Life Sciences

Social Sciences

BRIEF CALCULUS
AND ITS APPLICATIONS

Daniel D. Benice

Montgomery College

HOUGHTON MIFFLIN COMPANY   Boston   Toronto

Dallas   Geneva, Illinois   Palo Alto   Princeton, New Jersey

Cover design by Catherine Hawkes.
Cover photograph by Frank Siteman, Stock Boston.

SPONSORING EDITOR: *Maureen O'Connor*

DEVELOPMENT EDITOR: *Robert Hupp*

SENIOR PROJECT EDITOR: *Jean Andon*

PRODUCTION/DESIGN COORDINATOR: *Karen Rappaport*

MANUFACTURING COORDINATOR: *Sharon Pearson*

MARKETING MANAGER: *Mike Ginley*

Printed in the U.S.A.

Library of Congress Catalog Card Number: 92-72365

ISBN Numbers:

  Text: 0-395-61549-6
  Exam Copy: 0-395-66107-2
  Solutions Manual: 0-395-61553-4
  Student Solutions Manual: 0-395-61552-6
  Instructor's Resource Manual: 0-395-61550-X

1 2 3 4 5 6 7 8 9–DH–96 95 94 93 92

To my wife, Sylvia

CONTENTS

CHAPTER

5

EXPONENTIAL AND LOGARITHMIC FUNCTIONS   237

CHAPTER

INTEGRATION   295

PREFACE

Brief Calculus and Its Applications provides a one-semester introduction to calculus for students of business, economics, management, social sciences, life sciences, and other fields. The text offers a motivated, comprehensive, applications-oriented approach to the subject. Crafted with care and experience, this book offers instructors and students an ideal combination of content, level, writing style, and special features.

♦ *A textbook that students can read and understand* Explanations are carefully presented. The writing style makes calculus interesting and readily accessible to students. The theory and applications of calculus are explained in ways they can understand.

♦ *Solid, honest mathematics* Instructors will find the mathematics carefully done and at the right level. Definitions are mathematically accurate. Theorems are presented with convincing geometric or intuitive justification or with a formal proof when appropriate.

♦ *An abundance of outstanding exercises* The huge number of exercises in the book have been carefully constructed and gently graduated to enable students to gain confidence as they master skills and learn concepts. The variety of exercises will help students to understand and appreciate calculus.

  The number, kind, and arrangement of exercises provides the instructor with an opportunity to teach at the level he or she prefers. Furthermore, the instructor will be able to select exercises even when only part of a section is completed.

♦ *Realistic applications* A variety of interesting applications are used as examples and exercises to demonstrate real-world use of calculus. The opportunity to apply calculus concepts and skills will create student interest, involvement, and thinking. While the emphasis is on business, economics, and management, many other types of applications are given. Simplifications and assumptions are sometimes made in order to keep the mathematics within the scope of the course. An index of applications can be found on the inside covers of the book.

♦ *Illustrative examples* Many examples are included to illustrate calculus ideas and techniques. Consistently, you will see that extra step or explanation that helps to avoid confusion and misunderstanding.

♦ *Motivation and direction*   Calculus is exciting mathematics. This book offers students an adventure of discovery, complete with a built-in guide that tells them where they are going, what they will be doing, and why.

♦ *Reinforcement*   Concepts are studied from a variety of perspectives. The initial explanation is followed by examples, usually with accompanying comments or notes. Exercises focus first on small, central ideas and then proceed to combine the ideas into more complex exercises. Applications provide different settings in which to use the calculus concepts. The reinforcement continues in later sections, when concepts are reviewed in a new environment.

♦ *A valuable reference*   The format and organization of the text make it valuable as a reference. It is easy for students to locate concepts, definitions, theorems, and examples when they are studying or doing homework exercises.

♦ *Integrated problem solving*   The text flows naturally into problem-solving examples. Within the examples, great care is taken to discuss the question being asked or the problem being solved. Students are shown how to set up a problem or determine an equation and how to use the appropriate mathematical concepts. The relation of the final solution to the original problem is made clear. This approach will encourage students to become active participants and learn how to solve calculus problems. See Example 1, beginning on page 194. For a less extensive case, see Example 1 on page 160 and the lead-in text preceding it. Briefer still is Example 2 on page 182 and the sentence preceding it.

♦ *Accuracy*   The text is accurate. All exercises have been worked out by four mathematics professors: Joel W. Irish (University of Southern Maine), Robert Levine (Community College of Allegheny County), Michael Schramm (LeMoyne College), and myself. Furthermore, Robert Levine, Michael Schramm, and I have examined the entire book line-by-line for mathematical accuracy. Finally, in order to maximize continuity, I personally have written the answer section and the Exercise Library (mentioned under "Special Features").

♦ *Calculator use*   The use of calculators is incorporated in a natural, helpful way throughout the book. Within exercise sets, the symbol ▦ is used to identify problems specifically designed for calculators. (See pages 251 and 263, for example.) Although many classes will use calculators, those instructors who prefer not to use them will not encounter obstacles.

Special Features

Exercise Library for Graphing Calculators and Computers— ▦

The Exercise Library (blue-edged pages at the back of the book) contains a variety of optional *graphing calculator exercises* that can be used to supplement the study of calculus. The exercises can also be done using *computer software.* The symbol ⎌⌁ appears at the end of exercise sets throughout the book to direct you to the Exercise Library.

Such exercises add a new dimension to the study of calculus. The graphing calculator offers visual perspectives and problem-solving approaches that are unavailable using conventional techniques. You will find that the Exercise Library offers an exciting complement to the extensive exercise sets in the main body of the text.

Writing Exercises—W

Some exercises are marked with a large blue **W**. They are the *writing exercises,* which ask for explanations *in words.* These questions force students to think and to have a better grasp of the subject than might be attained otherwise. If a concept is not really understood, the student is forced to confront the realization. A sample of writing exercises can be seen on page 212.

These innovative questions address proposals from the NCTM, MAA, AMATYC, and others who support writing across the curriculum.

Notes ⌐ Note ⌐

Many "notes" have been strategically placed throughout the text. Their purpose is to help the reader by anticipating questions or problems, offering reinforcement or reminders, adding information, linking ideas, or making comparisons. To sample a variety of notes, look on pages 53, 135, 262, and 325.

Applications Examples

A variety of applications examples are given throughout the text. Not only do they illustrate how calculus can be used, but many of them also take the learning process a step further by providing the opportunity for students to be involved in problem solving. See Examples 2–4 on pages 76–78, and Example 3 on page 306.

A Section on Basic Economics Functions

All of the basic economics functions are presented and carefully explained in one section (Section 1.7). Cost, revenue, and profit functions are introduced. Fixed cost and breakeven quantity are explained. Price functions, demand equations, supply equations, and equilibrium are also covered.

Later, Section 3.4 is dedicated to presenting marginal analysis thoroughly and without interruption.

Practical Algebra Review

Chapter 1 includes a review of essential basic algebra. Elsewhere, algebra is included when needed so that students can work with the calculus. For example, in Chapter 5 the presentation of exponentials and logarithms provides what is needed, but does not cover everything one might learn about the subject by taking a precalculus course.

Supplements

The following supplements are available for use with the text:

♦ *Instructor's Resource Manual*   The Instructor's Resource Manual contains a printed test bank of all items in the computerized test generator, two chapter tests for each chapter, materials and exercises to facilitate the introduction and use of graphing calculators, and overhead transparency masters for selected examples from the text.

♦ *Computerized Test Generator*   The Computerized Test Generator contains more than 2000 test questions, organized by section to follow topics in the text. Over one third of these are applications questions. The instructor may choose between multiple choice and free-response answer formats.

♦ *Solutions Manual*   The Solutions Manual contains complete solutions to all exercises.

♦ *Student Solutions Manual*   The Student Solutions Manual contains complete solutions to all odd-numbered exercises.

♦ *Instructor's Even-numbered Answer Booklet*   The Answer Booklet contains answers to all even-numbered exercises, in an easy-to-carry format.

♦ *PC-81 Emulation Software*   PC-81 Emulation Software is available for the IBM PC (and compatibles). This powerful and compact package completely emulates the functionality of the popular TI-81 graphing calculator. The software is offered in cooperation with Texas Instruments.

♦ *Math Assistant Software*   Math Assistant Software is available for the Macintosh, Apple, and IBM PC (and compatibles). This package easily plots algebraic and trigonometric functions.

Content and Organization

The text offers comprehensive coverage of the calculus topics appropriate for this course. Since there is always the need to tailor a textbook to fit your specific course, we present here a guide that will help you to select the sections and topic sequence that will be best for your class.

The first five sections of Chapter 1 constitute a review of algebra and accordingly can be covered in class or omitted if the students are already familiar with the material. By contrast, Section 1.7 (Functions in Economics) introduces and explains key concepts that will be used in business applications throughout the book. The ideas in optional Section 1.6 are never formally applied elsewhere in the text; however, those students who become familiar with translation and reflection may choose to use them when they see the opportunity.

Since calculus cannot be understood without some knowledge of limits, Chapter 2 offers a pragmatic study of this important topic. If time constraints or personal preference demand that you get to the derivative quickly, cover only Section 2.1 (Introduction to Limits) and then go directly to Chapter 3 (Derivatives). Other sections of Chapter 2 can be covered as needed and as time permits. Another alternative is to present Sections 2.1 through 2.3 and return later as needed to limits at infinity (2.4) and infinite limits (2.5).

Chapter 3 introduces the derivative. You will want to present Sections 3.1–3.8 in sequence. Notice that Section 3.4 offers thorough coverage of the marginal concepts. Section 3.9 (Related Rates) can be omitted if it is not considered part of your course.

Chapter 4 presents applications of the derivative. The sections should be studied in sequence. Elasticity of demand (Section 4.6) may be omitted, although many business calculus courses consider the topic essential.

Chapter 5 presents the exponential and logarithmic functions. The sections should be covered in sequence. Section 5.5 (Some Additional Business Applications) can be omitted if desired.

Chapter 6 covers integration. Sections 6.1–6.4 should be taught in sequence. The three remaining sections offer applications and can be presented in any order. Note that in Section 6.5 the application of integration to finding the average value of a function and the volume of a solid of revolution are not essential to the continuity of the chapter or the text. Section 6.6 (Surplus) is an important business topic, but it can be omitted or studied later without loss of continuity.

Integration by substitution (Section 7.1) is an important extension of the integration ideas presented in Chapter 6. The remainder of the chapter includes standard topics that can be covered or omitted to fit the nature of your course.

Section 7.6 (Differential Equations) can be taught after 7.1, if desired.

Section 7.7 (Probability and Calculus) can also be covered after 7.1 or even without 7.1 if the integration in Example 2 is done by using an earlier formula rather than by substitution.

The sections of Chapter 8 (Multivariable Calculus) are arranged in a fairly standard order. However, you can rearrange the sequence or omit sections. The only restrictions are that Sections 8.1 and 8.2 must be presented first and that Section 8.3 must be covered before 8.5.

Many students taking this course have never studied trigonometry. Consequently, what seems like a simple review (Sections 9.1 and 9.2) may in fact serve as an introduction to trigonometry for some students. The sections of Chapter 9 should be covered in sequence.

Acknowledgments

I would like to express my genuine appreciation to the reviewers listed here. Their constructive comments have made a significant impact on the quality of this book. I am grateful for the support and encouragement they have offered.

Daniel D. Anderson, University of Iowa
Ronald Barnes, University of Houston, downtown campus
Margaret Russell Berkes, University of Vermont, Montpelier
George R. Bradley, Duquesne University
Michael J. Bradley, Merrimack College
Gabriel B. Costa, Seton Hall University
Sam Councilman, California State University
Preston Dinkins, Southern University
John Erbland, University of Hartford
Gerald K. Goff, Oklahoma State University
Kwang Chul Ha, Illinois State University
Gerald Higdon, Fitchburg State College
Joel W. Irish, University of Southern Maine
Thomas Judson, University of Portland
Donald LaTorre, Clemson University
Robert Levine, Community College of Allegheny County
Norman Martin, Northern Arizona University
Michael E. Mays, West Virginia University
Reginald Mazeres, Tennessee Technological University
Patsy N. Newman, Richard Bland College
R. Glen Powers, Western Kentucky University
Michael Schramm, LeMoyne College
Jean Shutters, Harrisburg Area Community College
Ronald Smith, Edison Community College
Kenneth W. Spackman, University of North Carolina, Wilmington
Robert F. Sutherland, Bridgewater State College
Arnold R. Vobach, University of Houston
Terry J. Walters, University of Tennessee at Chattanooga
Jan E. Wynn, Brigham Young University
Earl Zwick, Indiana State University

I would like to give special thanks to Gerald Higdon of Fitchburg State College for class-testing the completed manuscript. Personal thanks are due Robert F. Sutherland for creating the test generator and Joel W. Irish for preparing the solutions manual and student solutions manual.

It was a pleasure working with the professionals at Houghton Mifflin: Jean Andon (Senior Project Editor), Rob Hupp (Developmental Editor), Maureen O'Connor (Sponsoring Editor), Greg Tobin (Vice President, Editor in Chief), and Anne Wightman (Developmental Editor).

Daniel D. Benice

1

FUNCTIONS

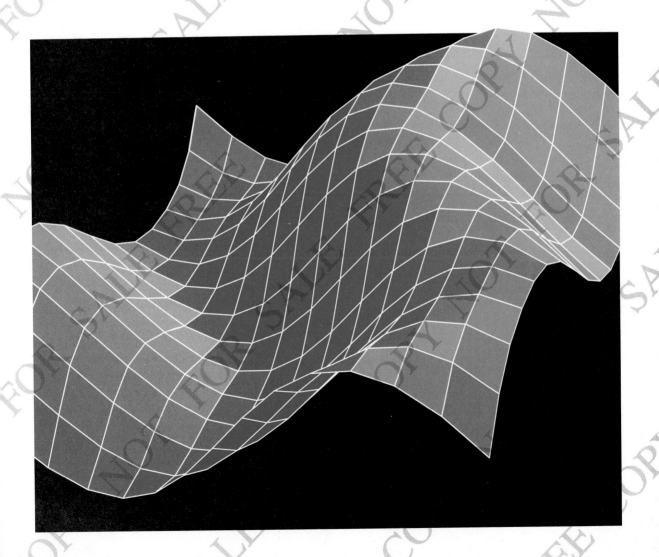

This chapter is intended to prepare you for a successful exploration of the ideas and applications of elementary calculus. We have included some algebra review, an introduction to functions and graphs, and a presentation of the functions used in business and economics. Since the material presented here will be used throughout the book, it is important that you become familiar with it.

1.1 | REAL NUMBERS

The study of elementary calculus requires a knowledge of the real number system. The real numbers can be considered points on a line. To every real number there corresponds one point. To every point there corresponds one real number. (See Figure 1.)

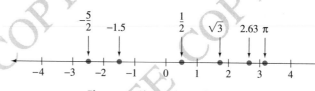

Figure 1 The real number line

Inequalities can be used to compare real numbers. The symbols used are > (greater than), < (less than), ≥ (greater than or equal to), and ≤ (less than or equal to). For example,

$$x > 3 \qquad (x \text{ is greater than 3})$$
$$y \le -2 \qquad (y \text{ is less than or equal to} -2)$$

In some applications it is useful to combine two inequalities in order to express an **interval**. For example,

$$2 < x < 5$$

combines the inequalities $2 < x$ and $x < 5$ and represents all real numbers between 2 and 5. The notation (2, 5) is used to denote such an **open interval** that excludes the endpoints. Graphically, the interval is shown as

The inequality $2 \le x \le 5$ expresses a **closed interval**, one in which the endpoints are included. The interval is denoted [2, 5] and is shown graphically as

The two intervals (2, 5) and [2, 5] and others are shown in Figure 2. A parenthesis is used to indicate that an endpoint is not included. A bracket is used to indicate that an endpoint is included. Intervals such as (2, 5] and [2, 5) are called **half-open intervals**. The symbol ∞ (infinity) is used to specify that the interval extends infinitely far to the right. Similarly, −∞ (minus infinity) is used to specify that an interval extends infinitely far to the left.

inequality	interval notation	graph
$2 < x < 5$	$(2, 5)$	
$2 \leq x \leq 5$	$[2, 5]$	
$2 < x \leq 5$	$(2, 5]$	
$2 \leq x < 5$	$[2, 5)$	
$x > 2$	$(2, \infty)$	
$x \leq 5$	$(-\infty, 5]$	

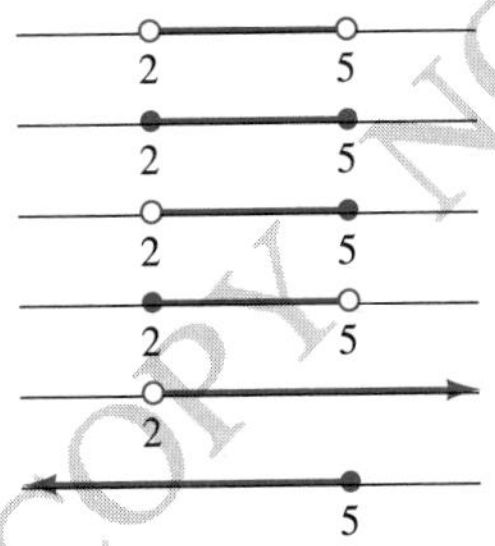

Figure 2 Intervals

This section concludes with a brief review of *linear inequalities*, which you will see have intervals for solutions. Some calculus problems require the solution of inequalities.

You should recall that linear inequalities are solved in much the same way as linear equations. But there is one key difference.

> If both sides of an inequality are multiplied or divided by a negative number, the direction of the inequality is reversed: > becomes <, and < becomes >.

EXAMPLE 1 Solve the inequality $3x + 4 > 15$.

SOLUTION Begin by adding -4 to both sides of the inequality $3x + 4 > 15$. The result is

$$3x > 11$$

Now, divide both sides by 3.

$$x > \frac{11}{3}$$

The solution is $x > 11/3$ or $(11/3, \infty)$. Any number greater than 11/3 will satisfy the original inequality. ◆

EXAMPLE 2   Solve the inequality $-3(2 + x) + x \geq 14$.

SOLUTION   When the expression on the left side is multiplied out, the inequality becomes

$$-6 - 3x + x \geq 14$$

or

$$-6 - 2x \geq 14$$

After adding 6 to both sides, we have

$$-2x \geq 20$$

Finally, dividing both sides by -2 yields

$$x \leq -10$$

Notice that division of both sides by a negative number (-2) resulted in a reversal of the direction of the inequality. The $\geq$ became $\leq$. The solution is $x \leq -10$ or $(-\infty, -10]$. ◆

1.1   Exercises

In Exercises 1–12, write each inequality using interval notation.

1. $5 \leq x \leq 9$

2. $-1 \leq x < 4$

3. $x \geq 6$

4. $3 < x \leq 8$

5. $x < 0$

6. $x > 0$

7. $x > -2$

8. $t \leq -4$

9. $t < \pi$

10. $t \geq \sqrt{2}$

11. $-\sqrt{3} < t < \sqrt{3}$

12. $-7 < t \leq 0$

In Exercises 13–24, write each interval as an inequality.

13. $[0, \infty)$

14. $(4, 19)$

15. $[1, 75]$

16. $(0, 100]$

17. $(-\infty, -2)$

18. $[7, \infty)$

19. $(-5, \infty)$

20. $(-\infty, 2)$

21. $[\pi, 7)$

22. $[-3, 7)$

23. $(-4, 19]$

24. $(2, \pi]$

In Exercises 25–34, draw the graph form of the given interval.

25. $(1, 4)$

26. $(3, 7)$

27. $[-1, 2]$

28. $(-6, 0]$

29. $(4, \infty)$

30. $(-\infty, -3)$

31. $(-\infty, 0]$

32. $[1, \infty)$

33. $[-5, -1)$

34. $[0, 2)$

Solve each linear inequality given in Exercises 35–48.

35. $2x + 1 > 17$

36. $3x + 1 < 25$

37. $5x - 1 \leq 29$

38. $4x - 3 \geq 33$

39. $3x \geq 0$

40. $x - 2 < 0$

41. $-2t > 14$

42. $-5y < 40$

43. $1 - 8x \leq 0$

44. $5 - 2x \geq 6$

45. $3(1 - x) + 2 > -1$

46. $8(1 - 2x) - 1 \leq 6$

47. $5(y + 1) < 13$

48. $4(t + 3) > 17$

1.2 | *SOME ALGEBRA REVIEW*

This section provides a brief *review* of some algebra concepts that will be used in the calculus presentation. The concepts included here should already be familiar to you.

Exponents

Recall the concept of exponent:

$$x^2 = x \cdot x \qquad x^3 = x \cdot x \cdot x \qquad x^4 = x \cdot x \cdot x \cdot x$$

In general,

Exponent

If x is any real number and n is a positive integer, then

$$x^n = x \cdot x \cdot x \cdots x \qquad n \text{ factors of } x$$

Here x is called the **base** and n is called the **exponent** or **power**.

Negative exponents are defined as follows:

Negative Exponents

If x is any nonzero real number and n is a positive integer, then

$$x^{-n} = \frac{1}{x^n} \qquad \frac{1}{x^{-n}} = x^n$$

The next example shows the value of changing negative exponents to positive exponents for calculation purposes.

EXAMPLE 1   Evaluate.   **(a)** 2^{-3} **(b)** $\dfrac{1}{4^{-2}}$

SOLUTION   **(a)** $2^{-3} = \dfrac{1}{2^3} = \dfrac{1}{8}$

(b) $\dfrac{1}{4^{-2}} = 4^2 = 16$ ◆

> **Note**
>
> Beginning in Chapter 3, you will see the occasional need to change from positive exponents to negative exponents in order to use calculus operations. An expression such as $1/x^4$ would be changed to x^{-4}.

Several important properties of exponents are given next.

Properties of Exponents

If x and y are real numbers and m and n are integers, then

$$x^0 = 1 \qquad\qquad x \neq 0$$

$$x^m \cdot x^n = x^{m+n}$$

$$(x^m)^n = x^{mn}$$

$$(x \cdot y)^m = x^m y^m$$

$$\left(\frac{x}{y}\right)^m = \frac{x^m}{y^m} \qquad y \neq 0$$

$$\frac{x^m}{x^n} = x^{m-n} \qquad x \neq 0$$

EXAMPLE 2 Use the properties of exponents to simplify each expression.

(a) $x^5 \cdot x^3$ **(b)** $(a^4)^6$ **(c)** $(2x)^4$ **(d)** $\left(\dfrac{a}{b}\right)^{10}$ **(e)** $\dfrac{x^{12}}{x^5}$

(f) $5x^0$

SOLUTION **(a)** $x^5 \cdot x^3 = x^{5+3} = x^8$

(b) $(a^4)^6 = a^{4 \cdot 6} = a^{24}$

(c) $(2x)^4 = 2^4 x^4 = 16x^4$

(d) $\left(\dfrac{a}{b}\right)^{10} = \dfrac{a^{10}}{b^{10}}$

(e) $\dfrac{x^{12}}{x^5} = x^{12-5} = x^7$

(f) $5x^0 = 5 \cdot x^0 = 5 \cdot 1 = 5$ ◆

Exponent notation can be extended to radicals. The idea that $\sqrt{x} \cdot \sqrt{x} = x$ suggests using $x^{1/2}$ as the exponent notation of $\sqrt{x}$, since $x^{1/2} \cdot x^{1/2} = x$ would be consistent with a known property of exponents.

> ## Exponent Form of Radicals
> $$x^{1/2} = \sqrt{x} \qquad\qquad x \geq 0$$
> $$x^{1/3} = \sqrt[3]{x} \qquad\qquad \text{all } x$$
> $$x^{1/n} = \sqrt[n]{x} \qquad\qquad x \geq 0 \text{ when } n \text{ even}$$
> $$x^{1/n} = \sqrt[n]{x} \qquad\qquad \text{all } x \text{ when } n \text{ odd}$$

Here is an example that demonstrates numerical computation.

EXAMPLE 3   Evaluate.   **(a)** $9^{1/2}$ **(b)** $8^{1/3}$

SOLUTION   **(a)** $9^{1/2} = \sqrt{9} = 3$

(b) $8^{1/3} = \sqrt[3]{8} = 2$ ◆

Any rational exponent of the form $x^{m/n}$ can be considered as $(x^{1/n})^m$ or as $(x^m)^{1/n}$.

> ## Rational Exponents
> $$x^{m/n} = (x^{1/n})^m \qquad \text{or} \qquad (x^m)^{1/n}$$
> $$x \geq 0 \text{ when } n \text{ is even}$$

The form $(x^{1/n})^m$ is usually easier to work with in numerical situations, because the number whose root you seek is usually smaller. This idea is demonstrated in the following example.

EXAMPLE 4   Evaluate $64^{3/2}$ two different ways. Compare.

SOLUTION $64^{3/2} = (64^{1/2})^3 = (8)^3 = 512$ (easier)

$64^{3/2} = (64^3)^{1/2} = (262{,}144)^{1/2} = 512$ (harder)   ◆

Negative fractional exponents are a natural extension.

EXAMPLE 5   Evaluate.   **(a)** $9^{-1/2}$ **(b)** $16^{-3/2}$

SOLUTION   **(a)** $9^{-1/2} = \dfrac{1}{9^{1/2}} = \dfrac{1}{3}$

(b) $16^{-3/2} = \dfrac{1}{16^{3/2}} = \dfrac{1}{(16^{1/2})^3} = \dfrac{1}{(4)^3} = \dfrac{1}{64}$ ◆

Decimal exponents such as $x^{1.2}$ and $x^{-.3}$ will appear in some settings. When necessary, you can change the decimals to fractions and deal with the fractional exponents.

EXAMPLE 6 Change to fractional exponents. (a) $x^{1.2}$ (b) $x^{-.3}$

SOLUTION

(a) $x^{1.2} = x^{6/5}$ since $1.2 = 1\frac{2}{10} = 1\frac{1}{5} = 6/5$

(b) $x^{-.3} = x^{-3/10}$ or $\dfrac{1}{x^{3/10}}$ ◆

Recall that numbers such as $\sqrt{2}$ and $\sqrt{7}$ are irrational numbers. This means that a fraction such as

$$\frac{5}{\sqrt{2}}$$

has an irrational denominator. You can make this denominator a rational number (that is, *rationalize the denominator*) by multiplying the denominator by $\sqrt{2}$. Of course, you must also multiply the numerator by $\sqrt{2}$ to avoid changing the value of the fraction. Here is the procedure.

$$\frac{5}{\sqrt{2}} = \frac{5}{\sqrt{2}} \cdot \frac{\sqrt{2}}{\sqrt{2}} = \frac{5\sqrt{2}}{2}$$

Quadratic Equations

An expression of the form $ax^2 + bx + c$, where $a \neq 0$, is called **quadratic** in x. Such expressions can sometimes be factored. For example,

$$x^2 + 7x + 10 = (x + 5)(x + 2)$$

Recall that in the factoring, x and x are selected to yield x^2 when multiplied. The $+5$ and $+2$ are chosen because their product is the $+10$ of the original expression and their sum is the $+7$ of the $+7x$.

EXAMPLE 7 Factor. (a) $2x^2 - 4x - 30$ (b) $3x^2 - 11x + 6$ (c) $x^2 - 49$
(d) $x^2 + 9x$

SOLUTION

(a) $2x^2 - 4x - 30 = 2(x^2 - 2x - 15)$
$\qquad\qquad\qquad = 2(x - 5)(x + 3)$

(b) $3x^2 - 11x + 6 = (3x - 2)(x - 3)$

(c) $x^2 - 49 = (x + 7)(x - 7)$

(d) $x^2 + 9x = x(x + 9)$ ◆

A *quadratic equation* can be solved by factoring if the quadratic expression can be factored.

EXAMPLE 8   Solve the quadratic equation $5x^2 + 2x - 3 = 0$.

SOLUTION   We begin by factoring the quadratic expression.

$$5x^2 + 2x - 3 = 0$$

becomes

$$(5x - 3)(x + 1) = 0$$

If either factor is zero, then the product will be zero. So set each factor equal to zero to solve the equation.

$$
\begin{array}{c|c}
5x - 3 = 0 & x + 1 = 0 \\
5x = 3 & x = -1 \\
x = 3/5 &
\end{array}
$$

The solutions of the quadratic equation are 3/5 and -1. ◆

If the quadratic expression cannot be factored or if you are having trouble factoring it, use the **quadratic formula** to solve the equation.

Quadratic Formula

If $ax^2 + bx + c = 0$ and $a \neq 0$, then

$$x = \frac{-b \pm \sqrt{b^2 - 4ac}}{2a}$$

The symbol $\pm$ is read "plus or minus."

EXAMPLE 9   Solve the quadratic equation $x^2 + 5x + 2 = 0$.

SOLUTION   The equation $x^2 + 5x + 2 = 0$ cannot be solved by factoring using integers. We will use the quadratic formula, with $a = 1$, $b = 5$, and $c = 2$.

$$x = \frac{-b \pm \sqrt{b^2 - 4ac}}{2a}$$

$$= \frac{-5 \pm \sqrt{25 - 4(1)(2)}}{2(1)}$$

$$= \frac{-5 \pm \sqrt{17}}{2} \qquad \text{(the two solutions)}$$

If desired, the fraction containing the two solutions can be split into two fractions in order to display each solution separately.

$$x = \frac{-5 + \sqrt{17}}{2}, \ x = \frac{-5 - \sqrt{17}}{2} \qquad \text{(alternative form)} \quad ◆$$

EXAMPLE 10 Solve the equation $x^2 - 4x = 1$.

SOLUTION We note that this is a quadratic equation. However, it looks different because the expression is not set equal to zero. *Whether we are solving by factoring or by formula, quadratic equations must be in an "equals zero" form before we can proceed.* In this case, we can add -1 to both sides to get the desired form. The result is

$$x^2 - 4x - 1 = 0$$

The expression cannot be factored using integers, so we will use the quadratic formula. Here $a = 1$, $b = -4$, and $c = -1$.

$$x = \frac{-(-4) \pm \sqrt{(-4)^2 - 4(1)(-1)}}{2(1)}$$

$$= \frac{4 \pm \sqrt{16 + 4}}{2}$$

$$= \frac{4 \pm \sqrt{20}}{2}$$

But $\sqrt{20}$ can be simplified, and the fraction will then reduce. Here is the process. First,

$$\sqrt{20} = \sqrt{4 \cdot 5}$$
$$= \sqrt{4}\sqrt{5}$$
$$= 2\sqrt{5}$$

So we have

$$x = \frac{4 \pm 2\sqrt{5}}{2}$$

Factoring a 2 out of the numerator yields

$$x = \frac{2(2 \pm \sqrt{5})}{2}$$

The *factor* of 2 in the numerator can be eliminated by division by the *factor* of 2 in the denominator (since $2 \div 2 = 1$). The result is the simplified form of the solution

$$x = 2 \pm \sqrt{5} \quad \blacklozenge$$

Quadratic expressions are examples of **polynomials**. A polynomial expression is a sum of terms (one or more) of the form ax^n, where a is a real number and n is a nonnegative integer. Examples of polynomials include $5x + 12$, $x^2 - 9x + 17$, $4x^3 - 2$, and x^6.

Rational Expressions (Fractions)

You will need to work with algebraic fractions in calculus. When the numerator and denominator expressions are polynomials, the fraction is called a **rational expression**. Such an expression is not defined when the denominator polynomial is equal to zero.

Fractions are *reduced* by factoring. Like factors can be eliminated by division.

EXAMPLE 11   Reduce $\dfrac{x^2 + 7x}{x^2 - x}$.

SOLUTION $\dfrac{x^2 + 7x}{x^2 - x} = \dfrac{x(x + 7)}{x(x - 1)} = \dfrac{x + 7}{x - 1}$ ◆

Fractions can be *added* or *subtracted* by obtaining a common denominator. The process is demonstrated in the next example.

EXAMPLE 12   Add $\dfrac{3}{x} + \dfrac{5}{x + 1}$.

SOLUTION $\dfrac{3}{x} + \dfrac{5}{x + 1} = \dfrac{3}{x} \cdot \dfrac{x + 1}{x + 1} + \dfrac{5}{x + 1} \cdot \dfrac{x}{x}$

$= \dfrac{3x + 3}{x(x + 1)} + \dfrac{5x}{x(x + 1)}$

$= \dfrac{8x + 3}{x(x + 1)}$ ◆

Fractions are *multiplied* by multiplying numerator by numerator and denominator by denominator. Reduce before you multiply.

EXAMPLE 13   Multiply $\dfrac{3}{x^2 - 1} \cdot \dfrac{x + 1}{7}$.

SOLUTION $\dfrac{3}{x^2 - 1} \cdot \dfrac{x + 1}{7} = \dfrac{3}{(x + 1)(x - 1)} \cdot \dfrac{(x + 1)}{7}$

$= \dfrac{3}{7x - 7}$ ◆

Division of fractions is accomplished by inverting the divisor fraction and changing the process to multiplication.

EXAMPLE 14   Divide $\dfrac{4}{x} \div \dfrac{3}{5x}$.

SOLUTION
$$\frac{4}{x} \div \frac{3}{5x} = \frac{4}{x} \cdot \frac{5x}{3}$$

$$= \frac{4}{\cancel{x}} \cdot \frac{5 \cdot \cancel{x}}{3}$$

$$= \frac{20}{3} \quad \blacklozenge$$

If the numerator or denominator of a fraction contains a fraction, then the entire expression is called a **complex fraction**. Here are three examples of complex fractions.

$$\frac{1 + \dfrac{1}{x}}{\dfrac{x}{y} - 2} \qquad \frac{\dfrac{x+1}{3}}{7} \qquad \frac{4}{\dfrac{x}{2}}$$

Such fractions can be simplified by determining the least common denominator of all the fractions within the complex fraction *and then* multiplying *each term* of the complex fraction by that common denominator. Consider

$$\frac{1 + \dfrac{1}{x}}{\dfrac{x}{y} - 2} \qquad \text{The fractions within are } \dfrac{1}{x} \text{ and } \dfrac{x}{y}$$

The least common denominator is xy, so multiply each term by xy. The result is

$$\frac{xy \cdot 1 + xy \cdot \dfrac{1}{x}}{xy \cdot \dfrac{x}{y} - xy \cdot 2} \qquad \text{or} \qquad \frac{xy + y}{x^2 - 2xy}$$

This process will always produce a simple (that is, noncomplex) fraction.

Equations containing fractions can be solved by multiplying both sides (all terms) by the least common denominator of the fractions present.

EXAMPLE 15   Solve the equation.

$$\frac{x+1}{x} + \frac{1}{3} = \frac{5}{x}$$

SOLUTION   Begin by multiplying both sides (all terms) by the least common denominator, $3x$.

$$3x \cdot \frac{x+1}{x} + 3x \cdot \frac{1}{3} = 3x \cdot \frac{5}{x}$$

which simplifies to

$$3(x + 1) + x = 3 \cdot 5$$

or

$$3x + 3 + x = 15$$
$$4x + 3 = 15$$
$$4x = 12$$
$$x = 3$$

The solution is 3, and it should be checked in the original equation. (Any time both sides of an equation are multiplied by an unknown quantity, the "solution" produced may be *extraneous*; that is, it may not be a solution of the original equation.) ◆

1.2 Exercises

In Exercises 1–18, use properties of exponents to simplify the expression. Do not leave negative exponents in your answer. Assume the variables are never equal to zero.

1. $x^{12} \cdot x^5$

2. $t^4 t^{10}$

3. $(b^7)^3$

4. $(y^5)^8$

5. $\dfrac{x^{14}}{x^8}$

6. $\dfrac{t^{19}}{t^7}$

7. $y^2 \cdot y^0$

8. $x^0 \cdot x^3$

9. $2 \cdot x^0$

10. $(2x)^0$

11. x^{-3}

12. y^{-7}

13. $\dfrac{7}{x^{-2}}$

14. $\dfrac{x^{-4}}{2}$

15. $\dfrac{t^6}{t^{15}}$

16. $\dfrac{x^4}{x^{18}}$

17. $\left(\dfrac{x^0}{y}\right)^4$

18. $\left(\dfrac{a}{b^0}\right)^3$

Evaluate each expression in Exercises 19–36. Do not use a calculator.

19. 3^{-2}

20. 2^{-4}

21. 5^{-3}

22. 6^{-2}

23. $49^{1/2}$

24. $25^{1/2}$

25. $27^{1/3}$

26. $64^{1/3}$

27. $16^{-1/2}$

28. $8^{-1/3}$

29. $1^{-1/3}$

30. $64^{-1/2}$

31. $4^{3/2}$

32. $16^{3/2}$

33. $27^{2/3}$

34. $64^{2/3}$

35. $9^{-3/2}$

36. $8^{-2/3}$

In Exercises 37–42, change each expression to one having fractional exponents and then evaluate it. Do not use a calculator.

37. 36^5

38. 64^5

39. $16^{1.5}$

40. $25^{1.5}$

41. $100^{-1.5}$

42. $4^{-1.5}$

In Exercises 43–48, use a calculator to evaluate each expression. Round the result to four decimal places. The steps, using $2^{1.4}$ as an example:

1. Enter the base (2).

2. Press the $\boxed{y^x}$ key.

3. Enter the exponent (1.4).

4. Press the $\boxed{=}$ key.

The result of this example is 2.6390.

43. $2^{3.5}$

44. $6^{1.9}$

45. $15^{2.6}$

46. $11^{4.2}$

47. $4.3^{.04}$

48. $12.9^{.01}$

Rationalize the denominator of each expression in Exercises 49–54.

49. $\dfrac{4}{\sqrt{3}}$

50. $\dfrac{7}{\sqrt{2}}$

51. $\dfrac{2}{\sqrt{5}}$

52. $\dfrac{3}{\sqrt{7}}$

53. $\dfrac{6}{\sqrt{3}}$

54. $\dfrac{12}{\sqrt{2}}$

Use factoring to solve each quadratic equation in Exercises 55–68.

55. $x^2 - 64 = 0$

56. $y^2 - 1 = 0$

57. $x^2 - 9x = 0$

58. $n^2 + 4n = 0$

59. $y^2 + 9y + 14 = 0$

60. $x^2 - x - 6 = 0$

61. $x^2 - 2x - 8 = 0$

62. $x^2 + 2x - 3 = 0$

63. $2t^2 - 6t - 20 = 0$

64. $2y^2 - 14y + 24 = 0$

65. $3x^2 - 10x - 8 = 0$

66. $2x^2 + 3x - 5 = 0$

67. $2x^2 + x = 6$

68. $3x^2 - 10x = -8$

Use the quadratic formula to solve each quadratic equation in Exercises 69–78.

69. $x^2 + 3x + 1 = 0$

70. $x^2 + 5x - 1 = 0$

71. $t^2 - 2t - 4 = 0$

72. $x^2 - 2x - 2 = 0$

73. $x^2 - 4x - 2 = 0$

74. $y^2 + 6y + 2 = 0$

75. $4y^2 + 10y - 5 = 0$

76. $3x^2 + 6x - 2 = 0$

77. $2x^2 + 6x = 3$

78. $2y^2 - 4y = 5$

Reduce each fraction in Exercises 79–86.

79. $\dfrac{x^2 - 9}{x + 3}$

80. $\dfrac{x + 1}{x^2 - 1}$

81. $\dfrac{x^2 - 4x}{x^2 - 16}$

82. $\dfrac{x^2 - 4}{x - 2}$

83. $\dfrac{2x - 10}{3x - 15}$

84. $\dfrac{2x^2 - 10x}{3x}$

85. $\dfrac{x}{x^2 + 5x}$

86. $\dfrac{x^2}{x^3}$

In Exercises 87–92, add or subtract the fractions as indicated.

87. $\dfrac{2}{x} + \dfrac{5}{3x}$

88. $\dfrac{1}{4x} + \dfrac{6}{x}$

89. $\dfrac{7}{x - 1} + \dfrac{1}{x}$

90. $\dfrac{3}{x - 4} + \dfrac{5}{x}$

91. $\dfrac{3}{2x} - \dfrac{5}{x + 2}$

92. $\dfrac{1}{x - 3} - \dfrac{2}{3x}$

In Exercises 93–100, multiply or divide as indicated.

93. $\dfrac{x + 1}{2x + 2} \cdot \dfrac{2x}{x - 1}$

94. $\dfrac{3}{7x} \cdot \dfrac{x^2}{4}$

95. $\dfrac{x}{x - 4} \div \dfrac{2}{5x - 20}$

96. $\dfrac{x}{x + 3} \div \dfrac{2x}{x + 3}$

97. $\dfrac{3x - 12}{3x} \cdot \dfrac{5x}{x^2 - 16}$

98. $\dfrac{5}{2x + 4} \cdot \dfrac{x^2 + 2x}{3x}$

99. $\dfrac{x^2 + x}{2} \div \dfrac{x^2 - 1}{8x}$

100. $\dfrac{x + 8}{8x} \div \dfrac{x + 2}{2x}$

In Exercises 101–104, simplify each complex fraction.

101. $\dfrac{\dfrac{x}{2} + x}{1 - \dfrac{x}{2}}$

102. $\dfrac{\dfrac{a}{b} + \dfrac{b}{a}}{7}$

103. $\dfrac{\dfrac{3}{x} + h}{x + \dfrac{1}{h}}$

104. $\dfrac{3x + \dfrac{2}{y}}{5 - \dfrac{1}{3}}$

In Exercises 105–108, solve each equation.

105. $\dfrac{x + 1}{3} + \dfrac{x}{2} = 7$

106. $\dfrac{x}{4} - 3 = \dfrac{x - 2}{2}$

107. $2 - \dfrac{1}{x} = \dfrac{x + 2}{x}$

108. $\dfrac{3}{x} + 4 = \dfrac{x + 1}{x}$

W 109. The quadratic formula is used to solve equations of the form $ax^2 + bx + c = 0$, where $a \neq 0$.
  (a) Looking at the equation $ax^2 + bx + c = 0$, explain why a cannot be 0.
  (b) Looking at the quadratic formula, explain why a cannot be 0.

1.3 | *INTRODUCTION TO FUNCTIONS*

You probably recall the geometry formula $A = \pi r^2$, which says that the area A of the region within a circle is equal to π times the square of the radius r.

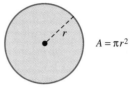

$A = \pi r^2$

The equation $A = \pi r^2$ defines a correspondence between two variables, r and A. For every nonnegative value of r there is a corresponding value of A. The formula $A = \pi r^2$ provides the rule of correspondence; it indicates how to compute the A value that corresponds to any particular r value supplied. The correspondence can be viewed as

$$r \xrightarrow{\;\pi r^2\;} A$$

$$0 \longrightarrow 0$$
$$1 \longrightarrow \pi \qquad \text{(approximately 3.14)}$$
$$2 \longrightarrow 4\pi \qquad \text{(approximately 12.56)}$$
$$.5 \longrightarrow .25\pi \qquad \text{(approximately .785)}$$

The area example, using $A = \pi r^2$, provides an introduction to the concept of **function**. Here the function is a rule that assigns one A value to each r value. (A is a function of r.) The equation $y = 2x + 1$ also defines a function; for every x value supplied there is one y value. (y is a function of x.)

$$x \xrightarrow{\;2x + 1\;} y$$

$$0 \longrightarrow 1$$
$$3 \longrightarrow 7$$
$$.5 \longrightarrow 2$$
$$-2 \longrightarrow -3$$

In general,

> **Function**
>
> A **function** is a rule of correspondence by which each element of one set (X) is assigned to exactly one element of the other set (Y).

If the elements of the first set are values of x and the elements of the second set are values of y, then we have a correspondence between x and y

$$x \longrightarrow y$$

The variable x is called the **independent variable**, and y is called the **dependent variable**.

The Swiss mathematician Leonhard Euler (1707–1783) suggested the use of function notation for certain applications, many of which arise in calculus. Functions are named by letters, with the letter f being the most popular.

If the function is named f and the independent variable is x, then the **function notation** $f(x)$ can be used instead of y. Thus, $y = 2x + 1$ becomes $f(x) = 2x + 1$. The $f(x)$ is read "f of x" or "f at x." It is the value of f at x. The relationship can be considered as follows:

$$x \xrightarrow{\ f\ } 2x + 1$$

$f(x) = 2x + 1$ is a rule that assigns to any number x the function value $2x + 1$. If $x = 3$, the function value is $2(3) + 1$, or 7. This can be seen as

$$3 \xrightarrow{\ f\ } 7$$

or as

$$f(3) = 7$$

Alternatively, if $f(x) = 2x + 1$, then $f(3) = 2(3) + 1 = 7$.

Note

When $y = 2x + 1$ is written in function notation as $f(x) = 2x + 1$, the *function* is f, not $f(x)$. Keep in mind that $f(x)$ is the *value* that corresponds to a particular x. $f(x) = 2x + 1$ *defines* or *gives* the function, but the function is f, not $f(x)$. At times we may choose to use an informal abbreviated statement such as "function $f(x) = \ldots$" rather than the more formal and complete statement "function *defined by* (or *given by*) $f(x) = \ldots$."

EXAMPLE 1   Let function f be defined by $f(x) = 5x^2 - 4x + 8$. Determine each value.

(a) $f(0)$ **(b)** $f(2)$ **(c)** $f(-2)$ **(d)** $f(x + 1)$ **(e)** $f(x - 2)$

SOLUTION   **(a)** $f(0) = 5(0)^2 - 4(0) + 8$

$\qquad\qquad = 0 - 0 + 8$

$\qquad\qquad = 8$

(b) $f(2) = 5(2)^2 - 4(2) + 8$

$\qquad\quad = 5 \cdot 4 - 8 + 8$

$\qquad\quad = 20$

(c) $f(-2) = 5(-2)^2 - 4(-2) + 8$

$\qquad\qquad = 5 \cdot 4 + 8 + 8$

$\qquad\qquad = 36$

(d) $f(x + 1) = 5(x + 1)^2 - 4(x + 1) + 8$

$\qquad\qquad\quad = 5(x^2 + 2x + 1) - 4x - 4 + 8$

$\qquad\qquad\quad = 5x^2 + 10x + 5 - 4x - 4 + 8$

$\qquad\qquad\quad = 5x^2 + 6x + 9$

(e) $f(x - 2) = 5(x - 2)^2 - 4(x - 2) + 8$

$\qquad\qquad\quad = 5(x^2 - 4x + 4) - 4x + 8 + 8$

$\qquad\qquad\quad = 5x^2 - 20x + 20 - 4x + 8 + 8$

$\qquad\qquad\quad = 5x^2 - 24x + 36$ ◆

EXAMPLE 2   Let function f be defined by $f(x) = 5x - 9$. Determine each value.

(a) $f(x + h)$ **(b)** $f(x + h) - f(x)$

SOLUTION   **(a)** To compute $f(x + h)$, replace x by $x + h$ in $f(x) = 5x - 9$.

$$f(x + h) = 5(x + h) - 9$$
$$= 5x + 5h - 9$$

(b) To compute $f(x + h) - f(x)$, subtract $f(x)$ from $f(x + h)$. We already know the value of $f(x + h)$ from part (a).

$$f(x + h) - f(x) = (5x + 5h - 9) - (5x - 9)$$
$$= 5x + 5h - 9 - 5x + 9$$
$$= 5h ◆$$

The letters used to name functions are often chosen to fit applications: R for revenue, C for cost, P for profit, v for velocity, etc.

EXAMPLE 3   ◆   **COST OF BOOKS**

If $C(x) = 12x$ gives the cost in dollars of x books, what is the cost of 5 books?

SOLUTION   Since $C(x) = 12x$ is the cost of x books, the cost of 5 books is $C(5)$.

$$C(5) = 12(5) = 60$$

We conclude that the cost of 5 books is \$60. ◆

EXAMPLE 4   ◆   **CONCENTRATION OF MEDICINE**

The concentration K of a particular medicine in the bloodstream t hours after being swallowed is

$$K(t) = \frac{.03t}{1 + t^2} \qquad t \geq 0$$

What is the concentration of the medicine after 2 hours?

SOLUTION   After 2 hours, the concentration is $K(2)$, namely,

$$K(2) = \frac{.03(2)}{1 + (2)^2}$$

$$= \frac{.06}{5}$$

$$= .012$$

After 2 hours the concentration of the medicine is .012 (or 1.2%). ◆

The set of all possible values of the independent variable x is called the **domain** of the function. (The corresponding set of all possible values of the dependent variable y is called the **range**.) For $f(x) = 2x + 1$, x can be any real number, since when any real number is used for x there will be a corresponding real number $f(x)$, or y. The domain of this function f includes all the real numbers.

If a function gives the distance traveled by a rocket for any time t, then t cannot be negative, since time cannot be negative. There will be further restriction on the domain of such a function, since the flight cannot go on indefinitely. If the flight lasts 20 seconds, then the domain would be t such that $0 \le t \le 20$, or [0, 20] using interval notation.

In addition to the nature of an application, there are other concerns that can restrict the domain of a function.

1. *Division by zero is not defined.* In view of this, any value of x that creates division by zero cannot be in the domain of a function. There would be no $f(x)$ corresponding to such an x.

2. *Square roots of negative numbers are not real numbers.* Thus, any value of x creating the square root of a negative number cannot be in the domain of a function. There would be no $f(x)$ corresponding to such an x.

EXAMPLE 5   Find the domain of $f(x) = \dfrac{1}{x - 4}$.

SOLUTION   If $x = 4$, division by zero results. Since division by zero is not defined, no f value is produced if 4 is used for x. Thus, 4 is not in the domain of f. This means that the domain of f is all the real numbers except 4. We can write this simply as $x \ne 4$. ◆

EXAMPLE 6   Find the domain of $g(x) = \sqrt{x - 1}$.

SOLUTION   If $x - 1$ is negative, the result is the square root of a negative number. Since the square root of a negative number is not a real number, no g value will be produced in such instances. This means that the domain is all x values for which $x - 1 \ge 0$. Solving this linear inequality yields $x \ge 1$. Thus, the domain of g is $x \ge 1$, or the interval [1, ∞). ◆

A **zero** of a function f is any real number x for which $f(x) = 0$.

EXAMPLE 7 Find the zeros of each function.

        (a) $f(x) = 5x - 20$ **(b)** $f(x) = x^2 + 5x - 14$

SOLUTION (a) Given $f(x) = 5x - 20$, it follows that $f(x) = 0$ when $5x - 20 = 0$. Solving $5x - 20 = 0$ yields $x = 4$. The (only) zero of this function is 4.

        (b) Given $f(x) = x^2 + 5x - 14$, it follows that $f(x) = 0$ when $x^2 + 5x - 14 = 0$. This is a quadratic equation that can be solved by factoring. The steps:

$$x^2 + 5x - 14 = 0$$
$$(x + 7)(x - 2) = 0$$
$$x = -7, \; x = 2$$

We conclude that the zeros of this function are -7 and 2. ◆

The section closes with a note on composition of functions. If f and g are functions, then the **composite functions** $f \circ g$ and $g \circ f$ are defined as follows:

$$(f \circ g)(x) = f(g(x))$$
$$(g \circ f)(x) = g(f(x))$$

The notation $f(g(x))$ is read "f of g of x." $f(g(x))$ is a function of a function.

EXAMPLE 8 If $f(x) = 5x - 2$ and $g(x) = x^2 + 1$, find **(a)** $(f \circ g)(x)$ **(b)** $(g \circ f)(x)$.

SOLUTION

(a) $(f \circ g)(x) = f(g(x))$
$= f(x^2 + 1)$
$= 5(x^2 + 1) - 2$
$= 5x^2 + 5 - 2$
$= 5x^2 + 3$

(b) $(g \circ f)(x) = g(f(x))$
$= g(5x - 2)$
$= (5x - 2)^2 + 1$
$= 25x^2 - 20x + 4 + 1$
$= 25x^2 - 20x + 5$ ◆

1.3   Exercises

In Exercises 1–10, find $f(0)$, $f(1)$, $f(2)$, and $f(-1)$ for the function defined.

1. $f(x) = 5x + 7$

2. $f(x) = 1 - 5x$

3. $f(x) = x^2 + 3x + 1$

4. $f(x) = 3x^2 - 6x + 4$

5. $f(x) = -x^2 + 5$

6. $f(x) = x^2 - 9x$

7. $f(x) = 6$

8. $f(x) = -2$

9. $f(x) = \sqrt{x + 1}$

10. $f(x) = \sqrt{x + 2}$

In Exercises 11–16, find $f(x + 2)$ and $f(x - 3)$ for the function defined.

11. $f(x) = x^2 - 3x + 7$ **12.** $f(x) = x^2 + 10x$

13. $f(x) = 4x^2 + 9x$ **14.** $f(x) = 3x^2 - 5x + 2$

15. $f(x) = \dfrac{x + 5}{x - 7}$ **16.** $f(x) = \dfrac{x - 6}{x + 1}$

In Exercises 17–20, find $f(x + h)$ and $f(x + h) - f(x)$ for the function defined.

17. $f(x) = 3x - 4$

18. $f(x) = 2x - 1$

19. $f(x) = -9x + 2$

20. $f(x) = -5x + 3$

21. (*Cost*) If $C(x) = 32x$ gives the cost in dollars of manufacturing x radios, what is the cost of making 8 radios?

22. (*Revenue*) Suppose that the total revenue a business receives from the sale of x bolts is

$$R(x) = 3x + \frac{1000}{x} \quad \text{cents}$$

What is the total revenue from the sale of 250 bolts?

23. (*Temperature*) After x seconds, the temperature of a metal plate undergoing a finishing process will be

$$T(x) = -2x^2 + 64x + 65 \quad \text{degrees Fahrenheit}$$

(a) What is the temperature after 10 seconds?
(b) What is the temperature at the beginning of the process?

24. (*Rocket flight*) A toy rocket is launched vertically upward with initial velocity of 200 feet per second. Its distance s (in feet) from the ground at any time t (in seconds) is $s(t) = -16t^2 + 200t$. How high is the rocket after 5 seconds?

25. (*Profit*) A manufacturer of telephones determines that the profit from producing and selling x telephones is $P(x) = .01x^2 + 60x - 500$ dollars. What is the manufacturer's profit on the production and sale of 1000 telephones?

26. (*Fence perimeter*) A rectangular fence is to be constructed so that its length is $3x + 2$ meters and its width is x meters. If P is the function that gives the perimeter, determine $P(x)$.

27. (*Bacteria growth*) A colony of 1000 bacteria is introduced to a growth-inhibiting environment. The number of bacteria present at any time t (hours) is given by

$$n(t) = 1000 + 20t + t^2$$

(a) How many bacteria are present after 1 hour?
(b) How many bacteria are present after 10 hours?

28. (*Balloon volume*) The volume of a spherical balloon can be expressed as a function of its radius. Specifically,

$$V(r) = \frac{4}{3}\pi r^3$$

where V is the volume and r is the radius. How much air is in a spherical balloon that is blown up to have a radius of 10 inches? (Use 3.14 for π and round your result to the nearest cubic inch.)

29. (*Medicine dosage*) Various methods are used to calculate the children's dosage of medicines, which is usually less than that for adults. One method gives the child's dosage as a function of age. Assuming that the drug is appropriate for children,

$$D(c) = \frac{c + 1}{24} \cdot a$$

where a = the adult dosage, c = the child's age in years, and D = the child's dosage.

(a) Write the formula for $D(c)$, assuming that the adult dosage of a particular drug is 400 milligrams.
(b) Compute $D(8)$, again assuming that the adult dosage is 400 milligrams.
W (c) Explain the meaning of $D(8)$, computed in part (b).

30. (*Air pollution*) A city's environmental advisors conclude that the amount of carbon monoxide in the air (in parts per million) is given by the function

$$f(x) = 1 + .003x^{1.5}$$

where x is the number of thousands of automobiles driven in the city.

(a) Determine the level of carbon monoxide when 100,000 cars are driven in the city. Use a calculator.
(b) What is the level of carbon monoxide when 1,000,000 cars are driven in the city?

In Exercises 31–48, determine the domain of each function.

31. $f(x) = x^2 + 5$

32. $f(x) = 1 - 5x$

33. $f(x) = \dfrac{1}{x}$

34. $g(x) = \dfrac{1}{x - 2}$

35. $f(x) = \sqrt{x - 2}$

36. $f(x) = \sqrt{x + 5}$

37. $f(x) = \dfrac{1}{x + 3}$

38. $g(x) = \sqrt{5x}$

39. $g(x) = \dfrac{1}{x(x - 1)}$

40. $f(x) = \dfrac{x}{x^2 + x}$

41. $f(x) = \sqrt{3x - 2}$

42. $g(x) = \sqrt{3 - 5x}$

43. $g(x) = (x + 1)^{1/2}$

44. $f(x) = x^3$

45. $g(x) = \dfrac{x}{x^2}$ **46.** $f(x) = \dfrac{x^3}{x}$

47. $f(x) = \dfrac{x + 3}{2x^2 - 9x - 5}$

48. $f(x) = \dfrac{x - 1}{3x^2 + 10x - 8}$

In Exercises 49–56, find the zeros of each function.

49. $f(x) = 2x + 6$ **50.** $f(x) = 3x - 2$

51. $f(x) = x^2 - 9$ **52.** $f(x) = 2x^2 - 50$

53. $f(x) = x^2 - 9x + 20$ **54.** $f(x) = x^2 - x - 12$

55. $f(x) = x^2 + 5x - 2$ **56.** $f(x) = x^2 - 7x + 3$

In Exercises 57–62, determine $(f \circ g)(x)$ and $(g \circ f)(x)$.

57. $f(x) = 3x + 1, g(x) = 7x$

58. $f(x) = 2x - 1, g(x) = x + 6$

59. $f(x) = x^2 + 2x, g(x) = x - 1$

60. $f(x) = x^2 - 7x + 10, g(x) = x + 4$

61. $f(x) = \dfrac{1}{x}, g(x) = 3x$

62. $f(x) = 2x, g(x) = \dfrac{7}{x}$

W 63. Do f and $f(x)$ mean the same thing? If not, explain the difference in meaning.

W 64. (*NUMBER OF MOSQUITOS*) What letter would you choose to name a function that represents the number of mosquitos in a region? Explain why you would select that letter.

1.4 | *LINEAR FUNCTIONS*

In elementary algebra, you obtained the graph of a *straight line* by determining points from the equation of the line. For example, to obtain the graph of $y = 2x + 1$, you can let x be 0, 1, and 2 (or any other real numbers) and then determine each corresponding y value from the equation.

x	$y = 2x + 1$	points
0	1	(0, 1)
1	3	(1, 3)
2	5	(2, 5)

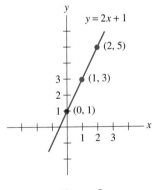

Figure 3

Once two or more points have been obtained, they can be plotted in the *xy* plane and a straight line can be passed through them. See Figure 3.

The graphs of equations such as $y = 3$ and $y = -1$ are *horizontal lines.* (See Figure 4.) In particular, the graph of $y = 0$ is the x axis. Note that the line $y = 3$ is horizontal because for every x value, the y value is 3. Points on the line include $(0, 3)$, $(1, 3)$, $(2, 3)$, and so on.

Horizontal Lines

If c is any real number, then

$$y = c$$

is the equation of a horizontal line.

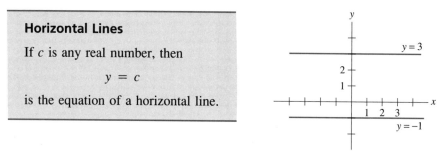

Figure 4 Horizontal lines

The graphs of equations such as $x = 4$ and $x = -3$ are *vertical lines.* (See Figure 5.) In particular, the graph of $x = 0$ is the y axis. Note that the line $x = 4$ is vertical because for every y value, the x value is 4. Points on the line include $(4, 0)$, $(4, 1)$, $(4, 2)$, and so on.

Vertical Lines

If c is any real number, then

$$x = c$$

is the equation of a vertical line.

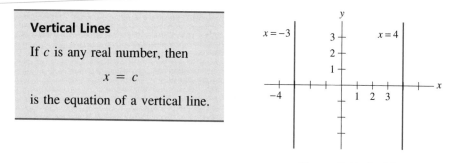

Figure 5 Vertical lines

The steepness or inclination or **slope** of a straight line can be formally defined and measured. In order to include the intuitive notion that the steeper the line, the greater the magnitude of its slope, the slope of a straight line is defined to be the change in y divided by the change in x between any two distinct points (x_1, y_1) and (x_2, y_2) of the line.

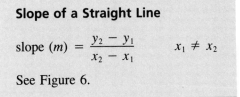

Slope of a Straight Line

$$\text{slope } (m) = \frac{y_2 - y_1}{x_2 - x_1} \qquad x_1 \neq x_2$$

See Figure 6.

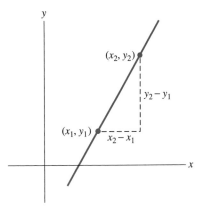

Figure 6 The slope of a line

In many instances, we will want to use the notation Δy (read as "delta y") for the change in y and Δx ("delta x") for the change in x. Using this notation, the definition of slope becomes

Slope (m) of a Line

$$m = \frac{\Delta y}{\Delta x} \qquad \Delta x \neq 0$$

EXAMPLE 1   Determine the slope of the line that passes through the given points and sketch a graph of the line.

(a) (2, 4) and (3, 1)   **(b)** (5, 2) and (8, 4)

SOLUTION **(a)** $m = \dfrac{\Delta y}{\Delta x} = \dfrac{y_2 - y_1}{x_2 - x_1} = \dfrac{1 - 4}{3 - 2} = \dfrac{-3}{1} = -3$

The slope of the line is -3. The graph is shown in Figure 7.

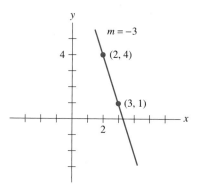

Figure 7

(b) $m = \dfrac{\Delta y}{\Delta x} = \dfrac{y_2 - y_1}{x_2 - x_1} = \dfrac{4 - 2}{8 - 5} = \dfrac{2}{3}$

The slope of the line is 2/3. The graph is shown in Figure 8.

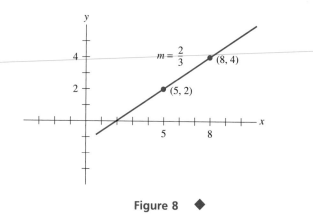

Figure 8 ◆

The example just completed demonstrates the following result.

Sign of the Slope

1. Lines with *negative slope* fall as they go from left to right.

2. Lines with *positive slope* rise as they go from left to right.

We will now digress briefly to present the concept of y intercept. Then we will be prepared to determine a special form for the equation of a straight line—a form involving both slope and y intercept.

The **y intercept** of a straight line is the point where the line crosses the y axis. (*Note*: Some mathematicians consider the y intercept to be merely the y coordinate of that point, since x is always 0 there.) The y intercept of the graph of any function is the point where the graph crosses the y axis. See Figure 9.

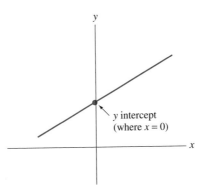

Figure 9 The y intercept

The y intercept of the line $y = 4x - 5$ is determined by letting $x = 0$ in the equation of the line. If $x = 0$, we have

$$y = 4(0) - 5 = 0 - 5 = -5$$

Thus, the y intercept of the line $y = 4x - 5$ is $(0, -5)$ or -5.

There are two very useful forms of the equation of a straight line. We can now proceed to determine both of them.

If a line with slope m passes through point (x_1, y_1) and if (x, y) can be any other point on the line, then by the definition of slope,

$$m = \frac{y - y_1}{x - x_1}$$

Multiplying both sides by $x - x_1$ yields the *point-slope form* of the equation of a straight line.

Equation of a Straight Line

Point-Slope Form

$$y - y_1 = m(x - x_1)$$

$(x_1, y_1) =$ a point on the line

$m =$ slope

Another form of the equation of a straight line can be obtained by considering $y - y_1 = m(x - x_1)$ with (x_1, y_1) being the y intercept. If $(0, b)$ is used to represent the y intercept, then $x_1 = 0$, $y_1 = b$, and we have

$$y - b = m(x - 0)$$

or

$$y - b = mx$$

Finally,

$$y = mx + b$$

This is the *slope-intercept form* of the equation of a straight line.

Equation of a Straight Line

Slope-Intercept Form

$$y = mx + b$$

$m = $ slope

$(0, b) = y$ intercept

EXAMPLE 2   Determine the slope and y intercept of each line.

   (a) $y = 7x - 3$ **(b)** $y = -x + 6$ **(c)** $2y - 3x = 10$

SOLUTION   **(a)** The equation $y = 7x - 3$ is already in the form $y = mx + b$, from which we can see that $m = 7$ and $b = -3$. Thus, the slope is 7 and the y intercept is -3 or $(0, -3)$.

   (b) The equation $y = -x + 6$ or $y = -1x + 6$ is in the form $y = mx + b$. Here $m = -1$ and $b = 6$. Thus, the slope is -1 and the y intercept is 6 or $(0, 6)$.

   (c) The equation $2y - 3x = 10$ is not in slope-intercept form ($y = mx + b$). However, if we add $3x$ to both sides and then divide both sides by 2, we have the desired form, namely,

$$y = \frac{3}{2}x + 5$$

   From this form it is clear that the slope is $3/2$ and the y intercept is 5 or $(0, 5)$. ◆

EXAMPLE 3   Determine the equation of the line having slope 2 and y intercept $(0, -6)$.

SOLUTION   Since the slope is 2, we know that $y = mx + b$ is $y = 2x + b$. Furthermore, given that the y intercept is $(0, -6)$, we know that $b = -6$. Thus, the equation of the line is

$$y = 2x - 6$$

A graph of the line is shown in Figure 10.

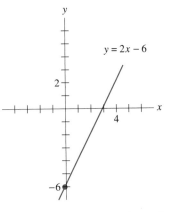

Figure 10 The line $y = 2x - 6$ ◆

The next two examples provide two different approaches to solving the same problem. Example 4 uses the slope-intercept form, whereas Example 5 uses the point-slope form.

EXAMPLE 4   Determine the equation of the line having slope 5 and passing through the point (3, 24). Use the slope-intercept form.

SOLUTION   Since the slope of the line is 5, we know that $y = mx + b$ is

$$y = 5x + b$$

Unlike in Example 3, we do not know the y intercept. But the value of b can be determined by using the fact that the point (3, 24) is on the line. Since (3, 24) is on the line, it must be true that together $x = 3$ and $y = 24$ satisfy the equation of the line. Substituting 3 for x and 24 for y into the equation $y = 5x + b$ will determine b.

$$24 = 5(3) + b$$
$$24 = 15 + b$$
$$b = 9$$

Now we know the equation of the line, namely,

$$y = 5x + 9 \quad ◆$$

EXAMPLE 5   Determine the equation of the line having slope 5 and passing through the point (3, 24). Use the point-slope form.

SOLUTION   Using the given point (3, 24) and slope 5, the point-slope form

$$y - y_1 = m(x - x_1)$$

becomes

$$y - 24 = 5(x - 3)$$

Rather than leave this unfinished equation, we will manipulate it into the more function-like form $y = mx + b$. First,

$$y - 24 = 5x - 15$$

Then

$$y = 5x + 9 \quad \blacklozenge$$

EXAMPLE 6 Determine the equation of the line that passes through the points $(2, 1)$ and $(1, 4)$.

SOLUTION To begin, compute the slope.

$$m = \frac{\Delta y}{\Delta x} = \frac{4 - 1}{1 - 2} = \frac{3}{-1} = -3$$

Since the slope is -3, the equation of the line will be of the form

$$y = -3x + b$$

The value of b can be determined by using either point, $(1, 4)$ or $(2, 1)$, in the equation $y = -3x + b$ in the same manner as in Example 4. Using $(2, 1)$, we have

$$1 = -3(2) + b$$
$$1 = -6 + b$$
$$b = 7$$

Thus, $b = 7$ and the equation of the line is

$$y = -3x + 7$$

A graph of the line is shown in Figure 11.

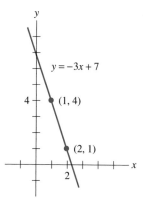

Figure 11 The line $y = -3x + 7$ $\blacklozenge$

EXAMPLE 7 ◆ *A SOCIAL WORKER'S STUDY*

A social worker has been studying data on child abuse that her colleagues have collected over a 3-year period. Based on the 3-year figures, she concludes that child abuse in her region is increasing linearly. Among the data is the fact that there were 230 known cases during the second year and 250 known cases during the third year.

(a) Determine the equation of the line ($y = mx + b$ form) that passes through the two data points (2, 230) and (3, 250).

(b) If child abuse continues to increase linearly according to the equation derived in part (a), how many cases will there be next year (the fourth year)?

SOLUTION **(a)** The equation of the line is of the form $y = mx + b$. Slope m can be computed as follows:

$$m = \frac{\Delta y}{\Delta x} = \frac{250 - 230}{3 - 2} = \frac{20}{1} = 20$$

Since the slope is 20, the line will have the form

$$y = 20x + b$$

The value of b can be determined by using either point, (2, 230) or (3, 250), in the equation $y = 20x + b$, as in Examples 4 and 6. Either way, the result is $b = 190$. Thus, the equation of the line is

$$y = 20x + 190$$

(b) To determine the number of cases in the fourth year (assuming that the pattern continues), let $x = 4$ in the equation $y = 20x + 190$.

$$y = 20(4) + 190$$
$$= 270$$

We conclude that there will be 270 cases next year, if child abuse continues to increase linearly according to the derived equation. ◆

From the intuitive notion of slope as a measure of steepness or inclination, it follows that

Parallel Lines

1. If two distinct lines have the same slope, then they are *parallel*.

2. If two lines are *parallel*, then they have the same slope.

The lines $y = 2x + 1$ and $y = 2x - 3$ are parallel; each line has a slope of 2. (See Figure 12.) By contrast, the lines $y = 3x + 2$ and $y = 2x + 1$ are not parallel, since their slopes are not the same.

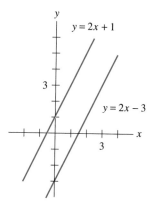

Figure 12 Parallel lines

There is also a relationship between the slopes of perpendicular lines, assuming neither is vertical.

> **Perpendicular Lines**
>
> **1.** If two lines are *perpendicular*, then their slopes are m and $-1/m$.
>
> **2.** Two lines with slopes m and $-1/m$ are *perpendicular*.

Here are three examples, the first of which is graphed in Figure 13.

perpendicular lines	their slopes
$y = 2x - 3$ $y = -\dfrac{1}{2}x + 5$	2 and $-1/2$
$y = -x + 6$ $y = x$	-1 and 1
$y = \dfrac{2}{3}x - 1$ $y = -\dfrac{3}{2}x - 4$	2/3 and $-3/2$

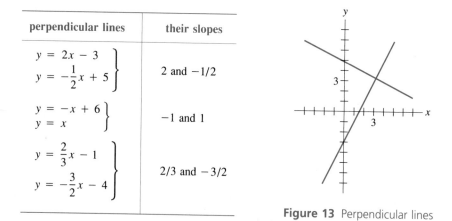

Figure 13 Perpendicular lines

Just as the *y* intercept of a straight line is the point where the line crosses the *y* axis, the **x intercept** of a straight line is the point where the line crosses the *x* axis. See Figure 14.

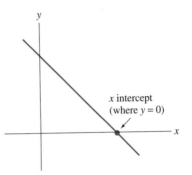

Figure 14 x intercept

To find the *x* intercept of the line $y = 5x - 4$, let $y = 0$ and solve for *x*. Solving $0 = 5x - 4$ yields $x = 4/5$. Thus, the *x* intercept of the line is (4/5, 0).

The section ends with a note about functions. We have been using equations of the form $y = mx + b$. Such equations define **linear functions**. Using function notation, we have $f(x) = mx + b$.

Linear Function f

$$f(x) = mx + b$$

$$m = \text{slope}$$

$$b = y \text{ intercept}$$

1.4   Exercises

In Exercises 1–8, obtain a few points and draw the graph of the given line.

1. $y = x + 2$

2. $y = x - 3$

3. $y = 3x - 2$

4. $y = 2x - 3$

5. $y = -x + 1$

6. $y = -x - 1$

7. $y = 5 - 2x$

8. $y = 1 - 3x$

In Exercises 9–14, determine the slope of the line that passes through the points and sketch a graph of the line.

9. (2, 4) and (6, 16)

10. (5, 2) and (7, 3)

11. (1, 5) and (3, 1)

12. (1, −3) and (4, 1)

13. (4, 9) and (7, 9)

14. (−2, 6) and (1, 11)

In Exercises 15–26, determine the slope and y intercept of each line.

15. $y = 5x + 3$ **16.** $y = 4x - 1$

17. $y = x - 9$ **18.** $y = -x + 2$

19. $y = 1 - 7x$ **20.** $y = 4 + 3x$

21. $y = 3$ **22.** $y = -2$

23. $y - 8x = 6$ **24.** $y + 2x = 7$

25. $5x + 2y - 8 = 0$ **26.** $3y - 8x + 1 = 0$

In Exercises 27–32, determine the equation of the line having the given slope and y intercept.

27. $m = -2, (0, 4)$ **28.** $m = 1, (0, 0)$

29. $m = 5, (0, -3)$ **30.** $m = -1, (0, -2)$

31. $m = 0, (0, -1)$ **32.** $m = 0, (0, 5)$

In Exercises 33–38, determine the equation of the line having the given slope and passing through the given point.

33. $m = 3, (1, 8)$ **34.** $m = 4, (2, 3)$

35. $m = -2, (5, -3)$ **36.** $m = -3, (-1, 4)$

37. $m = -1, (-3, 0)$ **38.** $m = 9, (0, -8)$

In Exercises 39–44, determine the equation of the straight line that passes through the given points.

39. $(4, 2)$ and $(6, 12)$ **40.** $(7, 5)$ and $(4, 2)$

41. $(1, 5)$ and $(2, 3)$ **42.** $(8, 9)$ and $(10, 3)$

43. $(3, 3)$ and $(7, -1)$ **44.** $(-5, 13)$ and $(1, 1)$

45. Determine the equation of the line that is parallel to the line $y = 3x + 5$ and passes through the point $(2, -4)$.

46. Determine the equation of the line that is perpendicular to the line $y = -\frac{1}{5}x$ and passes through the point $(1, 9)$.

47. Are the lines $y = x - 4$ and $y = -x + 4$ perpendicular or parallel or neither?

48. (CELSIUS/FAHRENHEIT) There is a linear relationship between temperature given in Celsius (°C) and in Fahrenheit (°F). Water freezes at 0°C or 32°F. Water boils at 100°C or 212°F. Consider the points $(0, 32)$ and $(100, 212)$, which are of the form (C, F).

(a) Using the points given, determine the slope.
(b) Determine the F intercept.
(c) Write the equation of the line.

49. (ENTOMOLOGY) Entomologists have found that the number of chirps per minute (y) made by a cricket depends on the temperature (x) in degrees Fahrenheit and is a linear relationship. Consider the points $(40, 0)$ and $(60, 80)$.

W (a) Explain the meaning of the point $(60, 80)$ in this setting.
(b) Find the equation of the line that describes the relationship.

50. (WATER TEMPERATURE) Suppose the temperature of the water in a swimming pool between 1 p.m. and 6 p.m. is a linear function of time. At 1 p.m. the water is 70°F, and at 6 p.m. the water is 85°F. Let t be the time in hours and y be the temperature.

(a) Using the information given, list the two known points of the form (t, y). Let $t = 0$ be 1 p.m.
(b) Determine the equation of the line (of form $y = mt + b$) on which the two points lie.
(c) Use the equation from part (b) to determine the water temperature at 2 p.m.
(d) At what time was the water temperature 80°?

51. (DEPRECIATION) Straight-line (linear) depreciation of equipment purchased by businesses is described by the equation

$$y = C - \frac{C - S}{n}t$$

where t is the time in years, y is the value of the asset after t years, n is the useful life in years, C is the original cost, and S is the scrap (resale) value of the asset.

(a) Your company purchases a machine for $3400. If the scrap value is $400 and the useful life is 15 years, determine the linear equation of form $y = mt + b$ that describes the machine's value at any time.
(b) What will be the value of the machine after 8 years?

52. (TAXI FARE) The taxi fare is $1.00, plus 50¢ per quarter mile. If F is the taxi fare and x is the number of quarter miles, find the linear equation that describes such taxi fares.

53. (APPRECIATION) A jewelry store guarantees its customers that the value of all diamonds bought from them will appreciate linearly and that purchasers can trade them in at any time at the appreciated value.

On a $2000 diamond, they guarantee an appreciation of $100 per year.

W (a) Consider $y = mx + b$. If x is the number of years since purchase and b is the original purchase price, what do m and y represent?

(b) Using the known m and b, write the linear equation that describes this situation.

(c) Use the linear equation determined in part (b) to find the guaranteed value of the diamond after 7 years.

(d) Use the linear equation obtained in part (b) to determine in how many years the guaranteed value of the diamond will be $3200.

W 54. (a) Find two points such that the slope of the line through them is undefined.

(b) Explain why the slope is undefined.

(c) What word would you use to describe the line through the two points [from part (a)]?

W 55. At the beginning of the section, it is stated that "to obtain the graph of $y = 2x + 1$, you can let x be 0, 1, and 2 (or any other real numbers) and then determine each corresponding y value." Explain why you can use *any real number* for x in this equation. Include the word "domain" in your explanation.

1.5 │ GRAPHS OF FUNCTIONS

In Section 1.4 the graph of a linear function was obtained by determining points from the equation. The points were then plotted, and a straight line was passed through them. This natural union of algebra and geometry that provides a geometric image of an algebraic equation was developed by the French mathematician René Descartes. His work on such "analytic geometry" was published in 1637 and helped lead to the development of calculus less than 50 years later. The rectangular coordinate system used for plotting points is also known as the *Cartesian* coordinate system—named for Descartes.

Figure 15 René Descartes (1596–1650)

When function notation is used, the linear equation $y = 2x + 1$ is written $f(x) = 2x + 1$. The $f(x)$ means the same as y, so for graphing purposes the points (x, y) become $(x, f(x))$, and $f(x)$ is the second coordinate when the first coordinate is x.

Consider the graph shown in Figure 16. Can you see from the graph that $f(4) = 7$; that is, when $x = 4$, the value of $f(x)$ is 7? [This follows from the fact that the point $(4, 7)$ is on the graph of f.] Also, note that $f(0) = 3$; that is, when $x = 0$, the value of y is 3. Finally, is it clear that $f(-5) = 0$?

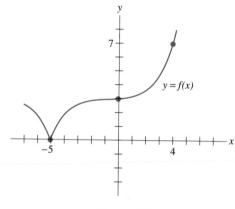

Figure 16

EXAMPLE 1 ◆ *TOWN POPULATION GROWTH*

The graph in Figure 17 shows the population of a small town over a 20-year period.

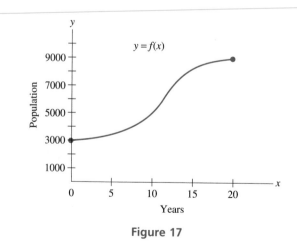

Figure 17

(a) Determine $f(10)$.

(b) Is $f(5) > f(15)$?

(c) Compute $f(20) - f(0)$. What does the result represent?

SOLUTION (a) Since $f(10)$ is the y value when $x = 10$, look for the y coordinate where $x = 10$. It appears from the graph that when $x = 10$, $y = 5000$. In other words, $f(10) = 5000$.

(b) From the graph, it appears that $f(5)$ is between 3000 and 4000, whereas $f(15)$ is more than 7000. Thus, $f(5)$ is not greater than $f(15)$. Alternatively, we can just look at the graph briefly to see that the curve is higher at $x = 15$ than it is at $x = 5$, which means that $f(15) > f(5)$.

(c) $f(20) - f(0) = 9000 - 3000 = 6000$, which is the increase in population during the 20-year period. ◆

The remainder of this section offers a variety of graphs of functions. Keep in mind that the **graph of a function** f is the set of all points $(x, f(x))$ satisfying the equation that defines the function. Ordinarily, we obtain only a few points of a graph from the equation. Then we draw the curve or line (the graph) through those points.

The definition of a function says that for every x there is *one y*. The graphical interpretation of this idea is considered in the **vertical line test**: If a vertical line crosses a curve in two or more places, then the curve is not the graph of a function. (After all, in that case there would be two or more y values corresponding to a particular x. See Figure 18.)

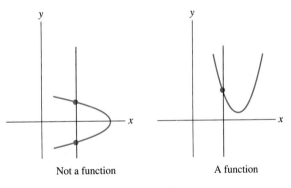

Not a function   A function

Figure 18 Vertical line test

Presented next are some special graphs—graphs of basic functions that will appear throughout the study of calculus.

> **Square Function**
>
> $$f(x) = x^2$$

The domain of the square function consists of all the real numbers. Using 0, 1, −1, 2, −2, etc. for x will yield some points. A smooth curve can then be passed through the points. The graph of $f(x) = x^2$ is an example of a *parabola*. See Figure 19.

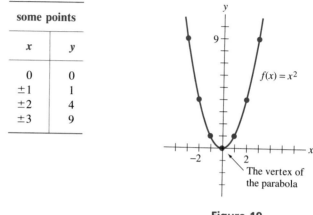

some points	
x	y
0	0
±1	1
±2	4
±3	9

The vertex of the parabola

Figure 19

The square function is one special example of a quadratic function. The graph of any quadratic function is a parabola.

Quadratic Functions

$$f(x) = ax^2 + bx + c \qquad a \neq 0$$

The x coordinate of the *vertex* of $f(x) = ax^2 + bx + c$ is

$$x = -\frac{b}{2a}$$

The vertex is the highest or lowest point on the graph of a quadratic function. It is the point where the graph turns around.

The statement about the x coordinate of the vertex can be proved easily using the calculus presented in Chapter 4. Furthermore, we can show that the parabola opens upward when $a > 0$ and downward when $a < 0$.

An easy way to graph a quadratic function is to use the vertex and at least two points on each side of it.

EXAMPLE 2   Sketch the graph of $f(x) = x^2 - 6x + 7$.

SOLUTION   Clearly f is a quadratic function. Here $a = 1$ and $b = -6$. Thus, the x coordinate of the vertex is

$$x = -\frac{b}{2a} = -\frac{-6}{2(1)} = 3$$

Since $f(3) = (3)^2 - 6(3) + 7 = -2$, the vertex is $(3, -2)$. By letting x be 1, 2, 4, and 5, we obtain two points on each side of the vertex, namely, $(1, 2)$, $(2, -1)$, $(4, -1)$, and $(5, 2)$. The graph is shown in Figure 20.

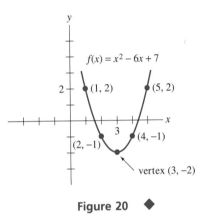

Figure 20 ◆

Next, consider the square root function.

Square Root Function
$$f(x) = \sqrt{x}$$

The domain of the square root function consists of all the nonnegative real numbers (that is, $x \geq 0$). Since the square root of a negative number is not a real number, no negative numbers can have corresponding functional values. Points and the graph of $f(x) = \sqrt{x}$ are shown in Figure 21.

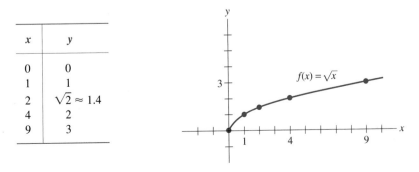

x	y
0	0
1	1
2	$\sqrt{2} \approx 1.4$
4	2
9	3

Figure 21

The numbers 0, 1, 4, and 9 were chosen for x because they are perfect squares. However, we could have chosen such numbers as 3, 5, 6, 7, and 8 and used approximation for the square root (as was done with 2). Note the use of the symbol $\approx$ to mean *approximately equal to*.

$\approx$ means "approximately equal to."

The cube function is presented next.

Cube Function

$$f(x) = x^3$$

The domain of the cube function consists of all the real numbers. See Figure 22 for some points and the graph.

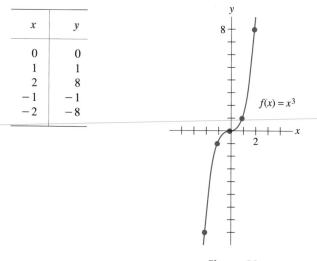

x	y
0	0
1	1
2	8
-1	-1
-2	-8

$f(x) = x^3$

Figure 22

The reciprocal function is presented next. The graph offers an example of a curve called a *hyperbola*.

Reciprocal Function

$$f(x) = \frac{1}{x}$$

The domain consists of all the real numbers except 0, since $x \neq 0$ or else division by zero would occur. The choice of numbers near 0 (that is, near the number for which f is not defined) leads to particularly helpful points. See Figure 23.

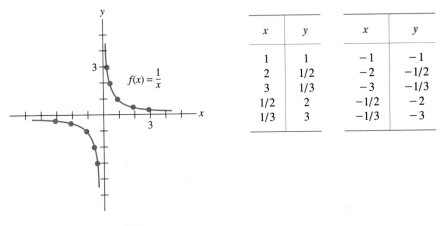

x	y	x	y
1	1	-1	-1
2	1/2	-2	$-1/2$
3	1/3	-3	$-1/3$
1/2	2	$-1/2$	-2
1/3	3	$-1/3$	-3

Figure 23

Recall the concept of **absolute value**, denoted $|\ |$. The absolute value of any real number is the magnitude of the number. Thus, $|+7| = 7$, $|-7| = 7$, and $|0| = 0$. The absolute value function is presented next, and its graph is shown in Figure 24.

Absolute Value Function

$$f(x) = |x|$$

The domain of the absolute value function consists of all the real numbers.

x	y	x	y
0	0	-1	1
1	1	-2	2
2	2	-3	3
3	3		

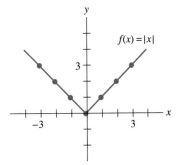

Figure 24 The absolute value function

Absolute value can be defined algebraically in a way that leads naturally to the use of a *two-part function*. (Two-part functions will be needed for the study of limits and continuity in Chapter 2.) We can say that

$$|x| = x \quad \text{when } x \geq 0$$
$$|x| = -x \quad \text{when } x < 0$$

The second statement above says that when x is a negative number (that is, when $x < 0$), $|x|$ will be the opposite signed number. For example, if $x = -6$, then $|x| = -x = -(-6) = 6$. Thus, we have the following formal definition of $|x|$.

$$|x| = \begin{cases} x & x \geq 0 \\ -x & x < 0 \end{cases}$$

To continue, let us graph the two-part function

$$f(x) = \begin{cases} x & x \geq 0 \\ -x & x < 0 \end{cases}$$

The notation of this function is understood to mean

$$f(x) = x \quad \text{when } x \geq 0$$
$$f(x) = -x \quad \text{when } x < 0$$

When using x values that are nonnegative, $f(x)$ is computed using $f(x) = x$. When using x values that are negative, $f(x)$ is computed using $f(x) = -x$. Some points and the graph are shown in Figure 25.

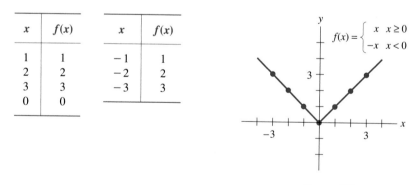

x	$f(x)$		x	$f(x)$
1	1		-1	1
2	2		-2	2
3	3		-3	3
0	0			

$$f(x) = \begin{cases} x & x \geq 0 \\ -x & x < 0 \end{cases}$$

Figure 25

EXAMPLE 3 Sketch the graph of

$$f(x) = \begin{cases} x + 3 & x < 0 \\ 4 & 0 \leq x \leq 7 \end{cases}$$

SOLUTION   Some points and the graph are shown in Figure 26.

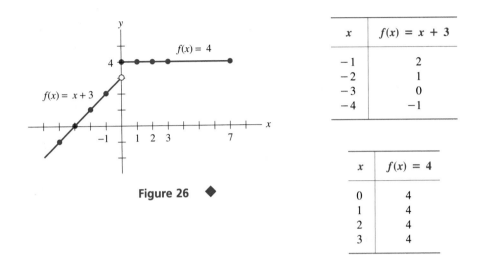

x	$f(x) = x + 3$
−1	2
−2	1
−3	0
−4	−1

x	$f(x) = 4$
0	4
1	4
2	4
3	4

Figure 26 ◆

The small circle at (0, 3) indicates that the left portion of the graph does not include (0, 3). After all, when $x = 0$, $f(x) = 4$. Accordingly, the right portion begins at (0, 4), as shown by the solid dot at (0, 4). The graph ends at (7, 4) because f is not defined for x beyond 7. For example, $f(8)$ is not defined.

EXAMPLE 4   Sketch the graph of

$$f(x) = \begin{cases} x^2 & x \le 0 \\ x + 1 & x > 0 \end{cases}$$

SOLUTION   Figure 27 shows some points and the graph of the function. The solid dot at (0, 0) indicates that (0, 0) is on the graph. The small circle at (0, 1) indicates that the right portion of the graph does not include the point (0, 1).

x	$f(x) = x^2$
0	0
−1	1
−2	4

x	$f(x) = x + 1$
1	2
2	3
3	4

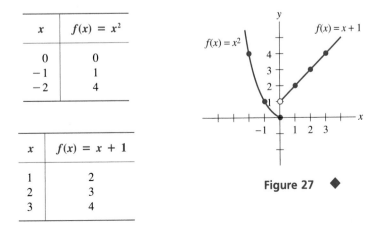

Figure 27 ◆

EXAMPLE 5 ◆ *PAYING VOLUNTEERS FOR AN EXPERIMENT*

A psychologist needs volunteers for an experiment. She offers to pay $8 per hour for volunteers who work up to 5 hours. Those who work more than 5 hours are paid $10 per hour for the additional hours. Let x represent the number of hours worked and write the function V that describes a volunteer's pay.

SOLUTION For x between 0 and 5, the pay is $8 per hour times the number of hours, x. Thus,

$$V(x) = 8x \qquad \text{for } 0 \le x \le 5$$

When x is greater than 5, the person makes $8 per hour for 5 hours ($40 total) plus $10 per hour for each hour above the 5 hours. The number of hours above 5 hours is $x - 5$ hours. So the earnings would be $40 + 10(x - 5)$ dollars for those who work more than 5 hours. The expression simplifies, and we have

$$V(x) = 10x - 10 \qquad \text{for } x > 5$$

The two parts can be combined to give the entire definition of V.

$$V(x) = \begin{cases} 8x & 0 \le x \le 5 \\ 10x - 10 & x > 5 \end{cases} \qquad ◆$$

1.5 Exercises

1. Answer the following questions based on the graph shown.

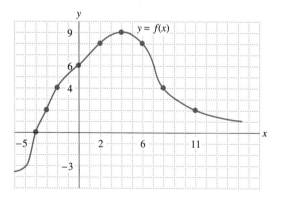

(a) What is y when $x = 6$?
(b) What is $f(0)$?
(c) Determine x for which $f(x) = 0$.
(d) Determine $f(11)$.
(e) Find x for which $f(x) = 8$.

2. Based on the graph shown in Exercise 1, label each of the following statements as true or false.
(a) $f(2) = 5$ (b) $f(-1) = 8$
(c) $f(4) = 9$ (d) $f(8) = 4$
(e) $f(-5) > 0$ (f) $f(-3) > 0$
(g) $f(13) = 5$ (h) $f(1) > f(3)$
(i) $f(10) < f(-1)$ (j) $f(0) \ge 6$

3. *(DISTANCE)* Answer the questions based on the following graph of a 1-hour boat ride.

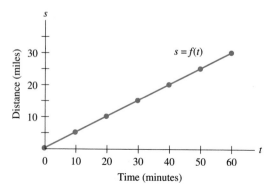

(a) Determine $s(60)$.

(b) How many miles were traveled in the first 10 minutes?

(c) For what t is $s(t) = 20$?

(d) Determine m and b in the linear function $s(t) = mt + b$.

(e) What is the domain of function s?

4. Answer the questions based on the graph shown.

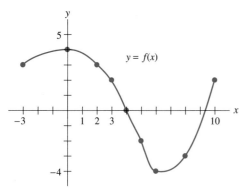

(a) What is the domain of f?

(b) What is the value of $f(5)$?

(c) For what x is $f(x) = 3$?

(d) What is the largest value that $f(x)$ can be?

(e) What is the smallest value that $f(x)$ can be?

5. *(MEMORIZATION)* Psychologists have conducted studies on the retention of memorized material. In the graph shown below, functions N and S both give the amount *forgotten* as a function of time t over a 2-week period after memorization has been completed. N represents nonsense words, and S represents the words of a song.

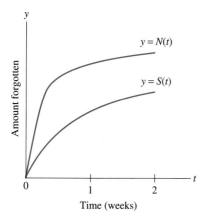

(a) For what value of t is $N(t) = S(t)$?

(b) Complete the sentence. $N(t) > S(t)$ for t _____.

W (c) Explain in nonmathematical words the meaning of the completed sentence in part (b).

6. *(STOCK MARKET)* Function J (graphed below) describes the Dow Jones Industrial Average (DJIA) during a day's trading session (10 a.m. to 4 p.m.). Here x is the time on a 24-hour clock, which means that 1 p.m. = 13, 2 p.m. = 14, 3 p.m. = 15, and 4 p.m. = 16.

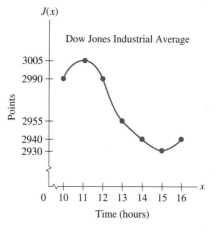

(a) Determine $J(16)$.

(b) Determine x for which $J(x) = 2955$.

(c) At what time did $J(x)$ reach its minimum value for the day?

(d) What was the amount of gain or loss in the DJIA for the day?

7. *(CITY POPULATION)* A sociologist has put together the graphs of two functions (shown below) that show the population growth in two different cities over a

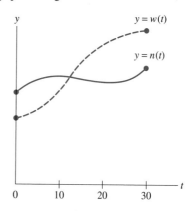

period of 30 years. Function n describes the growth in a selected northeastern U.S. city, and function w describes the growth in a particular western U.S. city.

(a) Which city had the greater population at the beginning?

W (b) Is it true that $w(15) > n(15)$? Explain.

W (c) Is it true that $n(5) < w(5)$? Explain.

8. Refer to the figure and complete the following statements by using inequalities involving x.

(a) $f(x) > g(x)$ when _____.

(b) $f(x) < g(x)$ when _____.

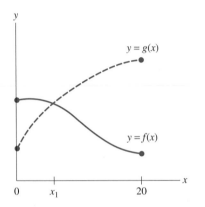

In Exercises 9–14, determine which curves are graphs of functions.

9.   10.

11.   12.

13.   14.

In Exercises 15–22, sketch the graph of the given quadratic function.

15. $f(x) = x^2 - 4x + 5$ 16. $f(x) = x^2 - 6x + 10$

17. $f(x) = x^2 + 6x + 6$ 18. $f(x) = x^2 + 8x + 10$

19. $y = -x^2 + 2x$ 20. $y = -x^2 - 4x - 6$

21. $y = 2x^2 - 8x + 5$ 22. $y = 2x^2 - 4x - 1$

In Exercises 23–32, sketch the graph of each two-part function.

23. $f(x) = \begin{cases} x^2 & x \geq 0 \\ 2 & x < 0 \end{cases}$ 24. $f(x) = \begin{cases} x & x \geq 2 \\ -1 & x < 2 \end{cases}$

25. $f(x) = \begin{cases} \sqrt{x} & x \geq 0 \\ x & x < 0 \end{cases}$ 26. $f(x) = \begin{cases} \sqrt{x} & x > 0 \\ x^3 & x \leq 0 \end{cases}$

27. $f(x) = \begin{cases} \dfrac{1}{x} & x > 0 \\ 1 & x \leq 0 \end{cases}$ 28. $f(x) = \begin{cases} 3 & x > 0 \\ |x| & x \leq 0 \end{cases}$

29. $f(x) = \begin{cases} 3 & x \geq 1 \\ -2 & x < 1 \end{cases}$ 30. $f(x) = \begin{cases} -x & x \geq 0 \\ x & x < 0 \end{cases}$

31. $f(x) = \begin{cases} x^3 & x \geq 0 \\ 2x & x < 0 \end{cases}$

32. $f(x) = \begin{cases} x + 4 & x \geq 0 \\ x^2 & x < 0 \end{cases}$

33. (WORKER'S PAY) Workers at a fast-food restaurant earn $5 per hour for the first 40 hours in a week and then $7.50 per hour for additional hours. Let x be the number of hours worked in a week and write the two-part function W that describes a worker's pay.

34. (TRAIN SPEED) A passenger train travels continuously for 10 hours. For the first 4 hours, the train travels at an average speed of 70 miles per hour. The remainder of the trip is at night, and the train goes an average of 58 miles per hour. Use t for time (in hours) and A for average speed (in miles per hour). Write the two-part function A that describes the train's average speed during the 10-hour ride.

W 35. Explain the meaning of $f(2) = 8$. You may use x and y in your explanation, but do not use f or $f(x)$.

W 36. Suppose that $f(x_1) > f(x_2)$. Does this mean that $x_1 > x_2$? Explain.

W 37. (INSECT POPULATION) Let function P give the insect population at any time t in months. Let $t = 0$ represent *now*. Explain the meaning of the expression $P(6) - P(0)$ in this setting.

W 38. Fill in the blank and then *explain* your answer.
"If function f has no real zero, it means that the graph of $y = f(x)$ never touches or crosses the _____ axis."

 The Exercise Library at the back of the book contains graphing calculator and computer exercises keyed to this section.

1.6 | *TRANSLATIONS AND REFLECTIONS (OPTIONAL)*

The expression $x^2 + 3$ is 3 units larger than x^2. It follows that the y values of $y = x^2 + 3$ will be 3 more than the y values of $y = x^2$—for the same x. In turn, the graph of $y = x^2 + 3$ will be the same shape *but 3 units above* the graph of $y = x^2$. See Figure 28. Similarly, the graph of $y = x^2 - 2$ will be the same shape *but 2 units below* the graph of $y = x^2$. See Figure 29.

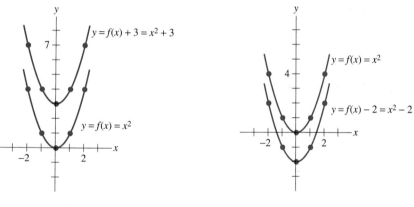

Figure 28   Figure 29

An awareness of such *vertical translations* (up or down) of familiar graphs can be helpful in drawing other graphs. For example, if you know the graph of $y = x^2$, then graphing $y = x^2 + 3$ is an easy matter. Simply draw the graph of $y = x^2$ and shift it up 3 units, point for point. See Figure 30. Similarly, the graph of $y = x^2 - 2$ can be drawn by shifting the graph of $y = x^2$ down 2 units.

Vertical Translation (Examples)

The graph of $y = f(x) + 3$ is 3 units above the graph of $y = f(x)$.

The graph of $y = f(x) - 2$ is 2 units below the graph of $y = f(x)$.

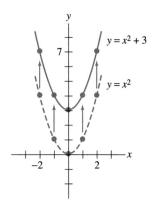

Figure 30 Obtaining the graph of $y = x^2 + 3$ by shifting the graph of $y = x^2$ up 3 units point for point

In general,

Vertical Translation ($c > 0$)

The graph of $y = f(x) + c$ is c units above the graph of $y = f(x)$.

The graph of $y = f(x) - c$ is c units below the graph of $y = f(x)$.

Next we will consider *horizontal translations*. Recall the square root function, $y = f(x) = \sqrt{x}$, the domain of which consists of all $x \geq 0$. The graph "begins" where $x = 0$. See Figure 31.

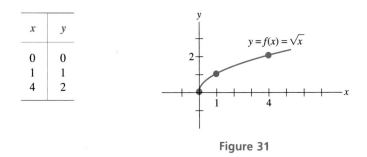

x	y
0	0
1	1
4	2

Figure 31

Consider the function $y = f(x - 1) = \sqrt{x - 1}$. The domain consists of all real numbers for which $x - 1 \geq 0$. In other words, $x \geq 1$. The graph "begins" where $x = 1$. See Figure 32.

x	y
1	0
2	1
5	2

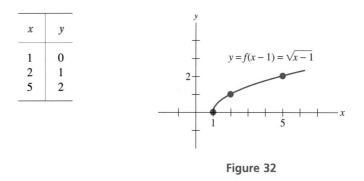

Figure 32

The graph of $y = f(x - 1) = \sqrt{x - 1}$ is point for point a shift to the right 1 unit of the graph of $f(x) = \sqrt{x}$.

To see a shift to the left, consider $y = f(x + 4) = \sqrt{x + 4}$. The domain of the function consists of the numbers $x \geq -4$. When $x = -4$, $y = 0$. The graph "begins" at $(-4, 0)$ and can be obtained by shifting the graph of $y = f(x) = \sqrt{x}$ to the left 4 units. See Figure 33.

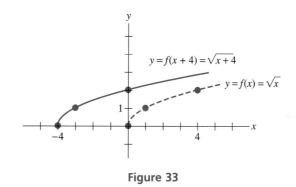

Figure 33

Similarly, the graph of $y = \sqrt{x + 1}$ can be obtained by shifting the graph of $y = \sqrt{x}$ to the left 1 unit.

Horizontal Translation (Examples)

The graph of $y = f(x - 1)$ is 1 unit to the right of the graph of $y = f(x)$.

The graph of $y = f(x + 4)$ is 4 units to the left of the graph of $y = f(x)$.

In general,

> ### Horizontal Translation ($c > 0$)
>
> The graph of $y = f(x - c)$ is c units to the right of the graph of $y = f(x)$.
>
> The graph of $y = f(x + c)$ is c units to the left of the graph of $y = f(x)$.

The concept of *reflection* is presented next. To compare the graphs of $y = x^2$ and $y = -x^2$, note first that points for the graph of $y = -x^2$ can be obtained from points for the graph of $y = x^2$ simply by changing the sign of the y coordinates. (After all, $-x^2$ means $-1 \cdot x^2$.)

x	$y = x^2$	$y = -x^2$
0	0	0
± 1	1	-1
± 2	4	-4
± 3	9	-9

The graphs are shown in Figure 34.

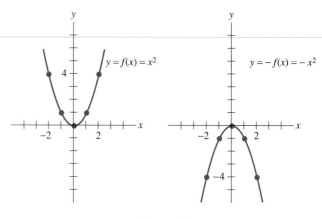

Figure 34

The sign change created by placing a minus in front of the x^2 resulted in a *reflection* of the graph across the x axis. In general,

> ### Reflection
>
> The graph of $y = -f(x)$ is a reflection across the x axis of the graph of $y = f(x)$.

1.6   Exercises

Use translations and reflections of the basic graphs $y = x^2$, $y = \sqrt{x}$, $y = x^3$, $y = |x|$, and $y = 1/x$ to sketch the graph of each function given.

1. $y = \sqrt{x} + 2$

2. $y = \sqrt{x} - 1$

3. $y = -\dfrac{1}{x}$

4. $y = (x - 1)^2$

5. $y = x^2 - 3$

6. $y = |x - 3|$

7. $y = -|x|$

8. $y = \dfrac{1}{x} + 1$

9. $y = -x^3$

10. $y = x^3 + 1$

11. $y = |x + 2|$

12. $y = |x| - 3$

13. $y = x^2 + 1$

14. $y = (x + 1)^3$

15. $y = \sqrt{x - 3}$

16. $y = \dfrac{1}{x + 2}$

1.7 | *FUNCTIONS IN ECONOMICS*

Functions that provide information about cost, revenue, and profit can be of great value to management. This section offers an introduction to the cost function (C), revenue function (R), and profit function (P) as well as a presentation of supply and demand concepts. We begin by establishing the notation for three important types of functions. Using x for the number of units produced or sold, we have

---

Cost, Revenue, Profit

$C(x) =$ The total *cost* of producing x units

$R(x) =$ The total *revenue* from the sale of x units

$P(x) =$ The total *profit* from the production and sale of x units

---

EXAMPLE 1 ◆ **COST OF PRODUCING RADIOS**

Assume that the cost of producing x radios is $C(x) = .4x^2 + 7x + 95$ dollars.

(a) Find the cost of producing 20 radios.

(b) Determine the cost of producing the 20th radio.

(c) Determine the cost of producing 0 radios.

SOLUTION   **(a)** The cost of producing 20 radios is $C(20)$.

$$C(20) = .4(20)^2 + 7(20) + 95$$
$$= 160 + 140 + 95$$
$$= 395$$

The cost of producing 20 radios is $395.

(b) The cost of producing the 20th radio can be determined by subtracting the cost of the first 19 radios from the cost of the first 20 radios. That cost is

$$C(20) - C(19) = (395) - [.4(19)^2 + 7(19) + 95]$$
$$= 395 - 372.40$$
$$= 22.60$$

The 20th radio costs $22.60 to produce.

(c) The cost of producing 0 radios is $C(0)$.

$$C(0) = .4(0)^2 + 7(0) + 95 = 95$$

The cost of producing 0 radios is $95.   ◆

The cost of producing no units [see Example 1, part (c)] is called the **fixed cost** or **overhead**. Such cost can vary from nearly zero to large amounts. Overhead can include such things as rent, tooling, training, insurance, equipment purchase, research, design, and other expenses that exist regardless of how many units are produced.

$$C(0) = \text{fixed cost or overhead}$$

A profit function P is sometimes given directly, but other times it may be necessary to determine profit as revenue minus cost.

$$\text{Profit} = \text{Revenue} - \text{Cost}$$
$$P(x) = R(x) - C(x)$$

EXAMPLE 2   ◆   *PROFIT ON THE MANUFACTURE AND SALE OF RADIOS*

It costs a manufacturer $C(x) = .4x^2 + 7x + 95$ dollars to produce x radios. They can be sold at $40 each; that is, revenue from the sale of x radios is $R(x) = 40x$ dollars.

(a) Determine the profit function.

(b) What is the profit on the manufacture and sale of 25 radios?

(c) What is the profit on the manufacture and sale of 2 radios?

SOLUTION   (a) Using $P(x) = R(x) - C(x)$, we have

$$P(x) = (40x) - (.4x^2 + 7x + 95)$$

which simplifies to

$$P(x) = -.4x^2 + 33x - 95$$

(b) The profit on the manufacture and sale of 25 radios is $P(25)$.

$$P(25) = -.4(25)^2 + 33(25) - 95 = 480$$

The company would have a profit of $480 on the manufacture and sale of 25 radios.

(c) The profit on the manufacture and sale of 2 radios is $P(2)$.

$$P(2) = -.4(2)^2 + 33(2) - 95 = -30.6$$

The *minus* (negative profit) indicates a *loss*. The company would *lose* $30.60 on the manufacture and sale of just 2 radios.   ◆

In parts (b) and (c) of Example 2, profit could have been computed by evaluating the cost and revenue functions separately. In (b), $R(25) = 1000$ and $C(25) = 520$. Then $P(25) = R(25) - C(25) = 480$, which demonstrates that $P(x) > 0$ when $R(x) > C(x)$. In (c), $R(2) = 80$ and $C(2) = 110.60$. Then $P(2) = R(2) - C(2) = -30.6$, which demonstrates that $P(x) < 0$ when $C(x) > R(x)$. In general,

Profit	when $R(x) > C(x)$
Loss	when $C(x) > R(x)$

It is natural to wonder what the profit will be when $R(x) = C(x)$. Since $P(x) = R(x) - C(x)$, clearly $P(x) = 0$ when $R(x) = C(x)$. The company will break even when $P(x) = 0$. Thus, x for which $R(x) = C(x)$ is called the *break-even quantity*. The point of intersection of the graphs of $y = R(x)$ and $y = C(x)$ is called the **break-even point**. See Figure 35. Knowing how many units must be sold in order to break even is important to management when they consider production and marketing of new products.

The next two examples show an algebraic approach to determining the break-even quantity.

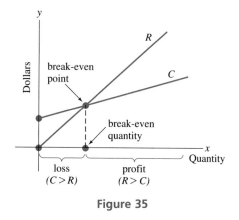

Figure 35

EXAMPLE 3 ◆ *Break-even Quantity*

A manufacturer of plastic containers for compact disks has the following profit function:

$$P(x) = .3x - 150 \quad \text{dollars}$$

where x is the number of CD containers produced and sold. How many containers must be made and sold in order to break even?

SOLUTION To break even, profit must be zero. That is,

$$.3x - 150 = 0$$

or

$$.3x = 150$$

This leads to

$$x = \frac{150}{.3} = 500$$

Thus, 500 CD containers must be manufactured and sold for this company to break even. Figure 36 shows a graph of the profit function.

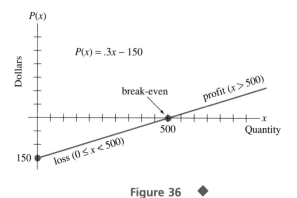

Figure 36 ◆

EXAMPLE 4 ◆ **BREAK-EVEN QUANTITY**

Assume that cost and revenue (in dollars) are given by

$$C(x) = 20x + 1250$$
$$R(x) = 50x - .1x^2$$

For what quantity x will this business break even?

SOLUTION It will break even when $R(x) = C(x)$; that is, when

$$50x - .1x^2 = 20x + 1250$$

or

$$.1x^2 - 30x + 1250 = 0$$

The quadratic equation can be solved by factoring (which you may find difficult), or by the quadratic formula as

$$x = \frac{30 \pm \sqrt{900 - 4(.1)(1250)}}{2(.1)}$$

$$= \frac{30 \pm 20}{.2}$$

which leads to

$$x = \frac{50}{.2} = 250$$

and

$$x = \frac{10}{.2} = 50$$

We conclude that *the business will break even when the quantity produced and sold is either 50 or 250 units.* (When $x < 50$ units, cost is greater than revenue, and a loss results. When $50 < x < 250$, revenue is greater than cost, and a profit results. When $x > 250$, cost is greater than revenue, and a loss results.) ◆

> *Note*
>
> In Example 4, the square root was $\sqrt{900 - 500}$ or $\sqrt{400}$, which is exactly 20. If instead it had been $\sqrt{300}$, then a calculator approximation (such as 17.3) would be used to continue the procedure in order to determine the quantity x.

Price Functions

Consumers know that often there is a relationship between the price of an item and the demand for it. When the price is high, the demand is low. When the price is lower, consumer demand is greater. The relationship between the price per unit p and the quantity demanded x may be given by a **demand equation**.

EXAMPLE 5 ◆ PRICE AND DEMAND

Assume that for some product, the equation

$$p = 80 - .2x \quad \text{dollars}$$

gives the relationship between the price per unit p and the quantity x demanded. If the price of this product is set at \$70 per unit, the quantity demanded is x such that

$$70 = 80 - .2x$$

Solving the equation for x yields

$$x = 50 \qquad \text{(quantity demanded)}$$

If the price is lowered to \$65, then we have

$$65 = 80 - .2x$$

or

$$x = 75 \qquad \text{(quantity demanded)}$$

This example shows the typical relationship between price and quantity demanded: the lower the price, the greater the demand. ◆

The relationship $p = 80 - .2x$ might have been stated as $p + .2x = 80$ or as $.2x = 80 - p$. However, the form $p = 80 - .2x$ was chosen in anticipation of using the equation as a **price function** p and using the price function to construct a total revenue function R.

Revenue

x = number of units (quantity)

p or $p(x)$ = price per unit

$R(x) = x \cdot p$ = total revenue from the sale of x units

EXAMPLE 6 ◆ CONSTRUCTING A REVENUE FUNCTION

Use the demand equation $p = 80 - .2x$ to construct the revenue function and then find the total revenue from the sale of 90 units.

SOLUTION The equation $p = 80 - .2x$ can be written as $p(x) = 80 - .2x$, using function notation to emphasize its price-function nature. Then

$$R(x) = x \cdot p(x)$$

becomes

$$R(x) = x(80 - .2x)$$

or

$$R(x) = 80x - .2x^2$$

The revenue from the sale of 90 units is then

$$R(90) = 80(90) - .2(90)^2$$
$$= 5580$$

The revenue function is $R(x) = 80x - .2x^2$, and the revenue from the sale of 90 units is $5580. ◆

The relationship between the price per unit paid to a supplier and the number of items being supplied may be given by a **supply equation**. Ordinarily, the quantity x supplied will be greater when the price p is higher and less when the price is lower.

EXAMPLE 7 ◆ **PRICE AND SUPPLY**

Assume that the equation

$$p = .02x + 3$$

gives the relationship between the price p (in dollars) per unit and the quantity x supplied.

(a) Find the number of units supplied when the price is $4.00 per unit.

(b) Find the number of units supplied when the price is $4.50 per unit.

SOLUTION **(a)** When the price is $4 per unit, we have

$$4 = .02x + 3$$

or

$$x = 50$$

50 units will be supplied at a price of $4 each.

(b) When the price is $4.50 per unit, we have

$$4.50 = .02x + 3$$

or

$$x = 75$$

75 units will be supplied when the price is $4.50 per unit. ◆

The supply equation $p = .02x + 3$ defines a price function. (We could write it as $p(x) = .02x + 3$ to emphasize the point.) The total cost function C can be constructed from the price function p.

Cost

x = number of units (quantity)

p or $p(x)$ = price per unit

$C(x) = x \cdot p$ = total cost to supply x units

EXAMPLE 8 ◆ *CONSTRUCTING A COST FUNCTION*

Use the supply equation $p = .02x + 3$ to construct the cost function and then find the total cost of supplying 85 units.

SOLUTION The equation $p = .02x + 3$ can be written as $p(x) = .02x + 3$, using function notation to emphasize its price-function nature. Then

$$C(x) = x \cdot p(x)$$

becomes

$$C(x) = x(.02x + 3)$$

or

$$C(x) = .02x^2 + 3x$$

The cost of supplying 85 units is then

$$C(85) = .02(85)^2 + 3(85)$$
$$= 399.50$$

The cost function is $C(x) = .02x^2 + 3x$ and the cost of supplying 85 units is $399.50. ◆

Note

Since price functions are used for both supply and demand, each time one arises there will be an indication of whether it is a supply equation or a demand equation. An alternative approach is to use D or $D(x)$ for demand and S or $S(x)$ for supply. The alternative avoids possible ambiguity.

$p = S(x)$ supply

$p = D(x)$ demand

The market for a product will be in a state of *equilibrium* when the quantity supplied (or produced) is equal to the quantity demanded. The point (x, p) for which equilibrium exists is called the **equilibrium point** and will be designated as (x_e, p_e). See Figure 37.

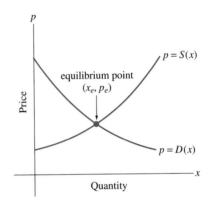

Figure 37 Supply and demand—and the equilibrium point

EXAMPLE 9 ◆ *SUPPLY, DEMAND, AND EQUILIBRIUM*

Suppose the demand equation for a product is $p = 17 - .2x$ (dollars) and the supply equation is $p = .4x + 8$ (dollars).

(a) Find the equilibrium quantity.

(b) Find the equilibrium price.

(c) Determine the equilibrium point.

SOLUTION **(a)** The quantity for which equilibrium will exist is the value of x for which supply and demand are equal. In other words, we seek x such that

$$.4x + 8 = 17 - .2x$$

or

$$.6x = 9$$

The result is

$$x = 15$$

The equilibrium quantity is 15.

(b) The price at market equilibrium is the p value corresponding to the equilibrium quantity, 15. Either equation (supply or demand) can be used to obtain p by using 15 for x. From the demand equation $p = 17 - .2x$, we have

$$p = 17 - .2(15)$$
$$= 17 - 3$$
$$= 14$$

The equilibrium price is $14.

(c) The equilibrium point (x_e, p_e) has as its coordinates the equilibrium quantity x_e and the equilibrium price p_e. Thus, based on parts (a) and (b), we have

$$(x_e, p_e) = (15, 14) \quad \blacklozenge$$

As a final note, using $p = S(x)$ and $p = D(x)$ for the supply and demand equations, we can write

$$C(x) = x \cdot p = x \cdot S(x)$$

and

$$R(x) = x \cdot p = x \cdot D(x)$$

In summary,

$$C(x) = x \cdot S(x)$$
$$R(x) = x \cdot D(x)$$

1.7   Exercises

1. *(COST)* Suppose the cost of producing x umbrellas is $C(x) = .1x^2 + 5x + 210$ dollars.
   (a) Find the cost of producing 25 umbrellas.
   (b) Find the cost of producing the 25th umbrella.
   (c) Determine the fixed cost or overhead.

2. *(COST)* Let $C(x) = .2x^2 + 17x + 95$ be the cost (in dollars) of making x wheel covers.
   (a) What is the cost of making 20 wheel covers?
   (b) Find the cost of making the 10th wheel cover.
   (c) What is the fixed cost?

3. *(COST)* If the cost of producing x lamps is given by $C(x) = 35x + 195$ dollars, how many lamps can be produced for $1000?

4. *(COST)* A calculator manufacturer determines that the cost to make each calculator is $3 and the fixed cost is $150. Determine the cost function, that is, the total cost of producing x calculators.

5. *(REVENUE)* A manufacturer of automobile batteries finds that the revenue from the sale of x batteries is $27x - .01x^2$ dollars.
   (a) How much revenue is derived from the sale of 100 batteries?
   (b) What is the revenue obtained from the sale of the 100th battery?

6. *(REVENUE)* A cigar box distributor has the revenue function

$$R(x) = 1.35x \text{ dollars}$$

where x is the number of boxes sold.

(a) How much revenue is obtained from selling 5 boxes?

(b) What is the revenue obtained from the sale of the 5th box?

(c) What is the revenue obtained from the sale of the 8th box?

7. *(PROFIT)* Consider that it costs a TV manufacturer $C(x) = .1x^2 + 150x + 1000$ dollars to produce x TV sets. The revenue from the sale of x TV sets is $R(x) = 280x$ dollars.

(a) Determine the profit function.

(b) What is the profit on the manufacture and sale of 50 TV sets?

8. *(PROFIT)* If revenue from the sale of x carpets is $R(x) = 90x$ and the cost to obtain the carpets is $C(x) = 50x + .03x^2$, determine the profit function.

9. *(PROFIT)* A tire maker can produce x tires at a cost of $29 + .02x$ dollars *per tire*. They can sell the tires at $54 each.

(a) Determine the cost function.

(b) Determine the revenue function.

(c) Determine the profit function.

10. *(PROFIT)* A manufacturer of felt-tip pens can produce x boxes of pens for $2.4 + .01x$ dollars *per box*. They can sell the pens at $3.59 per box.

(a) Determine the cost function.

(b) Determine the revenue function.

(c) Determine the profit function.

11. *(PROFIT)* If x barrels can be produced at a cost of $6 eacn and sold at a price of $15 - .02x$ dollars each, determine the profit function P.

12. *(PROFIT)* If x jackets can be produced at a cost of $.01x + 19$ dollars each and sold at a price of $50 each, determine the profit function P.

13. *(BREAK-EVEN QUANTITY)* A travel agent determines that her monthly profit on the sale of x dollars worth of airline tickets is $P(x) = .1x - 410$ dollars.

(a) What is her profit on $5800 in monthly airline ticket sales?

(b) How much is her loss if sales are only $2000?

(c) What must be her monthly sales in order to break even?

14. *(BREAK-EVEN QUANTITY)* The management of a publishing company informs the marketing department that

the profit function is $P(x) = .08x - 15{,}200$ dollars, where x is the number of dollars of sales.

W (a) How would management react to sales of $100,000?

(b) How many dollars of sales are needed to break even?

W 15. *(PROFIT)* Suppose you know the cost and revenue functions for a particular business. Explain how you would use $C(x)$ and $R(x)$ to determine their profit from the sale of the 75th unit. Use words rather than expressions or equations.

W 16. *(COST)* Let $C(x)$ be the cost of producing x limousines.

(a) What is the meaning of the expression $C(45) - C(43)$?

(b) What is the meaning of $C(2)$?

17. *(DEMAND)* Suppose the equation $p = 90 - .02x$ gives the relationship between the price (in dollars) per unit and the quantity x demanded. If the price is set at $52 per unit, what is the quantity demanded?

18. *(DEMAND)* If $p = -.04x + 72$ is the relationship between the price per unit (in dollars) and the quantity x demanded, what is the quantity demanded when the price is $50 per unit?

19. *(DEMAND AND REVENUE)* Use the demand equation $p = 50 - .1x$ dollars to construct the revenue function and then find the total revenue from the sale of 40 units.

20. *(DEMAND AND REVENUE)* Let the demand equation be $p = 34 - .3x$ dollars.

(a) Determine $R(x)$.

(b) Determine $R(15)$.

21. *(SUPPLY)* Assume that the equation $p = .3x + 17$ gives the relationship between the price (in dollars) per unit and the quantity x supplied. If the price is set at $65 per unit, what quantity will be supplied?

22. *(SUPPLY)* Let $p = 5 + .04x$ be the relationship between the price (in dollars) per unit and the quantity (x) supplied. If the price is set at $73 per unit, what quantity will be supplied?

23. *(SUPPLY AND COST)* Assume the supply equation is $p = 24 + .4x$ dollars.

(a) Determine the cost function.

(b) Find $C(20)$.

(c) What is the cost of the 10th unit?

24. (*SUPPLY AND COST*) What is the cost of producing 100 units if the supply equation is $p = 8 + .01x$ dollars?

(*EQUILIBRIUM*) In Exercises 25–30, use the given demand equation and supply equation to determine (a) the equilibrium quantity (b) the equilibrium price (c) the equilibrium point. Assume p is in dollars.

25. demand: $p = 20 - .3x$; supply: $p = .1x + 8$

26. demand: $p = 74 - .08x$; supply: $p = .02x + 3$

27. demand: $p = 100 - .1x$; supply: $p = 52$

28. demand: $p = 104$; supply: $p = .5x + 14$

29. demand: $p = 47 - .2x$; supply: $p = 1 + .03x$

30. demand: $p = 22 - .04x$; supply: $p = .2x + 4$

(*EQUILIBRIUM*) In Exercises 31–34, use the given supply and demand functions to determine the quantity and price at which equilibrium occurs. Assume the monetary unit is dollars.

31. $S(x) = 2x + 43$; $D(x) = 160 - x$

32. $S(x) = .04x + 10$; $D(x) = 38 - .03x$

33. $S(x) = 5 + .3x$; $D(x) = 29$

34. $S(x) = x + 1$; $D(x) = 91 - .2x$

(*SUPPLY, DEMAND, AND PROFIT*) For each supply and demand function given in Exercises 35–38, determine the cost, revenue, and profit functions.

35. $S(x) = .2x + 11$; $D(x) = 90 - .4x$

36. $S(x) = .05x + 3$; $D(x) = 17$

37. $S(x) = 7x + 10$; $D(x) = 5 + \frac{3}{x}$

38. $S(x) = 32$; $D(x) = 150 - .2x$

39. (*SUPPLY, DEMAND, AND PROFIT*) If $S(x) = .02x + 30$ dollars and $D(x) = 50 - .01x$ dollars, what is the profit on the production and sale of 20 units?

40. (*SUPPLY, DEMAND, AND PROFIT*) If $S(x) = 15 + .01x$ dollars and $D(x) = 21 - .03x$ dollars, what is the profit on the production and sale of 100 units?

W 41. (*EQUILIBRIUM*) Once the equilibrium quantity x is determined, then the equilibrium price can be determined from either the supply equation or the demand equation. Why doesn't it matter which of the two equations is used?

The Exercise Library at the back of the book contains graphing calculator and computer exercises keyed to this section.

Chapter List *Important terms and ideas*

open interval
closed interval
half-open interval
quadratic formula
function
independent variable
dependent variable
domain
range
zero of a function
composition of functions
linear function
slope
y intercept

point-slope form
slope-intercept form
x intercept
graph of a function
vertical line test
square function
quadratic function
vertex of a parabola
square root function
cube function
reciprocal function
absolute value function
two-part function

vertical translation (optional)
horizontal translation (optional)
reflection (optional)
cost function
revenue function
profit function
overhead
fixed cost
break-even point
demand equation
price function
supply equation
equilibrium point

Review Exercises for Chapter 1

1. Write the inequality $1 \le x < 7$ using interval notation.

2. Write the inequality $t \ge 0$ using interval notation.

3. Solve the linear inequality $4(x - 2) \le 3$ and write the answer in interval notation.

4. Solve the linear inequality $10 - 4x > 15$ and write the answer in interval notation.

5. If $f(x) = \dfrac{3x^2}{1 + x}$, find $f(0)$, $f(2)$, and $f(-2)$.

W 6. Comment on the calculation of $f(-1)$, where f is the function defined in Exercise 5.

7. If $f(x) = 2\sqrt{x - 3}$, find $f(4)$, $f(7)$, and $f(8)$.

W 8. Comment on the calculation of $f(0)$, where f is the function defined in Exercise 7.

9. If $f(x) = 3x^2$, find $f(x + 1)$ and $f(x + h)$.

10. If $g(x) = x^2 - x$, find $g(x + 2)$ and $g(x + h)$.

In Exercises 11–14, determine the domain of each function.

11. $f(x) = \dfrac{x}{2x - 1}$

12. $f(x) = \dfrac{1 - x}{x}$

13. $g(x) = \sqrt{x + 9}$

14. $g(x) = x^3$

15. Find the zeros of $f(x) = 3x^2 - 27$.

16. Find the zeros of $g(x) = 1 - 3x$.

17. If $f(x) = x^2 - 3$ and $g(x) = x + 1$, find $(f \circ g)(x)$.

18. If $f(x) = 4x^2$ and $g(x) = 5x - 19$, find $(g \circ f)(x)$.

19. Find the slope of the line that passes through the points $(-1, 2)$ and $(1, 5)$ and sketch the graph.

20. Determine two points that lie on the line given by $y = -2x + 3$.

21. Find the slope and y intercept of the line given by $y = 5x - 1$.

22. Find the equation of the line having slope 3 and y intercept 11.

23. Determine the equation of the line having slope 6 and passing through the point $(2, -7)$.

24. Determine the equation of the line parallel to the line $y = -2x + 5$ and passing through the point $(4, 0)$.

25. Write the equation of the line that crosses the y axis where y is 6 and crosses the x axis where x is 2.

26. What is the slope of a line that is perpendicular to the line $y = 7x - 2$?

27. Sketch the graph of the quadratic function given by $f(x) = x^2 - 2x + 7$.

28. Sketch the graph of the quadratic function given by $y = -x^2 + 6x - 5$.

In Exercises 29–32, sketch the graph of each two-part function.

29. $f(x) = \begin{cases} 2 & x \le 0 \\ -x & x > 0 \end{cases}$

30. $f(x) = \begin{cases} x & x < 0 \\ x^2 & x \ge 0 \end{cases}$

31. $g(x) = \begin{cases} x & x < 1 \\ \sqrt{x} & x \ge 1 \end{cases}$

32. $g(x) = \begin{cases} |x| & x < 0 \\ 2x & x \ge 0 \end{cases}$

33. (*RENTAL CAR COST*) An executive rents a car for one day. The cost of the rental is $26 plus 30¢ per mile driven. Let x be the number of miles driven and y be the total rental car bill. Express y as a function of x, using $y = mx + b$.

34. (*PLUMBER'S BILL*) A plumber charges $30 to come to the house plus $50 per hour once there. Let x be the number of hours the plumber works and y be the total bill. Express y as a function of x, using $y = mx + b$.

35. (*BACTERIA CULTURE*) Suppose $n(t) = 300 + 12t + t^2$ gives the number of bacteria present in a lab culture at any time t, where t is in hours.
   (a) How many bacteria were present at the beginning, when the culture was started?
   W (b) Compute $n(5)$ and tell what the result means.

36. (*REVENUE*) Suppose the revenue from the sale of x bags of pretzels is $.75x$ dollars.
   (a) What is the revenue when 110 bags of pretzels are sold?
   (b) What is the revenue from the sale of the last (110th) bag?

37. (*Cost, revenue, and profit*) It costs a stereo manufacturer $C(x) = .1x^2 + 170x + 900$ dollars to
produce x stereo units. The stereos can be sold for
$300 each.
   (a) Determine the revenue function R.
   (b) Determine the profit function P.
W **(c)** Compute $P(5)$. Interpret the result.
   (d) What is the cost of making 10 stereos?
   (e) What is the cost of making the 10th stereo?

38. (*Cost, revenue, and profit*) If flower pots can be
produced at a cost of $1.25 each and sold at a price
of $2 - .01x$ dollars each, determine the profit
function P. Note that x is the number of flower pots.

39. (*Equilibrium*) Let the demand equation be given by
$p = 25 - .1x$ and the supply equation be given by
$p = 1 + .02x$. Find the equilibrium point.

2

AN INTRODUCTION
TO LIMITS

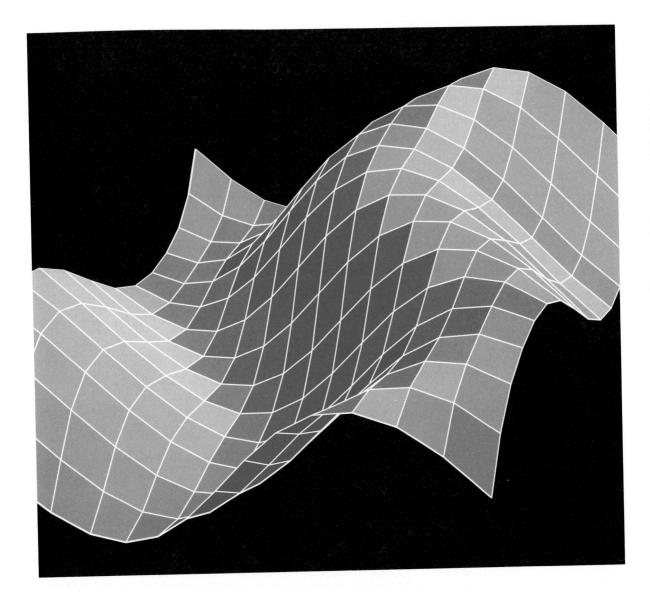

We begin with some historical notes on the invention of calculus by Isaac Newton and Gottfried Wilhelm Leibniz. We then proceed with the study of calculus, beginning with an introduction to the concept of limit. The limit ideas presented here will provide a foundation on which other calculus concepts will be built.

2.1 | *INTRODUCTION TO LIMITS*

Newton and Leibniz

The Englishman Isaac Newton and the German Gottfried Wilhelm Leibniz are the mathematicians credited with inventing calculus. They worked separately from each other and without exchanging ideas. Newton invented calculus in 1665, but he took more than 20 years to publish his results. Consequently, Leibniz published his own development of calculus first. Furthermore, Leibniz developed notation that was considered superior to the notation that Newton used.

Figure 1 Isaac Newton (1642–1727)

Figure 2 Gottfried Wilhelm Leibniz (1646–1716)

Calculus-like methods had been developed before Newton and Leibniz. As problems of physics arose, solutions were found. But no one saw the pattern or underlying mathematics relating the nature of the various problems and solutions. Simon Stevin (1548–1620) used a calculus-like method to determine the force due to water pressure on a vertical dam. Johannes Kepler (1571–1630) used calculus-like methods in his investigation of the motion of planets. And there were others, including Galileo (1564–1642), Cavalieri (1598–1647), Torricelli (1608–1647), Fermat (1601–1665), Huygens (1629–1695), Wallis (1616–1703), and Barrow (1630–1677).

The invention of calculus had a great impact on technology as well as on the development of mathematics. Years later, calculus applications were found in a variety of nonengineering areas, including business and economics, biology, medicine, sociology, and psychology. Calculus can be used to:

1. Determine the average speed at which blood flows through an artery.

2. Select the least expensive dimensions for a container.

3. Calculate how high a rocket will travel.

4. Find the production level that will maximize a company's profit.

Calculus developed as two separate branches—integral calculus and differential calculus. The original problems that led to the invention of differential calculus did not appear to resemble the problems solved by integral calculus. However, the two branches of calculus were linked together by Newton and by Leibniz in a theorem now known as the *Fundamental Theorem of Calculus*.

You will be introduced to differential calculus in Chapter 3 and integral calculus in Chapter 6. The Fundamental Theorem is presented in Section 6.4.

Both differential calculus and integral calculus use the basic concept and notation of *limit*, as you will see in Chapters 3 and 6 and elsewhere in the text. Because of their importance to the study of calculus, Chapter 2 is devoted to the presentation of limits as part of the necessary background for understanding differential and integral calculus.

Introduction to Limits

This section offers a practical introduction to *limits*, including the background needed to understand and appreciate many of the ideas of calculus. Our approach to the subject will be intuitive.

To begin, consider the expression

$$\frac{x^2 + x - 12}{x - 3}$$

Suppose we want to know what will happen to the value of this expression as the value of x gets nearer and nearer to 3. We could study the expression by evaluating it, using x values nearer and nearer to 3. Here is such a study:

x	$\dfrac{x^2 + x - 12}{x - 3}$
2.9	6.9
2.95	6.95
2.99	6.99
2.999	6.999

and

x	$\dfrac{x^2 + x - 12}{x - 3}$
3.1	7.1
3.05	7.05
3.01	7.01
3.001	7.001

The first table shows x "approaching" 3 from the left, through values less than 3. The second table shows x approaching 3 from the right, through values greater than 3. In each instance you can see that as x gets closer to 3, the value of the expression gets closer to 7. We say

"The limit of $\dfrac{x^2 + x - 12}{x - 3}$, as x approaches 3, is 7."

In mathematical notation, we write

$$\lim_{x \to 3} \frac{x^2 + x - 12}{x - 3} = 7$$

Notice the use of an *arrow* to show what number x is approaching and the *lim* shorthand for the word limit.

Using function notation, we can say that if

$$f(x) = \frac{x^2 + x - 12}{x - 3}$$

then

$$\lim_{x \to 3} f(x) = 7$$

Also, $\lim_{x \to 3} f(x) = 7$ means that as x gets closer and closer to 3, $f(x)$ gets closer and closer to 7. In general,

---

Limit

For any function f,

$$\lim_{x \to a} f(x) = L$$

means that as x gets closer and closer to a, $f(x)$ gets closer and closer to L.

This definition of limit is intuitive. There is a more formal definition, one that is used in courses for mathematicians and engineers.

Note

In the example leading up to the definition of limit, we considered $\lim\limits_{x \to 3} f(x)$ by letting x get closer and closer to 3 from both sides. We saw that the limit as x approaches 3 from the left (through values less than 3) is the same as the limit as x approaches 3 from the right (through values greater than 3). In general, *the limit exists only if the limits from both the left and the right are equal.*

Although it may be quite natural to wonder what happens when x is equal to 3 in the function we considered, that is *not* the "limit" concern. *The limit as $x \to 3$ deals with x values approaching 3 without actually being equal to 3.* The particular function and expression used here are not even defined at $x = 3$, for if 3 is substituted for x, division by zero will arise, and division by zero is not defined. Thus, we have seen that

$$\lim_{x \to 3} \frac{x^2 + x - 12}{x - 3} = 7$$

even though

$$\frac{x^2 + x - 12}{x - 3}$$

is not defined at 3.

Often the expression in question *is* defined at the number which the variable is approaching. Consider, for example,

$$\lim_{x \to 2} 3x$$

Two tables can be used to study this limit.

x	$3x$		x	$3x$
1.9	5.7		2.1	6.3
1.95	5.85		2.05	6.15
1.99	5.97		2.01	6.03
1.999	5.997		2.001	6.003

From the tables it appears that

$$\lim_{x \to 2} 3x = 6$$

and 6 is the number you would obtain by simply substituting 2 for x in the expression $3x$. In other words, in this example,

$$\lim_{x \to 2} 3x = 3(2) = 6$$

The example suggests that *sometimes limits can be evaluated by making the obvious substitution.* Here are two more examples.

EXAMPLE 1   Evaluate $\lim_{x \to 3} (2x + 1)$ by making the substitution for x.

SOLUTION   Using substitution, we obtain

$$\lim_{x \to 3} (2x + 1) = 2(3) + 1 = 7$$

Of course, we could make tables using x values that get closer and closer to 3. The tables would show that $2x + 1$ gets closer and closer to 7. ◆

EXAMPLE 2   ◆   *DISTANCE TRAVELED BY A TRAIN*

The distance traveled by a train that is going 60 miles per hour for t hours is $d(t) = 60t$. Make the substitution for t to evaluate

$$\lim_{t \to 3} d(t)$$

and interpret the answer.

SOLUTION
$$\lim_{t \to 3} d(t) = \lim_{t \to 3} 60t$$
$$= 60 \cdot 3$$
$$= 180$$

As the time t gets closer and closer to 3 hours, the distance traveled d approaches 180 miles.   ◆

It is natural to wonder when you can evaluate limits by this simple "plugging in" type of substitution and when you cannot. The answer is based on limit theorems (presented next) and continuity (the topic of the next section).

Limit Theorems

Our approach to limits has been rather informal. However, at this point a brief look at limit theorems will help you to understand the evaluation of limits. In later chapters, limit theorems will be used occasionally in the calculus operations of differentiation and integration.

Limit Theorems

If a, c, and n are real numbers, then

1. $\displaystyle\lim_{x \to a} c = c$

2. $\displaystyle\lim_{x \to a} x = a$

3. $\displaystyle\lim_{x \to a} [c \cdot f(x)] = c \cdot \lim_{x \to a} f(x)$

4. $\displaystyle\lim_{x \to a} [f(x) + g(x)] = \lim_{x \to a} f(x) + \lim_{x \to a} g(x)$

5. $\displaystyle\lim_{x \to a} [f(x) - g(x)] = \lim_{x \to a} f(x) - \lim_{x \to a} g(x)$

6. $\displaystyle\lim_{x \to a} [f(x) \cdot g(x)] = \lim_{x \to a} f(x) \cdot \lim_{x \to a} g(x)$

7. $\displaystyle\lim_{x \to a} \frac{f(x)}{g(x)} = \frac{\displaystyle\lim_{x \to a} f(x)}{\displaystyle\lim_{x \to a} g(x)} \qquad (\lim_{x \to a} g(x) \neq 0)$

8. $\displaystyle\lim_{x \to a} [f(x)]^n = [\lim_{x \to a} f(x)]^n$

If n indicates an even root, then the limit of $f(x)$ must be nonnegative.

In words, the limit theorems say:

1. The limit of a constant is that constant.

2. The limit of x as x approaches a is a.

3. The limit of a constant times a function is equal to the constant times the limit of the function.

4. The limit of a sum is equal to the sum of the limits.

5. The limit of a difference is equal to the difference of the limits.

6. The limit of a product is equal to the product of the limits.

7. The limit of a quotient is equal to the quotient of the limits (provided the denominator is nonzero).

8. The limit of the nth power of a function is the nth power of the limit of the function.

EXAMPLE 3 Use limit theorems to evaluate each limit.

(a) $\displaystyle\lim_{x \to 3} 6$ (b) $\displaystyle\lim_{x \to 2} (4x + 5)$ (c) $\displaystyle\lim_{x \to 1} \sqrt{10 - x}$

SOLUTION **(a)** By limit theorem 1,

$$\lim_{x \to 3} 6 = 6$$

(b) $\lim_{x \to 2} (4x + 5) = \lim_{x \to 2} 4x + \lim_{x \to 2} 5$ limit theorem 4

$$= \lim_{x \to 2} 4 \cdot \lim_{x \to 2} x + \lim_{x \to 2} 5 \qquad \text{limit theorem 6}$$

$$= (4)(2) + 5 \qquad \text{limit theorems 1 and 2}$$

$$= 13$$

(c) $\lim_{x \to 1} \sqrt{10 - x} = \lim_{x \to 1} (10 - x)^{1/2}$ exponent notation

$$= [\lim_{x \to 1} (10 - x)]^{1/2} \qquad \text{limit theorem 8}$$

$$= [\lim_{x \to 1} 10 - \lim_{x \to 1} x]^{1/2} \qquad \text{limit theorem 5}$$

$$= (10 - 1)^{1/2} \qquad \text{limit theorems 1 and 2}$$

$$= 9^{1/2} = 3 \qquad \blacklozenge$$

The limit theorems and Example 3 suggest that *many limits can indeed be evaluated by the "plugging in" type of substitution used in Examples 1 and 2.* But some limits cannot be evaluated in that manner. This concern is addressed in the next section, which presents the concept of continuity.

French mathematician Augustin Louis Cauchy (1789–1857) was a leader in establishing a formal theory of limits and a rigorous treatment of calculus in general.

Figure 3 Augustin Louis Cauchy (1789–1857).
Photo courtesy of Smithsonian Institution Libraries.

2.1   Exercises

Find each limit in Exercises 1–18 by using limit theorems. (See Example 3.)

1. $\lim\limits_{x \to 1} (x + 7)$

2. $\lim\limits_{x \to 7} (x - 5)$

3. $\lim\limits_{x \to 4} 5x$

4. $\lim\limits_{x \to -3} 8x$

5. $\lim\limits_{x \to 5} (4x - 1)$

6. $\lim\limits_{x \to 2} (6x + 1)$

7. $\lim\limits_{t \to -3} t^2$

8. $\lim\limits_{u \to -1} u^4$

9. $\lim\limits_{x \to 0} \dfrac{x + 2}{9}$

10. $\lim\limits_{x \to 5} \dfrac{2}{3x}$

11. $\lim\limits_{x \to 8} \sqrt{x + 1}$

12. $\lim\limits_{x \to 16} \sqrt{4x}$

13. $\lim\limits_{x \to 2} x^{-3}$

14. $\lim\limits_{x \to -1} x^{-7}$

15. $\lim\limits_{x \to -1} (x^3 + x)$

16. $\lim\limits_{x \to 2} (1 - x^3)$

17. $\lim\limits_{x \to 0} \dfrac{5x}{1 + x}$

18. $\lim\limits_{x \to 9} \dfrac{x}{1 + x}$

In Exercises 19–30, attempt to evaluate each limit by using the substitution approach (plugging in). If that method fails, write "cannot determine by substitution" and then proceed to *use a calculator* to create tables in order to determine the limit.

19. $\lim\limits_{x \to 2} \dfrac{x^2 - 4}{x - 2}$

20. $\lim\limits_{x \to -3} \dfrac{x^2 - 9}{x + 3}$

21. $\lim\limits_{x \to 0} \dfrac{x}{3}$

22. $\lim\limits_{x \to 0} \dfrac{7x}{2}$

23. $\lim\limits_{t \to 2} \dfrac{t^2 - 7t + 10}{t - 2}$

24. $\lim\limits_{t \to 4} \dfrac{t^2 - 7t + 10}{t - 5}$

25. $\lim\limits_{x \to 5} \dfrac{2x^2 - 13x + 15}{x - 5}$

26. $\lim\limits_{x \to 2} \dfrac{x^2 + 3x - 10}{x^2 + 6x - 16}$

27. $\lim\limits_{t \to 3} \dfrac{t^2 - 9}{t^2 + 9}$

28. $\lim\limits_{x \to 1} \dfrac{1 - 8x}{\sqrt{x}}$

29. $\lim\limits_{x \to -1} \dfrac{x^2 + 6x + 5}{x^2 + x}$

30. $\lim\limits_{x \to 4} \dfrac{x^2 - 16}{x^2 - 4x}$

Find each limit in Exercises 31–34.

31. $\lim\limits_{x \to 1} 7$

32. $\lim\limits_{x \to 2} 15$

33. $\lim\limits_{x \to 0} (-6)$

34. $\lim\limits_{x \to -1} 4$

35. (*AIRPLANE TRAVEL*) The distance $d(t)$ traveled by an airplane going 400 miles per hour for t hours is $d(t) = 400t$.
 (a) Evaluate $\lim\limits_{t \to 2.5} d(t)$.
 W (b) Interpret your answer to part (a).

36. (*GEOMETRY*) The area of a circular region of radius r is $A = \pi r^2$. (See figure.)

 (a) Find $\lim\limits_{r \to 0} A$.
 W (b) Explain the meaning of the limit in part (a) in light of this geometry setting.

37. (*PRICE FUNCTION*) Ordinarily, the price p of an item depends on the quantity x that is supplied or demanded. But occasionally, the price $p(x)$ is constant regardless of x. This is especially true for such consumer items as newspapers. Assume that newspapers sell for 25¢ each.
 (a) Determine the price function p. Use x for the quantity of newspapers.
 (b) Find $\lim\limits_{x \to 2000} p(x)$.

38. (*GEOMETRY*) Let $P(x)$ be the perimeter of the triangle shown in the figure.

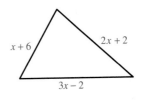

 (a) Find $\lim\limits_{x \to 7} P(x)$.

W **(b)** Explain the meaning of the limit in part (a) in light of this geometry setting.

39. (*NUMBER OF CHIRPS*) The number of chirps per minute made by a cricket depends on the temperature. If x is the temperature in degrees Fahrenheit, and $x \geq 40$, then the number of chirps per minute is $4x - 160$. What will be the number of chirps per minute on a cold night when the temperature approaches 40° Fahrenheit?

40. (*GEOMETRY*) If x represents the length of the side of a cube, then the volume of the cube is x^3. Suppose the length of each side approaches zero. What does the volume approach?

41. (*TELEPHONE CALL*) When you make a telephone call, the phone rings for 2 seconds, is quiet for 4 seconds, rings again for 2 seconds, and so on. The figure below shows the graph of the sound-level function S based on the sound from a phone call in which the caller hangs up after three rings. Here A represents the sound level in a room when the phone is quiet and B represents the sound level when the phone is ringing.

(a) Find $\lim_{t \to 4} S(t)$.

(b) Find $\lim_{t \to 7} S(t)$.

Consideration of a limit such as

$$\lim_{t \to 6} S(t)$$

introduces some concerns. The limit does not exist. After all, if t is approached using values less than 6 (that is, from the left), the $S(t)$ values obtained are different from those obtained using an approach through values greater than 6 (that is, from the right). These concerns are examined in the next two sections—Continuity (2.2) and One-Sided Limits (2.3).

The Exercise Library at the back of the book contains graphing calculator and computer exercises keyed to this section.

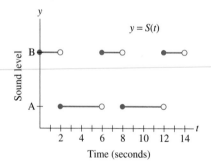

Three rings of a telephone

2.2 | CONTINUITY

We turn to graphs in order to see why some limits cannot be found directly by substitution. To begin, let f be the function defined by $f(x) = 2x + 1$ and consider the limit of $f(x)$ as x approaches 3.

$$\lim_{x \to 3} f(x) = \lim_{x \to 3} (2x + 1) = 7$$

The graph of $y = 2x + 1$ shows that as x gets closer and closer to 3, the y values get closer and closer to 7. See Figure 4.

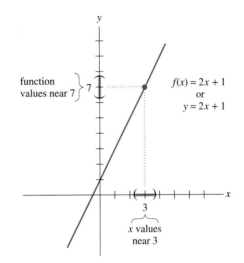

Figure 4 The function values are near 7 when the *x* values are near 3.

Notice that there is no break or jump in the graph of $f(x) = 2x + 1$. Functions whose graphs have no break in them are called **continuous functions**. If the limit of $f(x)$ as x approaches some number a is $f(a)$, then there can be no break in the graph at $x = a$. This means that when the limit of $f(x)$ as x approaches a is $f(a)$, the function is continuous at a. These observations are restated below, to provide the definition of what it means to say that **function f is continuous at a.**

Function f Continuous at a

If f is *continuous at a*, then $\lim_{x \to a} f(x) = f(a)$.

If $\lim_{x \to a} f(x) = f(a)$, then f is *continuous at a*.

The definition includes three conditions that must be met for f to be continuous at a.

1. $\lim_{x \to a} f(x)$ must exist.

2. $f(a)$ must be defined.

3. The limit (in #1) must be equal to $f(a)$.

Note

A function is **continuous on an open interval** (*a*, *b*) if it is continuous at every number in the interval. This idea will be pursued in Exercises 56–63.

Because many of the examples and exercises of the previous section involved functions that were continuous, the limits could be determined by substituting for *x* the value that *x* was approaching. The following table serves to summarize some of those examples. Notice that in each instance the limit of *f*(*x*) as *x* approaches *a* is the same as *f*(*a*), which means that *f* is continuous at *a*.

function f	$\lim\limits_{x \to a} f(x)$	$f(a)$
$f(x) = 3x$	$\lim\limits_{x \to 2} 3x = 6$	$f(2) = 6$
$f(x) = 2x + 1$	$\lim\limits_{x \to 3} (2x + 1) = 7$	$f(3) = 7$
$f(x) = 15$	$\lim\limits_{x \to 2} 15 = 15$	$f(2) = 15$
$f(x) = x^3 + x$	$\lim\limits_{x \to -1} (x^3 + x) = -2$	$f(-1) = -2$
$f(x) = \dfrac{5x}{1 + x}$	$\lim\limits_{x \to 0} \dfrac{5x}{1 + x} = 0$	$f(0) = 0$

Some types of functions are continuous at every number in their domain. Limits of such functions can always be determined by the substitution approach. *Polynomial functions* are continuous at every real number. *Rational functions* are continuous at every real number, except at numbers for which the denominator is zero.

For contrast, consider the following limit:

$$\lim_{x \to 3} \frac{x^2 + x - 12}{x - 3}$$

In the previous section, an attempt to determine this limit by substituting 3 for *x* led to division by zero. Specifically, using 3 for *x* will result in 0/0, which is undefined. The function

$$f(x) = \frac{x^2 + x - 12}{x - 3}$$

is not continuous at 3 because *f*(3) is not defined. That is precisely why we cannot evaluate this limit simply by using 3 for *x* in *f*(*x*). However, the limit can be evaluated without using a calculator. In fact, the next example shows the use of *factoring* to determine the limit when 0/0 results from the substitution approach.

EXAMPLE 1   Find the limit.

$$\lim_{x \to 3} \frac{x^2 + x - 12}{x - 3}$$

SOLUTION   As suggested above, we begin by factoring.

$$\lim_{x \to 3} \frac{x^2 + x - 12}{x - 3} = \lim_{x \to 3} \frac{(x - 3)(x + 4)}{(x - 3)}$$

The factors $(x - 3)$ can be eliminated by division as long as x is not 3. (If x is 3, then we would be dividing by zero.) And we know that x is not 3 because the limit as $x \to 3$ indicates that x is *approaching* 3, getting closer and closer to 3, but is not actually 3. The limit can be simplified as

$$\lim_{x \to 3} \frac{(x - 3)(x + 4)}{(x - 3)} = \lim_{x \to 3} (x + 4)$$
$$= 7$$

Thus,

$$\lim_{x \to 3} \frac{x^2 + x - 12}{x - 3} = 7 \quad \blacklozenge$$

Consider the following two ideas, which are based on the function and results of Example 1.

1. For this function, $\lim_{x \to a} f(x) \neq f(a)$. Specifically, we have $\lim_{x \to 3} f(x) \neq f(3)$, since $\lim_{x \to 3} f(x) = 7$ and $f(3)$ is not defined. Thus, this function is *not continuous* at 3. We say it is **discontinuous** at 3.

2. Since the expression

$$\frac{x^2 + x - 12}{x - 3}$$

reduces to

$$x + 4$$

when $x \neq 3$, the two expressions are in fact equal to each other for all values of x except 3. Hence the graph of

$$f(x) = \frac{x^2 + x - 12}{x - 3}$$

is the same as the graph of

$$f(x) = x + 4 \qquad (x \neq 3)$$

It is the graph of the line $y = x + 4$, with a point missing at $x = 3$. It is a straight line with a hole in it at $(3, 7)$. Visually, since the graph has a break (hole) in it at $x = 3$, the function is discontinuous there. See Figure 5.

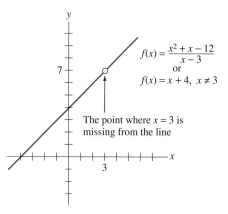

$$f(x) = \frac{x^2 + x - 12}{x - 3}$$
or
$$f(x) = x + 4, \ x \neq 3$$

The point where $x = 3$ is missing from the line

Figure 5 Function *f* is discontinuous at 3

Here is an application involving a discontinuity.

EXAMPLE 2   ◆   *FLASH PHOTOGRAPHY*

Consider what you see when someone takes a photograph using a flash. The level of light in the room is constant. Suddenly, the flash goes off and creates more light for an instant. Then the light in the room returns immediately to the preflash level. See Figure 6 and note that the graphed function is discontinuous at t_1.

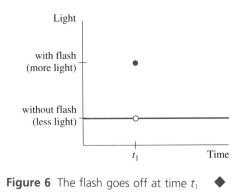

Figure 6 The flash goes off at time t_1 ◆

The next example provides an everyday application that involves a function with discontinuities.

EXAMPLE 3   ◆   *POSTAL RATES FOR FIRST-CLASS MAIL*

The function that describes first-class mail postage rates has discontinuities. The cost of first-class mail (1992) is 29¢ for the first ounce (or less) and then 23¢

for each additional ounce above that. The graph for up to 4 ounces is shown. (See Figure 7.)

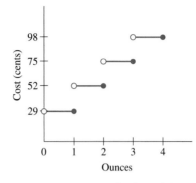

Figure 7 Postal rates for first-class mail

Notice the discontinuities at 1, 2, and 3 ounces, where the graph jumps. ◆

Note

In Example 3, the discontinuities at 1, 2, and 3 are obvious from the jumps in the graph. From a limit perspective, the function is not continuous at 1, 2, and 3 because the limit does not exist. In each case, the limit when approaching from the left is different from the limit when approaching from the right. This concern will be pursued more formally when one-sided limits are studied in the next section.

EXAMPLE 4 ◆ *ADAPTATION OF THE EYE TO DARKNESS*

Sensory psychologists have studied the adaptation of the human eye to light and dark. Adaptation begins with the cones of the eyes. After about 8 minutes the rods take over the adaptation. The graph of the adaptation function shows a discontinuity at the time ($t = 8$ minutes) when the rods take over. See Figure 8.

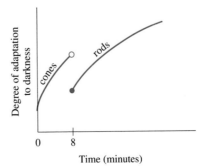

Figure 8 Adaptation of the human eye to darkness

Once again we have a situation in which the limit does not exist (at 8) because the left-hand and right-hand limits are different. ◆

2.2   Exercises

Evaluate each limit in Exercises 1–12 by using the substitution approach (plugging in). Keep in mind that the method will work because each function given here is continuous at the number that x is approaching.

1. $\lim_{x \to 10} (3x - 2)$

2. $\lim_{x \to 4} (x^3 - 7)$

3. $\lim_{x \to -3} (1 - 9x + x^2)$

4. $\lim_{x \to 0} (12 - x^2)$

5. $\lim_{x \to 5} \dfrac{x - 3}{x + 2}$

6. $\lim_{x \to 7} \dfrac{x + 6}{x - 3}$

7. $\lim_{x \to 4} \dfrac{x - 4}{x}$

8. $\lim_{x \to -3} \dfrac{3 + x}{x}$

9. $\lim_{x \to 2} \dfrac{1 - x}{x^2 + 1}$

10. $\lim_{x \to 4} \dfrac{x - 1}{1 + 3x^2}$

11. $\lim_{x \to -4} \sqrt{x^2 + 9}$

12. $\lim_{x \to 40} \sqrt{1 + 2x}$

Attempt to evaluate each limit in Exercises 13–26 by the substitution approach. If 0/0 is produced, simplify the expression by factoring and then try again.

13. $\lim_{x \to 4} \dfrac{x^2 - x - 12}{x - 4}$

14. $\lim_{x \to 3} \dfrac{x^2 + 2x - 15}{x - 3}$

15. $\lim_{x \to 0} \dfrac{x^2}{x}$

16. $\lim_{x \to 5} \dfrac{x^2 + 8x + 15}{x + 3}$

17. $\lim_{x \to -2} \dfrac{x^2 - 4}{x + 2}$

18. $\lim_{x \to 2} \dfrac{x^2 - 4}{x - 2}$

19. $\lim_{x \to 1} \dfrac{x^2 + 1}{x + 1}$

20. $\lim_{x \to 0} \dfrac{3x}{x}$

21. $\lim_{x \to 3} \dfrac{x^2}{x}$

22. $\lim_{x \to 3} \dfrac{x^2 - 9}{x - 3}$

23. $\lim_{x \to 2} \dfrac{x - 2}{x + 3}$

24. $\lim_{x \to 1} \dfrac{x - 1}{x + 1}$

25. $\lim_{x \to 1} \dfrac{3x - 3}{x^2 - 1}$

26. $\lim_{x \to 2} \dfrac{x^2 - 4}{2x - 4}$

Sketch a graph of each function in Exercises 27–34. Use the limit and continuity information obtained from the exercise referenced.

27. $f(x) = \dfrac{x^2 - x - 12}{x - 4}$ See Exercise 13.

28. $f(x) = \dfrac{x^2 + 2x - 15}{x - 3}$ See Exercise 14.

29. $f(x) = \dfrac{x^2 - 4}{x + 2}$ See Exercise 17.

30. $f(x) = \dfrac{x^2 - 4}{x - 2}$ See Exercise 18.

31. $f(x) = \dfrac{3x}{x}$ See Exercise 20.

32. $f(x) = \dfrac{2x}{x}$

33. $f(x) = \dfrac{x^2}{x}$ See Exercise 15.

34. $f(x) = \dfrac{3x^2}{x}$

In Exercises 35–40, each function is discontinuous at the given value of x because f is not defined there. In each case, attempt to obtain $f(x)$ for the given value of x and indicate why the attempt fails.

35. $f(x) = \sqrt{x - 4}$ at $x = 3$

36. $f(x) = \dfrac{x + 5}{x - 2}$ at $x = 2$

37. $f(x) = \dfrac{x - 1}{x + 1}$ at $x = -1$

38. $f(x) = \dfrac{x^2 + 3x - 28}{x - 4}$ at $x = 4$

39. $f(x) = \dfrac{x^2 + 8x}{x + 8}$ at $x = -8$

40. $f(x) = \dfrac{x^2}{x}$ at $x = 0$

In Exercises 41–44, indicate for what x the graphed function is discontinuous.

41.   **42.**

43.   **44.**

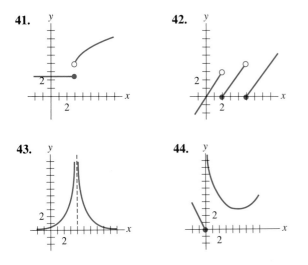

45. (*SOCIAL WORK*) A social worker begins a new job in January 1991 (1/91) and is given a specified workload. Since it is based on the number of cases assigned, the workload may be several thousand hours. Many of the cases will be completed during the year. Additional cases are given to the social worker at the beginning of each successive year (1/92, 1/93, and so on).
  (a) From the graph, determine when discontinuities occur.
W (b) From a workload perspective, explain why the discontinuities occur.

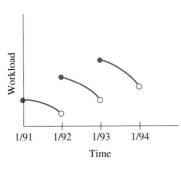

46. (*BANKING*) In the banking industry, a failed bank or savings and loan association is often acquired by a healthy institution. The graph below shows the assets of an acquiring institution as a function of time.

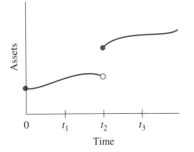

  (a) When (for what value of t) does the stronger institution acquire the failed institution?
W (b) Explain the reason for the discontinuity at t_2 (in terms of the banking situation).

47. (*POSTAL INSURANCE*) The post office allows customers to insure valuable mail. The rates are as follows: 75¢ for coverage up to and including $50, $1.60 for coverage above $50 and up to and including $100, $2.40 for coverage above $100 up to and including $200, and $3.50 for coverage above $200 and up to and including $300. Sketch a graph of this function and indicate where the discontinuities are found.

48. (*COMPOUND INTEREST*) $1000 is invested at 8% per year compounded quarterly (that is, every three months, or four times per year). This means that the investment begins with $1000, at 3 months it becomes $1020, at 6 months it is $1040, at 9 months it is $1061, and at one year it is $1082. Sketch a graph of this function, one that gives the amount of money at any time during the first year. Indicate where the discontinuities are to be found.

49. (*LONG DISTANCE CALL*) A person-to-person phone call from Chicago to Los Angeles costs $3.50 for the first minute and then 25¢ per minute after that. Sketch a graph showing the total cost of a phone call for all times up to five minutes.

50. (*PLUMBER'S CHARGES*) A plumber charges $40 to come to your house. In addition, he charges $30 per half hour (or fraction of a half hour) of work once he arrives. Graph the function showing the plumber's charges for every period of time up to 3 hours.

51. (*CAR RENTAL COST*) Suppose a rental car costs $25 per day, with no charge for mileage. Graph the function showing the cost for as many as 5 days.

W 52. Suppose that f is not defined at a. Does this mean that $\lim\limits_{x \to a} f(x)$ does not exist? Explain.

W 53. If $\lim\limits_{x \to 2} f(x) = 5$, is it necessarily true that $f(2) = 5$? Explain.

W 54. Given the graph of a function, how can you tell if it is not continuous at some number such as 7?

W 55. You are given a function f and asked to determine (without a graph) if it is continuous at 4. Explain the procedure.

As noted earlier, a function is *continuous on an open interval* (*a*, *b*) if it is continuous at every number in the interval. In Exercises 56–63, examine each function and interval and indicate whether the function is continuous on the interval and, if it is not, where in the interval it is discontinuous.

56. $f(x) = 1 - x^3$ on $(-\infty, \infty)$

57. $f(x) = x^2 + 8x + 1$ on $(-\infty, \infty)$

58. $f(x) = \dfrac{3}{x - 4}$ on $(0, 30)$

59. $f(x) = \dfrac{1}{x^2}$ on $(-3, 10)$

60. $g(x) = \dfrac{x}{2x}$ on $(-10, 5)$

61. $g(x) = \dfrac{3x}{x}$ on $(-\infty, \infty)$

62. $f(x) = \sqrt{x - 1}$ on $(0, 1)$

63. $f(x) = \sqrt{x}$ on $(-\infty, 0)$

Evaluate each limit in Exercises 64–67.

64. $\lim\limits_{x \to 3} \dfrac{x^3 - 9x}{x^2 - 3x}$

65. $\lim\limits_{x \to -2} \dfrac{x^2 + 8x + 12}{4 - x^2}$

66. $\lim\limits_{x \to 1} \dfrac{x^3 - 1}{x - 1}$

67. $\lim\limits_{x \to -2} \dfrac{x + 2}{x^3 + 8}$

Evaluate each limit in Exercises 68–69. *Hint:* In Exercise 68, rationalize the denominator by multiplying the denominator and numerator by $\sqrt{x} + 2$. In Exercise 69, rationalize the numerator by multiplying both the numerator and denominator by $\sqrt{x} + 1$.

68. $\lim\limits_{x \to 4} \dfrac{x - 4}{\sqrt{x} - 2}$

69. $\lim\limits_{x \to 1} \dfrac{\sqrt{x} - 1}{x - 1}$

Evaluate each limit in Exercises 70–71. You will need to simplify the complex fraction in each case.

70. $\lim\limits_{x \to 4} \dfrac{\dfrac{1}{4} - \dfrac{1}{x}}{x - 4}$

71. $\lim\limits_{x \to 3} \dfrac{\dfrac{1}{x} - \dfrac{1}{3}}{x - 3}$

2.3 | ONE-SIDED LIMITS

As you know, $\lim\limits_{x \to a} f(x)$ is the value that $f(x)$ approaches as x gets closer and closer to a. With this in mind, consider $\lim\limits_{x \to 2} f(x)$ for the function graphed in Figure 9.

From the graph, you can see that as x approaches 2 from the left (that is, via numbers less than 2), $f(x)$ approaches 3. In other words, coming from the left, the closer x gets to 2, the closer $f(x)$ gets to 3. By contrast, the graph also shows that as x approaches 2 from the right (via numbers greater than 2), $f(x)$

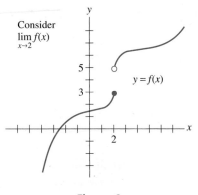

Figure 9

approaches 5. The limit is different depending on the side from which x is approaching 2. This situation suggests the following:

1. The graph will have a break at $x = 2$, and thus f is discontinuous at 2.

2. $\lim\limits_{x \to 2} f(x)$ does not exist in the usual sense.

The *left-hand limit* and *right-hand limit* can be considered separately as special limits having their own notation.

$$\text{The left-hand limit:} \quad \lim_{x \to 2^-} f(x) = 3$$

$$\text{The right-hand limit:} \quad \lim_{x \to 2^+} f(x) = 5$$

The raised minus specifies "from the left." The raised plus specifies "from the right."

Here are some additional examples of *one-sided limits*.

EXAMPLE 1 For the function graphed in Figure 10, determine the one-sided limits $\lim\limits_{x \to 3^-} f(x)$ and $\lim\limits_{x \to 3^+} f(x)$.

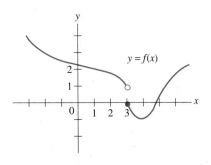

Figure 10

SOLUTION   A study of the graph shows that

$$\lim_{x \to 3^-} f(x) = 1$$

and

$$\lim_{x \to 3^+} f(x) = 0 \quad \blacklozenge$$

EXAMPLE 2   ◆   *POSTAL RATES FOR FIRST CLASS MAIL*

Consider the first-class postage example graphed in Figure 11.

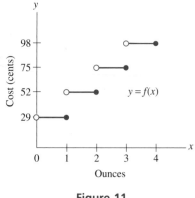

Figure 11

Determine $\lim_{x \to 1} f(x)$.

SOLUTION   From the graph we can see that

$$\lim_{x \to 1^-} f(x) = 29$$

$$\lim_{x \to 1^+} f(x) = 52$$

Since the left-hand limit and the right-hand limit are different, it follows that

$$\lim_{x \to 1} f(x) \text{ does not exist} \quad \blacklozenge$$

EXAMPLE 3   Determine $\lim_{x \to 0} \sqrt{x}$.

SOLUTION   At first glance this limit may look simple, since you could easily believe that the limit is zero. However, *the limit is not zero.* Although it is true that

$$\lim_{x \to 0^+} \sqrt{x} = 0$$

it is also true that

$$\lim_{x \to 0^-} \sqrt{x} \text{ does not exist}$$

Why? Because approaching from the left in this instance means using numbers less than zero (negative numbers) for x, and $\sqrt{x}$ is not defined for such numbers. *A function must be defined for all of the values of x along the approach, or else the limit does not exist.* Figure 12 shows the graph of $f(x) = \sqrt{x}$, which is defined only for $x \geq 0$.

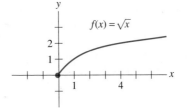

Figure 12 A function defined only for $x \geq 0$

Our conclusion:

$$\lim_{x \to 0^+} \sqrt{x} = 0$$

$$\lim_{x \to 0^-} \sqrt{x} \text{ does not exist}$$

$$\lim_{x \to 0} \sqrt{x} \text{ does not exist} \quad \blacklozenge$$

In general,

> $$\lim_{x \to a} f(x) \text{ exists if both } \lim_{x \to a^-} f(x) \text{ and } \lim_{x \to a^+} f(x)$$
>
> exist and are equal.

In Example 1, both one-sided limits existed, but they were different. The same thing was true in Example 2. In Example 3, one of the one-sided limits failed to exist. (A look back at the first limit presented in Section 2.1 shows an instance where both one-sided limits existed and were the same.)

EXAMPLE 4   Determine $\lim_{x \to 0^-} f(x)$ and $\lim_{x \to 0^+} f(x)$ for the function defined by

$$f(x) = \begin{cases} x^2 & x \leq 0 \\ x + 3 & x > 0 \end{cases}$$

SOLUTION   In this two-part function, the value of $f(x)$ is computed as x^2 when $x \leq 0$, but as $x + 3$ when $x > 0$. Thus,

$$\lim_{x \to 0^-} f(x) = \lim_{x \to 0^-} x^2 = 0$$

$$\lim_{x \to 0^+} f(x) = \lim_{x \to 0^+} (x + 3) = 3$$

Since the left-hand and right-hand limits are different, we conclude that

$$\lim_{x \to 0} f(x) \quad \text{does not exist}$$

Figure 13 shows the graph of f.

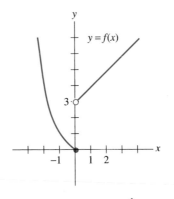

Figure 13   ◆

EXAMPLE 5   Determine $\lim_{x \to 1^-} f(x)$ and $\lim_{x \to 1^+} f(x)$ for the function defined by

$$f(x) = \begin{cases} 2x + 1 & x \leq 1 \\ 4 - x & x > 1 \end{cases}$$

SOLUTION   Here

$$\lim_{x \to 1^-} f(x) = \lim_{x \to 1^-} (2x + 1) = 3$$

$$\lim_{x \to 1^+} f(x) = \lim_{x \to 1^+} (4 - x) = 3$$

Since the left- and right-hand limits are the same, 3, the limit as x approaches 1 is 3.

$$\lim_{x \to 1} f(x) = 3$$

Furthermore, since $f(1) = 3$, we have $\lim_{x \to 1} f(x) = f(1)$. In other words, the function is continuous at 1. The graph is shown in Figure 14.

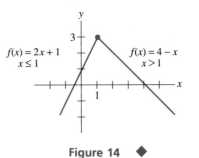

Figure 14   ◆

Note

Recall that for f to be continous at a we need

$$\lim_{x \to a} f(x) = f(a)$$

In other words, the following three conditions must be met:

1. $\lim_{x \to a} f(x)$ must exist.

2. f must be defined at a.

3. The limit (in #1) must be equal to $f(a)$.

In Example 5, the conditions are met at $x = 1$, so f is continuous at 1. In Example 4, $\lim_{x \to 0} f(x)$ fails to exist because the left- and right-hand limits are different. Thus, f is *not continuous at* 0, because the limit fails to exist.

2.3 Exercises

Evaluate each one-sided limit in Exercises 1–6. (*Note:* Some limits may not exist.) Be sure to refer to the proper graph in each case.

1.(a) $\lim_{x \to 2^-} f(x)$.

(b) $\lim_{x \to 2^+} f(x)$.

2.(a) $\lim_{x \to 0^-} g(x)$.

(b) $\lim_{x \to 0^+} g(x)$.

3.(a) $\lim_{x \to 1^-} h(x)$.

(b) $\lim_{x \to 1^+} h(x)$.

4.(a) $\lim_{x \to 4^-} k(x)$.

(b) $\lim_{x \to 4^+} k(x)$.

5.(a) $\lim_{x \to 5^-} m(x)$.

(b) $\lim_{x \to 5^+} m(x)$.

6.(a) $\lim_{x \to 4^-} n(x)$.

(b) $\lim_{x \to 4^+} n(x)$.

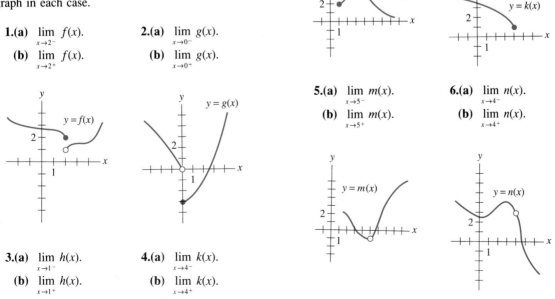

In Exercises 7–14, explain why each limit does not exist.

7. $\lim\limits_{x \to 0^-} \sqrt{2x}$

8. $\lim\limits_{x \to 4^-} \sqrt{x - 4}$

9. $\lim\limits_{x \to 1^+} \sqrt{1 - x}$

10. $\lim\limits_{x \to 5^+} \sqrt{5 - x}$

11. $\lim\limits_{x \to 0} \sqrt{3x}$

12. $\lim\limits_{x \to 2} \sqrt{x - 2}$

13. $\lim\limits_{x \to -2} \sqrt{x + 2}$

14. $\lim\limits_{x \to 3} \sqrt{3 - x}$

Evaluate each one-sided limit in Exercises 15–22.

15.(a) $\lim\limits_{x \to 0^-} f(x)$
  (b) $\lim\limits_{x \to 0^+} f(x)$
  $$f(x) = \begin{cases} x + 2 & x \leq 0 \\ x - 1 & x > 0 \end{cases}$$

16.(a) $\lim\limits_{x \to 0^-} g(x)$
  (b) $\lim\limits_{x \to 0^+} g(x)$
  $$g(x) = \begin{cases} 3x & x \leq 0 \\ x^2 & x > 0 \end{cases}$$

17.(a) $\lim\limits_{x \to 2^-} h(x)$
  (b) $\lim\limits_{x \to 2^+} h(x)$
  $$h(x) = \begin{cases} 1 - x & x \leq 2 \\ x^2 - 5 & x > 2 \end{cases}$$

18.(a) $\lim\limits_{x \to 4^-} j(x)$
  (b) $\lim\limits_{x \to 4^+} j(x)$
  $$j(x) = \begin{cases} 5x - 1 & x < 4 \\ 4x + 3 & x \geq 4 \end{cases}$$

19.(a) $\lim\limits_{x \to 0^-} k(x)$
  (b) $\lim\limits_{x \to 0^+} k(x)$
  $$k(x) = \begin{cases} x^2 & x < 0 \\ \sqrt{x} & x \geq 0 \end{cases}$$

20.(a) $\lim\limits_{x \to 1^-} m(x)$
  (b) $\lim\limits_{x \to 1^+} m(x)$
  $$m(x) = \begin{cases} \dfrac{x + 1}{x} & x \leq 1 \\ x^2 + 1 & x > 1 \end{cases}$$

21.(a) $\lim\limits_{x \to 3^-} n(x)$
  (b) $\lim\limits_{x \to 3^+} n(x)$
  $$n(x) = \begin{cases} 0 & x < 3 \\ x & x \geq 3 \end{cases}$$

22.(a) $\lim\limits_{x \to -4^-} p(x)$
  (b) $\lim\limits_{x \to -4^+} p(x)$
  $$p(x) = \begin{cases} x^2 & x \leq -4 \\ 2x & x > -4 \end{cases}$$

In Exercises 23–30, find the limit if it exists or indicate that it does not exist. The functions used here (f, g, h, j, k, m, n, and p) are the functions of Exercises 15–22.

23. $\lim\limits_{x \to 0} f(x)$

24. $\lim\limits_{x \to 0} g(x)$

25. $\lim\limits_{x \to 2} h(x)$

26. $\lim\limits_{x \to 4} j(x)$

27. $\lim\limits_{x \to 0} k(x)$

28. $\lim\limits_{x \to 1} m(x)$

29. $\lim\limits_{x \to 3} n(x)$

30. $\lim\limits_{x \to -4} p(x)$

31. (*POSTAL INSURANCE*) Function M (shown in the figure) gives the cost of postal insurance for coverage up to $200.

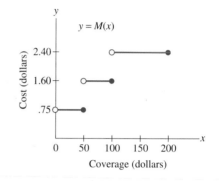

(a) Determine $M(120)$.
(b) Determine $M(100)$.
(c) Determine $\lim\limits_{x \to 50^+} M(x)$, if it exists.
(d) Determine $\lim\limits_{x \to 50^-} M(x)$, if it exists.
(e) Determine $\lim\limits_{x \to 50} M(x)$, if it exists.
(f) Is M continuous at 50?
W (g) Is M continuous at 120? Explain.

32. (*COMPUTER RENTAL COST*) A computer rental company charges $30 per week to rent their basic PC. If $C(x)$ is the cost of renting the PC for x weeks, then the graph of $y = C(x)$ for $0 \leq x \leq 4$ is shown in the figure.

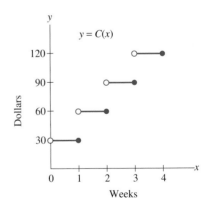

(a) Determine $\lim\limits_{x \to 2^-} C(x)$, if it exists.
(b) Determine $\lim\limits_{x \to 2^+} C(x)$, if it exists.

(c) Determine $\lim\limits_{x \to 2} C(x)$, if it exists.

W (d) Is C continuous at 2? Explain.

33. (HOME LOAN RATES) The following graph shows the interest rates offered by a lender who examines the market every 30 days for possible rate changes. The function is called H. The variable t represents the number of days since the beginning of the year.

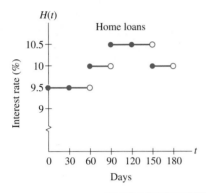

(a) Determine $H(60)$.
(b) Determine $H(100)$.
(c) Is H discontinuous at 30?
(d) Find $\lim\limits_{t \to 60^-} H(t)$.
(e) Find $\lim\limits_{t \to 60^+} H(t)$.
(f) Is H continuous at 60?

34. (PARKING FEE) A downtown parking lot charges $2 for the first hour (or portion of the first hour) and then $1 per hour (or portion of an hour). The maximum fee for a day is $8. Let p be the function that gives the parking fee in terms of time t, where t is in hours.
(a) Sketch a graph of function p.
(b) At what values of t is p discontinuous?
(c) What is the value of $p(1/2)$?
(d) What is the value of $p(12)$?

W In Exercises 35–38, explain why the given function is *continuous* at $x = a$. Be sure to determine the limit as x approaches a and the value of $f(a)$ and use them in your explanation. (You may want to read Example 5 and the note that follows it.)

35. $f(x) = \begin{cases} 3x + 2 & x \le 3 \\ 17 - 2x & x > 3 \end{cases}$
at $x = 3$

36. $f(x) = \begin{cases} 1 + x & x \le 7 \\ 4x - 20 & x > 7 \end{cases}$
at $x = 7$

37. $f(x) = \begin{cases} x^2 + 1 & x \le 5 \\ 3x & x > 5 \end{cases}$
at $x = 2$

38. $f(x) = \begin{cases} \sqrt{x} & 0 \le x \le 4 \\ x^3 & x > 4 \end{cases}$
at $x = 3$

W In Exercises 39–42, explain why the function is *not* *continuous* at $x = a$. (You may want to read the note that follows Example 5.)

39. $f(x) = \begin{cases} 5x + 1 & 0 \le x \le 2 \\ x + 14 & 2 < x \le 5 \end{cases}$
at $x = 2$

40. $f(x) = \begin{cases} 4 + 2x & x \le 0 \\ 9 - x & x > 0 \end{cases}$
at $x = 0$

41. $f(x) = \begin{cases} 3x + 2 & x \le 4 \\ x^2 - 3 & x > 4 \end{cases}$
at $x = 4$

42. $f(x) = \begin{cases} x^3 & -5 \le x \le -1 \\ x^2 & x > -1 \end{cases}$
at $x = -1$

W 43. Suppose $\lim\limits_{x \to 3^-} f(x) = 5$ and $\lim\limits_{x \to 3} f(x) = 5$. Is it possible that $\lim\limits_{x \to 3^+} f(x) = 4$? Explain.

The Exercise Library at the back of the book contains graphing calculator and computer exercises keyed to this section.

2.4 | *LIMITS AT INFINITY*

Sometimes we are concerned with the behavior of a function f as the magnitude of the variable increases without bound (that is, the magnitude becomes infinitely large). The limits studied in such instances are called **limits at infinity** and are written as

$$\lim_{x \to \infty} f(x)$$

and

$$\lim_{x \to -\infty} f(x)$$

The notation $x \to \infty$ can be read "as x increases without bound" or "as x tends toward infinity." Similarly, $x \to -\infty$ is read "as x decreases without bound" or "as x tends toward minus infinity."

Consider the function $f(x) = 1/x$. The larger x becomes, the closer $1/x$ gets to zero. Here is a table that illustrates this statement.

x	$\dfrac{1}{x}$
2	.5
10	.1
100	.01
1000	.001

The graph of $f(x) = 1/x$ also shows this — that as x becomes larger, $1/x$ gets closer to 0. See Figure 15.

We say that the limit of $1/x$ as x increases without bound is 0. This is written

$$\lim_{x \to \infty} \frac{1}{x} = 0$$

In a similar manner, we can determine that

$$\lim_{x \to -\infty} \frac{1}{x} = 0$$

Note that $1/x$ is *never equal to zero*. The limits above merely describe the behavior of $1/x$ as the magnitude of x increases without bound. They describe the tendency of $1/x$ toward zero as x tends toward infinity.

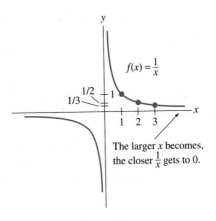

Figure 15

Consider next the function defined by

$$f(x) = \frac{1}{x^n} \qquad n \text{ is a positive integer}$$

Compare $1/x^n$ and $1/x$, noting that the magnitude of x^n is larger than the magnitude of x when x is approaching infinity or minus infinity. It then follows that

$$\lim_{x \to \infty} \frac{1}{x^n} = 0 \qquad n \text{ is a positive integer}$$

and

$$\lim_{x \to -\infty} \frac{1}{x^n} = 0 \qquad n \text{ is a positive integer}$$

If the function is $f(x) = c/x^n$, where c is a finite constant, then the limit as x increases or decreases without bound will still be zero. Consider that c/x^n is $c \cdot 1/x^n$, and $1/x^n$ will be zero in the limit, so that c/x^n will be $c \cdot 0$, or 0 in the limit. This important result is stated below.

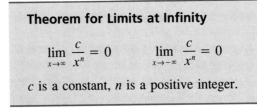

Theorem for Limits at Infinity

$$\lim_{x \to \infty} \frac{c}{x^n} = 0 \qquad \lim_{x \to -\infty} \frac{c}{x^n} = 0$$

c is a constant, n is a positive integer.

EXAMPLE 1 Evaluate each limit.

(a) $\displaystyle\lim_{x \to \infty} \frac{1}{x^8}$ **(b)** $\displaystyle\lim_{x \to -\infty} \frac{500}{x^3}$ **(c)** $\displaystyle\lim_{x \to \infty} \frac{-30}{x^2}$

SOLUTION   These limits can be determined by using the Theorem for Limits at Infinity.

$$\textbf{(a)} \lim_{x \to \infty} \frac{1}{x^8} = 0 \qquad \textbf{(b)} \lim_{x \to -\infty} \frac{500}{x^3} = 0 \qquad \textbf{(c)} \lim_{x \to \infty} \frac{-30}{x^2} = 0 \quad \blacklozenge$$

EXAMPLE 2   Determine $\lim_{x \to \infty} \dfrac{3x + 2}{4x + 9}$.

SOLUTION   This limit does not appear to fit the c/x^n form under consideration. However, if each term of the numerator and denominator is divided by x, the expression will consist only of constants and terms that are of the form c/x^n. Then we can apply the limit theorems (from Section 2.1), which do hold for limits at infinity.

$$\lim_{x \to \infty} \frac{3x + 2}{4x + 9} = \lim_{x \to \infty} \frac{\dfrac{3x}{x} + \dfrac{2}{x}}{\dfrac{4x}{x} + \dfrac{9}{x}}$$

$$= \lim_{x \to \infty} \frac{3 + \dfrac{2}{x}}{4 + \dfrac{9}{x}}$$

$$= \frac{\lim_{x \to \infty} \left(3 + \dfrac{2}{x} \right)}{\lim_{x \to \infty} \left(4 + \dfrac{9}{x} \right)}$$

$$= \frac{\lim_{x \to \infty} 3 + \lim_{x \to \infty} \dfrac{2}{x}}{\lim_{x \to \infty} 4 + \lim_{x \to \infty} \dfrac{9}{x}}$$

$$= \frac{3 + 0}{4 + 0}$$

$$= \frac{3}{4}$$

Note that in the limit, $2/x$ tends to 0 and $9/x$ tends to 0. $\blacklozenge$

EXAMPLE 3   Determine $\lim_{x \to -\infty} \dfrac{x^3 + 7x^2 - 8x + 1}{5x^3 - 3x^2 + 9x - 17}$.

SOLUTION   As an extension of the approach used in Example 2, divide each term of the numerator and denominator by *the highest power of x present in the denominator*, namely x^3. The resulting expression will consist only of constants and terms of the form c/x^n—each of which is zero in the limit, according to the Theorem for Limits at Infinity. Thus,

$$\lim_{x \to -\infty} \frac{x^3 + 7x^2 - 8x + 1}{5x^3 - 3x^2 + 9x - 17} = \lim_{x \to -\infty} \frac{1 + \dfrac{7}{x} - \dfrac{8}{x^2} + \dfrac{1}{x^3}}{5 - \dfrac{3}{x} + \dfrac{9}{x^2} - \dfrac{17}{x^3}}$$

$$= \frac{1 + 0 - 0 + 0}{5 - 0 + 0 - 0}$$

$$= \frac{1}{5} \quad \blacklozenge$$

EXAMPLE 4   Determine $\displaystyle\lim_{x \to \infty} \frac{7x^2 - 8x + 2}{4x^3 + 9x^2 - 8x + 1}$.

SOLUTION   Begin by dividing each term of the numerator and denominator by the highest power of x in the denominator, namely x^3.

$$\lim_{x \to \infty} \frac{7x^2 - 8x + 2}{4x^3 + 9x^2 - 8x + 1} = \lim_{x \to \infty} \frac{\dfrac{7}{x} - \dfrac{8}{x^2} + \dfrac{2}{x^3}}{4 + \dfrac{9}{x} - \dfrac{8}{x^2} + \dfrac{1}{x^3}}$$

$$= \frac{0 - 0 + 0}{4 + 0 - 0 + 0}$$

$$= \frac{0}{4}$$

$$= 0 \quad \blacklozenge$$

> **Note**
>
> The technique of dividing each term by a power of x is used *only* for limits at infinity. This approach will not help in the evaluation of other kinds of limits.

In the next section, we will see that with limits at infinity such as those of Examples 2, 3, and 4, if the degree of the numerator is greater than the degree of the denominator, then the limit of the entire expression as x approaches infinity or minus infinity will be infinite.

A limit at infinity such as $\displaystyle\lim_{x \to \infty} (1/x) = 0$ describes the behavior of the function $f(x) = 1/x$ as x tends toward infinity. The function values approach 0 as x tends toward infinity. The graph in Figure 16 shows how the curve tends

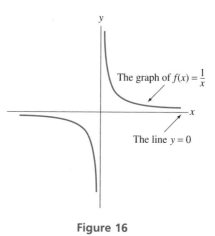

Figure 16

toward the line $y = 0$ as x approaches infinity. We say that the x axis (the horizontal line $y = 0$) is an *asymptote* for the graph of the function defined by $f(x) = 1/x$. In general,

> The line $y = L$ is a **horizontal asymptote** for the graph of $y = f(x)$ if either limit at infinity is the number L.

EXAMPLE 5   Determine the horizontal asymptote for the graph of each function.

(a) $f(x) = \dfrac{3x + 2}{4x + 9}$ **(b)** $f(x) = \dfrac{x^2 + 7x - 18}{x^3 + 5}$

SOLUTION   **(a)** In Example 2 the limit of $f(x)$ as $x \to \infty$ was determined to be 3/4.

$$\lim_{x \to \infty} \frac{3x + 2}{4x + 9} = \frac{3}{4}$$

Thus, $y = 3/4$ is a horizontal asymptote for the graph of the function.

(b) Consider the limit of $f(x)$ as $x \to \infty$.

$$\lim_{x \to \infty} f(x) = \lim_{x \to \infty} \frac{x^2 + 7x - 18}{x^3 + 5}$$

$$= \lim_{x \to \infty} \frac{\dfrac{x^2}{x^3} + \dfrac{7x}{x^3} - \dfrac{18}{x^3}}{\dfrac{x^3}{x^3} + \dfrac{5}{x^3}}$$

$$= \lim_{x \to \infty} \frac{\dfrac{1}{x} + \dfrac{7}{x^2} - \dfrac{18}{x^3}}{1 + \dfrac{5}{x^3}}$$

$$= \frac{0 + 0 - 0}{1 + 0} = \frac{0}{1} = 0$$

Since the limit of $f(x)$ as $x \to \infty$ is 0, we conclude that $y = 0$ is a horizontal asymptote for the graph of this function. ◆

EXAMPLE 6 ◆ *LIMIT ON PROFIT*

Suppose that profit from the sale of x units is $P(x) = 2000 - \dfrac{300}{x}$ dollars for $x \geq 1$. Considering the limit of $P(x)$ as x increases without bound, we find that

$$\lim_{x \to \infty} P(x) = \lim_{x \to \infty} \left(2000 - \frac{300}{x} \right)$$

$$= \lim_{x \to \infty} 2000 - \lim_{x \to \infty} \frac{300}{x}$$

$$= 2000 - 0$$

$$= 2000$$

The graph of function P shows profit beginning at \$1700 for $x = 1$ unit and tending toward \$2000 as x increases without bound. (See Figure 17.) Note that the line $y = 2000$ is a horizontal asymptote.

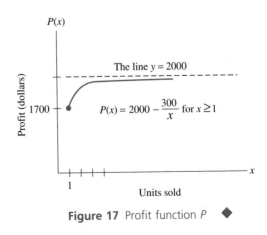

Figure 17 Profit function P ◆

2.4   Exercises

In Exercises 1–4, complete the tables to determine the limits. Use a calculator.

1.

$$\lim_{x \to \infty} \frac{x + 1}{x}$$

x	$\dfrac{x + 1}{x}$
100	
1000	
10,000	
1,000,000	

2.

$$\lim_{x \to \infty} \frac{x}{1 + x^2}$$

x	$\dfrac{x}{1 + x^2}$
100	
1000	
10,000	
100,000	

3.

$$\lim_{x \to -\infty} \frac{1 + 3x}{2x}$$

x	$\dfrac{1 + 3x}{2x}$
-100	
-1000	
-10,000	
-1,000,000	

4.

$$\lim_{x \to -\infty} \frac{1 - 5x}{2x}$$

x	$\dfrac{1 - 5x}{2x}$
-100	
-1000	
-10,000	
-1,000,000	

Use the Theorem for Limits at Infinity to evaluate each limit in Exercises 5–12.

5. $\displaystyle \lim_{x \to \infty} \frac{1}{x}$

6. $\displaystyle \lim_{x \to \infty} \frac{1}{x^2}$

7. $\displaystyle \lim_{x \to \infty} \frac{-20}{x^4}$

8. $\displaystyle \lim_{x \to -\infty} \frac{170}{x^3}$

9. $\displaystyle \lim_{x \to \infty} \frac{1000}{x^2}$

10. $\displaystyle \lim_{x \to \infty} \frac{-1}{x^6}$

11. $\displaystyle \lim_{x \to -\infty} \left(-\frac{1}{x^4}\right)$

12. $\displaystyle \lim_{x \to \infty} \left(-\frac{2}{x^5}\right)$

Use the Theorem for Limits at Infinity to evaluate each limit in Exercises 13–28.

13. $\displaystyle \lim_{x \to \infty} \frac{3x + 2}{5x - 4}$

14. $\displaystyle \lim_{x \to \infty} \frac{7x + 1}{2x + 6}$

15. $\displaystyle \lim_{x \to \infty} \frac{2x^2 + 8x + 6}{x^2 - 3x + 1}$

16. $\displaystyle \lim_{x \to \infty} \frac{5x^2 - 17}{x^2 + 2}$

17. $\displaystyle \lim_{x \to \infty} \frac{9 + 3x - 2x^2}{5 - x + x^2}$

18. $\displaystyle \lim_{x \to \infty} \frac{23 + 5x^2}{7 - 3x - 12x^2}$

19. $\displaystyle \lim_{x \to -\infty} \frac{5x^2 + 3x - 1}{x^3 - 9x^2 + 4}$

20. $\displaystyle \lim_{x \to -\infty} \frac{2x^3 + 5x - 7}{x^3 + 6x^2 + x}$

21. $\displaystyle \lim_{x \to \infty} \frac{x^3 - 15}{2x^3 + x^2 - 3x + 1}$

22. $\displaystyle \lim_{x \to \infty} \frac{5x^2 - 19}{1 + x + x^3}$

23. $\displaystyle \lim_{x \to -\infty} \frac{x^3 + x^2}{x^3 - x^2}$

24. $\displaystyle \lim_{x \to -\infty} \frac{4x - 3}{2x^2 - 5x + 1}$

25. $\lim\limits_{x\to\infty} \dfrac{1 - x}{1 + 2x}$

26. $\lim\limits_{x\to\infty} \dfrac{1 + x^2}{1 - x^2}$

27. $\lim\limits_{x\to\infty} \dfrac{1 + 3x}{x^2 - 5x + 2}$

28. $\lim\limits_{x\to\infty} \dfrac{5x + x^2}{x - x^2}$

Determine the horizontal asymptote for the graph of each function defined in Exercises 29–38.

29. $f(x) = \dfrac{2x + 1}{x - 4}$

30. $f(x) = \dfrac{8x - 7}{2x + 3}$

31. $f(x) = \dfrac{x}{1 - x}$

32. $f(x) = \dfrac{8x}{1 + 4x}$

33. $f(x) = \dfrac{x - 1}{x}$

34. $f(x) = \dfrac{2x - 3}{1 - x}$

35. $f(x) = \dfrac{x^2 + 3}{x^3 - 1}$

36. $f(x) = \dfrac{3x^2 - 7x + 1}{x - 5x^3}$

37. $f(x) = \dfrac{x^2 - 8x + 2}{3x^2 + 6x - 5}$

38. $f(x) = \dfrac{1 - 16x^2}{2x^2 - 3}$

39. (*AVERAGE COST*) A company makes soccer balls. When x balls are made, the average cost per ball, $\bar{C}(x)$, is

$$\bar{C}(x) = \frac{1500 + 12x}{x} \text{ dollars} \qquad \text{for } x \geq 1$$

When 100 balls are made, the average cost per ball is $\bar{C}(100)$, or $27. When 1000 balls are made, the average cost per ball drops to $13.50. As more and more soccer balls are made, the average cost per ball will continue to drop and tend toward a particular low average cost value. Determine that value.

40. (*PROFIT*) If the profit from the sale of x units is

$$P(x) = \frac{1400x - 250}{x} \text{ dollars} \qquad \text{for } x \geq 1$$

what is the limit of the profit as the quantity sold increases without bound?

41. (*ANIMAL HEIGHT*) Suppose that the function h defined by

$$h(t) = \frac{5t - 2}{t} \text{ feet} \qquad t \geq 1$$

gives the approximate height of a particular animal after t years. Toward what height does the adult animal tend?

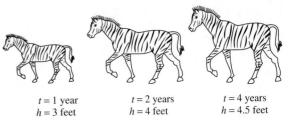

$t = 1$ year
$h = 3$ feet

$t = 2$ years
$h = 4$ feet

$t = 4$ years
$h = 4.5$ feet

42. (*TREE HEIGHT*) After t years, the height of a particular type of tree is given by

$$f(t) = \frac{20t - 15}{2t} \text{ feet} \qquad t \geq 1$$

Toward what height does the tree grow?

43. (*DRUG ABSORPTION*) The amount of a drug that remains in a person's bloodstream t hours after being injected is given by

$$f(t) = \frac{.15t}{1 + t^2}$$

(a) Find $\lim\limits_{t\to\infty} f(t)$.

W (b) Use nonmathematical words to explain the meaning of the result obtained in part (a). Explain it in terms of the application.

W 44. In the limit statement

$$\lim_{x\to\infty} \frac{c}{x^n} = 0$$

why can't n be a *negative* integer?

45. (*WEISS' LAW*) According to Weiss' law, the intensity of electric current needed to excite muscle and nerve tissue depends on how long the current flows to the tissue. The longer the duration of the current flow, the less current is needed to excite the tissue. The relationship is given by

$$I(t) = \frac{a}{t} + b$$

where I is the current's intensity, t is the time (duration) of current flow, and a and b are constants. Determine the (theoretically) smallest possible current that would be needed to excite the tissue.

46. (*GEOMETRY*) The number of degrees d in each interior angle of a regular polygon of n sides is

$$d = \frac{180n - 360}{n}$$

For example, in an equilateral triangle ($n = 3$ sides), the number of degrees in each angle is 60, calculated from

$$d = \frac{180(3) - 360}{3} = 60$$

In a square ($n = 4$ sides), each angle is 90°. In a regular pentagon ($n = 5$ sides), each angle is 108°.

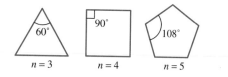

$n = 3$ $\qquad$ $n = 4$ $\qquad$ $n = 5$

(a) Determine the number of degrees in each interior angle of a regular hexagon ($n = 6$ sides).

(b) The number of degrees d in each interior angle of a regular polygon gets larger as n gets larger. Find the limit, that is, the number of degrees toward which the angles tend as n gets larger and larger.

W 47. Limits at infinity are quite different from ordinary limits in which the variable x approaches a finite number a. Nevertheless, in a way they do resemble one-sided limits. Explain.

The Exercise Library at the back of the book contains graphing calculator and computer exercises keyed to this section.

2.5 | *INFINITE LIMITS*

When considering $\lim_{x \to a} f(x)$, it may happen that $f(x)$ increases without bound (that is, becomes infinite) as x approaches a. In such instances, the limit does not exist in the usual sense. However, this situation can be described by writing

$$\lim_{x \to a} f(x) = \infty$$

Such a limit is called an **infinite limit**. A study of the graph shown in Figure 18 should help you to understand this kind of limit.

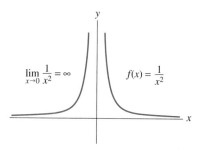

$$\lim_{x \to 0} \frac{1}{x^2} = \infty \qquad f(x) = \frac{1}{x^2}$$

Figure 18

One incidental observation that can be made from the graph is that function f is not continuous at 0, since there is a break in the graph at $x = 0$. Note that the closer x gets to 0, the larger the functional values become. A table has been prepared to demonstrate that as x gets closer and closer to 0, $1/x^2$ gets larger and larger.

x	$\dfrac{1}{x^2}$
1	1
.5	4
.1	100
.01	10,000
.001	1,000,000

Approaching 0 from the left, the x values would be -1, $-.5$, $-.1$, $-.01$, and $-.001$. The corresponding values of $1/x^2$ would be the same as those shown in the table.

If instead function f was defined to be $f(x) = -1/x^2$, then the graph would appear as shown in Figure 19. Again the magnitude of $f(x)$ increases without bound (becomes infinite) as x approaches 0. However, this time the values of $f(x)$ are negative; the values of $f(x)$ approach $-\infty$.

$$\lim_{x \to 0} \left(-\frac{1}{x^2} \right) = -\infty$$

Next, consider the graph of $f(x) = 1/x$, which suggests two interesting one-sided limits. The graph is shown in Figure 20.

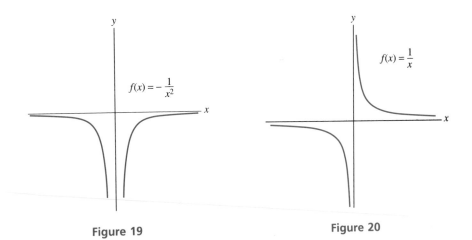

$$f(x) = -\frac{1}{x^2}$$

$$f(x) = \frac{1}{x}$$

Figure 19

Figure 20

The limits are:

$$\lim_{x \to 0^+} \frac{1}{x} = \infty$$

$$\lim_{x \to 0^-} \frac{1}{x} = -\infty$$

Numerically, you can interpret the limit $\lim_{x \to 0^+} (1/x) = \infty$ as follows: If the number 1 is divided by positive numbers that get closer and closer to zero, the results obtained will get larger and larger. In fact, for any positive constant c,

$$\lim_{x \to 0^+} \frac{c}{x} = \infty \qquad (c > 0)$$

If c is a negative constant, then the result will be negative:

$$\lim_{x \to 0^+} \frac{c}{x} = -\infty \qquad (c < 0)$$

Based on the numerical interpretation of these two limits, we will use the notation "appr zero(+)" to indicate the behavior of the denominator x as $x \to 0^+$ for the following arithmetic.

$$\frac{\text{positive constant}}{\text{appr zero(+)}} = \infty$$

$$\frac{\text{negative constant}}{\text{appr zero(+)}} = -\infty$$

EXAMPLE 1   Determine each limit. Use the ideas developed above.

(a) $\lim_{x \to 2^+} \dfrac{1}{x - 2}$ 　　(b) $\lim_{x \to 0} \dfrac{-35}{x^2}$

SOLUTION   (a) $\lim_{x \to 2^+} \dfrac{1}{x - 2} = \dfrac{\text{positive constant}}{\text{appr zero(+)}} = \infty$

When $x \to 2^+$, the x values are greater than 2; thus expression $x - 2$ will approach zero via positive values.

(b) $\lim_{x \to 0} \dfrac{-35}{x^2} = \dfrac{\text{negative constant}}{\text{appr zero(+)}} = -\infty$

Here it does not matter whether x approaches 0 from the left or the right, because when x is squared the result will be positive in either case. ◆

$$\lim_{x\to\infty} \frac{4x^3 + 5}{x + 1} = \lim_{x\to\infty} \frac{\dfrac{4x^3}{x} + \dfrac{5}{x}}{\dfrac{x}{x} + \dfrac{1}{x}}$$

$$= \lim_{x\to\infty} \frac{4x^2 + \dfrac{5}{x}}{1 + \dfrac{1}{x}}$$

As $x \to \infty$, the expressions $5/x$ and $1/x$ tend to 0 and $4x^2$ tends to ∞ (an example of cx^n as $x \to \infty$). This means that the numerator of the fraction is tending toward ∞ while the denominator tends toward 1. Thus, the limit is ∞.

$$\lim_{x\to\infty} \frac{4x^3 + 5}{x + 1} = \frac{\infty}{1} = \infty \quad \blacklozenge$$

In general,

> When the degree of the numerator is greater than the degree of the denominator, the limit at infinity will be infinite (either ∞ or $-\infty$).

2.5 Exercises

Determine each limit in Exercises 1–12.

1. $\lim\limits_{x\to 0} \dfrac{2}{x^2}$

2. $\lim\limits_{x\to 0} \dfrac{100}{x^2}$

3. $\lim\limits_{x\to 0} \dfrac{-3}{x^2}$

4. $\lim\limits_{x\to 0} \dfrac{-4}{x^2}$

5. $\lim\limits_{x\to 1^+} \dfrac{1}{x - 1}$

6. $\lim\limits_{x\to 2^+} \dfrac{-3}{x - 2}$

7. $\lim\limits_{x\to 3^+} \dfrac{-6}{x - 3}$

8. $\lim\limits_{x\to 4^+} \dfrac{15}{x - 4}$

9. $\lim\limits_{x\to 4^-} \dfrac{11}{x - 4}$

10. $\lim\limits_{x\to 5^-} \dfrac{2}{x - 5}$

11. $\lim\limits_{x\to 1^-} \dfrac{-3}{x - 1}$

12. $\lim\limits_{x\to 2^-} \dfrac{-2}{x - 2}$

Determine each limit in Exercises 13–22.

13. $\lim\limits_{x\to 2^+} \dfrac{5}{2 - x}$

14. $\lim\limits_{x\to 3^+} \dfrac{17}{3 - x}$

15. $\lim\limits_{x\to 1^+} \dfrac{-9}{1 - x}$

16. $\lim\limits_{x\to 2^-} \dfrac{3}{2 - x}$

17. $\lim\limits_{x\to 4^-} \dfrac{1}{4 - x}$

18. $\lim\limits_{x\to 2^+} \dfrac{-1}{2 - x}$

19. $\lim\limits_{x\to 2^-} \dfrac{-3}{2 - x}$

20. $\lim\limits_{x\to 1^-} \dfrac{-2}{1 - x}$

21. $\lim\limits_{x\to 1^-} \dfrac{1}{1 - x}$

22. $\lim\limits_{x\to 3^-} \dfrac{5}{3 - x}$

Determine each limit in Exercises 23–30.

23. $\lim\limits_{x \to 3^+} \dfrac{5}{(x - 3)^2}$ **24.** $\lim\limits_{x \to 5^+} \dfrac{1}{(5 - x)^2}$

25. $\lim\limits_{x \to 2^+} \dfrac{1}{(2 - x)^2}$ **26.** $\lim\limits_{x \to 1^+} \dfrac{2}{(x - 1)^2}$

27. $\lim\limits_{x \to 1^-} \dfrac{3}{(x - 1)^2}$ **28.** $\lim\limits_{x \to 2^-} \dfrac{-1}{(x - 2)^2}$

29. $\lim\limits_{x \to 4^-} \dfrac{-2}{(x - 4)^2}$ **30.** $\lim\limits_{x \to 2^-} \dfrac{4}{(x - 2)^2}$

Determine each limit in Exercises 31–38.

31. $\lim\limits_{x \to -1^+} \dfrac{2}{x + 1}$ **32.** $\lim\limits_{x \to -3^-} \dfrac{4}{x + 3}$

33. $\lim\limits_{x \to -1^-} \dfrac{2}{x + 1}$ **34.** $\lim\limits_{x \to -5^+} \dfrac{3}{x + 5}$

35. $\lim\limits_{x \to -2^-} \dfrac{-3}{2 + x}$ **36.** $\lim\limits_{x \to -2^+} \dfrac{-5}{2 + x}$

37. $\lim\limits_{x \to -3^+} \dfrac{-1}{3 + x}$ **38.** $\lim\limits_{x \to -1^-} \dfrac{-2}{1 + x}$

Determine the vertical asymptote for the graph of each function defined in Exercises 39–48.

39. $f(x) = \dfrac{4}{x - 5}$ **40.** $f(x) = \dfrac{1}{7 - x}$

41. $f(x) = \dfrac{2}{x^2}$ **42.** $f(x) = \dfrac{2}{x - 6}$

43. $f(x) = \dfrac{1}{x + 2}$ **44.** $f(x) = \dfrac{-1}{x^3}$

45. $f(x) = \dfrac{-3}{(x - 1)^2}$ **46.** $f(x) = \dfrac{2}{1 - x}$

47. $f(x) = \dfrac{1}{(x + 4)^2}$ **48.** $f(x) = \dfrac{3}{(x - 4)^2}$

Evaluate each limit in Exercises 49–54.

49. $\lim\limits_{x \to \infty} 3x^2$ **50.** $\lim\limits_{x \to \infty} \dfrac{x^5}{x^3 + 7x}$

51. $\lim\limits_{x \to \infty} \dfrac{x^3 - 6x^2 + 3}{5x^2 + x - 1}$ **52.** $\lim\limits_{x \to \infty} \dfrac{x^2 + 8x - 2}{x^4 + 6x + 1}$

53. $\lim\limits_{x \to \infty} \dfrac{5x - 2}{7x + 1}$ **54.** $\lim\limits_{x \to \infty} 8$

The Exercise Library at the back of the book contains graphing calculator and computer exercises keyed to this section.

Chapter List   *Important terms and ideas*

limit   discontinuity   horizontal asymptotes
continuous   one-sided limits   infinite limits
discontinuous   limits at infinity   vertical asymptotes

Review Exercises for Chapter 2

In Exercises 1–12, evaluate each limit.

1. $\lim\limits_{x \to 4} x^{-3}$ **2.** $\lim\limits_{n \to 3} \sqrt{n^2 + 16}$ **5.** $\lim\limits_{t \to -2} 17$ **6.** $\lim\limits_{x \to 3} \dfrac{x - 3}{x + 2}$

3. $\lim\limits_{x \to 6} \dfrac{\sqrt{37 - 2x}}{x + 7}$ **4.** $\lim\limits_{m \to 2} \pi$ **7.** $\lim\limits_{x \to 5} \dfrac{x^2 - 25}{2x - 10}$ **8.** $\lim\limits_{x \to 2} \sqrt{\dfrac{5x^2}{x + 2}}$

9. $\lim\limits_{x \to 1} \dfrac{3x}{1 + (x - 1)^2}$

10. $\lim\limits_{x \to -1} \dfrac{x^2 + x}{1 + x}$

11. $\lim\limits_{x \to 7} \dfrac{14 - 2x}{7x}$

12. $\lim\limits_{x \to 0} \dfrac{x^3}{x^2}$

Evaluate each one-sided limit in Exercises 13–16.

13. $\lim\limits_{x \to 1^-} f(x)$

14. $\lim\limits_{x \to 1^+} f(x)$

$f(x) = \begin{cases} 4x & x \le 1 \\ 2x + 1 & x > 1 \end{cases}$

15. $\lim\limits_{x \to 0^-} g(x)$

16. $\lim\limits_{x \to 0^+} g(x)$

$g(x) = \begin{cases} 3x & x < 0 \\ \sqrt{x + 1} & x \ge 0 \end{cases}$

In Exercises 17–20, evaluate each limit at infinity.

17. $\lim\limits_{x \to \infty} \dfrac{5x - 19}{7x + 2}$

18. $\lim\limits_{x \to \infty} \dfrac{30}{1 + x^{10}}$

19. $\lim\limits_{x \to \infty} \dfrac{3x^2}{x^3 - x^2 + x}$

20. $\lim\limits_{x \to \infty} \dfrac{5 + x^2}{5 + x^3}$

Determine each limit in Exercises 21–24.

21. $\lim\limits_{x \to 5^-} \dfrac{2}{x - 5}$

22. $\lim\limits_{x \to 5^+} \dfrac{2}{x - 5}$

23. $\lim\limits_{x \to 0^+} \dfrac{3}{2x^5}$

24. $\lim\limits_{x \to 3^-} \dfrac{-4}{x - 3}$

In Exercises 25–30, determine the limit, if it exists.

25. $\lim\limits_{x \to 0} f(x)$

26. $\lim\limits_{x \to 2} f(x)$

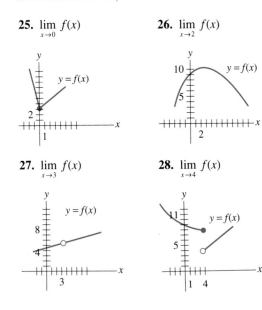

27. $\lim\limits_{x \to 3} f(x)$

28. $\lim\limits_{x \to 4} f(x)$

29. $\lim\limits_{x \to 2} g(x)$

30. $\lim\limits_{x \to 5} g(x)$

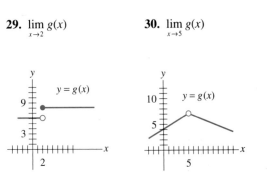

31. Answer the questions based on the graph of f shown in the figure below.
(a) Is f continuous at 0?
(b) Is f continuous at -3?
(c) Does $\lim\limits_{x \to 2} f(x)$ exist?
(d) Is f continuous at 2?
(e) Does $\lim\limits_{x \to 6} f(x)$ exist?
(f) Is f continuous at 6?
(g) Does $\lim\limits_{x \to 4} f(x)$ exist?
(h) Is f continuous at 4?

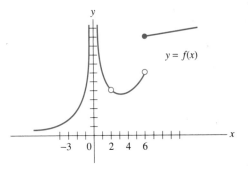

32. Answer the questions based on the graph of g shown in the figure on the next page.
(a) Does $\lim\limits_{x \to 0} g(x)$ exist?
(b) Is g continuous at 0?
(c) Does $\lim\limits_{x \to 2} g(x)$ exist?
(d) Is g continuous at 2?
(e) Does $\lim\limits_{x \to 5} g(x)$ exist?
(f) Is g continuous at 5?
(g) Is g continuous at 1?
(h) Is g continuous at 10?

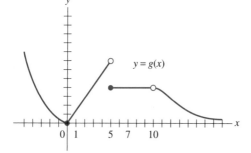

33. (*PSYCHOLOGY EXPERIMENT*) A psychologist studies the time required to complete a devised task. He finds that as people gain experience, they complete the task in less time; in fact,

$$T(n) = 4 + \frac{6}{n}$$

where n is the number of times they perform the task and T is the time (in minutes) it takes them to complete the task. Based on the formula and your knowledge of limits, what is the (theoretically) shortest time needed to complete the task?

34. (*PRICE TENDENCY*) The price of oak chairs being supplied to a wholesaler is given by

$$p = \frac{40}{x} + 120 \quad \text{dollars}$$

where p is the price per chair and x is the number of chairs supplied. Toward what amount is the price tending as the number of chairs supplied increases?

3

DERIVATIVES

The study of limits provided new directions and applications. Familiar algebra and graphs were used in ways that you probably had not anticipated.

As we continue the presentation of calculus, the opportunity will arise to use limits in the study of an operation called differentiation. You will see how limits occur in the basic definitions and throughout the development of this key calculus topic. Limits will be combined with the familiar idea of slope of a straight line. This will set the stage for defining differentiation and developing basic rules for carrying it out. A variety of applications will be included, all based on various interpretations of the differentiation process.

3.1 INTRODUCTION TO THE DERIVATIVE

Before proceeding to a formal introduction to the derivative, we consider some background information on the nature of the derivative. Simply stated, *the derivative is a rate of change.* Three examples are given next to offer some preliminary insight into the notion of rate of change.

1. Consider an outbreak of the flu. A function specifies the number of people sick with flu at any particular time. The derivative of the function indicates the *rate* at which illness due to flu is spreading at any particular time.

2. Suppose that a function gives the average cost per unit of producing x units. The derivative of the function yields information about when the average cost per unit is increasing and when it is decreasing.

3. A function may describe the motion of a rocket, giving the distance traveled for any time t. The derivative of this function is the rate of change of distance with respect to time—the velocity. From the derivative we can determine the velocity at any instant desired.

The slope of a straight line can be considered to be a rate of change. It is the rate of change of y with respect to x.

$$\text{slope } (m) \; = \; \frac{\Delta y}{\Delta x}$$

The slope of a line is a number that specifies the change in y compared with the change in x in going from point to point on the line. For any particular line, the slope is constant for the entire line. See Figure 1.

But what happens when the function under consideration is not linear? In order to consider the rate of change in such instances, we might choose to extend

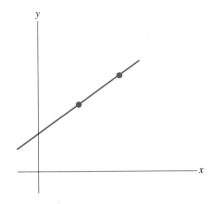

Figure 1 A linear function (straight line); constant slope and constant rate of change

the notion of slope to curves other than straight lines. Consider, for example, the graph of the function f defined by $f(x) = x^2$, shown in Figure 2.

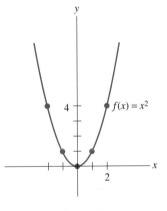

Figure 2

How do we talk about the slope of a curve that is not a straight line? *The slope of a curve at any point is considered to be the slope of the line tangent to (that is, touching) the curve at that point. Such a line has the same steepness as the curve at that point.* For example, the slope of the parabola $f(x) = x^2$ at $(1, 1)$ is the slope of the line tangent to the curve at $(1, 1)$. See Figure 3.

The slope of a straight line can be determined from two points on that line by computing the change in the y coordinates (Δy) and the change in the x coordinates (Δx). The slope is

$$m = \frac{\Delta y}{\Delta x}$$

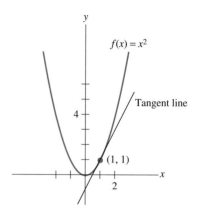

Figure 3 The tangent line at (1, 1)

Proceeding in a similar fashion for a parabola or other type of curve, we can begin our search for the slope of the tangent line by using two points on the curve. Let $(x, f(x))$ be any point through which we want the tangent line. To obtain a second point on the curve, add to the x coordinate the nonzero amount Δx. This means that the x coordinate of the second point is $x + \Delta x$, and the second point itself is $(x + \Delta x, f(x + \Delta x))$. The two points and the line through them are shown in Figure 4.

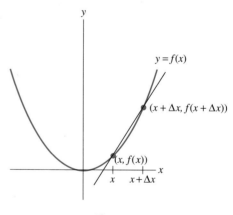

Figure 4

A true tangent line to the graph of $y = f(x)$ at $(x, f(x))$ would touch the curve at only one point, namely at $(x, f(x))$. Yet the line in our drawing passes through the curve at *two* points. Such a line is called a *secant line*. Using the notation m_{sec} to mean "the slope of the secant line," we have

$$m_{\text{sec}} = \frac{\Delta y}{\Delta x} = \frac{f(x + \Delta x) - f(x)}{(x + \Delta x) - x}$$

which simplifies to

$$m_{\text{sec}} = \frac{f(x + \Delta x) - f(x)}{\Delta x}$$

Notice from the drawing in Figure 5 that as Δx gets smaller, the secant line comes closer and closer to being the tangent line. Thinking in terms of limits, we can say that as Δx approaches 0, the secant line approaches the tangent line.

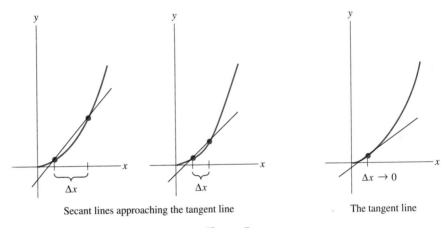

Secant lines approaching the tangent line   The tangent line

Figure 5

Our analysis suggests that the slope of the tangent line is the limit of the slope of the secant line as Δx approaches 0. Thus, using the notation m_{tan} to mean "the slope of the tangent line," we make the following definition:

Slope of the Tangent Line

The *slope of the tangent line* to the graph of f at any point $(x, f(x))$ is

$$m_{\text{tan}} = \lim_{\Delta x \to 0} \frac{f(x + \Delta x) - f(x)}{\Delta x}$$

provided this limit exists.

The **tangent line** to the graph of f at any point $(x, f(x))$ can now be defined; it is the line through the point $(x, f(x))$ and having slope

$$m = \lim_{\Delta x \to 0} \frac{f(x + \Delta x) - f(x)}{\Delta x}$$

provided the limit exists.

EXAMPLE 1   Determine the slope of the line tangent to the graph of $f(x) = x^2$ at any point $(x, f(x))$.

SOLUTION   Using

$$m_{\tan} = \lim_{\Delta x \to 0} \frac{f(x + \Delta x) - f(x)}{\Delta x}$$

with $f(x) = x^2$ yields

$$m_{\tan} = \lim_{\Delta x \to 0} \frac{(x + \Delta x)^2 - x^2}{\Delta x}$$

$$= \lim_{\Delta x \to 0} \frac{x^2 + 2x(\Delta x) + (\Delta x)^2 - x^2}{\Delta x}$$

$$= \lim_{\Delta x \to 0} \frac{2x(\Delta x) + (\Delta x)^2}{\Delta x}$$

At this point, Δx can be factored out of the numerator.

$$m_{\tan} = \lim_{\Delta x \to 0} \frac{\Delta x(2x + \Delta x)}{\Delta x}$$

The factors of Δx in the numerator and denominator can be eliminated by division, since $\Delta x \neq 0$.

$$m_{\tan} = \lim_{\Delta x \to 0} (2x + \Delta x)$$

Finally, taking the limit as $\Delta x \to 0$, we obtain

$$m_{\tan} = 2x$$

Thus, the slope of the line tangent to the graph of $f(x) = x^2$ at any point $(x, f(x))$ is $2x$. Specifically, the slope of the tangent line at $(1, 1)$ is $2(1)$, or 2. The slope of the tangent line at $(-2, 4)$ is $2(-2)$, or -4. See Figure 6.

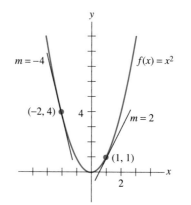

Figure 6 Slopes of tangent lines to the graph of $y = x^2$ ◆

Later in this chapter, as well as in Chapter 4, you will see that the limit

$$\lim_{\Delta x \to 0} \frac{f(x + \Delta x) - f(x)}{\Delta x}$$

has many applications. In view of this, it is given a special name and notation. It is called the **derivative** of the function. The definition:

Derivative

The **derivative** of function f with respect to x is the function f' defined by

$$f'(x) = \lim_{\Delta x \to 0} \frac{f(x + \Delta x) - f(x)}{\Delta x}$$

provided this limit exists.

The *domain* of f' consists of all x for which this limit exists. When the derivative exists, the function is said to be **differentiable**. The process of obtaining the derivative is called **differentiation**. To **differentiate** a function means to determine its derivative.

A variety of notations are used to indicate the derivative.

Notations for the Derivative

$f'(x)$	read as "f prime of x"
y'	read as "y prime"
$\dfrac{dy}{dx}$	read as either "the derivative of y with respect to x" or "dee y dee x"
$\dfrac{d}{dx} f(x)$	read as "the derivative of $f(x)$ with respect to x" or as "dee f dee x"
$D_x y$	read as "the derivative of y with respect to x"
$D_x f(x)$	read as "the derivative of $f(x)$ with respect to x"

The third notation listed, dy/dx, is the original notation of Leibniz.

Keep in mind that in working with functions of x, we are differentiating *with respect to x*. With functions of t, we would be differentiating with respect to t and the notation would be $f'(t)$, dy/dt, etc. If we had $y = f(u)$, then the differentiation would be with respect to u and the notation would be $f'(u)$, dy/du, etc. (Examples using letters other than x will appear throughout the text, beginning in the next section.)

As a final note on notation, consider that

$$\frac{f(x + \Delta x) - f(x)}{\Delta x}$$

is the same as

$$\frac{\Delta y}{\Delta x}$$

which means that the derivative can also be considered as

$$\lim_{\Delta x \to 0} \frac{\Delta y}{\Delta x}$$

That is,

$$\frac{dy}{dx} = \lim_{\Delta x \to 0} \frac{\Delta y}{\Delta x}$$

provided the limit exists.

The *dy/dx* notation is particularly popular, and consequently it will be used extensively, beginning in the next section.

Here are several examples in which the definition of the derivative is used to obtain $f'(x)$.

Note

Δx is one number. Think of it as a single symbol. Keep in mind that Δx *does not mean* Δ *times* x.

EXAMPLE 2   Let $f(x) = x^3$. Find the derivative.

SOLUTION   Using the definition of the derivative, we have

$$f'(x) = \lim_{\Delta x \to 0} \frac{f(x + \Delta x) - f(x)}{\Delta x}$$

$$= \lim_{\Delta x \to 0} \frac{(x + \Delta x)^3 - x^3}{\Delta x}$$

$$= \lim_{\Delta x \to 0} \frac{x^3 + 3x^2(\Delta x) + 3x(\Delta x)^2 + (\Delta x)^3 - x^3}{\Delta x}$$

$$= \lim_{\Delta x \to 0} \frac{3x^2(\Delta x) + 3x(\Delta x)^2 + (\Delta x)^3}{\Delta x}$$

$$= \lim_{\Delta x \to 0} \frac{\Delta x[3x^2 + 3x(\Delta x) + (\Delta x)^2]}{\Delta x}$$

$$= \lim_{\Delta x \to 0} [3x^2 + 3x(\Delta x) + (\Delta x)^2]$$

$$= 3x^2$$

We see that if $f(x) = x^3$, then $f'(x) = 3x^2$. ◆

Note

We have been using Δx to show the change in the independent variable x and to display the connection between the slope, the derivative, and the notation for derivatives. Other uses for Δx appear throughout the remainder of this chapter.

1. In Section 3.3, the Δx notation will smooth the transition from average rate of change to instantaneous rate of change.

2. The discussion of marginal analysis (Section 3.4) is clarified by using the Δx notation.

3. The chain rule (Section 3.6) can be seen intuitively using the "delta" notation.

4. In order to define differentials (Section 3.10), Δx notation must be used along with the slopes of the secant line and the tangent line.

Some people prefer to replace Δx by the letter h when calculating the derivative from the definition. In view of this, the next example offers a comparison. Your instructor may indicate a preference for Δx or for h.

EXAMPLE 3   (a) Determine $f'(x)$ if $f(x) = x^2 - 5x + 1$. Use the Δx notation.

(b) Determine $f'(x)$ if $f(x) = x^2 - 5x + 1$. Use h instead of Δx.

SOLUTION   (a) Using the definition of derivative, we have

$$f'(x) = \lim_{\Delta x \to 0} \frac{f(x + \Delta x) - f(x)}{\Delta x}$$

To continue, obtain $f(x + \Delta x)$ by replacing *every* x by $x + \Delta x$.

$$f'(x) = \lim_{\Delta x \to 0} \frac{[(x + \Delta x)^2 - 5(x + \Delta x) + 1] - (x^2 - 5x + 1)}{\Delta x}$$

$$= \lim_{\Delta x \to 0} \frac{x^2 + 2x(\Delta x) + (\Delta x)^2 - 5x - 5(\Delta x) + 1 - (x^2 - 5x + 1)}{\Delta x}$$

$$= \lim_{\Delta x \to 0} \frac{2x(\Delta x) + (\Delta x)^2 - 5(\Delta x)}{\Delta x}$$

$$= \lim_{\Delta x \to 0} \frac{\Delta x(2x + \Delta x - 5)}{\Delta x}$$

$$= \lim_{\Delta x \to 0} (2x + \Delta x - 5)$$

$$= 2x - 5$$

Using d/dx to mean "the derivative with respect to x of," we can write

$$\frac{d}{dx}(x^2 - 5x + 1) = 2x - 5$$

(b) Using the definition of derivative, with h instead of Δx, we have

$$f'(x) = \lim_{h \to 0} \frac{f(x + h) - f(x)}{h}$$

To continue, obtain $f(x + h)$ by replacing *every* x by $x + h$.

$$f'(x) = \lim_{h \to 0} \frac{[(x + h)^2 - 5(x + h) + 1] - (x^2 - 5x + 1)}{h}$$

$$= \lim_{h \to 0} \frac{(x^2 + 2xh + h^2 - 5x - 5h + 1) - (x^2 - 5x + 1)}{h}$$

$$= \lim_{h \to 0} \frac{2xh + h^2 - 5h}{h}$$

$$= \lim_{h \to 0} \frac{h(2x + h - 5)}{h}$$

$$= \lim_{h \to 0} (2x + h - 5)$$

$$= 2x - 5$$

Using d/dx to mean "the derivative with respect to x of," we can write

$$\frac{d}{dx}(x^2 - 5x + 1) = 2x - 5 \quad \blacklozenge$$

EXAMPLE 4   Find $f'(x)$, if $f(x) = \dfrac{1}{x}$.

SOLUTION   Using the definition of the derivative, we have

$$f'(x) = \lim_{\Delta x \to 0} \frac{f(x + \Delta x) - f(x)}{\Delta x}$$

$$= \lim_{\Delta x \to 0} \frac{\dfrac{1}{x + \Delta x} - \dfrac{1}{x}}{\Delta x}$$

Since the expression is a complex fraction involving denominators $x + \Delta x$ and x, multiply all three terms of the complex fraction by the common denominator

$(x + \Delta x)(x)$. This will change the complex fraction to an ordinary fraction and make it easier to work with.

$$f'(x) = \lim_{\Delta x \to 0} \frac{\dfrac{1}{x + \Delta x} \cdot (x + \Delta x)(x) - \dfrac{1}{x} \cdot (x + \Delta x)(x)}{\Delta x (x + \Delta x)(x)}$$

$$= \lim_{\Delta x \to 0} \frac{x - (x + \Delta x)}{(\Delta x)(x + \Delta x)(x)}$$

$$= \lim_{\Delta x \to 0} \frac{-\Delta x}{(\Delta x)(x + \Delta x)(x)}$$

Next, divide out the factors of Δx in the numerator and denominator of this fraction.

$$f'(x) = \lim_{\Delta x \to 0} \frac{-1}{(x + \Delta x)(x)}$$

As $\Delta x \to 0$, $x + \Delta x \to x + 0$, or x. Thus,

$$f'(x) = \frac{-1}{(x)(x)} = \frac{-1}{x^2}$$

In conclusion,

$$\frac{d}{dx}\left(\frac{1}{x}\right) = -\frac{1}{x^2}$$

This result can also be expressed using negative exponents, as

$$\frac{d}{dx}(x^{-1}) = -1 \cdot x^{-2}$$

This latter form will be used in the next section to obtain an important generalization.   ◆

In the next section we will begin to develop some rules that will make it much easier to determine derivatives of functions. Nevertheless, use of the definition of the derivative will remain important, because we will need it whenever new types of functions arise. This will happen when we study exponential functions, logarithmic functions, and trigonometric functions, among others.

As a final note, consider this theorem.

> If a function f is differentiable at $x = a$,
> then it is continuous at $x = a$.

After all, if f is differentiable at a, then a tangent line can be drawn to the graph at $(a, f(a))$. This means there can be no break (discontinuity) at $(a, f(a))$, and therefore f must be continuous at a. (See Figure 7.)

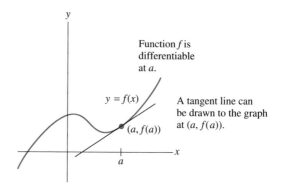

Figure 7 If f is differentiable at a, then it is continuous at a.

In Section 4.1 we will examine some graphs that lead to the following conclusion: *If function f is continuous at x = a, it is not necessarily differentiable at x = a.*

3.1 Exercises

For each function given in Exercises 1–20, use the *definition* of the derivative to obtain $f'(x)$.

1. $f(x) = 5x + 1$

2. $f(x) = 3x - 4$

3. $f(x) = x^2 + 3$

4. $f(x) = x^2 - 8$

5. $f(x) = x^2 - 4x + 2$

6. $f(x) = x^2 + 3x - 7$

7. $f(x) = 3x^2 + 7x$

8. $f(x) = 2x^2 - 5x$

9. $f(x) = 6$

10. $f(x) = 2$

11. $f(x) = x^3 + 2$

12. $f(x) = x^3 - 1$

13. $f(x) = x^3 + x^2 + x$

14. $f(x) = x^3 - 4x^2 + 2$

15. $f(x) = \dfrac{2}{x}$

16. $f(x) = \dfrac{5}{x}$

17. $f(x) = \dfrac{1}{3x}$

18. $f(x) = -\dfrac{1}{x}$

19. $f(x) = \dfrac{1}{x + 1}$

20. $f(x) = \dfrac{1}{x - 1}$

In Exercises 21–26, find the slope of the line tangent to the graph of the function f at any point $(x, f(x))$. Then find the slope at the specific point given.

21. $f(x) = x^2 + 6x$; the point $(2, 16)$

22. $f(x) = x^2 - 3x + 5$; the point $(4, 9)$

23. $f(x) = x^3 - 9$; the point $(4, 55)$

24. $f(x) = \dfrac{1}{x}$; the point $(1/2, 2)$

25. $f(x) = 4x - 1$; the point $(3, 11)$

26. $f(x) = 9$; the point $(2, 9)$

27. Determine the equation of the tangent line mentioned in Exercise 21.

28. Determine the equation of the tangent line mentioned in Exercise 24.

Obtain the derivative f' for each function defined in Exercises 29–34.

29. $f(x) = x^4$

30. $f(x) = x^5$

31. $f(x) = \sqrt{x}$ *Hint:* After using the definition of the derivative, you will be unable to proceed until you have rationalized the numerator of the expression. Multiply both numerator and denominator by $\sqrt{x + \Delta x} + \sqrt{x}$ in order to rationalize the numerator.

32. $f(x) = \sqrt{x + 3}$

33. $f(x) = \dfrac{1}{\sqrt{x}}$

34. $f(x) = \dfrac{1}{x^2}$

35. Explain why we can eliminate the Δx factors by division in going from

$$\lim_{\Delta x \to 0} \frac{\Delta x (2x + \Delta x)}{\Delta x}$$

to

$$\lim_{\Delta x \to 0} (2x + \Delta x)$$

After all, division by zero is not defined. Isn't this division by zero?

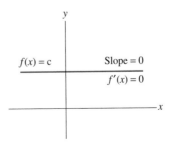 The Exercise Library at the back of the book contains graphing calculator and computer exercises keyed to this section.

3.2 | BASIC RULES FOR DIFFERENTIATION

As a beginning, we have obtained derivatives of functions by using the definition of the derivative, namely,

$$f'(x) = \lim_{\Delta x \to 0} \frac{f(x + \Delta x) - f(x)}{\Delta x}$$

Because using the definition of the derivative can be a lengthy and difficult procedure, rules that simplify differentiation have been developed. This section presents some of these rules.

Functions such as $f(x) = 2$ and $f(x) = -5$ are called *constant functions* because they have the same value for every x. The graph of any constant function $f(x) = c$ is a horizontal straight line. The slope of any such line is zero, which means that the derivative $f'(x)$ is zero. Thus, the derivative of a constant is zero. See Figure 8.

y

$f(x) = c$ Slope $= 0$
                    $f'(x) = 0$

x

Figure 8

The Derivative of a Constant Is Zero

$$\frac{d}{dx}(c) = 0$$

where c is a constant.

EXAMPLE 1   Included here are four examples of the rule that the derivative of a constant is zero.

SOLUTION   **(a)** $\dfrac{d}{dx}(15) = 0$ **(b)** If $f(t) = 1$, then $f'(t) = 0$

(c) $\dfrac{d}{dx}(-6) = 0$ **(d)** If $y = \sqrt{3}$, then $\dfrac{dy}{dx} = 0$ ◆

In the first section we obtained the following results, which are restated below to make the pattern more obvious.

$$\frac{d}{dx}(x^2) = 2x^1$$

$$\frac{d}{dx}(x^3) = 3x^2$$

$$\frac{d}{dx}(x^{-1}) = -1x^{-2}$$

These three derivatives are examples of the **power rule** for derivatives of powers of x.

Power Rule

$$\frac{d}{dx}(x^n) = nx^{n-1}$$

where n is a real number.

The power rule will be proved formally at the end of Section 5.4.
Here are some examples that demonstrate the use of the power rule to differentiate functions we have not previously differentiated.

EXAMPLE 2   Find the derivative of $y = x^5$.

SOLUTION   By the power rule,

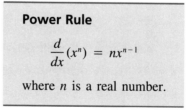

$$\frac{dy}{dx} = \frac{d}{dx}(x^5) = 5x^4$$ ◆

EXAMPLE 3   If $y = \sqrt{x}$, determine dy/dx.

SOLUTION   In order to apply the power rule, we need a *numerical exponent* representation for $\sqrt{x}$. So we write $\sqrt{x}$ as $x^{1/2}$. Now

$$y = x^{1/2}$$

and

$$\frac{dy}{dx} = \frac{1}{2}x^{-1/2} = \frac{1}{2x^{1/2}} = \frac{1}{2\sqrt{x}} \quad \blacklozenge$$

EXAMPLE 4   Obtain the derivative of $f(t) = \dfrac{1}{t^2}$.

SOLUTION   In order to apply the power rule, rewrite $1/t^2$ as t^{-2}. Now the expression is indeed a power of t. As originally given, the expression was 1 *divided by* a power of t. The derivative of

$$f(t) = t^{-2}$$

is

$$f'(t) = -2t^{-3} = \frac{-2}{t^3} = -\frac{2}{t^3} \quad \blacklozenge$$

The next rule specifies how to find the derivative of a constant times a function, when the derivative of the function is known.

$$\frac{d}{dx}[c \cdot f(x)] = c \cdot \frac{d}{dx}f(x)$$

where c is a constant.

In words, *the derivative of a constant times a function is that constant times the derivative of the function.* Here is an opportunity to see some "formal" mathematics. Proof of this derivative rule follows readily from the definition of the derivative, as shown next.

$$\frac{d}{dx}[c \cdot f(x)] = \lim_{\Delta x \to 0} \frac{c \cdot f(x + \Delta x) - c \cdot f(x)}{\Delta x}$$

$$= \lim_{\Delta x \to 0} \frac{c[f(x + \Delta x) - f(x)]}{\Delta x}$$

$$= \lim_{\Delta x \to 0} c \cdot \frac{f(x + \Delta x) - f(x)}{\Delta x}$$

$$= c \cdot \lim_{\Delta x \to 0} \frac{f(x + \Delta x) - f(x)}{\Delta x}$$

$$= c \cdot \frac{d}{dx}f(x)$$

The rule says, in effect, that a constant can be moved out in front of the differentiation process.

EXAMPLE 5 Find $\dfrac{dy}{dx}$. **(a)** $y = 4x^6$ **(b)** $y = \dfrac{x^3}{5}$ **(c)** $y = -2x^5$ **(d)** $y = -\dfrac{3}{x^7}$

SOLUTION **(a)** $\dfrac{dy}{dx} = \dfrac{d}{dx}(4x^6) = 4 \cdot \dfrac{d}{dx}(x^6) = 4 \cdot 6x^5 = 24x^5$

(b) $\dfrac{dy}{dx} = \dfrac{d}{dx}\left(\dfrac{x^3}{5}\right) = \dfrac{d}{dx}\left(\dfrac{1}{5}x^3\right) = \dfrac{1}{5} \cdot \dfrac{d}{dx}(x^3) = \dfrac{1}{5} \cdot 3x^2 = \dfrac{3x^2}{5}$

(c) $\dfrac{dy}{dx} = \dfrac{d}{dx}(-2x^5) = -2 \cdot \dfrac{d}{dx}(x^5) = -2 \cdot 5x^4 = -10x^4$

(d) $\dfrac{dy}{dx} = \dfrac{d}{dx}\left(-\dfrac{3}{x^7}\right) = \dfrac{d}{dx}(-3x^{-7}) = -3 \cdot \dfrac{d}{dx}(x^{-7}) = -3\,(-7x^{-8})$

$\qquad = 21x^{-8} = \dfrac{21}{x^8}$ ◆

Perhaps you can see the shortcut in differentiating expressions such as those of the previous example, that is, expressions of the form cx^n.

$$\frac{d}{dx}(cx^n) = n \cdot cx^{n-1}$$

Thus, for example,

$$\frac{d}{dx}(8x^4) = 4 \cdot 8x^3 = 32x^3$$

$$\frac{d}{dt}(3t^9) = 9 \cdot 3t^8 = 27t^8$$

Another special case is worth noting here. Since $y = cx$ is the equation of a straight line with slope c, it follows that

$$\frac{d}{dx}(cx) = c$$

This result can also be seen as a special case of the derivative of cx^n, as shown next.

$$\frac{d}{dx}(cx) = \frac{d}{dx}(cx^1) = 1 \cdot cx^0 = 1 \cdot c \cdot 1 = c$$

Here are some examples of this special case.

$$\frac{d}{dx}(5x) = 5$$

$$\frac{d}{dx}(\pi x) = \pi$$

$$\frac{d}{dx}(x) = 1$$

$$\frac{d}{dt}(9t) = 9$$

The next rule will enable us to combine the separate rules already developed.

Sum and Difference Rules

$$\frac{d}{dx}[f(x) + g(x)] = \frac{d}{dx}f(x) + \frac{d}{dx}g(x)$$

$$\frac{d}{dx}[f(x) - g(x)] = \frac{d}{dx}f(x) - \frac{d}{dx}g(x)$$

In words, *the derivative of a sum is the sum of the derivatives* and *the derivative of a difference is the difference of the derivatives*. A sum or difference can be differentiated term by term. If you do not feel that this rule is true intuitively, then consider the following proof, which uses the definition of the derivative.

$$\frac{d}{dx}[f(x) + g(x)] = \lim_{\Delta x \to 0} \frac{[f(x + \Delta x) + g(x + \Delta x)] - [f(x) + g(x)]}{\Delta x}$$

$$= \lim_{\Delta x \to 0} \frac{[f(x + \Delta x) - f(x)] + [g(x + \Delta x) - g(x)]}{\Delta x}$$

$$= \lim_{\Delta x \to 0} \left[\frac{f(x + \Delta x) - f(x)}{\Delta x} + \frac{g(x + \Delta x) - g(x)}{\Delta x} \right]$$

$$= \lim_{\Delta x \to 0} \frac{f(x + \Delta x) - f(x)}{\Delta x} + \lim_{\Delta x \to 0} \frac{g(x + \Delta x) - g(x)}{\Delta x}$$

$$= \frac{d}{dx}f(x) + \frac{d}{dx}g(x)$$

EXAMPLE 6   Differentiate $y = x^{10} + 3x$.

SOLUTION   The differentiation will make use of the sum rule.

$$\frac{dy}{dx} = \frac{d}{dx}(x^{10} + 3x)$$

$$= \frac{d}{dx}(x^{10}) + \frac{d}{dx}(3x) \qquad \text{by the sum rule}$$

$$= 10x^9 + 3 \qquad \text{after differentiating} \quad \blacklozenge$$

The sum rule can be extended to sums of three or four or more functions, as demonstrated in the next example.

EXAMPLE 7 Determine the derivative of $y = x^3 + 4x^2 - 7x + 1$.

SOLUTION

$$\frac{dy}{dx} = \frac{d}{dx}(x^3 + 4x^2 - 7x + 1) \qquad \text{(step 1)}$$

$$= \frac{d}{dx}(x^3 + 4x^2) + \frac{d}{dx}(-7x + 1) \qquad \text{(step 2)}$$

$$= \frac{d}{dx}(x^3) + \frac{d}{dx}(4x^2) + \frac{d}{dx}(-7x) + \frac{d}{dx}(1) \qquad \text{(step 3)}$$

$$= 3x^2 + 8x - 7$$

The second step above was included to show you why the sum rule can be extended to more than two terms. In the future, you should go directly from step 1 to step 3. ◆

EXAMPLE 8 **(a)** Find the slope of the line tangent to the graph of $y = x^2 + 5x - 3$ at the point $(3, 21)$.

(b) Find the equation of the tangent line mentioned in part (a).

SOLUTION **(a)** Since the slope of the tangent line is the derivative, we will obtain dy/dx.

$$m_{\text{tan}} = \frac{dy}{dx} = 2x + 5$$

At $(3, 21)$, $m_{\text{tan}} = 2(3) + 5 = 11$; the slope of the tangent line is 11.

(b) The line we seek has the form $y = mx + b$. We know that the slope m is 11. Thus,

$$y = 11x + b$$

Since the line passes through the point $(3, 21)$, it follows that $x = 3$ and $y = 21$ must satisfy the equation of the line. We have then

$$21 = 11(3) + b$$
$$21 = 33 + b$$
$$b = -12$$

The equation of the tangent line is

$$y = 11x - 12 \quad ◆$$

EXAMPLE 9 Given $f(x) = x^2 + 4\sqrt{x}$, determine $f'(1)$.

SOLUTION We seek a specific value of the derivative function f'. First obtain $f'(x)$. Then evaluate f' at $x = 1$.

$$f(x) = x^2 + 4\sqrt{x}$$

or

$$f(x) = x^2 + 4x^{1/2}$$

So

$$f'(x) = 2x + \frac{1}{2} \cdot 4x^{-1/2}$$

or

$$f'(x) = 2x + \frac{2}{\sqrt{x}}$$

Finally, substituting 1 for x yields

$$f'(1) = 2(1) + \frac{2}{\sqrt{1}} = 4 \quad \blacklozenge$$

3.2 Exercises

Determine the derivative of each function defined in Exercises 1–20. Do not leave negative exponents in answers. Also, recall that $\sqrt[n]{x} = x^{1/n}$.

1. $f(x) = x^4$

2. $f(x) = x^{80}$

3. $f(x) = x^{-2}$

4. $y = x^{-10}$

5. $g(x) = 16$

6. $g(x) = -8$

7. $y = x^{3/2}$

8. $y = x^{2/3}$

9. $y = x^{-2/3}$

10. $y = x^{-3/2}$

11. $f(x) = \dfrac{1}{x^5}$

12. $f(t) = \dfrac{1}{t^{11}}$

13. $y = \sqrt[3]{x}$

14. $y = \sqrt[4]{x}$

15. $f(x) = 30\sqrt{x}$

16. $f(x) = 10\sqrt[3]{x}$

17. $y = \dfrac{x^4}{4}$

18. $f(x) = \dfrac{x^5}{3}$

19. $y = \dfrac{10}{\sqrt{t}}$

20. $f(x) = \pi$

In Exercises 21–30, obtain each derivative.

21. $D_x(1 - x^3)$

22. $D_x(x^2 + 3x - 1)$

23. $\dfrac{d}{dx}\left(\dfrac{1}{x} + \sqrt{2}\right)$

24. $\dfrac{d}{dx}\left(5 - \dfrac{4}{x}\right)$

25. $\dfrac{d}{dx}\left(x - \dfrac{3}{x^2}\right)$

26. $\dfrac{d}{dx}\left(x^2 - \dfrac{6}{x^3}\right)$

27. $D_x(\sqrt{x} - 2)$

28. $D_x(5 + \sqrt[3]{x})$

29. $D_x(x^3 - x^{-3})$

30. $D_x(x^2 - x^{-2})$

Find dy/dx in Exercises 31–38.

31. $y = x^2 - 5x + 19$

32. $y = x^3 + 6x - 3$

33. $y = x^{3/2} + 4x^2$

34. $y = x^{4/3} - 16x^2$

35. $y = 8\sqrt{x} + 16x - 3$

36. $y = 6\sqrt[3]{x} - 25x^2$

37. $y = \dfrac{2}{\sqrt{x}} - 3 + \pi^2$

38. $y = 1 - \dfrac{6}{\sqrt[3]{x}}$

Find $D_x y$ in Exercises 39–44.

39. $y = 1 - x^4 + 3x^6$

40. $y = 2x^{10} + 5x^9 + 13x^5$

41. $y = 1 - x^{-5}$

42. $y = x^{-3} + 4$

43. $y = 11 + 6\sqrt[3]{x}$

44. $y = 10\sqrt{x} + 13x - 1$

Find $f'(x)$ in Exercises 45–52.

45. $f(x) = x^3 - 6x^2 + 4x - 1$

46. $f(x) = x^7 + 5x^4 - 16x$

47. $f(x) = 8x^{7/4} + 6x^{5/3} - 9$

48. $f(x) = 12x^{5/3} + 3x^{4/3}$

49. $f(x) = 3\sqrt{x} + 5x^2$

50. $f(x) = 4\sqrt[3]{x} + 2x^3$

51. $f(x) = \dfrac{4}{x} - 2x - 5$

52. $f(x) = 18 - \dfrac{5}{x^2}$

Determine each of the derivatives specified in Exercises 53–60.

53. $f'(3)$ when $f(x) = x^2 + 8x + 4$

54. $f'(1)$ when $f(x) = 5x^2 - 16x + 2$

55. $f'(-1)$ when $f(x) = 4x^3 - 7x^2 + 8x - 12$

56. $f'(4)$ when $f(x) = 1 - x - x^2 - x^3$

57. $f'(0)$ when $f(x) = 10x^{4/3} + 6x - 13$

58. $f'(-2)$ when $f(x) = \dfrac{5}{x^2}$

59. $f'(9)$ when $f(x) = 1 + \sqrt{x}$

60. $f'(8)$ when $f(x) = 6\sqrt[3]{x}$

In Exercises 61–68, find the slope of the line tangent to the graph of the given function at the given point.

61. $y = x^2 + 4x + 1$ at $(1, 6)$

62. $y = x^2 - 5x + 12$ at $(4, 8)$

63. $y = 1 - x^3$ at $(-2, 9)$

64. $y = x^4$ at $(-2, 16)$

65. $y = 8\sqrt{x}$ at $(4, 16)$

66. $y = \sqrt[3]{x}$ at $(1, 1)$

67. $y = \dfrac{1}{x}$ at $(1/3, 3)$

68. $y = 2 - \dfrac{1}{x}$ at $(1, 1)$

69. Determine the equation of the tangent line mentioned in Exercise 64.

70. Determine the equation of the tangent line mentioned in Exercise 65.

71. Use the *definition* of the derivative to prove that if $f(x) = c$, then $f'(x) = 0$.

72. Use the *definition* of the derivative to prove that

$$\frac{d}{dx}[f(x) - g(x)] = \frac{d}{dx}f(x) - \frac{d}{dx}g(x)$$

W 73. Consider the function $f(x) = 1/x$. In view of the *definition* of the derivative, explain why $f'(x)$ does not exist when $x = 0$.

W 74. Complete the sentence. For any function f, if f is not defined at some number a, then f' _____.

W 75. When finding a specific value of a derivative, such as $f'(3)$ or $f'(1)$, we differentiate the function first *and then* substitute the 3 or the 1 for x. Why don't we substitute the 3 or the 1 at the beginning and then differentiate the function?

W 76. Suppose $f'(x) = g'(x)$. Does this necessarily mean that $f(x) = g(x)$? Explain.

77. Answer each question based on the graphs of f and g shown in the figure below. You will need to use the fact that the derivative of a function is the slope of the tangent line.

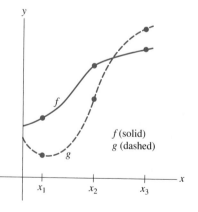

(a) Which is greater, $f'(x_1)$ or $g'(x_1)$?
(b) Which is greater, $f'(x_2)$ or $g'(x_2)$?
(c) Which is greater, $f'(x_3)$ or $g'(x_3)$?

78. Answer each question based on the graphs of f and g shown in the figure below.

(a) Which is greater, $f'(x_1)$ or $g'(x_1)$?
(b) Which is greater, $f'(x_2)$ or $g'(x_2)$?
(c) Which is greater, $f'(x_3)$ or $g'(x_3)$?

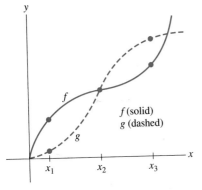

f (solid)
g (dashed)

3.3 | *RATES OF CHANGE*

Applications that follow from interpreting the derivative as a rate of change are presented next. Additional rate-of-change applications will be given in Sections 3.7, 3.9, and 4.3, and elsewhere in the text.

To set the stage for the rate-of-change interpretation, we begin with three examples of **average rate of change**.

EXAMPLE 1   ◆   *AVERAGE SPEED*

If Janice walks 7 miles in 2 hours, what is her average speed?

SOLUTION   We shall use the letter s to represent *distance*, not speed. This is a common convention in mathematics.

Average speed is defined to be the change in distance Δs divided by the change in time Δt.

$$\text{average speed} = \frac{\Delta s}{\Delta t}$$

In this instance,

$$\frac{\Delta s}{\Delta t} = \frac{7 \text{ miles}}{2 \text{ hours}} = 3.5 \text{ miles per hour}$$

Her average speed is 3.5 miles per hour.   ◆

EXAMPLE 2 ◆ *AVERAGE CHANGE IN PROFIT*

From 1980 to 1985 a company's annual profit increased from $20,000 to $180,000. What was the average change in profit per year during that time?

SOLUTION The average change in profit per year is the change in profit ΔP divided by the change in years Δt.

$$\frac{\Delta P}{\Delta t} = \frac{\$180,000 - \$20,000}{1985 - 1980} = \frac{\$160,000}{5 \text{ years}} = \$32,000 \text{ per year}$$

As you can now see, from 1980 to 1985 the company's profit increased at the rate of $32,000 per year. ◆

The "average rate of change" concept suggested in Examples 1 and 2 can be applied to a function $y = f(x)$, as shown in the next example.

EXAMPLE 3 What is the average rate of change of $y = x^2$ from $x = 1$ to $x = 5$?

SOLUTION The average rate of change is the change in y divided by the change in x, namely

$$\frac{\Delta y}{\Delta x} = \frac{(5)^2 - (1)^2}{5 - 1} = \frac{25 - 1}{4} = \frac{24}{4} = 6 \quad ◆$$

The $\Delta y/\Delta x$ in Example 3 should remind you of slope. The slope of a line is an example of a rate of change — the rate of change of y with respect to x. A slope of 3 means that the rate of change of y with respect to x is 3; that is, there is a 3-unit change in y for each 1-unit change in x.

For $y = f(x)$, the derivative

$$f'(x) = \lim_{\Delta x \to 0} \frac{\Delta y}{\Delta x}$$

gives the slope of the tangent line at any point (x, y). So it can also be said that $f'(x)$ or dy/dx gives the rate of change of y with respect to x at any *point (x, y)*. It is the *rate of change at a particular value of x*, rather than an average rate of change over an interval. Consequently, the derivative is considered to be the **instantaneous rate of change** of y with respect to x.

> If $y = f(x)$, then dy/dx is the *instantaneous rate of change* of y with respect to x, or the instantaneous rate of change of $f(x)$ with respect to x.

> **Note**
>
> The instantaneous rate of change is often called simply the **rate of change**.

EXAMPLE 4 ◆ *VELOCITY OF A FALLING OBJECT*

A ball dropped from the top of a cliff will fall such that the distance it has traveled after t seconds is

$$s(t) = -16t^2 \text{ feet}$$

(a) What is the *average* velocity for the first 3 seconds?

(b) How fast is the ball traveling at 3 seconds?

SOLUTION **(a)** Just as in Example 1, the *average* speed is $\Delta s/\Delta t$. Here it is the average speed between $t = 0$ and $t = 3$.

$$\frac{\Delta s}{\Delta t} = \frac{s(3) - s(0)}{3 - 0}$$

$$= \frac{[-16(3)^2] - [-16(0)^2]}{3}$$

$$= \frac{-144}{3}$$

$$= -48 \text{ feet per second}$$

The average speed is 48 feet per second. The *minus* indicates that the ball is traveling in a *downward* direction. The *velocity* (which includes speed and direction) is -48 feet per second. If the ball were traveling upward, then the velocity would be positive.

(b) The velocity at 3 seconds is an *instantaneous* velocity. It is the velocity at a specific time, when $t = 3$. So we need to evaluate ds/dt at $t = 3$. Since $s = -16t^2$,

$$v = \frac{ds}{dt} = -32t \text{ feet per second}$$

At $t = 3$,

$$\frac{ds}{dt} = -32(3) = -96 \text{ feet per second}$$

The ball is traveling 96 feet per second (downward) after 3 seconds. ◆

For future reference, we shall note here that **instantaneous velocity** is defined as the rate of change of distance with respect to time.

Instantaneous Velocity

$$v = \frac{ds}{dt}$$

where

$$v = \text{velocity}$$
$$s = \text{distance}$$
$$t = \text{time}$$

EXAMPLE 5   Let $y = 5x^3 - x^2 + 8x + 1$. Determine the rate of change of y with respect to x when x is 4.

SOLUTION   The rate of change of y with respect to x is dy/dx. Since we are given that $y = 5x^3 - x^2 + 8x + 1$, it follows that

$$\frac{dy}{dx} = 15x^2 - 2x + 8$$

Specifically, when $x = 4$,

$$\frac{dy}{dx} = 15(4)^2 - 2(4) + 8 = 240$$

Thus, when $x = 4$, the rate of change of y with respect to x is 240.   ◆

EXAMPLE 6   ◆   *VELOCITY OF A RACING CAR*

The driver of an experimental racing car enters his car and begins a test run. During the first 6 seconds, the distance s (in feet) of the car from the starting point is

$$s = 14t^2 - \frac{1}{3}t^3 \qquad 0 \le t \le 6$$

where t is the number of seconds the car has been moving. (See Figure 9.) What is the velocity of the car after 4 seconds?

SOLUTION   The velocity at any particular time is the value of ds/dt at that time. Thus we seek ds/dt for $t = 4$. From

$$s = 14t^2 - \frac{1}{3}t^3$$

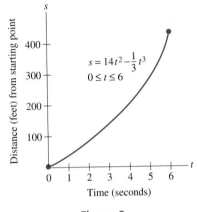

Figure 9

we determine that

$$v = \frac{ds}{dt} = 28t - t^2$$

Specifically, for $t = 4$,

$$v = \frac{ds}{dt} = 28(4) - (4)^2 = 96$$

We conclude that the velocity of the car after 4 seconds is 96 feet per second (which is, by the way, approximately 65 miles per hour). ◆

EXAMPLE 7 ◆ *CITY GOVERNMENT EXPENDITURES*

Figure 10 shows the city government expenditures during the administrations of three mayors, each of whom promised to reduce expenditures.

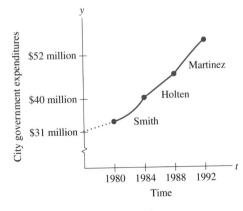

Figure 10

(a) Did any of the three mayors actually reduce the city government expenditures?

(b) Which mayor can claim to have reduced the *rate* of government expenditure?

SOLUTION **(a)** The *y* coordinate represents expenditures. From the graph you can see that the *y* coordinate (expenditures) increased throughout each mayor's term. We conclude that none of the mayors reduced expenditures.

(b) The *rate* of government expenditures is the derivative, *dy/dt*—the slope of the curve. By following the curve from before 1980, we can see that the slope of the curve (that is, the rate of expenditure) increased during Smith's term. When Holten took over in 1984, the rate of expenditure was very high. (Note the steep slope at the end of Smith's term.) The rate of expenditure during Holten's term is considerably less than it was when she took office in 1984. (The slope of her portion of the curve is much less than that of Mayor Smith.) When Martinez took over in 1988, the rate of expenditure increased again (as displayed by the increased slope of the curve).

We conclude that only Holten can claim that as Mayor she reduced the rate of government expenditures. ◆

Note on Marginal Analysis

It seems natural to apply the "instantaneous rate of change" concept to functions that describe cost, revenue, and profit. This application, called *marginal analysis*, is a particularly important business application, one that deserves full attention when studied. Consequently, it will be presented separately and unhurriedly in the next section.

3.3   Exercises

Determine the *average rate of change* that is requested in Exercises 1–12.

1. If Bill hikes 7 miles in 4 hours, what is his average speed during the hike?

2. (*INVENTORY*) At the end of November, an automobile manufacturer's inventory consisted of 160,000 cars. At the end of March, the inventory consisted of 220,000 cars. What was the average rate of increase in inventory per month during this period?

3. A breeder begins with 24 birds. A year later she has 36 birds. A year after that she has 62 birds. What is the average rate of increase of her bird population over the 2-year period?

4. Assume that the college is 5 miles from your home. It's a nice day, so you decide to walk home after your last class. If you leave the campus at 3 p.m. and arrive home at 5 p.m., what would be your average speed during the walk?

5. *(CORPORATE PROFIT)* In 6 years a corporation's annual profit increased from $10,000 to $130,600. What was the average rate of increase in profit per year during this 6-year period?

6. *(CORPORATE PROFIT)* In 5 years a corporation's annual profit decreased from $750,000 per year to $40,000 per year. Determine the average rate of decrease in profit per year over the 5-year period.

7. *(POPULATION DECLINE)* In 1970 the population of Buffalo, New York was 460,000. In 1980 it was 355,000. What was the average rate of decrease in the population per year between 1970 and 1980?

8. *(POPULATION GROWTH)* In 1970 the population of Austin, Texas was 250,000. In 1980 it was 340,000. What was the average rate of increase in the population per year between 1970 and 1980?

9. *(STOCK MARKET)* The Dow Jones Industrial Average gained 162 points over the last 3 days. What was the average rate of gain per day?

10. Susan has been dieting for 15 months and has lost 24 pounds. What is her average weight loss per month?

11. *(FALLING OBJECT)* A ball is dropped from the top of a 400-foot-tall building and falls such that its distance from the ground after t seconds is $s = -16t^2 + 400$. What is the average velocity of the ball for the first 4 seconds?

12. *(FALLING OBJECT)* Assume that the ball of Exercise 11 is dropped from the top of a different building and falls according to $s = -16t^2 + 450$. What is the average velocity of the ball during the first 5 seconds?

13. Let $y = x^2 - 6x + 2$. Determine the rate of change of y with respect to x when x is 4.

14. Let $f(x) = 1 - x^2 + x^3$. Determine the rate of change of $f(x)$ with respect to x when x is 10.

15. Let $f(t) = 4\sqrt{t} - 3$. Determine the rate of change of $f(t)$ with respect to t when t is 25.

16. *(ROCKET VELOCITY)* A toy rocket is shot straight up from the ground and travels so that its distance s (in feet) from the ground after t seconds is given by $s = 200t - 16t^2$. What is the velocity of the rocket after 2 seconds?

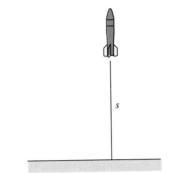

17. *(FALLING OBJECT)* A brick comes loose from near the top of a building and falls such that its distance s (in feet) from the street (after t seconds) is given by $s = 150 - 16t^2$. How fast is the brick falling after 3 seconds?

18. *(VELOCITY OF A CAR)* A racing car begins a short test run and travels according to $s = 16t^2 - \frac{2}{3}t$, where s is the distance traveled in feet and t is the time in seconds. What is the velocity of the car after 3 seconds?

19. *(BACTERIA GROWTH)* Two bacteria cultures are used to test the effect of a growth inhibitor. The number of bacteria n in the culture after t hours is given by

$$n = 1000 + 100t + 20t^2 \quad \text{(no inhibitor used)}$$
$$n = 1000 + 200t - 10t^2 \quad \text{(inhibitor used)}$$

Compare the rates of growth of the two cultures after 5 hours and after 10 hours.

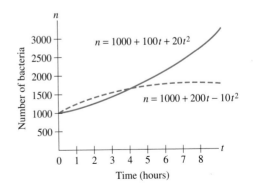

20. *(BACTERIA GROWTH)* A colony of 1000 bacteria is introduced to a growth-inhibiting environment and grows according to the formula

$$n = 1000 + 20t + t^2$$

where n is the number of bacteria present at any time t (t is measured in hours).

(a) According to the formula, how many bacteria are present at the beginning?

(b) What is the rate of growth of the bacteria at any time t?

(c) What is the rate of growth of the bacteria after 3 hours?

(d) How many bacteria are present after 3 hours?

21. (SALES) Suppose that in June a chain of stores had combined daily sales of ice cream cones given by

$$s = -.01x^2 + .48x + 50$$

where s is the number of hundreds of ice cream cones sold and x is the day of the month.

(a) How many ice cream cones were sold by the chain on June 3?

(b) At what rate were sales changing on June 10?

(c) At what rate were sales changing on June 28?

(d) On what day was the rate of change of sales equal to 10 cones per day?

22. (FLU OUTBREAK) Suppose that when a flu outbreak strikes, the number of people n that are sick with flu at a particular time t (within a month of the outbreak) is given by $n = 100t^2 - 2t^3$. Time t is measured in days and is the number of days after the start of the outbreak.

(a) How many people will be sick with flu after 20 days?

(b) At what rate is illness due to flu increasing after 20 days?

23. (EYE PUPIL SIZE) The intensity of light that enters the eye depends on the radius of the pupil. As the pupil increases in size, the amount of light entering the eye increases. Specifically, $I = kr^2$, where I is the intensity of light, r is the radius of the pupil, and k is some constant. Determine the (instantaneous) rate of change of intensity with respect to radius.

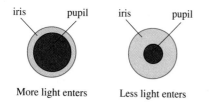

iris   pupil   iris   pupil

More light enters   Less light enters

24. (VOLUME) The volume V of a spherical balloon with radius r is $V = \frac{4}{3}\pi r^3$. As air is blown into the balloon,

both the radius and the volume of the balloon increase. Determine the rate of change of the volume with respect to the radius when the radius is 10 centimeters.

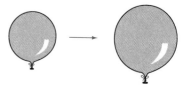

25. (AREA) Determine the rate of change of the area A of a circle with respect to its radius r. The area of a circle is $A = \pi r^2$.

26. (TEMPERATURE) Determine the rate of change of temperature in degrees Fahrenheit (F) with respect to the temperature in degrees Celsius (C). The relationship between F and C is $F = 1.8C + 32$.

27. (RISING BALLOON) A weather balloon is released and rises vertically according to

$$s(t) = t^2 + t + 4 \qquad 0 \le t \le 15$$

where s is its distance (in feet) from the ground after t seconds.

(a) Determine the velocity of the balloon after 1 second and after 5 seconds.

(b) When is the balloon 24 feet above the ground?

(c) How fast is the balloon traveling when it is 24 feet above the ground?

28. (GROWTH OF YEAST CULTURE) Consider the rate of change in the number of yeast cells growing in a culture (see figure). Three times are marked in the illustration: t_1, t_2, and t_3.

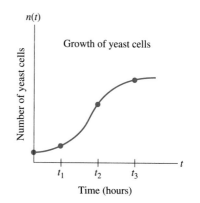

(a) At which of the three times (t_1, t_2, t_3) is the rate of growth the greatest?

(b) At which of the three times (t_1, t_2, t_3) is the rate of growth the least?

(c) True or false: $n(t_3) > n(t_1)$?

(d) True or false: $n'(t_1) > n'(t_2)$?

(e) True or false: $n'(t_3) < n'(t_2)$?

W 29. (GROSS NATIONAL PRODUCT) The figure below shows the graphs of two estimates of future GNP growth.

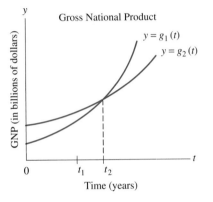

(a) Which function (g_1 or g_2) shows the greater rate of increase of GNP at time t_1? Explain.

(b) Consider the rates of increase of GNP at time t_2. Are the rates of increase the same for both g_1 and g_2? Explain.

30. (LEARNING / RECALL) Psychologists have found that previous learning interferes with recall. The more groups of items people are required to learn, the smaller the percentage that they can recall. A graph of three recall functions r_1, r_2, and r_3 is shown in the figure.

Here t is time and y is the percentage of items recalled. Function r_3 includes the most groups; r_1 includes the least.

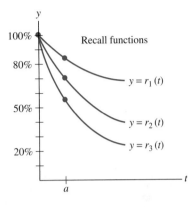

(a) Consider $r_1(a)$, $r_2(a)$, and $r_3(a)$. Which number is the largest and which is the smallest?

(b) Consider the absolute value of $r_1'(a)$, $r_2'(a)$, and $r_3'(a)$. Which number is the largest and which is the smallest?

W 31. Compare the instantaneous rate of change and the average rate of change for a linear function such as $f(x) = 10x + 25$.

W 32. Compare the instantaneous rate of change and the average rate of change for a constant function such as $f(x) = 50$.

The Exercise Library at the back of the book contains graphing calculator and computer exercises keyed to this section.

3.4 | *MARGINAL ANALYSIS*

The management of a manufacturing operation is naturally concerned about the total cost of maintaining a particular level of production. In other words, they want to know the cost $C(x)$ of producing x units. Furthermore, when a particular level of production is being maintained, it is important to know the cost of producing one additional unit. For example, if they are already making

100 TV sets, what will it cost to make one more—the 101st TV? Such information assists management in making decisions about production.

The rate-of-change interpretation of the derivative leads to a calculus application here. If $C(x)$ is the total cost of producing x units, then $C'(x)$ is the rate of change of the total cost and gives the *approximate* cost of producing one additional unit. $C'(x)$ is called the **marginal cost**.

Marginal Cost

$C(x)$ = the total cost of producing x units

$C'(x)$ = *marginal cost*, the approximate cost of producing the next unit

---

Ordinarily, $C'(x)$ is a good approximation of the exact cost of producing one more unit, the $(x + 1)$st unit. In other words, $C'(x) \approx C(x + 1) - C(x)$. Consider that the exact cost of one more unit is $\Delta C / \Delta x$ with $\Delta x = 1$ and the marginal cost is dC/dx. The two values are equal when C is a linear function; otherwise dC/dx only approximates $\Delta C/\Delta x$. (Graphically, dC/dx is the slope of the tangent line, whereas $\Delta C/\Delta x$ is the slope of the secant line.) Consider that

$$C(x + 1) - C(x) = \frac{C(x + 1) - C(x)}{1} \approx \lim_{\Delta x \to 0} \frac{C(x + \Delta x) - C(x)}{\Delta x} = C'(x)$$

since $\Delta x \to 0$ is approximated by $\Delta x = 1$.

In this section, you will see that the marginal cost function and other marginal functions are relatively easy to obtain and use.

EXAMPLE 1 ◆ *MARGINAL COST*

Suppose the cost of producing x units is $C(x) = 100 + 30x - x^2$ dollars (for $0 \leq x \leq 12$). Determine the marginal cost when $x = 9$.

SOLUTION The marginal cost is

$$C'(x) = 30 - 2x$$

For $x = 9$, we have

$$C'(9) = 30 - 2(9) = 12$$

The marginal cost when $x = 9$ is \$12. This means that after 9 units have been produced, the cost of producing the next unit (the tenth unit) will be *approximately* \$12.

By the way, the *exact* cost of producing the tenth unit can be computed as $C(10) - C(9)$, a method used in Chapter 1.

$$C(10) - C(9) = [100 + 30(10) - (10)^2] - [100 + 30(9) - (9)^2]$$
$$= 300 - 289$$
$$= \$11$$

Figure 11 shows a graph of the marginal cost $C'(x)$ and the exact cost of the next unit for x between 0 and 11 units.

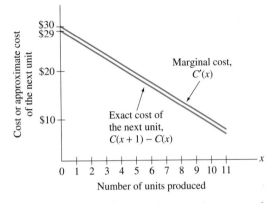

Figure 11 Comparison of marginal cost and exact cost for the cost function of Example 1 ◆

Note

It is natural to wonder why we would want the approximate cost of the tenth unit when the exact cost is relatively easy to determine. The answer to this question will emerge as we study various applications, beginning later in this section with a relationship involving marginal cost, marginal revenue, and marginal profit. In a later chapter, you will learn how to use the marginal cost function to construct the cost function C.

EXAMPLE 2 ◆ MARGINAL COST

A table manufacturer determines that the marginal cost is always increasing. They decide to stop production when the marginal cost reaches $110, because they cannot make a profit when it costs more than $110 to make each table. Assuming that the cost function is

$$C(x) = x^2 + 80x + 100$$

how many tables should they make?

SOLUTION Since the cost is given as $C(x) = x^2 + 80x + 100$, it follows that the marginal cost $C'(x)$ is

$$C'(x) = 2x + 80$$

The marginal cost will be \$110 when $C'(x)$ is 110, so solve

$$110 = 2x + 80$$

The result is $x = 15$. Thus, they should make 15 tables. ◆

Marginal revenue and **marginal profit** are defined in much the same manner as marginal cost.

Marginal Revenue

$R(x)$ = the total revenue from the sale of x units

$R'(x)$ = *marginal revenue*, the approximate revenue from the sale of the next unit

---

Marginal Profit

$P(x) = R(x) - C(x)$, the total profit from the production and sale of x units

$P'(x) = R'(x) - C'(x)$, *marginal profit*, the approximate profit from the production and sale of the next unit

---

Economists sometimes use the notations MC for marginal cost, MR for marginal revenue, and MP for marginal profit.

$MC(x) = C'(x)$ = marginal cost

$MR(x) = R'(x)$ = marginal revenue

$MP(x) = P'(x)$ = marginal profit

EXAMPLE 3 ◆ *MARGINAL COST, MARGINAL REVENUE, AND MARGINAL PROFIT*

Let the cost of producing x tires be $C(x) = 300 + 40x - .02x^2$ dollars and let the revenue from the sale of x tires be $R(x) = 100x - .03x^2$ dollars.

(a) Determine marginal cost.

(b) Determine marginal revenue.

(c) Determine MR(50) and tell what it means.

(d) Determine marginal profit.

(e) For what value of x is the marginal cost equal to the marginal revenue, and what is the marginal profit in that instance?

SOLUTION **(a)** $MC(x) = C'(x) = 40 - .04x$. The marginal cost is $40 - .04x$ dollars.

(b) $MR(x) = R'(x) = 100 - .06x$. The marginal revenue is $100 - .06x$ dollars.

(c) Since $MR(x) = 100 - .06x$, it follows that

$$MR(50) = 100 - .06(50) = \$97$$

This means that once 50 tires have been sold, the revenue to be obtained from the sale of the next tire (the 51st) is approximately \$97.

(d) $P(x) = R(x) - C(x)$
$$= (100x - .03x^2) - (300 + 40x - .02x^2)$$
$$= -.01x^2 + 60x - 300$$

Differentiating yields

$$MP(x) = P'(x) = -.02x + 60$$

The marginal profit is $-.02x + 60$ dollars.

(e) Marginal cost is $40 - .04x$ and marginal revenue is $100 - .06x$. They are equal when $x = 3000$, as shown by solving the equation

$$40 - .04x = 100 - .06x$$

In steps,

$$.02x = 60$$
$$x = 3000$$

Marginal profit is then $P'(3000) = -.02(3000) + 60 = 0$. Thus, the marginal profit is zero when $x = 3000$ tires. This should not be surprising, since $P'(x) = R'(x) - C'(x)$ and $R'(x) - C'(x) = 0$ when marginal cost and marginal revenue are equal. ◆

The last result of Example 3 is worth noting for future reference.

> Marginal profit is zero when
> marginal revenue equals marginal cost.

If marginal revenue and marginal cost are equal, then the revenue from the sale of the next item is the same as the cost of producing that next item. Consequently, there is no profit on the next item, which means that the marginal profit is zero.

3.4 Exercises

(*MARGINAL COST*) Determine the marginal cost function in Exercises 1–8.

1. $C(x) = 50 - .1x^2$

2. $C(x) = 80 - .2x^2$

3. $C(x) = 1000 + 150x - x^2$

4. $C(x) = 500 + 40x - x^2$

5. $C(x) = 90x + .02x^2$

6. $C(x) = 120x + .03x^2$

7. $C(x) = 1200 + 3x - .001x^3$

8. $C(x) = 1000 + 6x - .04x^3$

9. (*MARGINAL COST*) Suppose the cost of producing x units is given by $C(x) = 150 + 40x - x^2$ dollars, for $0 \le x \le 16$.
  (a) Determine the marginal cost when $x = 10$ units.
  (b) Determine the exact cost of the 11th unit.
  W **(c)** What is the meaning of MC(10)?

10. (*MARGINAL COST*) Suppose the cost of producing x units is given by $C(x) = 200 + 15x - .5x^2$ dollars, for $0 \le x \le 12$.
  (a) Determine the marginal cost when $x = 7$ units.
  (b) Determine the exact cost of the 8th unit.
  W **(c)** What is the meaning of MC(7)?

11. (*LIMITING PRODUCTION*) A chair maker plans to stop production if and when the marginal cost reaches $70. If the cost function is $C(x) = .5x^2 + 50x + 90$ dollars, how many chairs should she make?

12. (*LIMITING PRODUCTION*) A small manufacturer of hats will stop production if and when the marginal cost reaches $31. If the cost of making x hats is given by $C(x) = .5x^2 + 5x + 120$ dollars, how many hats will this manufacturer produce?

13. (*LIMITING PRODUCTION*) Redo Exercise 12 assuming that the cost function is $C(x) = x^2 + 3x + 75$.

(*MARGINAL REVENUE*) Determine the marginal revenue function in Exercises 14–19.

14. $R(x) = 20x + .1x^2$ **15.** $R(x) = 50x + .2x^2$

16. $R(x) = 300x - .001x^2$ **17.** $R(x) = 400x - .003x^2$

18. $R(x) = .002x^2 + .4x$ **19.** $R(x) = .001x^2 + .7x$

20. (*MARGINAL REVENUE*) Suppose revenue from the sale of x liter bottles of cola is $R(x) = .70x + .001x^2$ dollars.
  (a) Determine the revenue from the sale of 1 bottle, 2 bottles, and 10 bottles.
  (b) Find the marginal revenue.
  W **(c)** Determine MR(10) and tell what it means.

21. (*MARGINAL REVENUE*) Let the revenue from the sale of x compact disc players be $R(x) = 400x - .01x^2$ dollars, for $0 \le x \le 20,000$.
  (a) Find the revenue from the sale of 1 player, 10 players, and 100 players.
  (b) Determine the marginal revenue.
  W **(c)** Determine MR(1000) and tell what it means.

(*MARGINAL PROFIT*) Determine the marginal profit function in Exercises 22–27.

22. $P(x) = .01x^2 + 5x - 100$

23. $P(x) = .02x^2 + 9x - 72$

24. $P(x) = 50x - .03x^2$

25. $P(x) = 40x - .01x^2$

26. $P(x) = 100x + .1x^2 - .01x^3$

27. $P(x) = 200x + .3x^2 - .001x^3$

28. (*PROFIT AND LOSS*) A compact disc (CD) maker says the company's profit on the manufacture and sale of x CDs is $P(x) = .02x^2 + 10x - 300$ dollars.
  (a) What is the company's total profit on 50 units?
  (b) How much money does the company lose if they make and sell only 10 units?
  (c) Determine the marginal profit function.

29. (*PROFIT AND LOSS*) A potato chip maker says the company's profit on the manufacture and sale of x 8-ounce bags of chips ($0 \le x \le 500$) is $P(x) = .0005x^2 + x - 160$ dollars.
  (a) Determine the total profit on 200 bags.
  (b) How much will the company lose if they make and sell only 50 bags?
  (c) Find the marginal profit function.

(*MARGINAL PROFIT*) Determine the marginal profit that corresponds to the given revenue and cost functions in Exercises 30–34.

30. $R(x) = 150x - .02x^2$; $C(x) = 200 + 15x - .01x^2$

31. $R(x) = 95x - .01x^2$; $C(x) = 140 + 10x - .02x^2$

32. $R(x) = 100x + .01x^2$; $C(x) = 140 + 90x + .01x^2$

33. $R(x) = 80x + .001x^2$; $C(x) = 210 + 70x + .001x^2$

34. $R(x) = .002x^2 + .3x$; $C(x) = 30 + 25x - .001x^2$

(*MARGINAL REVENUE / MARGINAL COST*) In Exercises 35–39, determine the number of units x for which marginal revenue is equal to marginal cost.

35. $R(x) = 70x - .02x^2$; $C(x) = 500 + 40x - .01x^2$

36. $R(x) = 55x - .03x^2$; $C(x) = 250 + 30x - .02x^2$

37. $R(x) = 300x - .02x^2$; $C(x) = 100x$

38. $R(x) = 240x - .02x^2$; $C(x) = 210x$

39. $R(x) = 150x + .01x^2$; $C(x) = 120x + .03x^2$

(*MARGINAL PROFIT*) In Exercises 40–45, determine the number of units x for which marginal profit is zero.

40. $R(x) = 100x - .02x^2$; $C(x) = 200 + 50x - .01x^2$

41. $R(x) = 80x - .01x^2$; $C(x) = 130 + 90x - .02x^2$

42. $R(x) = 5x + .002x^2$; $C(x) = 12x + .001x^2$

43. $R(x) = 7x + .001x^2$; $C(x) = 5x + .003x^2$

44. $R(x) = 14x - .02x^3$; $C(x) = 42 + 2x - .01x^3$

45. $R(x) = 81x - .03x^3$; $C(x) = 20 + 27x - .01x^3$

46. (*MARGINAL PROFIT*) A manufacturer of telephones determines that the cost of producing x telephones is $C(x) = 500 + 15x - .01x^2$ dollars and the revenue received from the sale of x telephones is $R(x) = 75x - .02x^2$ dollars.
(a) Determine the marginal profit function.
(b) What production level results in a marginal profit of zero?

47. (*MARGINAL PROFIT*) A tire manufacturer has found that the cost of making x tires is given by the function $C(x) = 240 + 64x - .02x^2$ dollars and the revenue received from the sale of x tires is $R(x) = 90x - .03x^2$ dollars.

(a) Find the marginal profit function.
(b) What production level results in a marginal profit of zero?

48. (*MARGINAL PROFIT*) Suppose that a manufacturer of knives determines that the cost to make x knives is $8 - .01x$ dollars *per knife*. If x knives can be sold for a total of $20x - .04x^2$ dollars, determine the marginal profit when the number of knives manufactured and sold is 50.

49. (*MARGINAL PROFIT*) A bicycle manufacturer estimates that the company can price their bicycles at $p = 140 - .02x$ dollars each, where x is the number sold. The cost of producing x bicycles is $900 + 80x - .01x^2$ dollars. Determine the marginal profit when 20 bicycles are made.

50. (*MARGINAL PROFIT*) Suppose each tennis racquet costs $15 to make. If x racquets can be sold at $70 - .03x$ dollars each, find the marginal profit function.

51. (*MARGINAL PROFIT*) It will cost a manufacturer $600 - .02x$ dollars each to make x stereo units. The price at which each unit can be sold is given by $p = 1000 - .04x$ dollars.
(a) Determine the marginal revenue function.
(b) Determine the marginal profit when $x = 150$.
(c) Determine the exact profit on the 151st stereo unit.

52. (*MARGINAL PROFIT*) A company plans to produce x units of its product. The fixed costs of production are $300. The variable costs are $20 per unit produced. The relationship between the number of units sold x and the price per unit are given by the demand equation $x = 5000 - 100p$.
(a) Determine the revenue function R by using $R(x) = xp$.
(b) Determine the marginal cost function.
(c) Determine the marginal profit function.

W 53. (*INCOME TAX*) Consider the following tax schedule. (i) If your taxable income is $0–$7999, there is no tax due. (ii) If your taxable income is $8000–$19,999, the tax due is 16% of taxable income. (iii) If your taxable income is $20,000 or more, the tax due is $3200 + 26% of the taxable income above $20,000. Explain why the tax rates 16% and 26% in this setting are sometimes called "marginal" rates. (The figure on the next page shows a graph of tax due as a function of taxable income.)

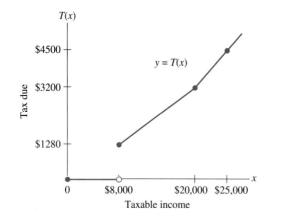

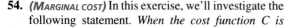

Taxable income

linear, the marginal cost $C'(x)$ is the same as the exact cost of producing the next unit. Consider the general linear cost function $C(x) = mx + b$. Compute the exact cost of the tenth item, namely, $C(10) - C(9)$. Then compute $C'(9)$. Both will have the same value. The computations can be repeated for the fifth unit or the eighth unit or any other unit, and the result will be the same, as you can see once you have done the calculations for the tenth unit.

The Exercise Library at the back of the book contains graphing calculator and computer exercises keyed to this section.

54. *(MARGINAL COST)* In this exercise, we'll investigate the following statement. *When the cost function C is*

3.5 | THE PRODUCT AND QUOTIENT RULES

The rules for finding the derivative of a product and the derivative of a quotient are not as simple or as straightforward as the rules developed and used in Section 3.2. In view of this, we introduce a notation that will simplify the appearance of the rules and make them easier to apply.

If a function is to be considered as the product or quotient of two functions, we will call the two functions u and v. This means we will have

$$u \cdot v \qquad \text{product of two functions; } u = \text{first, } v = \text{second}$$

and

$$\frac{u}{v} \qquad \text{quotient of two functions; } u = \text{numerator, } v = \text{denominator}$$

The derivative of a product is *not* the product of the derivatives. To see this, consider $y = x^3$. Clearly $dy/dx = 3x^2$. If we consider x^3 as the product $x^2 \cdot x$, then the product of the derivatives is $2x \cdot 1$, or $2x$, which is incorrect.

> *Note*
>
> The derivative of a product is *not* the product of the derivatives.

The product rule is stated next.

Since it is far from obvious that the product rule is true in general, we will use the definition of the derivative to prove the product rule.

PROOF   Using the definition of the derivative, we can begin with

$$\frac{d}{dx}[u(x)v(x)] = \lim_{\Delta x \to 0} \frac{u(x + \Delta x)v(x + \Delta x) - u(x)v(x)}{\Delta x}$$

At this point we must do something to make the numerator look more like an expression that will eventually lead to $uv' + vu'$. Specifically, we will add "zero" to the numerator. But zero will be chosen in the useful form

$$-u(x)v(x + \Delta x) + u(x)v(x + \Delta x)$$

The reason for this choice will become apparent as the proof progresses. As a result of adding this form of zero to the numerator, the limit that represents the derivative of product $u(x)v(x)$ becomes

$$\lim_{\Delta x \to 0} \frac{u(x + \Delta x)v(x + \Delta x) - u(x)v(x + \Delta x) + u(x)v(x + \Delta x) - u(x)v(x)}{\Delta x}$$

The quotient can be split into two quotients and then into two limits, as we anticipate a limit that will yield vu' and a limit that will yield uv'.

$$\lim_{\Delta x \to 0} \frac{u(x + \Delta x)v(x + \Delta x) - u(x)v(x + \Delta x)}{\Delta x} +$$

$$\lim_{\Delta x \to 0} \frac{u(x)v(x + \Delta x) - u(x)v(x)}{\Delta x}$$

Next, $v(x + \Delta x)$ can be factored out of the first numerator and $u(x)$ can be factored out of the second numerator. The result is

$$\lim_{\Delta x \to 0} \frac{[u(x + \Delta x) - u(x)]v(x + \Delta x)}{\Delta x} + \lim_{\Delta x \to 0} \frac{u(x)[v(x + \Delta x) - v(x)]}{\Delta x}$$

Perhaps you can see in the first limit the part that will become $u'(x)$ and in the second limit the part that will become $v'(x)$. We proceed in that direction by using the limit theorem that says "the limit of the product is the product of the limits." The result:

$$\lim_{\Delta x \to 0} \frac{u(x + \Delta x) - u(x)}{\Delta x} \cdot \lim_{\Delta x \to 0} v(x + \Delta x) + \lim_{\Delta x \to 0} u(x) \cdot \lim_{\Delta x \to 0} \frac{v(x + \Delta x) - v(x)}{\Delta x}$$

The first limit above is the derivative of $u(x)$, namely, $u'(x)$. The second limit is simply $v(x)$. The third limit is $u(x)$. The fourth limit is the derivative of $v(x)$, namely, $v'(x)$. Thus, we have

$$\frac{d}{dx}[u(x)v(x)] = u'(x)v(x) + u(x)v'(x)$$

which can also be written with the $u(x)v'(x)$ term first, as

$$\frac{d}{dx}[u(x)v(x)] = u(x)v'(x) + v(x)u'(x)$$

EXAMPLE 1   Find $f'(x)$ for $f(x) = (3x + 5)(x^2 - 7x)$. Use the product rule.

SOLUTION   Function f is clearly a product. The factors are $u = 3x + 5$ and $v = x^2 - 7x$. Using the product rule yields

$$f'(x) = (3x + 5) \cdot \frac{d}{dx}(x^2 - 7x) + (x^2 - 7x) \cdot \frac{d}{dx}(3x + 5)$$
$$= (3x + 5)(2x - 7) + (x^2 - 7x)(3)$$

The expression can be simplified in steps. Begin by multiplying.

$$f'(x) = 6x^2 - 21x + 10x - 35 + 3x^2 - 21x$$
$$= 9x^2 - 32x - 35$$

Thus, $f'(x) = 9x^2 - 32x - 35$.   ◆

In this instance, it is natural to ask why we don't just multiply out the original expression *before* differentiating. Then the sum rule could be used, which seems simpler. Actually, you could do that, and indeed it may even be simpler. However, in the next section there will be products that are difficult to multiply out, such as $(4x + 1)^3(x^2 - 3)^5$. The idea in this section is to learn the product rule by using it now. Then, when you really need it, you will know it.

EXAMPLE 2   Find the derivative of $y = (2x + 1)\left(1 - \dfrac{1}{x}\right)$. Use the product rule.

SOLUTION   Before using the product rule, rewrite $1/x$ as x^{-1} to prepare for the differentiation process.

$$\frac{dy}{dx} = \frac{d}{dx}[(2x + 1)(1 - x^{-1})]$$

$$= (2x + 1) \cdot \frac{d}{dx}(1 - x^{-1}) + (1 - x^{-1}) \cdot \frac{d}{dx}(2x + 1)$$

$$= (2x + 1)(x^{-2}) + (1 - x^{-1})(2)$$

$$= (2x + 1)\left(\frac{1}{x^2}\right) + \left(1 - \frac{1}{x}\right)(2)$$

$$= \frac{2x + 1}{x^2} + 2 - \frac{2}{x}$$

All three terms can be combined by using the common denominator x^2.

$$\frac{dy}{dx} = \frac{2x + 1}{x^2} + \frac{2x^2}{x^2} - \frac{2x}{x^2}$$

$$= \frac{2x^2 + 1}{x^2} \quad \blacklozenge$$

Next we consider the quotient rule. The derivative of a quotient is *not* the quotient of the derivatives. To see this, consider $y = x^3$. Clearly, $dy/dx = 3x^2$. If we consider x^3 as the quotient x^4/x, then the quotient of the derivatives is $4x^3/1$, or $4x^3$, which is incorrect.

Here then is the quotient rule. The proof is similar to the proof of the product rule and is suggested as an exercise with hint. (See Exercise 58.)

Quotient Rule

If $u = u(x)$ and $v = v(x)$ are differentiable functions and $v(x) \neq 0$, then

$$\frac{d}{dx}\left(\frac{u}{v}\right) = \frac{v \cdot u' - u \cdot v'}{v^2}$$

or

$$\frac{d}{dx}\left(\frac{u}{v}\right) = \frac{v \cdot \dfrac{du}{dx} - u \cdot \dfrac{dv}{dx}}{v^2}$$

The derivative of a quotient of two functions is the denominator times the derivative of the numerator *minus* the numerator times the derivative of the denominator—all divided by the denominator squared.

Note

The derivative of a quotient is *not* the quotient of the derivatives.

EXAMPLE 3   Find the derivative of $f(x) = \dfrac{5x - 1}{1 + 2x}$.

SOLUTION   The function is clearly a quotient. The numerator (u) is $5x - 1$. The denominator (v) is $1 + 2x$.

$$f'(x) = \frac{(1 + 2x) \cdot \dfrac{d}{dx}(5x - 1) - (5x - 1) \cdot \dfrac{d}{dx}(1 + 2x)}{(1 + 2x)^2}$$

$$= \frac{(1 + 2x)(5) - (5x - 1)(2)}{(1 + 2x)^2}$$

$$= \frac{5 + 10x - 10x + 2}{(1 + 2x)^2}$$

$$= \frac{7}{(1 + 2x)^2} \quad \blacklozenge$$

Note

1. It is better *not* to consider an expression such as $\dfrac{x^2}{9}$ to be a quotient. Instead, consider that

$$\frac{x^2}{9} = \frac{1}{9}x^2$$

and thus

$$\frac{d}{dx}\left(\frac{x^2}{9}\right) = \frac{d}{dx}\left(\frac{1}{9}x^2\right) = 2 \cdot \frac{1}{9}x = \frac{2}{9}x \quad \text{or} \quad \frac{2x}{9}$$

Of course, the expression $x^2/9$ *could* be considered to be a quotient, and the quotient rule could be used, but the differentiation and simplification are more complicated. (See Exercise 59).

2. An expression such as

$$\frac{1}{x^2}$$

should be rewritten as

$$x^{-2}$$

so that the power rule can be used. The original expression should *not* be considered a quotient.

EXAMPLE 4 Find the derivative of $f(x) = \dfrac{3x^2 - 4x}{x^2 + 7x + 1}$.

SOLUTION The function f is a quotient, with the numerator (u) being $3x^2 - 4x$ and the denominator (v) being $x^2 + 7x + 1$.

$$f'(x) = \frac{(x^2 + 7x + 1)(6x - 4) - (3x^2 - 4x)(2x + 7)}{(x^2 + 7x + 1)^2}$$

$$= \frac{6x^3 - 4x^2 + 42x^2 - 28x + 6x - 4 - (6x^3 + 21x^2 - 8x^2 - 28x)}{(x^2 + 7x + 1)^2}$$

$$= \frac{6x^3 + 38x^2 - 22x - 4 - (6x^3 + 13x^2 - 28x)}{(x^2 + 7x + 1)^2}$$

$$= \frac{25x^2 + 6x - 4}{(x^2 + 7x + 1)^2} \quad \blacklozenge$$

3.5 Exercises

Use the product rule to determine the derivative of each function in Exercises 1–18.

1. $y = (x + 1)(x - 2)$ **2.** $y = (x - 7)(x + 4)$

3. $f(x) = (5x - 3)(2x + 1)$ **4.** $f(x) = (3x - 5)(4x - 2)$

5. $y = (t^2 + 6)(1 + t^2)$ **6.** $y = (t^3 - 1)(t^2 + 3)$

7. $f(x) = (4x - 3)x^3$ **8.** $y = (3 + x^2)x^4$

9. $f(t) = t^2(1 + t)$ **10.** $f(t) = t^3(t + 5)$

11. $g(x) = (4x + 1)\left(1 + \dfrac{1}{x}\right)$

12. $y = (1 + 2x)\left(2 + \dfrac{1}{x}\right)$ **13.** $y = \left(\dfrac{2}{x} + 3\right)(6x - 2)$

14. $f(x) = \left(\dfrac{4}{x} + 1\right)(12x - 1)$

15. $y = (x^3 + 2)(1 - x)$ **16.** $y = (x^2 + 5)(1 - 2x)$

17. $y = (1 - x^3)(1 + x^2)$ **18.** $y = (1 - x^4)(1 + x^3)$

Use the quotient rule to determine the derivative of each function in Exercises 19–36.

19. $y = \dfrac{x}{1 + x}$ **20.** $y = \dfrac{x - 2}{4x + 1}$

21. $y = \dfrac{2x + 1}{x + 4}$ **22.** $y = \dfrac{5x + 1}{2x - 1}$

23. $f(x) = \dfrac{2x}{3x + 1}$ **24.** $f(x) = \dfrac{x^2}{2x - 1}$

25. $f(t) = \dfrac{1 - t}{t - 1}$ **26.** $f(t) = \dfrac{t - 3}{3 - t}$

27. $y = \dfrac{1 - x}{1 + x}$ **28.** $y = \dfrac{2 + x}{x - 2}$

29. $s = \dfrac{1 - t^2}{t^2 + 1}$ **30.** $s = \dfrac{1 + 9t}{3t^2 + t}$

31. $s(t) = \dfrac{4t^2 + t}{1 + 3t}$ **32.** $s(t) = \dfrac{t^2 + t + 1}{t + 1}$

33. $y = \dfrac{x^2 + 5x - 3}{x + 4}$ **34.** $f(x) = \dfrac{x^2 - 7x + 1}{x - 2}$

35. $y = \dfrac{4 + 2x}{x^{3/2}}$ **36.** $y = \dfrac{6x - 2}{x^{5/2}}$

Use appropriate rules to determine the derivative of each function in Exercises 37–46.

37. $y = 1 - x - x^2$ **38.** $y = \dfrac{1 + x}{1 - x}$

39. $y = (1 - x)x^2$ **40.** $f(x) = x^{-5} + x^{-1}$

41. $f(x) = \dfrac{1}{x} + 7x - \dfrac{x^2}{3}$ **42.** $f(t) = \dfrac{t}{7} + 10\sqrt{t} - 19$

43. $f(t) = (t^2 - 8)(t^3 + 1)$ **44.** $y = x^2(1 - 10x)$

45. $s = \dfrac{t^2 + 3t + 5}{1 - t}$ **46.** $y = 9x^2 - 6x + 14$

In Exercises 47–50, find the slope of the line tangent to the graph of the function at the given point.

47. $f(x) = (x^3 - 4)(1 + 2x)$ at $(3, 161)$

48. $f(x) = x^2(1 + 3x^2)$ at $(2, 52)$

49. $y = \dfrac{x^2 - 4}{x + 8}$ at $(4, 1)$

50. $y = \dfrac{x - 2}{x^2 + 3x + 1}$ at $(-1, 3)$

In Exercises 51–54, determine the rate of change of y with respect to x for the given value of x.

51. $y = \dfrac{x^2 + 3}{x + 1}$ when $x = 1$

52. $y = \dfrac{x + 4}{x - 1}$ when $x = 2$

53. $y = (1 - x)(1 - x^2)$ when $x = 2$

54. $y = (1 + 3x)(1 - x)$ when $x = 3$

55. (*VELOCITY*) An inventor has created a space toy that travels so that after t seconds its distance in inches from a starting point is

$$s = \frac{4t^2 + 6}{t + 1} \qquad 1 \le t \le 30$$

What is the velocity of the toy after 9 seconds?

56. (*MARGINAL COST*) Find the marginal cost function that corresponds to the cost function

$$C(x) = \frac{100 + 80x + x^2 + x^3}{x + 2}$$

57. (*PROFIT*) A company's profit (in dollars) from the sale of x VCRs is approximately

$$P(x) = \frac{10x^2 - 50x}{x + 1}$$

(a) If they sell only 1 VCR, what is the profit? Any comment?
(b) If they sell 5 VCRs, what is the profit?
(c) What is the profit on the sale of 10 VCRs?
(d) Determine the marginal profit function.

58. *Quotient Rule.* Use the definition of the derivative to prove the quotient rule,

$$\frac{d}{dx}\left[\frac{u(x)}{v(x)}\right] = \frac{v(x)u'(x) - u(x)v'(x)}{[v(x)]^2}$$

Hint: Once you have applied the definition of the derivative to the function $u(x)/v(x)$, it will be necessary to follow the lead of the proof of the product rule. Insertion of an appropriate form of zero is needed. Your choice should be based on what must be added in order to create the desired numerator, $v(x)u'(x) - u(x)v'(x)$.

59. Near the end of the section, we recommended avoiding the use of the quotient rule in finding the derivative of $x^2/9$. Now, use the quotient rule to determine the derivative of $x^2/9$. You will see that it is indeed more complicated this way.

60. (a) Use your knowledge of differentiation to obtain the derivative of

$$y = \frac{5}{x} \qquad y = \frac{6}{x} \qquad y = -\frac{3}{x}$$

(b) The three examples you worked out in part (a) are specific cases of the general formula

$$\boxed{\begin{array}{c} \dfrac{d}{dx}\left(\dfrac{c}{x}\right) = -\dfrac{c}{x^2} \\[2mm] c = \text{constant} \end{array}}$$

Use the general formula to obtain the derivative of

$$y = \frac{12}{x} \qquad y = -\frac{2}{x}$$

W 61. Explain why it is better to write $1/x^2$ as x^{-2} to prepare for differentiation. Why not just use the quotient rule to differentiate $f(x) = 1/x^2$?

W 62. Explain how the derivative of

$$y = \frac{x^2 + 1}{x}$$

can be obtained easily without using the quotient rule or the product rule.

The Exercise Library at the back of the book contains graphing calculator and computer exercises keyed to this section.

3.6 | *THE CHAIN RULE*

We already have a rule for obtaining derivatives of powers of x, namely,

$$\frac{d}{dx}x^n = nx^{n-1}$$

For example,

$$\frac{d}{dx}x^8 = 8x^7$$

But we do not yet have a rule for obtaining derivatives of powers of *functions of* x. This means that we cannot yet easily differentiate expressions such as

$$(x^2 + 1)^8 \qquad \text{or} \qquad (2x - 3)^8$$

The rule we have is for differentiating powers of x, not powers of $x^2 + 1$, or powers of $2x - 3$, or powers of other functions of x.

In the presentation that follows we will be using $u(x)$, or "u" for short, as the name of the expression (such as $x^2 + 1$ or $2x - 3$) that is raised to a power. This means that we need a rule for obtaining dy/dx when $y = f(u)$. To begin, we return to the definition of the derivative.

$$\frac{dy}{dx} = \lim_{\Delta x \to 0} \frac{\Delta y}{\Delta x}$$

Consider that since u is a function of x, a change in x (called Δx) creates a change in u (called Δu). If we multiply $\Delta y/\Delta x$ by 1 in the form of $\Delta u/\Delta u$, we have

$$\frac{dy}{dx} = \lim_{\Delta x \to 0} \left(\frac{\Delta y}{\Delta x} \cdot \frac{\Delta u}{\Delta u} \right)$$

$$= \lim_{\Delta x \to 0} \left(\frac{\Delta y}{\Delta u} \cdot \frac{\Delta u}{\Delta x} \right) \qquad \text{after rearranging denominators}$$

$$= \left(\lim_{\Delta x \to 0} \frac{\Delta y}{\Delta u} \right) \left(\lim_{\Delta x \to 0} \frac{\Delta u}{\Delta x} \right) \qquad \text{by using a limit theorem}$$

As $\Delta x \to 0$, Δu also approaches zero, so we can change the notation $\Delta x \to 0$ to $\Delta u \to 0$ in the limit of $\Delta y/\Delta u$. Then we have

$$\frac{dy}{dx} = \lim_{\Delta u \to 0} \frac{\Delta y}{\Delta u} \cdot \lim_{\Delta x \to 0} \frac{\Delta u}{\Delta x}$$

Each limit on the right side of the equation is a derivative. The first is dy/du; the second is du/dx. Thus, we now have the result

$$\frac{dy}{dx} = \frac{dy}{du} \cdot \frac{du}{dx}$$

This important result is known as the **chain rule**.

Chain Rule

If y is a function of u and u is a function of x, then y is a function of x, and

$$\frac{dy}{dx} = \frac{dy}{du} \cdot \frac{du}{dx}$$

Our "proof" or development of this rule assumed that whenever $\Delta x \neq 0$, then $\Delta u \neq 0$ also. However, it is possible that for some function u, $\Delta u = 0$ even though $\Delta x \neq 0$. In such a case we cannot introduce $\Delta u / \Delta u$, since division by zero is not defined. A more elaborate formal proof is required to prove the chain rule for the case when Δu is zero.

Considering the derivative as a rate of change leads to another way of seeing what is happening in the chain rule. Suppose that $dy/du = 5$ and $du/dx = 3$. This means that the rate of change of y with respect to u is 5 and the rate of change of u with respect to x is 3. In other words, y is changing 5 times as fast as u and u is changing 3 times as fast as x. It follows that y is changing 15 times as fast as x. In other words, $dy/dx = 5 \cdot 3 = 15$. This result is an example of the chain rule,

$$\frac{dy}{dx} = \frac{dy}{du} \cdot \frac{du}{dx}$$

Now let us apply the chain rule.

EXAMPLE 1 Differentiate $y = (x^2 + 1)^8$.

SOLUTION $y = (x^2 + 1)^8$ can be considered $y = (u)^8$, where $u = x^2 + 1$. Using the chain rule,

$$\frac{dy}{dx} = \frac{dy}{du} \cdot \frac{du}{dx}$$

Here $y = u^8$ and $u = x^2 + 1$, so that $dy/du = 8u^7$ and $du/dx = 2x$. As a result,

$$\frac{dy}{dx} = \frac{dy}{du} \cdot \frac{du}{dx}$$

$$= 8u^7 \cdot 2x$$

Next, replacing u by $x^2 + 1$ produces the final result, one expressed in terms of x (that is, without u in it).

$$\frac{dy}{dx} = 8(x^2 + 1)^7 \cdot 2x$$

$$= 16x(x^2 + 1)^7 \quad \blacklozenge$$

Alternatively, the chain rule can be stated using composition of functions.

$$[f(g(x))]' = f'(g(x)) \cdot g'(x)$$

Note

In practice, the chain rule is often applied rather mechanically, without actually substituting u into the written work. If you examine the function and the derivative obtained in the preceding example, you will be able to see how this can be done.

$$\text{Function:} \quad y = (x^2 + 1)^8$$

$$\text{Derivative:} \quad \frac{dy}{dx} = 8(x^2 + 1)^7 \cdot 2x$$

The $8(x^2 + 1)^7$ portion of the derivative comes from the power rule. The factor $2x$ is the derivative of the function $x^2 + 1$ that was raised to the power.

Here are two more examples that show the use of the chain rule without actually substituting u into the written work.

$$\frac{d}{dx}(7x + 2)^{10} = 10(7x + 2)^9 \cdot 7$$

$$= 70(7x + 2)^9$$

$$\frac{d}{dx}(1 - x^4)^{3/2} = \frac{3}{2}(1 - x^4)^{1/2} \cdot (-4x^3)$$

$$= -6x^3(1 - x^4)^{1/2}$$

All of the chain rule applications shown thus far are examples of the **general power rule**.

General Power Rule

If u is a differentiable function of x and n is a real number, then

$$\frac{d}{dx}u^n = nu^{n-1} \cdot \frac{du}{dx}$$

In function notation, the general power rule appears as follows for $f(x) = [u(x)]^n$:

$$f'(x) = n[u(x)]^{n-1} \cdot u'(x)$$

EXAMPLE 2 Determine the derivative of $\quad y = \sqrt{x^2 + 4x - 1}$.

SOLUTION First, change the radical form of the square root to the numerical exponent 1/2.

$$y = (x^2 + 4x - 1)^{1/2}$$

Next, apply the general power rule.

$$\frac{dy}{dx} = \frac{1}{2}(x^2 + 4x - 1)^{-1/2}(2x + 4)$$

The calculus is now complete; however, we must simplify the result. The steps are:

$$\frac{dy}{dx} = \frac{1}{2}(2x + 4)(x^2 + 4x - 1)^{-1/2}$$

$$= (x + 2)(x^2 + 4x - 1)^{-1/2}$$

$$= \frac{x + 2}{(x^2 + 4x - 1)^{1/2}} \quad \text{or} \quad \frac{x + 2}{\sqrt{x^2 + 4x - 1}}$$

Either of these last two forms is mathematically correct and simplified. Some people prefer exponents, others prefer radicals. In particular, the example *began* with a radical, so it would seem fitting that the final form of the derivative might also have a radical.   ◆

EXAMPLE 3 Find the derivative of $f(x) = \dfrac{1}{(x^2 + 5)^3}$.

SOLUTION Although the quotient rule could be used here, it is simpler to rewrite the expression by using negative exponents. Then the general power rule applies immediately.

$$f(x) = \frac{1}{(x^2 + 5)^3} = (x^2 + 5)^{-3}$$

Now, by the general power rule,

$$f'(x) = -3(x^2 + 5)^{-4}(2x)$$

or

$$f'(x) = \frac{-6x}{(x^2 + 5)^4}$$

If you are curious, try using the quotient rule instead. You will see that it is more difficult. (See Exercise 58.)   ◆

EXAMPLE 4 Differentiate $f(x) = \dfrac{(2x + 1)^5}{3x + 1}$.

SOLUTION Using the quotient rule, we obtain as a first step

$$f'(x) = \frac{(3x + 1) \cdot \dfrac{d}{dx}(2x + 1)^5 - (2x + 1)^5 \cdot \dfrac{d}{dx}(3x + 1)}{(3x + 1)^2}$$

Next, keeping in mind that the general power rule applies when differentiating $(2x + 1)^5$, we obtain

$$f'(x) = \frac{(3x + 1) \cdot 5(2x + 1)^4 \cdot 2 - (2x + 1)^5 \cdot 3}{(3x + 1)^2}$$

This expression can be simplified by factoring $(2x + 1)^4$ from each term of the numerator.

$$f'(x) = \frac{(2x + 1)^4[(3x + 1) \cdot 5 \cdot 2 - (2x + 1) \cdot 3]}{(3x + 1)^2}$$

or

$$f'(x) = \frac{(2x + 1)^4(30x + 10 - 6x - 3)}{(3x + 1)^2}$$

Combining like terms within the parentheses produces the final form of the derivative, namely,

$$f'(x) = \frac{(2x + 1)^4(24x + 7)}{(3x + 1)^2} \quad \blacklozenge$$

EXAMPLE 5 Find dy/dx when $y = (4x + 1)^3(x^2 - 3)^5$.

SOLUTION Using the product rule, we obtain as a first step

$$\frac{dy}{dx} = (4x + 1)^3 \cdot \frac{d}{dx}(x^2 - 3)^5 + (x^2 - 3)^5 \cdot \frac{d}{dx}(4x + 1)^3$$

Next, keeping in mind that the general power rule applies when differentiating $(x^2 - 3)^5$ and $(4x + 1)^3$, we have

$$\frac{dy}{dx} = (4x + 1)^3 \cdot 5(x^2 - 3)^4 \cdot 2x + (x^2 - 3)^5 \cdot 3(4x + 1)^2 \cdot 4$$

At this point you can factor out $(4x + 1)^2$ and $(x^2 - 3)^4$ and 2. The result is

$$\frac{dy}{dx} = 2(4x + 1)^2(x^2 - 3)^4[(4x + 1) \cdot 5 \cdot x + (x^2 - 3) \cdot 3 \cdot 2]$$

or

$$\frac{dy}{dx} = 2(4x + 1)^2(x^2 - 3)^4(20x^2 + 5x + 6x^2 - 18)$$

Finally,

$$\frac{dy}{dx} = 2(4x + 1)^2(x^2 - 3)^4(26x^2 + 5x - 18) \quad \blacklozenge$$

3.6 Exercises

Simplify all derivatives in the exercises of this section.

Differentiate each function in Exercises 1–18.

1. $y = (x^2 + 3)^5$

2. $y = (x^2 + 1)^4$

3. $y = (3x)^5$

4. $y = (2x)^7$

5. $y = (3x + 4)^{1/3}$

6. $y = (9x)^{1/3}$

7. $s = (1 - t^4)^{-6}$

8. $s = (t^3 + 2)^{-10}$

9. $y = \sqrt{4x + 1}$

10. $y = \sqrt{10x - 7}$

11. $y = (1 - 6x)^{4/3}$

12. $y = (1 + 8x)^{3/2}$

13. $f(x) = (x^4 + x^2 + 1)^4$

14. $f(x) = (x^2 + 7x - 2)^5$

15. $f(t) = \dfrac{1}{(t^2 + 1)^6}$

16. $f(t) = \sqrt{1 - t}$

17. $y = \dfrac{1}{\sqrt{6x + 5}}$

18. $y = 3 + \sqrt{1 - 2t^2}$

Determine the derivative in Exercises 19–42.

19. $y = x^2(5x - 2)^4$

20. $y = x^3(3x + 4)^6$

21. $y = (x + 3)(2x + 1)^3$

22. $y = (2x + 1)^4(x^2 + 3)$

23. $y = (x^2 + 1)^4(x - 2)$

24. $y = (3x - 1)(1 - x^2)^3$

25. $y = (5x - 2)^2(x^2 + 7)^3$

26. $y = (3x + 1)^3(1 + 2x)^2$

27. $y = (2t + 3)^4(t - 7)^3$

28. $y = (1 + 4t)^3(t^2 + 3)^4$

29. $y = \dfrac{(x + 4)^3}{x + 1}$

30. $y = \dfrac{(x - 1)^5}{x + 2}$

31. $y = \dfrac{(2t + 3)^4}{t - 2}$

32. $y = \dfrac{(4t + 1)^6}{t - 3}$

33. $y = \dfrac{(1 + x^2)^5}{3x - 2}$

34. $y = \dfrac{(x^3 + 2)^4}{5x + 1}$

35. $y = (2x - 5)(2x + 1)^{3/2}$

36. $y = (6x - 1)^{4/3}(x^2 + 3)$

37. $y = (x^3 + 1)^{4/3}(1 + 2x)$

38. $y = (8x + 5)(1 + 2x)^{3/2}$

39. $y = 2x\sqrt{2x + 1}$

40. $y = x^2\sqrt{4x + 1}$

41. $y = x^2(x^2 + 1)^{1/2}$

42. $y = 4x(1 + x)^{1/2}$

Find dy/dx in Exercises 43–46.

43. $y = \left(\dfrac{x - 1}{x + 1}\right)^4$

44. $y = \left(\dfrac{x^2 + 1}{x + 1}\right)^3$

45. $y = \sqrt{\dfrac{2x + 1}{x - 1}}$

46. $y = \sqrt{\dfrac{1 - 2x}{1 + x}}$

In Exercises 47–50, find the slope of the line tangent to the graph of the function at the given point.

47. $f(x) = (x^2 - 3x + 1)^4$ at $(2, 1)$

48. $f(x) = 3x(x^2 + 1)^3$ at $(0, 0)$

49. $y = \sqrt{4x + 1}$ at $(2, 3)$

50. $y = \dfrac{1}{(1 + 2x)^2}$ at $(-1, 1)$

In Exercises 51–54, determine the rate of change of y with respect to x at the given value of x.

51. $y = (1 + x^2)^4$ when $x = 1$

52. $y = (5x - 1)^{3/2}$ when $x = 2$

53. $y = \sqrt[3]{6x - 4}$ when $x = 2$

54. $y = (10x + 1)^{-1}$ when $x = 0$

55. (*MARGINAL PROFIT*) An upstart insurance company has determined that its profit from the sale of x "units" of auto insurance is approximately P dollars, where

$$P(x) = 100(x^2 - 1)^{1/2} \qquad \text{for } x \geq 2$$

Determine the marginal profit function.

56. (*SPEED*) A windup toy car rolls around, returns to its starting point, and then stops. Its distance s from the starting point at any time t during its 10-second run is

$$s = \frac{2}{9}(10 - t)^2 t^{3/2}$$

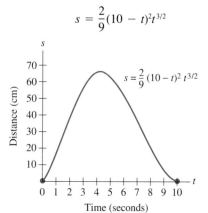

where s is in centimeters and t is in seconds. (See figure.) What is the speed of the car after 1 second?

57. (POLLUTANT CONCENTRATION) A factory has dumped pollutants into a river that passes next to it. The relative concentration is considered to be 100% at the spot where the pollutants enter the river. The relative concentration then decreases until the pollutants virtually disappear about 21.54 miles downstream. The *percent* of relative concentration (p) is given by

$$p = \sqrt{10,000 - x^3}$$

where x is the distance in miles downstream from the factory ($0 \le x \le 21.54$).

(a) Using the function given, determine the relative concentration of pollutants at distances of 0 miles, 10 miles, and 20 miles from the factory. Use a calculator.

(b) Determine the rate of change of this relative concentration at 0 miles, 10 miles, and 20 miles from the factory. Use a calculator.

58. Use the quotient rule to find $f'(x)$ for the function f defined in Example 3.

W 59. Explain why we need the chain rule. Include two functions such as $y = x^3$ and $y = (x^2 + 1)^3$ in your explanation. Be sure to focus on *why* the differentiation is different rather than on *how* it is different.

Determine dy/dx in Exercises 60–63.

60. $y = (2x + 1)^4 (x - 2)^3 \sqrt{1 + 4x}$

61. $y = (x + 1)^2 (2x - 3)^5 (x^2 + 1)^8$

62. $y = \dfrac{(1 + x)^3 (2x + 3)^4}{5x - 1}$

63. $y = \dfrac{(x^2 + 3)^5 (x + 1)^2}{2x + 7}$

The Exercise Library at the back of the book contains graphing calculator and computer exercises keyed to this section.

3.7 | HIGHER-ORDER DERIVATIVES

We proceed next to find the derivative of the derivative of a function. The "second derivative," as it is called, will be used here in applications involving acceleration of a moving object and rate of change of marginal cost and marginal revenue. In Chapter 4 the second derivative will be used to assist curve sketching. Other uses of the second derivative and higher-order derivatives appear in the study of differential equations and infinite series.

To begin, consider the function defined by

$$f(x) = x^5 + 3x^4 - x^2 + 9x - 22$$

Differentiation yields

$$f'(x) = 5x^4 + 12x^3 - 2x + 9$$

Since f' is itself a function that can be differentiated, we will differentiate it. The result is called the **second derivative** and is denoted by f''. Here

$$f''(x) = 20x^3 + 36x^2 - 2$$

Similarly, the derivative of f'' is the **third derivative** and is denoted f'''. Here

$$f'''(x) = 60x^2 + 72x$$

The fourth derivative is denoted $f^{(4)}$, the fifth derivative is $f^{(5)}$, and so on.

Since the second derivative is the derivative of the first derivative, in dy/dx notation we have the second derivative appearing as

$$\frac{d}{dx}\left(\frac{dy}{dx}\right)$$

which is abbreviated to

$$\frac{d^2y}{dx^2}$$

Next is a summary table showing notations used to denote various derivatives.

Derivative Notations

first derivative	y'	f'	$\dfrac{dy}{dx}$	$D_x y$
second derivative	y''	f''	$\dfrac{d^2y}{dx^2}$	$D_x^2 y$
third derivative	y'''	f'''	$\dfrac{d^3y}{dx^3}$	$D_x^3 y$
fourth derivative	$y^{(4)}$	$f^{(4)}$	$\dfrac{d^4y}{dx^4}$	$D_x^4 y$
nth derivative	$y^{(n)}$	$f^{(n)}$	$\dfrac{d^ny}{dx^n}$	$D_x^n y$

EXAMPLE 1   If $y = 8x^{3/2}$, determine y''.

SOLUTION

$$y = 8x^{3/2}$$

$$y' = \frac{3}{2} \cdot 8x^{1/2} = 12x^{1/2}$$

$$y'' = \frac{1}{2} \cdot 12x^{-1/2} = 6x^{-1/2} = \frac{6}{x^{1/2}} \quad \text{or} \quad \frac{6}{\sqrt{x}} \quad \blacklozenge$$

EXAMPLE 2   Let $f(x) = x^2 + 7x - 1$. Find $f^{(5)}(x)$.

SOLUTION

$$f(x) = x^2 + 7x - 1$$
$$f'(x) = 2x + 7$$
$$f''(x) = 2$$
$$f'''(x) = 0$$
$$f^{(4)}(x) = 0$$
$$f^{(5)}(x) = 0$$

Notice that for $f(x) = x^2 + 7x - 1$, all derivatives from f''' on are equal to zero. ◆

EXAMPLE 3   Let $f(x) = x^4 - 4x^2 + x - 2$. Find $f''(1)$.

SOLUTION $f''(1)$ is the second derivative evaluated at $x = 1$. First determine $f''(x)$. Then let $x = 1$ in $f''(x)$.

$$f(x) = x^4 - 4x^2 + x - 2$$
$$f'(x) = 4x^3 - 8x + 1$$
$$f''(x) = 12x^2 - 8$$

Now

$$f''(1) = 12(1)^2 - 8 = 4$$

Thus, $f''(1) = 4$. ◆

Recall from Section 3.3 that the derivative of a function can be interpreted as the instantaneous rate of change of the function with respect to the variable. f' is the (instantaneous) rate of change of f with respect to x. It follows that

> *The second derivative of a function is the* (instantaneous) *rate of change of the first derivative* with respect to the variable.

Thus, f'' is the rate of change of f' with respect to x.

EXAMPLE 4   ◆   **RATE OF CHANGE OF MARGINAL COST**

If the cost of producing x items is $C(x) = 800 + 50x - .04x^2$ dollars, determine the rate of change of the marginal cost when 35 items are produced.

SOLUTION   Since $C(x) = 800 + 50x - .04x^2$, the marginal cost $C'(x)$ is given by

$$C'(x) = 50 - .08x$$

The rate of change of the marginal cost is the derivative of the marginal cost, that is, $C''(x)$.

$$C''(x) = -.08$$

The function C'' gives the rate of change of marginal cost for any level of production x. In this example, $C''(x)$ is constant; $C''(x) = -.08$ for all x. Specifically, $C''(35) = -.08$. Thus, the rate of change of the marginal cost when 35 items are produced is $-.08$ dollars per item per item. In other words, marginal cost is decreasing (which accounts for the *minus*) at the rate of 8¢ per item per item. ◆

Acceleration a is the instantaneous rate of change of velocity v with respect to time t. That is,

$$a = \frac{dv}{dt}$$

Using s for distance and the fact that $v = ds/dt$, we can obtain the alternative form

$$a = \frac{dv}{dt} = \frac{d}{dt}\left(\frac{ds}{dt}\right) = \frac{d^2s}{dt^2}$$

In conclusion,

$$a = \frac{dv}{dt} = \frac{d^2s}{dt^2}$$

EXAMPLE 5   ◆   *ACCELERATION OF A FALLING OBJECT*

In Example 4 of Section 3.3, a ball is dropped and travels such that the distance it has traveled after t seconds is $s = -16t^2$ feet. Find the acceleration of the ball at any time t.

SOLUTION   From $s = -16t^2$, we obtain

$$v = \frac{ds}{dt} = -32t$$

and

$$a = \frac{dv}{dt} = -32$$

The acceleration of the ball is -32 feet per second per second. The minus sign specifies that the direction of the acceleration is downward, toward the ground. This acceleration is due to the constant pull of gravity on the ball, which is why the acceleration is a constant.   ◆

3.7   Exercises

Determine the specified derivative for each function in Exercises 1–8.

1. Find $f''(x)$ for $f(x) = x^4 - 10x^2$

2. Find $f''(x)$ for $f(x) = 1 - 3x + x^2 + 10x^3$

3. Find y''' for $y = x^3 - 7x^2 + 100x + 1$

4. Find y''' for $y = x^{1/2}$

5. Find $f^{(4)}(x)$ for $f(x) = x^4$

6. Find $f^{(4)}(x)$ for $f(x) = x^5$

7. Find $D_x^3 y$ for $y = x^{5/3}$

8. Find $D_x^3 y$ for $y = 1 - x^{3/2}$

Find y'' for each function in Exercises 9–14.

9. $y = 9x$

10. $y = 1 - 2x$

11. $y = x^{-4} + x$

12. $y = \dfrac{1}{x^5}$

13. $y = \dfrac{x}{1 + x}$

14. $y = \dfrac{1 - x}{1 + x}$

Find d^2y/dx^2 for each function in Exercises 15–18.

15. $y = x^3 - 5x^2 + 10x - 8$ **16.** $y = x^4 - 7x + 2$

17. $y = (3x - 1)^5$ **18.** $y = (1 - 2x)^6$

Obtain $D_x^2 y$ for each function in Exercises 19–22.

19. $y = x^{5/2}$

20. $y = x^{3/4}$

21. $y = \dfrac{1}{\sqrt{x}}$

22. $y = \dfrac{4}{x}$

In Exercises 23–28, determine the specified derivative.

23. $g(x) = x^3 - 8x^2 + x - 15$;   find $g''(x)$

24. $s = \dfrac{3}{t - 1}$;   find s''

25. $y = \dfrac{10}{t^4}$;   find d^2y/dt^2

26. $C(x) = 100 - .03x^2$;   find $C''(x)$

27. $s = -16t^2 + 96t + 108$;   find d^2s/dt^2

28. $y = 5x^2 - 3x + 17$;   find y'''

Determine $f''(4)$ for each function in Exercises 29–34.

29. $f(x) = x^3 - 5x^2 + x - 3$

30. $f(x) = x^3 + 2x^2 - 10x + 16$

31. $f(x) = \dfrac{1}{x}$

32. $f(x) = \dfrac{1}{x^2}$

33. $f(x) = 10\sqrt{x}$

34. $f(x) = \dfrac{1}{\sqrt{x}}$

In Exercises 35–44, determine the indicated rate of change.

35. The rate of change of $f'(x)$ when $f(x) = x^3 - 7x^2$

36. The rate of change of $f'(x)$ when $f(x) = 1 - x + x^2$

37. The rate of change of $g'(x)$ when $g(x) = 1 + \dfrac{10}{x}$

38. The rate of change of $g'(x)$ when $g(x) = 5 - \dfrac{12}{x^2}$

39. The rate of change of $f'(x)$ when $f(x) = 6x^{3/2}$ and $x = 1$

40. The rate of change of $f'(x)$ when $f(x) = 9x^{4/3}$ and $x = 1$

41. The rate of change of $f'(t)$ when $f(t) = \sqrt{8t}$ and $t = 2$

42. The rate of change of dy/dx for $y = \dfrac{1 + x}{x}$ when $x = 1$

43. The rate of change of dy/dx for $y = \dfrac{1 - x}{x}$ when $x = 1$

44. The rate of change of y' for $y = (1 + 2x)^4$

(*COST/MARGINAL COST*) For each cost function in Exercises 45–48, determine the rate of change of the marginal cost when 10 items are produced.

45. $C(x) = 300 + 40x - .06x^2$

46. $C(x) = 500 + 30x - .03x^2$

47. $C(x) = 200 + 30x - \dfrac{100}{x}$

48. $C(x) = 400 + 70x - \dfrac{1000}{x^2}$

(*REVENUE/MARGINAL REVENUE*) For each revenue function in Exercises 49–52, determine the rate at which marginal revenue is changing when 20 items are sold.

49. $R(x) = 50x - .01x^2$ **50.** $R(x) = 80x - .001x^3$

51. $R(x) = 20x - \dfrac{20}{x}$ **52.** $R(x) = 30x + \dfrac{40}{x}$

53. (*ACCELERATION*) A ball is thrown straight up from the ground and travels such that its distance from the ground at any time t is $s = -16t^2 + 80t$ feet. Determine its acceleration at any time t.

54. (*ACCELERATION*) What would be the acceleration of the ball mentioned in Exercise 53, if the distance function is $s = -16t^2 + 200t$?

55. (*ACCELERATION*) An object travels according to $s = t^3 - t^2 + 3t$, where s is in feet and t is in seconds. Determine the function that describes its acceleration.

56. (*ACCELERATION*) Redo Exercise 55 using $s = .5t^2 + 6t$.

57. (*VELOCITY / ACCELERATION*) A ball travels on a hilly path according to

$$s = t^2 - 2t + \frac{4}{t}$$

where s is its distance in centimeters from the beginning point and t is the number of seconds it has been rolling. (Note: $t > 0$.)
(a) Determine a formula for the speed of the ball at any time t.
(b) Determine the magnitude of its acceleration at any time t.
(c) Determine the magnitude of its acceleration when $t = 1$.
(d) Determine the magnitude of its acceleration after 2 seconds.

58. (*VELOCITY / ACCELERATION*) Redo Exercise 57 using the function

$$s = t^2 - 3t - \frac{8}{t}$$

Determine the second derivative of each function given in Exercises 59–62.

59. $f(x) = \dfrac{x^3 + 1}{x^2 + 1}$

60. $f(x) = \dfrac{1 + x^2}{1 - x^2}$

61. $y = 5x \sqrt{1 + x^2}$

62. $y = x (3x + 1)^6$

63. (*VELOCITY / ACCELERATION*) A ball thrown upward from the surface of the moon will go higher and take longer to return than a ball thrown upward from the surface of the earth, assuming the same initial velocity. Assume that the ball on the moon travels according to

$$s = -2.65t^2 + 106t$$

where s is the distance from the surface (in feet) and t is time (in seconds).
(a) Determine the velocity at any time t.
(b) Determine the acceleration at any time t.
W (c) Does your answer to part (b) seem consistent with the statement that the ball will go higher and take longer to return than it would on earth? Explain.
(d) How long does it take the ball to reach the ground?

64. Consider the third derivative of y with respect to x as an instantaneous rate of change and fill in the blanks. d^3y/dx^3 is the instantaneous rate of change of _____ with respect to _____.

W 65. (*PROFIT*) Explain the meaning of $P''(x)$, if $P(x)$ is a profit function.

66. For the quadratic polynomial function $f(x) = x^2$, we find that $f'''(x) = 0$. For $g(x) = x^{-2}$, we find that $g'''(x) \neq 0$.
(a) Verify the two statements.
W (b) For $g(x) = x^{-2}$, how many more derivatives (beyond g''') must we obtain before the resulting function will be zero (as it was with f)? Explain.

The Exercise Library at the back of the book contains graphing calculator and computer exercises keyed to this section.

3.8 | IMPLICIT DIFFERENTIATION

Up until now, every function of x that we have differentiated has been given in an explicit manner, such as

$$y = x^2 + 3x - 2$$
$$y = (x + 3)^2(1 + x^2)^3$$
$$f(x) = \sqrt{x^3 - 6}$$

In other words, the variable y has been written explicitly in terms of the other variable x. Always y, or $f(x)$, has been on one side of the equation and the expression in x has been on the other side.

But it can happen that the function to be differentiated is not presented this way. For example,

$$3xy - 2 = 0$$

and

$$y^3 - x^2 + xy = 3$$

are two relationships between x and y in which y is not given explicitly as a function of x. When x and y appear together on the same side of the equation, as they do in these examples, the equation is said to define an *implicit function*. A relationship between x and y is implied, but it is not given explicitly, as it was in the three functions noted at the very beginning of this section.

The equation $3xy - 2 = 0$ can be readily manipulated to give y explicitly in terms of x. The steps are shown next.

$$3xy - 2 = 0$$
$$3xy = 2$$
$$y = \frac{2}{3x}$$

By contrast, the equation $y^3 - x^2 + xy = 3$ cannot be so readily manipulated.

While there is nothing "wrong" with having equations in which y is given implicitly as a function of x, the procedure for determining dy/dx is different in such cases. The process used to determine the derivative is called **implicit differentiation**, and it uses the chain rule in a very interesting way. Recall the general power rule, with u being a function of x:

$$\frac{d}{dx} u^n = n \cdot u^{n-1} \cdot \frac{du}{dx}$$

In a setting such as $y^3 - x^2 + xy = 3$, y can be considered as a function of x, just as u is a function of x in the general power rule stated above. In view of this, we can differentiate y^3 by using the general power rule, with u changed to y.

$$\frac{d}{dx} y^n = n \cdot y^{n-1} \cdot \frac{dy}{dx}$$

Specifically,

$$\frac{d}{dx} y^3 = 3y^2 \cdot \frac{dy}{dx}$$

Similarly, xy can be differentiated by using the product rule.

$$\frac{d}{dx}(xy) = x \cdot \frac{d}{dx}(y) + y \cdot \frac{d}{dx}(x)$$

$$= x \cdot \frac{dy}{dx} + y \cdot 1$$

$$= x\frac{dy}{dx} + y$$

Now we will combine all of these ideas in an example that shows how to determine the derivative dy/dx when given an equation such as $y^3 - x^2 + xy = 3$, in which y is given implicitly as a function of x.

EXAMPLE 1 Determine dy/dx if $y^3 - x^2 + xy = 3$.

SOLUTION Since $y^3 - x^2 + xy$ is equal to 3, it follows that the derivative of $y^3 - x^2 + xy$ is equal to the derivative of 3. That is, from

$$y^3 - x^2 + xy = 3$$

we obtain

$$\frac{d}{dx}(y^3 - x^2 + xy) = \frac{d}{dx}(3)$$

or

$$\frac{d}{dx}(y^3) - \frac{d}{dx}(x^2) + \frac{d}{dx}(xy) = \frac{d}{dx}(3)$$

In work done just before this example, we had already determined that

$$\frac{d}{dx}y^3 = 3y^2\frac{dy}{dx}$$

and

$$\frac{d}{dx}(xy) = x\frac{dy}{dx} + y$$

Also, we can see readily that

$$\frac{d}{dx}x^2 = 2x$$

and

$$\frac{d}{dx}(3) = 0$$

Putting all of this together yields

$$3y^2\frac{dy}{dx} - 2x + x\frac{dy}{dx} + y = 0$$

At this point the "calculus" is finished. We need only do some algebraic manipulation to solve for dy/dx. Begin by adding $2x$ and $-y$ to both sides of the equation. This will leave only dy/dx terms on the left side.

$$3y^2 \frac{dy}{dx} + x\frac{dy}{dx} = 2x - y$$

Next, factor out dy/dx.

$$(3y^2 + x)\frac{dy}{dx} = 2x - y$$

Finally, divide by $3y^2 + x$. The result is the desired derivative, dy/dx.

$$\frac{dy}{dx} = \frac{2x - y}{3y^2 + x} \quad \blacklozenge$$

While implicit differentiation is not a difficult concept, it is surely different from what you are used to doing. Here is another example.

EXAMPLE 2 Find dy/dx if $xy^4 + y - 2y^{3/2} = x^3 - 6$.

SOLUTION Begin by indicating the term-by-term differentiation. From

$$xy^4 + y - 2y^{3/2} = x^3 - 6$$

we obtain

$$\frac{d}{dx}(xy^4) + \frac{d}{dx}(y) - \frac{d}{dx}(2y^{3/2}) = \frac{d}{dx}(x^3) - \frac{d}{dx}(6)$$

Differentiate, keeping in mind that xy^4 is a product.

$$x \cdot 4y^3 \cdot \frac{dy}{dx} + y^4 \cdot 1 + \frac{dy}{dx} - 3y^{1/2} \cdot \frac{dy}{dx} = 3x^2 - 0$$

or

$$4xy^3 \frac{dy}{dx} + y^4 + \frac{dy}{dx} - 3y^{1/2}\frac{dy}{dx} = 3x^2$$

Add $-y^4$ to both sides to isolate the dy/dx terms.

$$4xy^3 \frac{dy}{dx} + \frac{dy}{dx} - 3y^{1/2}\frac{dy}{dx} = 3x^2 - y^4$$

Factor out dy/dx.

$$(4xy^3 - 3y^{1/2} + 1)\frac{dy}{dx} = 3x^2 - y^4$$

Finally, divide both sides by the coefficient of dy/dx.

$$\frac{dy}{dx} = \frac{3x^2 - y^4}{4xy^3 - 3\sqrt{y} + 1} \quad \blacklozenge$$

The procedure for determining dy/dx by implicit differentiation is summarized next.

Implicit Differentiation

If the equation contains x and y,

1. Differentiate both sides (all terms) of the equation with respect to x.

2. Collect all terms containing dy/dx on one side and all other terms on the other side.

3. Factor out dy/dx from all terms that contain it.

4. Solve for dy/dx by dividing both sides by the coefficient of dy/dx.

EXAMPLE 3   Find the slope of the tangent line to the ellipse given by $3x^2 + y^2 = 12$ at the point $(1, 3)$. See Figure 12.

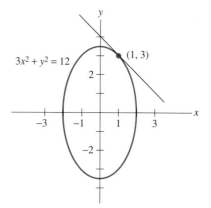

Figure 12 The tangent line at $(1, 3)$

SOLUTION   The slope of the tangent line is given by dy/dx. So we must find dy/dx and evaluate it at $(1, 3)$. From

$$3x^2 + y^2 = 12$$

we obtain

$$\frac{d}{dx}(3x^2) + \frac{d}{dx}(y^2) = \frac{d}{dx}(12)$$

$$6x + 2y\frac{dy}{dx} = 0$$

$$2y\frac{dy}{dx} = -6x$$

$$\frac{dy}{dx} = \frac{-6x}{2y} = -\frac{3x}{y}$$

At (1, 3), we have

$$\frac{dy}{dx} = \frac{-3(1)}{(3)} = -1$$

Thus, the slope of the tangent line to the ellipse at (1, 3) is −1. ◆

In the next section, the concept of implicit differentiation will be applied to situations involving rates of change.

3.8   Exercises

Obtain dy/dx by implicit differentiation in Exercises 1–22.

1. $x^2 + y^2 = 3$

2. $3x^2 + 2y^2 = 6$

3. $y^3 - x^2 = 6x$

4. $x^4 - y^2 = 2x - 5$

5. $x^3 + y^3 = 9$

6. $y^5 = x^4 + 6$

7. $y^2 = x^2 + 7x - 4$

8. $y^2 = x^3 - 10x^2$

9. $x^2 + y^2 - y^3 - x = 0$

10. $y^4 - x^3 + y^2 - 10 = 0$

11. $2x^{3/2} + 2y^{3/2} = 15$

12. $3x^{4/3} + 3y^{4/3} = 1$

13. $y^2 - 4x^{5/2} + x = 0$

14. $x^2 + 4y^{3/2} = y$

15. $x^2y^2 - x - 3y = 0$

16. $x^2y^2 - 2x + 7y = 0$

17. $xy^3 + 5y = 3x$

18. $xy^2 + x = 4y$

19. $xy - 7 = 0$

20. $xy + 4 = 0$

21. $8xy + x = 9$

22. $7xy + y = 2$

In Exercises 23–28, use implicit differentiation to determine dy/dx. Then find the *slope* of the tangent line to the curve at the specified point.

23. $x^2 + y^3 - y = 7$ at (1, 2)

24. $x^3 + y^2 + x = 19$ at (2, 3)

25. $y^2 = 3x^2 + 6x + 4$ at (−1, 1)

26. $y^3 = 28 - x - x^2$ at (4, 2)

27. $10xy + x + 45 = 0$ at (5, −1)

28. $y - 3xy = 22$ at (4, −2)

29. Determine the equation of the line tangent to the circle $x^2 + y^2 = 25$ at the point (3, 4). See Figure on next page.

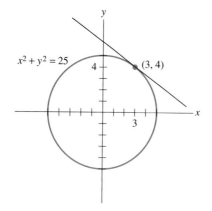

$x^2 + y^2 = 25$ (3, 4)

30. Determine the equation of the line tangent to the ellipse $x^2 + 4y^2 = 25$ at the point $(3, -2)$.

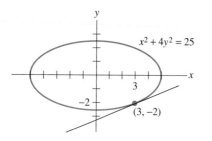

$x^2 + 4y^2 = 25$

$(3, -2)$

Obtain dy/dx by implicit differentiation in Exercises 31–34.

31. $(y - 3)^3 = (x + 5)^2$ **32.** $x + y = \sqrt{2x + 3y}$

33. $x\sqrt{y} + y = 7x$ **34.** $\dfrac{x}{y} + y = x$

35. If you were asked to determine dy/dx, would you know when implicit differentiation is needed and when it is not? Think about it and then answer "yes" or "no" to whether implicit differentiation should be used to determine dy/dx in each of the four equations given.

(a) $y = x^3 + x$ **(b)** $x^3y = 1 + xy^2$

(c) $y = x^2 - 3xy^2$ **(d)** $x^2 - 9x = y$

W 36. Explain your answer to Exercise 35(c).

W 37. Explain your answer to Exercise 35(d).

The Exercise Library at the back of the book contains graphing calculator and computer exercises keyed to this section.

3.9 | RELATED RATES

The concept of implicit differentiation can be applied to situations involving rates of change. In most of our applications there will be two variables, and each of them will be a function of some third variable. For example, the variables may be x and y, and each of them may be a function of time t. As we shall see, differentiating such an equation with respect to t will introduce a relationship between dx/dt and dy/dt. In other words, we will have a relationship between rates of change—**related rates**.

We begin with a preliminary example in order to provide some orientation.

EXAMPLE 1   Suppose that x and y are functions of t and that $x^3 + 5x^2 - 2y = 19$. Differentiate this equation with respect to t.

SOLUTION   As a beginning step, we have

$$\frac{d}{dt}(x^3 + 5x^2 - 2y) = \frac{d}{dt}(19)$$

or

$$\frac{d}{dt}(x^3) + \frac{d}{dt}(5x^2) - \frac{d}{dt}(2y) = \frac{d}{dt}(19)$$

To complete the differentiation, we must obtain the four derivatives suggested in this last equation. And they are as follows:

$$\frac{d}{dt}(19) = 0 \qquad \text{(The derivative of a constant is zero.)}$$

$$\frac{d}{dt}(2y) = 2 \cdot \frac{d}{dt}(y) = 2\frac{dy}{dt}$$

$$\frac{d}{dt}(5x^2) = 10x \cdot \frac{d}{dt}(x) = 10x\frac{dx}{dt}$$

$$\frac{d}{dt}(x^3) = 3x^2 \cdot \frac{dx}{dt} \qquad \text{(Using the general power rule)}$$

Putting all this together, we see that differentiation yields the equation

$$3x^2\frac{dx}{dt} + 10x\frac{dx}{dt} - 2\frac{dy}{dt} = 0 \quad \blacklozenge$$

The remaining examples of this section present applications in which related rates arise naturally.

EXAMPLE 2 ◆ *JAPANESE BEETLE INFESTATION*

A Japanese beetle infestation is spreading from the center of a small rural town. (See Figure 13.) Since the beetles fly off in all directions, the region they cover

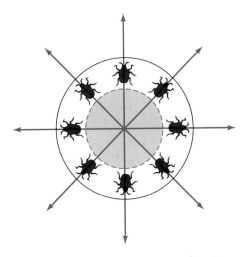

Figure 13 Spread of a beetle infestation

is circular. The radius of the circular region of infestation is increasing at the rate of 1.5 miles per year. Determine the rate of change of the area of infestation when the radius is 4 miles.

SOLUTION The area of the circular region of infestation is given by

$$A = \pi r^2$$

where r is the radius of the circle. We want to determine the rate of change of area A with respect to time t. This means we want the derivative of A with respect to t, dA/dt. We are given the rate of change of the radius r with respect to time t; that is, we know from the statement of the problem that dr/dt is 1.5. In view of all this, and realizing that both A and r are functions of time t, we differentiate $A = \pi r^2$ with respect to t.

$$\frac{d}{dt}(A) = \frac{d}{dt}(\pi r^2)$$

$$= \pi \frac{d}{dt}(r^2)$$

$$= \pi \cdot 2r \frac{dr}{dt}$$

That is,

$$\frac{dA}{dt} = 2\pi r \frac{dr}{dt}$$

We are seeking dA/dt when $r = 4$, and we already know that dr/dt is 1.5. Putting these known values into our equation for dA/dt produces

$$\frac{dA}{dt} = 2\pi(4)(1.5)$$

or

$$\frac{dA}{dt} = 12\pi \quad \text{square miles per year}$$

Using 3.14 as an approximation for π, we can determine that 12π is approximately 38. This means that the area of infestation is increasing at the rate of approximately 38 square miles per year. ◆

EXAMPLE 3 ◆ *INCREASING PRODUCTION LEVEL*

A manufacturer of tennis balls decides to increase production at the rate of 30 packages per day (time t will be in days). Total revenue from the sale of all x packages produced is

$$R = 2.14x - .0001x^2 \quad \text{dollars}$$

Determine the rate of change of revenue with respect to time when the daily production level is 1500 packages.

SOLUTION The rate of change of revenue with respect to time is dR/dt, which can be obtained by differentiating

$$R = 2.14x - .0001x^2$$

with respect to t. (Such differentiation is reasonable because both revenue R and quantity x are functions of time t.) The derivative is

$$\frac{dR}{dt} = 2.14\frac{dx}{dt} - .0002x\frac{dx}{dt}$$

A daily production level of 1500 means that $x = 1500$. Production increasing at the rate of 30 packages per day means that $dx/dt = 30$. Using 1500 for x and 30 for dx/dt yields

$$\frac{dR}{dt} = 2.14\,(30) - .0002\,(1500)(30) = 55.20$$

We conclude that the manufacturer's revenue is increasing at the rate of \$55.20 per day. ◆

EXAMPLE 4 ◆ *KITE FLYING*

A kite is flying 150 feet high, where the wind causes it to move horizontally at the rate of 5 feet per second. In order to maintain the kite at a height of 150 feet, the person must allow more string to be let out. (See Figure 14.) At what rate is the string being let out when the length of the string already out is 250 feet?

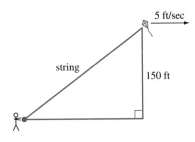

Figure 14 Flying a kite

SOLUTION We will use x for the length of the horizontal side of the right triangle and z for the length of the hypotenuse. (If the vertical side were not a constant 150 feet, we would use y to represent its length.) See Figure 15.

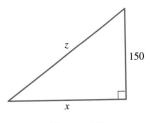

Figure 15

As the kite moves horizontally, both x and z increase. See Figure 16.

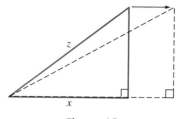

Figure 16

The rate of change of x is dx/dt, and it is known to be 5 feet per second (from the statement of the problem). The rate of change of z is dz/dt and is not yet known. In fact, dz/dt is the rate at which the string is being let out; it is the rate we seek. If we can find a relationship between x and z, then we can differentiate it and thus introduce dx/dt and dz/dt. We would then have an equation from which we could determine the value of dz/dt.

The Pythagorean theorem provides a relationship between x and z. From the right triangle with hypotenuse z and sides x and 150, we obtain

$$z^2 = x^2 + 150^2$$

or

$$z^2 = x^2 + 22{,}500$$

Differentiating with respect to t yields

$$2z\frac{dz}{dt} = 2x\frac{dx}{dt}$$

From the statement of the problem, we know that $dx/dt = 5$ and the string length is $z = 250$. The value of x can be determined from the right triangle by using the Pythagorean theorem. (See Figure 17.)

$$250^2 = x^2 + 150^2$$
$$x^2 = 250^2 - 150^2$$
$$x^2 = 40{,}000$$
$$x = \sqrt{40{,}000} = 200$$

Figure 17

Now substitute 200 for x, as well as 250 for z and 5 for dx/dt. The result:

$$2(250)\left(\frac{dz}{dt}\right) = 2(200)(5)$$

or

$$\frac{dz}{dt} = \frac{2(200)(5)}{2(250)} = 4 \quad \text{feet per second}$$

The string is being let out at the rate of 4 feet per second. ◆

When you encounter related rates problems, you may want to use the procedure suggested next.

Steps in Solving Related Rates Problems

1. Make a drawing of the situation, if possible.

2. Use letters to represent the variables involved in the situation—say x, y, etc.

3. Identify all rates of change given and those to be determined. Use calculus notation dx/dt, dy/dt, etc., to represent them.

4. Determine an equation that both (a) involves the variables of step 2 *and* (b) will involve the derivatives of step 3, when differentiated.

5. Differentiate (by implicit differentiation) the equation of step 4.

6. Substitute all known values into the differentiated equation. (The original equation of step 4 may be needed to calculate an unknown value.)

7. Use algebraic manipulation, if necessary, to solve for the desired unknown rate or quantity.

3.9 Exercises

1. **(GYPSY MOTH INFESTATION)** A gypsy moth infestation is spreading from the center of a small rural town. The radius of the circular region of infestation is increasing at the rate of .75 mile per year. Determine the rate of change of the area of infestation when the radius is 3 miles.

2. **(POND RIPPLE)** A boy walking on a bridge across a pond stops in the middle. He then drops a rock into the water, causing a circular ripple. If the ripple moves away from the center in such a way that its radius is increasing at the rate of 2.5 feet per second, determine how fast the area within the ripple is changing when the radius is 9 feet.

3. **(OIL SPILL)** Suppose an oil spill has taken the form of a circular region and its area is increasing at the rate

of 100 square meters per hour. At what rate is the radius of the region increasing when the radius is 200 meters?

4. **(MELTING ICE)** A large cube of ice is melting in such a way that the length s of a side is decreasing at the rate of 2 centimeters per minute. Find the rate of decrease in volume when each side is 100 centimeters. (Use $V = s^3$.)

5. **(TUMOR GROWTH)** Assuming that a tumor is approximately spherical in shape, its volume is approximately

$$V = \frac{4}{3}\pi r^3$$

The radius r of a tumor growing in an animal is

increasing at a rate of 1.25 millimeters per month. Determine how fast the volume of the tumor is increasing when the radius is 10 millimeters.

6. (*TUMOR REMISSION*) Refer to Exercise 5. Suppose the animal is in remission and the radius of the tumor is decreasing at the rate of 1.5 millimeters per month. How fast is the volume of the tumor decreasing when its radius is 12 millimeters?

7. (*INFLATING A BALLOON*) Consider a spherical balloon $\left(V = \frac{4}{3}\pi r^3 \right)$ that is being inflated by helium at the rate of 4 cubic feet per minute. At what rate is the radius increasing when the radius is 2 feet?

8. (*INFLATING A BALLOON*) A child is blowing air into a spherical balloon. The radius of the balloon is increasing at the rate of 2 centimeters per second. What is the rate of change of the volume of the balloon when its radius is 3 centimeters?

9. (*INCREASING AREA*) The length of each side of a square is increasing at the rate of 3 millimeters per second. Determine the rate of increase of the area when the sides are each 24 millimeters.

10. (*SPINNING A WEB*) Assume that a spider is spinning a circular web and suppose that the radius of the web is changing at the rate of 8 millimeters per hour.
   (a) Find the rate at which the area of the web is increasing when the radius is 50 millimeters.
   (b) Find the rate at which the circumference $(C = 2\pi r)$ is increasing when the radius is 50 millimeters.

11. (*MANUFACTURING*) A manufacturer of baseball bats decides to increase production at the rate of 50 bats per day. The total cost of producing x bats is

$$C = 8x + 350 \quad \text{dollars}$$

Determine the rate of change of cost with respect to time.

12. (*MANUFACTURING*) A calculator manufacturer will be increasing production at the rate of 400 calculators per week. Total revenue from the sale of all x calculators produced is

$$R = 18x - .0002x^2 \quad \text{dollars}$$

Determine the rate of change of revenue with respect to time when the weekly production level is 8000 calculators.

13. (*BOTTLE PRODUCTION*) A bottle producer is increasing its production at the rate of 250 bottles per week. The selling price p per bottle is given by the weekly demand function

$$p = 12 - .00015x \quad \text{cents}$$

where x is the number of bottles produced and sold.
   (a) Determine the weekly revenue function R.
   (b) Find the rate of change of revenue with respect to time when the weekly production level is 10,000 bottles.

14. (*PRODUCTION*) A company is increasing production of its product at the rate of 20 units per day. The selling price p per unit is given by the daily demand function

$$p = 340 - .003x \quad \text{dollars}$$

where x is the number of units produced and sold. Determine the rate of change of revenue with respect to time when the daily production level is 3000 units.

15. (*SLIDING LADDER*) A 30-foot ladder is leaning against a building. Suppose the ladder is sliding down the wall in such a way that the bottom of the ladder is moving away from the wall at the rate of 2 feet per second. At what rate is the top of the ladder sliding down the wall when the top of the ladder is 24 feet above the ground?

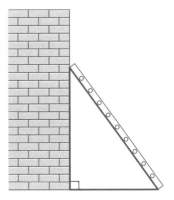

16. (*SLIDING LADDER*) A 20-foot ladder is leaning against a building. Suppose the ladder is sliding down the wall in such a way that the top of the ladder is moving down the wall at the rate of 1.5 feet per second. At what rate is the bottom of the ladder moving away from the wall when it is 16 feet from the wall?

17. *(SAND PILE FORMATION)* Sand poured at the rate of 8 cubic feet per minute is forming a pile in the shape of a cone (see figure).

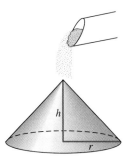

(a) The volume of a cone is $V = \frac{1}{3}\pi r^2 h$, where r is the radius of the base and h is the height. Suppose the height is always the same as the radius in this instance ($h = r$). Make a substitution into the volume equation to produce V as a function of r (that is, with no h in the equation).

(b) Use the assumption and the result of part (a) to determine how fast the radius of the base is increasing when the pile is 10 feet high.

18. *(AIRPLANE FLIGHT)* An airplane is flying at a height of 5 miles and traveling at a speed of 400 miles per hour toward Denver. When its distance (z in the figure) from Denver is 13 miles, how fast is it approaching Denver?

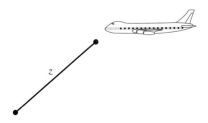

19. *(ROCKET LAUNCH)* A rocket is launched straight up. There is an observation station (see "dot" in figure) 7 miles from the launch site. At what rate is the distance between the rocket and the station increasing when the rocket is 24 miles high and traveling at 2000 miles per hour?

20. *(AUTO TRAVEL)* A city is surrounded by a circular interstate highway with a radius of 10 miles. The path cars travel on it is thus described by $x^2 + y^2 = 100$. Assume that when cars are at the point (8, 6), their x (east-west) coordinate is changing at the rate of -33 miles per hour. How fast is their y (north-south) coordinate changing? (See figure.)

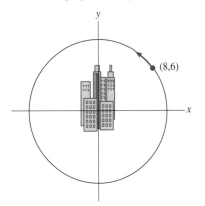

21. *(BEETLE INFESTATION)* Consider Example 2, the beetle infestation.

(a) What are the two related rates—in symbols?

W (b) What are the two related rates—in words?

(c) What equation shows how the two rates of change are related?

3.10 | *DIFFERENTIALS*

The derivative, *dy/dx*, has been defined as

$$\frac{dy}{dx} = \lim_{\Delta x \to 0} \frac{\Delta y}{\Delta x}$$

Since *dy/dx* is the slope of the tangent line and $\Delta y/\Delta x$ is the slope of the secant line, it follows that *for small values of* Δx,

$$\frac{dy}{dx} \approx \frac{\Delta y}{\Delta x}$$

This idea of approximation can be pursued by separating the notation *dy/dx* into two parts, *dy* and *dx*, called **differentials**. Before actually stating the definition of the differentials *dy* and *dx*, let us first see some justification.

By considering *dy* and *dx* to be separable quantities, we can rewrite

$$\frac{dy}{dx} = f'(x)$$

as

$$dy = f'(x)\, dx$$

Figure 18 shows a drawing of the secant and tangent lines to the graph of $y = f(x)$.

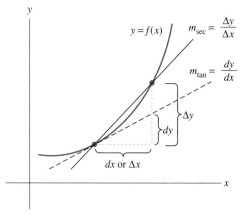

Figure 18 The secant and tangent lines to the graph of a function

The drawing shows that dy $\approx$ Δy *and suggests that dx* $=$ Δx. *We use this geometric justification to define dx. Here then are the definitions of dy and dx.*

> **Differentials**
>
> Let $y = f(x)$ be a function whose derivative exists. Then
>
> $$dx = \Delta x$$
> $$dy = f'(x)\, dx$$
> $$(dy = f'(x)\, \Delta x)$$
>
> It is understood that dx is a small, nonzero real number.

The remainder of the section will offer an opportunity to become familiar with differential notation and to see the use of differentials to obtain approximations. The notation of differentials will be used extensively in our study of integration (Chapter 6).

EXAMPLE 1   Let $y = x^2$. Determine dy.

SOLUTION   *One approach*:

From $y = f(x) = x^2$, we readily obtain

$$f'(x) = 2x$$

Since

$$dy = f'(x)\, dx$$

we have

$$dy = 2x\, dx$$

Another approach:

The derivative is

$$\frac{dy}{dx} = 2x$$

Since dy and dx can be considered separate quantities, we can multiply both sides of this equation by dx. The result is

$$dy = 2x\, dx \quad \blacklozenge$$

EXAMPLE 2   Find dy if $y = (x^2 - 7)^4$.

SOLUTION Differentiating

$$y = (x^2 - 7)^4$$

yields

$$\frac{dy}{dx} = 4(x^2 - 7)^3 \cdot 2x$$

or

$$\frac{dy}{dx} = 8x(x^2 - 7)^3$$

Finally,

$$dy = 8x(x^2 - 7)^3 \, dx \quad \blacklozenge$$

Since $dy \approx \Delta y$, the differential dy gives an approximation to the change in y (that is, Δy) corresponding to a small change in x.

EXAMPLE 3 Let $f(x) = 4x^2 - 9$.

(a) Use differentials to find dy (the approximate change in y) when x changes from 6 to 6.01.

(b) Determine the exact change in y (that is, Δy) when x changes from 6 to 6.01.

SOLUTION (a) First, obtain the derivative.

$$f'(x) = 8x$$

Then, using

$$dy = f'(x) \, dx$$

we have

$$dy = 8x \, dx$$

The x value begins at 6 and changes to 6.01. This means that $x = 6$ and $dx = .01$. Thus,

$$dy = 8(6)(.01) = .48$$

The change in y is approximately .48.

(b) The exact change, Δy, is $f(6.01) - f(6)$.

$$f(6.01) - f(6) = [4(6.01)^2 - 9] - [4(6)^2 - 9]$$
$$= 135.4804 - 135$$
$$= .4804$$

Notice that differentials provide a quick and good approximation—.48 versus .4804. $\blacklozenge$

EXAMPLE 4   ◆   **REVENUE ASSOCIATED WITH ADVERTISING**

The revenue of many corporations is affected by advertising. Suppose that for a particular corporation the function defined below (for $0 \leq x \leq 40$) gives the revenue in *thousands* of dollars that is associated with an expenditure of *x hundred* dollars on advertising.

$$R(x) = -.25x^2 + 20x + 100$$

If the amount being spent on advertising now stands at $1500, use differentials to determine the approximate increase in revenue associated with a $100 increase in advertising.

SOLUTION   The approximate increase in revenue will be the value of dR for $x = 15$ and $dx = 1$. (An x value of 15 represents $1500 and $dx = 1$ represents $100.)

$$\frac{dR}{dx} = -.5x + 20$$

$$dR = (-.5x + 20)\, dx$$
$$= [-.5(15) + 20]\,(1)$$
$$= 12.5$$

Thus, the approximate increase in revenue associated with a $100 increase in advertising is $12,500.

Exercise 25 asks you to show that the exact increase in revenue is $12,250.   ◆

3.10   Exercises

In Exercises 1–10, determine dy.

1. $y = x^3$

2. $y = x^4$

3. $y = x^2 + 3$

4. $y = 3x + 1$

5. $y = x^2 + 5x - 1$

6. $y = x^3 + 7x^2 + 5$

7. $y = (x^2 - 3)^5$

8. $y = (5x + 2)^4$

9. $y = (1 - 9x)^4$

10. $y = (5 - x^2)^3$

In Exercises 11–16, determine dy for the given values of x and dx. (It is understood that $y = f(x)$.)

11. $f(x) = 5x^2$; $x = 4$, $dx = .02$

12. $f(x) = x^2 - 7x$; $x = 3$, $dx = .01$

13. $f(x) = \sqrt{x^2 + 9}$; $x = 4$, $dx = .01$

14. $f(x) = \sqrt{3x}$; $x = 12$, $dx = .03$

15. $f(x) = \dfrac{2x}{x + 1}$; $x = 3$, $dx = .001$

16. $f(x) = \dfrac{x}{x - 1}$; $x = 2$, $dx = .002$

In Exercises 17–22, use differentials to approximate the change in y corresponding to the given change in x. (It is understood that $y = f(x)$.)

17. $f(x) = 3x^2 + 1$; x changes from 5 to 5.01

18. $f(x) = x^3 - 7$; x changes from 4 to 4.02

19. $f(x) = 2x - 5\sqrt{x}$; x changes from 9 to 9.015

20. $f(x) = x + \sqrt{x}$; x changes from 16 to 16.01

21. $f(x) = x - \dfrac{20}{x}$; x changes from 4 to 3.98

22. $f(x) = \dfrac{5x}{x + 1}$; x changes from 3 to 2.99

23. (*ADVERTISING*) Redo Example 4 assuming that the amount now spent on advertising stands at $2300 and a $150 increase is planned.

24. (*REVENUE*) Determine the *exact* increase in revenue for the situation described in Exercise 23.

25. (*REVENUE*) Show that the *exact* increase in revenue for the situation described in Example 4 is $12,250.

26. (*ADVERTISING*) Suppose the revenue in thousands of dollars associated with an expenditure of x hundred dollars on advertising is

$$R(x) = -.5x^2 + 18x + 70 \qquad 0 \le x \le 18$$

Use differentials to determine the approximate decrease in revenue associated with cutting the advertising expenditure from $1100 to $1000.

27. (*DEMAND*) Consider the demand equation given by $p = 20 - .4\sqrt{x}$, where x is the quantity demanded and p is the price per unit. Use differentials to approximate the change in price that would cause the quantity demanded to increase from 100 to 101 units.

28. (*SUPPLY*) Consider the supply equation given by $p = 6 + .01\sqrt{x + 25}$, where x is the quantity supplied and p is the price per unit. Use differentials to approximate the change in price that would cause the quantity supplied to increase from 144 to 145.

29. (*BACTERIA GROWTH*) After t hours, the number of bacteria in a laboratory culture is given by

$$n = 6t^2 + 200$$

Use differentials to approximate the change in the number of bacteria when t changes from 5 hours to 5 hours and 3 minutes.

30. (*COST CUTTING*) A fast food restaurant serves soft drinks in a cylindrical cup that has a radius of 1.5 inches. The volume of soft drink that this cup can hold (using $V = \pi r^2 h$, with $r = 1.5$) is $V = 2.25\pi h$ cubic inches. Ordinarily, the restaurant fills the cup to a height of 8 inches. Suppose that instead they decide to fill the cup to a height of only $7\frac{1}{2}$ inches. Use differentials to approximate the number of cubic inches of soft drink the restaurant saves on each serving.

Chapter List   *Important terms and ideas*

slope of the tangent line
tangent line
derivative
differentiable
differentiation
differentiate
rules for differentiation
power rule
sum rule

difference rule
average rate of change
instantaneous rate of change
instantaneous velocity
marginal cost
marginal revenue
marginal profit
product rule

quotient rule
chain rule
general power rule
second derivative
acceleration
implicit differentiation
related rates
differentials

Review Exercises for Chapter 3

In Exercises 1–6, find the derivative of each function.

1. $f(x) = 2\pi$ **2.** $f(t) = 20t$

3. $f(x) = \dfrac{x^7}{7}$ **4.** $f(x) = \dfrac{2x^7}{5}$

5. $f(t) = \dfrac{1}{4\sqrt{t}}$ **6.** $f(x) = \dfrac{1}{6\sqrt[3]{x}}$

In Exercises 7–8, find the indicated derivative.

7. $\dfrac{d}{dx}(x^2 + 3x)$ **8.** $\dfrac{d}{dx}(7x^3 - x^2)$

Find dy/dx in Exercises 9–12.

9. $y = 6x^{2/3} - 14x$ **10.** $y = 15 - 9x^{1/3}$

11. $y = 1 + x^{-1}$ **12.** $y = 17 - x^{-2}$

In Exercises 13–16, find $D_x y$.

13. $y = x - x^9$ **14.** $y = 5x - x^5 + 2x^7$

15. $y = \dfrac{3}{\sqrt[3]{x}}$ **16.** $y = 15 - \dfrac{4}{\sqrt{x}}$

In Exercises 17–18, find $f'(x)$.

17. $f(x) = x^{5/3} + 14$ **18.** $f(x) = x^{8/5} + 19$

Differentiate each function in Exercises 19–24.

19. $y = (8x)^{1/2}$ **20.** $y = (1 + 4x)^{1/2}$

21. $f(t) = \sqrt[3]{1 + t^2}$ **22.** $g(t) = 1 + \sqrt{t^2 + 1}$

23. $y = \dfrac{1}{\sqrt[3]{1 + x^3}}$ **24.** $y = \dfrac{1}{(x^3 - 2)^4}$

In Exercises 25–28, determine dy/dx.

25. $y = \dfrac{(2x - 1)^4}{x^2 + 4}$ **26.** $y = \dfrac{(7x + 1)^3}{x^2 + 1}$

27. $y = (x^2 + 7)^3(x - 5)^4$ **28.** $y = (2x - 3)^4(3x + 1)^2$

In Exercises 29–30, find y''.

29. $y = \dfrac{2x}{x + 3}$ **30.** $y = x(x^2 + 1)^5$

In Exercises 31–32, use implicit differentiation to find dy/dx.

31. $xy^4 + x^2 - y = 1$ **32.** $y^2 - 3xy + 4y = 0$

In Exercises 33–34, use differentials to approximate the change in y corresponding to the given change in x. (It is understood that $y = f(x)$.)

33. $f(x) = x + 7\sqrt{x}$; x changes from 4 to 4.01

34. $f(x) = \dfrac{3x}{x + 4}$; x changes from 3 to 3.02

35. (RATE OF CHANGE) Determine the rate of change of $f(t) = (t^2 + 5)^3$ with respect to t when $t = 2$.

36. (VELOCITY) A ball is dropped from the top of a tall building and falls such that its distance s (in feet) from the ground after t seconds is $s = 200 - 16t^2$. How fast is the ball falling after 2 seconds?

37. (MARGINAL COST) Let the cost of producing x units be $C(x) = 400 + 35x - .01x^2$ dollars.
    (a) Determine the marginal cost when x is 10.
    (b) Determine the exact cost of producing the 11th unit.
    (c) Find the rate of change of the marginal cost.

38. (MARGINAL PROFIT) If $MR(x) = 140 - .04x$ and $MC(x) = 12 - .02x$, for what value of x is marginal profit zero?

39. (MARGINAL PROFIT) If $R(x) = 6x + .002x^2$ and $C(x) = 150 + 10x + .001x^2$, determine the marginal profit function.

40. (MARGINAL PROFIT) If $C(x) = 300 + 40x - .01x^2$ and $R(x) = 70x - .02x^2$, determine the rate of change of marginal profit.

41. (MARGINAL COST) The cost to manufacture x tables is $C(x) = x^2 + 80x + 150$. At what production level does the marginal cost equal $150?

4

ADDITIONAL APPLICATIONS OF THE DERIVATIVE

Applications of calculus to business, science, and industry are widespread. Our examples and exercises have been chosen to provide a feeling for how the derivative can be used to solve real problems when the situation can be represented by or approximated by a function.

Given a few intuitive guidelines, you will be able to look at the graph of a function and see where the function is increasing, where it is decreasing, where it is maximum, and where it is minimum. This means that you will be able to "read" the graph of a profit function to see where profit is increasing, where it is decreasing, and how to obtain the maximum profit. You will be able to look at the graph of a temperature function and see when a metal is being heated or cooled. Learning how to interpret graphs will be an important experience, one filled with practical applications. A study of how derivatives apply to graphs will also enable you to make many of the same determinations without the benefit of a graph.

4.1 | INCREASING AND DECREASING, GRAPHS, AND CRITICAL NUMBERS

A function is said to be **increasing** when its graph rises as it goes from left to right. A function is **decreasing** when its graph falls as it goes from left to right. See Figure 1.

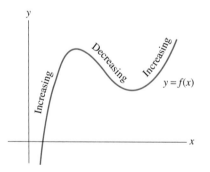

Figure 1 A graph with increasing and decreasing portions

The increasing/decreasing concept can be associated with the slope of the tangent line. After all, the slope of the tangent line to a curve will be positive when the curve is rising and negative when it is falling. Figure 2 shows an illustration.

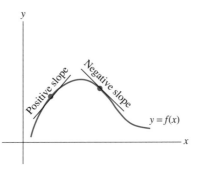

Figure 2 The slope of the tangent line

Since $f'(x)$ is the slope of the tangent line, it follows that if $f'(x) > 0$, then function f is increasing, and if $f'(x) < 0$, then f is decreasing.

Increasing/Decreasing

1. *At a point* (at which f is defined)

   (a) If $f'(a) > 0$, then f is increasing at $x = a$.

   (b) If $f'(a) < 0$, then f is decreasing at $x = a$.

2. *On an interval* (on which f is defined)

   (a) If $f'(x) > 0$ for all x in an interval, then f is increasing on the interval.

   (b) If $f'(x) < 0$ for all x in an interval, then f is decreasing on the interval.

EXAMPLE 1 Consider the function defined by $f(x) = x^2 - 8x + 7$. Where is f increasing and where is it decreasing?

SOLUTION The derivative of $f(x) = x^2 - 8x + 7$ is $f'(x) = 2x - 8$. Thus, $f(x)$ is increasing when $2x - 8 > 0$, that is, when $x > 4$. It can also be said that $f(x)$ is increasing when x is in the interval $(4, \infty)$. Similarly, $f(x)$ is decreasing when $2x - 8 < 0$, that is, when $x < 4$. We can also say that $f(x)$ is decreasing when x is in the interval $(-\infty, 4)$. Figure 3 shows a graph of f. Note that it is decreasing for $x < 4$ and increasing for $x > 4$.

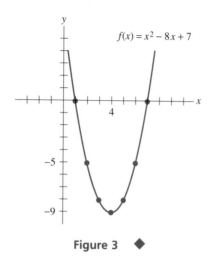

Figure 3 ◆

Although the increasing/decreasing idea was derived by studying graphs, applications of the increasing/decreasing concept extend far beyond graphs. The next four examples offer such applications.

EXAMPLE 2 ◆ *HEATING AND COOLING*

Let $T(x) = -3x^2 + 60x + 70$ be the temperature after x seconds of a metal tray undergoing a chemical finishing process. Determine when the metal is cooling.

SOLUTION The metal is cooling when the temperature $T(x)$ is decreasing. Since a function is decreasing when its derivative is less than zero, $T(x)$ is decreasing when $T'(x) < 0$. From

$$T(x) = -3x^2 + 60x + 70$$

we obtain

$$T'(x) = -6x + 60$$

$T'(x)$ is less than zero when $-6x + 60$ is less than zero. Solving the inequality $-6x + 60 < 0$ leads to

$$x > 10$$

We conclude that the metal is cooling after 10 seconds. (It is heated during the first 10 seconds and cooled after that. Notice that $T'(x) > 0$ for $x < 10$ shows heating for the first 10 seconds.)   ◆

EXAMPLE 3 ◆ *GLUCOSE TOLERANCE TEST*

A glucose tolerance test is used to determine if a person is diabetic. If a certain amount of glucose is eaten or drunk, a normal person's blood sugar will rise somewhat, then fall back to a normal range within 2 hours. A diabetic's blood sugar will start high and continue to get higher throughout the 2-hour interval. See Figure 4.

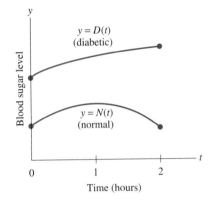

Figure 4 Glucose tolerance test

(a) For what values of t in [0, 2] is $D'(t) > 0$?

(b) For what values of t in [0, 2] is $N'(t) > 0$?

(c) Is $N'(t) > D'(t)$ on [1, 2]? Explain.

SOLUTION **(a)** $D'(t)$ is the slope of the curve given by $y = D(t)$. The slope is positive throughout the interval [0, 2]. Thus, $D'(t) > 0$ for all t in the interval [0, 2]. The function D is increasing throughout the interval.

(b) $N'(t)$ is the slope of the curve given by $y = N(t)$. From the graph, it appears that the slope is positive throughout the interval [0, 1]. (We can't be sure just where the slope becomes negative, but it looks like it changes from positive to negative at about $t = 1$.) Thus, $N'(t) > 0$ for all t in the interval [0, 1]. The function is increasing there.

(c) No, $N'(t)$ is not greater than $D'(t)$ on [1, 2]. The graph of N has a negative slope on [1, 2], whereas the graph of D has a positive slope on [1, 2]. Thus, $N'(t) < D'(t)$ on [1, 2]. ◆

EXAMPLE 4 ◆ *TELEPHONE MANUFACTURER'S PROFIT*

Suppose that $P(x) = -.01x^2 + 60x - 500$ is the profit from the manufacture and sale of x telephones. Is profit increasing or decreasing when 100 phones have been sold?

SOLUTION The derivative is

$$P'(x) = -.02x + 60$$

When $x = 100$, we have

$$P'(100) = -.02(100) + 60 = 58$$

Thus, $P'(100) > 0$, which means that profit is increasing when 100 phones have been sold.   ◆

EXAMPLE 5   ◆   PROFIT, MARGINAL REVENUE AND MARGINAL COST

Show that profit $P(x)$ is increasing when marginal revenue is greater than marginal cost.

SOLUTION   Begin with profit expressed as revenue minus cost.

$$P(x) = R(x) - C(x)$$

Differentiating will introduce marginal revenue $R'(x)$ and marginal cost $C'(x)$.

$$P'(x) = R'(x) - C'(x)$$

The idea that "marginal revenue is greater than marginal cost" translates into $R'(x) > C'(x)$. Adding $-C'(x)$ to both sides of this inequality leads to the inequality $R'(x) - C'(x) > 0$. Since $P'(x)$ is the same as $R'(x) - C'(x)$, we have, by substitution, $P'(x) > 0$. And $P'(x) > 0$ means that $P(x)$ is increasing. Thus, we have shown that profit is increasing when marginal revenue is greater than marginal cost.

For another perspective, note that if marginal revenue exceeds marginal cost, then the revenue generated by the sale of the next unit will be more than the cost of making the next unit. As a result, there will be a (positive) profit on the production and sale of the next unit. Thus, marginal profit is positive when marginal revenue is greater than marginal cost.   ◆

Assuming that the graph of a function is continuous at the point where the function changes from increasing to decreasing, that point is called a **relative maximum point**. This concept is illustrated in the two graphs of Figure 5. Similarly, if the graph is continuous at the point where a function changes from decreasing to increasing, that point is called a **relative minimum point**. This idea is illustrated in the two graphs of Figure 6. The points are called *relative* maximum or minimum points because they are the highest or lowest points compared with other points "nearby." There may be higher or lower points

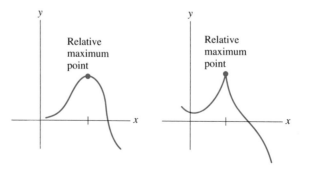

Figure 5 Relative maximum points

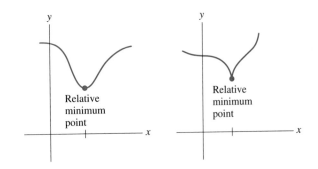

Figure 6 Relative minimum points

elsewhere on the graph, however. Figure 7 shows a graph with a relative maximum point and higher points elsewhere, away from the relative maximum point.

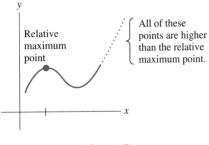

Figure 7

The graph will be "rounded" at the relative maximum or minimum point if $f'(x) = 0$ *there.* The slope of the tangent line is zero at such points. The tangent line is horizontal. See Figure 8. *The graph will be "pointed" at the relative maximum or minimum point when* $f'(x)$ *is not defined there.* The slope of the tangent line is undefined at such points. The tangent line either is vertical or else

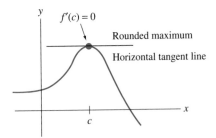

Figure 8 A "rounded" relative maximum

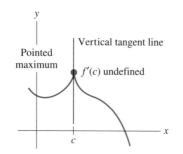

Figure 9 A "pointed" relative maximum

does not exist. See Figure 9. Exercise 89 offers an opportunity to consider an example of a function with a pointed relative maximum where the tangent line does not exist. Both Exercise 89 and the graph shown in Figure 9 suggest the following result.

If function f is continuous at $x = a$, it is not necessarily differentiable at $x = a$.

The derivative will fail to exist where there is a vertical tangent (slope undefined) or where there is a sharp corner or abrupt turn in the graph.

In order to sketch graphs of functions and to solve various applied problems, it will be very helpful to be able to determine relative maximum and relative minimum points. This means you will be looking for x such that $f'(x) = 0$ or $f'(x)$ is undefined. Consequently, the following definition will be useful.

Critical Number

Suppose f is defined at some real number c; that is, $f(c)$ exists. If $f'(c) = 0$ or $f'(c)$ is undefined, then c is called a **critical number** of function f.

The critical numbers are the candidates (the only *possible* numbers) that can lead to relative maximum and relative minimum points. Only when $f'(x)$ is zero or undefined can the graph of the function have a relative maximum or minimum point. (Recall that if $f'(x) > 0$, the graph is increasing, and if $f'(x) < 0$, it is decreasing. The graph cannot have a relative maximum or minimum point where it is increasing or decreasing.)

EXAMPLE 6 Determine the critical numbers of $f(x) = x^3 + 3x^2 - 24x + 17$.

SOLUTION   The function is defined for every real number x. The derivative of the function is $f'(x) = 3x^2 + 6x - 24$, and it is never undefined. But $f'(x)$ may be zero. Specifically, $f'(x) = 0$ when

$$3x^2 + 6x - 24 = 0$$

or, after dividing both sides by 3,

$$x^2 + 2x - 8 = 0$$

Next

$$(x + 4)(x - 2) = 0$$

or

$$x = -4, 2$$

Thus, $f'(x) = 0$ when $x = -4$ or $x = 2$. The numbers -4 and 2 are the critical numbers of the function $f(x) = x^3 + 3x^2 - 24x + 17$.   ◆

> *Note*
>
> If the quadratic expression of Example 6 could not be factored, then the quadratic formula would have been needed to solve the equation.

> *Note*
>
> If the roots of the quadratic equation being solved had contained the square root of a negative number, then there would have been no critical numbers derived from the equation. That is because all of the functions here are functions of real numbers, and the square root of a negative number is not a real number.

EXAMPLE 7   Find the critical numbers of $f(x) = \sqrt[3]{x}$.

SOLUTION   From

$$f(x) = \sqrt[3]{x} = x^{1/3}$$

we obtain

$$f'(x) = \frac{1}{3}x^{-2/3} = \frac{1}{3x^{2/3}}$$

The derivative is not defined at 0, since the denominator of $f'(x)$ is 0 when $x = 0$. Yet function f is defined at 0. Thus, 0 is a critical number of the function.

Considering $f'(x) = 0$ for other possible critical numbers, we see that $f'(x)$ always has 1 in the numerator and thus can never be zero.

$$\frac{1}{3x^{2/3}} \quad \text{can never be zero}$$

We conclude that there are no additional critical numbers. This means that 0 is the only critical number of f. ◆

Note

Keep in mind that a fractional expression is equal to zero only when the numerator is zero and the denominator is not zero.

$$\frac{\textbf{zero}}{\textbf{nonzero}} = \textbf{zero}$$

EXAMPLE 8   Determine the critical numbers of the function defined by

$$f(x) = \frac{x^2}{x - 3}$$

SOLUTION   Upon differentiating, we obtain

$$f'(x) = \frac{(x - 3)(2x) - x^2(1)}{(x - 3)^2}$$

$$= \frac{2x^2 - 6x - x^2}{(x - 3)^2}$$

or

$$f'(x) = \frac{x^2 - 6x}{(x - 3)^2}$$

For critical numbers, note that $f'(x) = 0$ when the numerator $x^2 - 6x$ is zero.

$$x^2 - 6x = 0$$
$$(x)(x - 6) = 0$$
$$x = 0, x = 6$$

We now have $f'(0) = 0$ and $f'(6) = 0$, *and* the original function f is defined at 0 and at 6. This means that 0 and 6 are critical numbers of the function.

We must also consider where $f'(x)$ is undefined. A quick look at the derivative

$$f'(x) = \frac{x^2 - 6x}{(x - 3)^2}$$

shows that $f'(x)$ is undefined when $x = 3$, since division by zero would be implied then, and division by zero is not defined. *However*, since the original function f is also undefined at 3, the number 3 cannot be a critical number. The only critical numbers of the function are 0 and 6. ◆

In the next section we will formalize some tests for determining whether a particular critical number obtained is associated with a relative maximum point, a relative minimum point, or neither.

4.1 Exercises

In Exercises 1–10, study each graph to determine the intervals on which the function f is (a) increasing and (b) decreasing. Use interval notation.

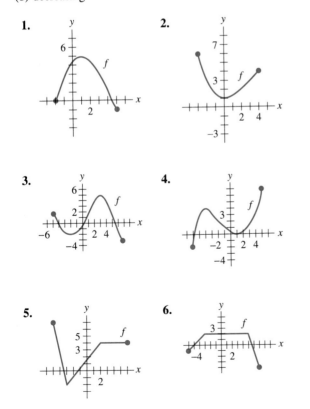

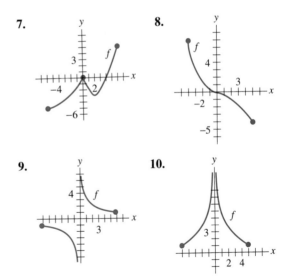

Determine where each function in Exercises 11–20 is increasing and where it is decreasing. Write your answers in inequality notation.

11. $f(x) = x^2 - 6x + 19$ **12.** $f(x) = x^2 - 14x + 100$

13. $g(x) = x^2 + 5x + 3$ **14.** $g(x) = x^2 + 4x + 20$

15. $f(x) = 1 - 4x^2$ **16.** $f(x) = 5x^2 + 42$

17. $m(x) = 10x - x^2$ **18.** $m(x) = 12x - 3x^2$

19. $f(x) = 50 + 6x - .02x^2$

20. $f(x) = 200 + 30x - .01x^2$

Determine where each function in Exercises 21–32 is increasing and where it is decreasing. Write your answers in inequality notation.

21. $f(x) = \dfrac{1}{x}$ Hint: $x^2 > 0$ for $x > 0$ and for $x < 0$.

22. $f(x) = \dfrac{1}{x^2}$ **23.** $f(x) = x^{3/2}$

24. $f(x) = x^{5/2}$ **25.** $f(x) = \sqrt{x}$

26. $f(x) = \sqrt[3]{x}$ **27.** $f(x) = \dfrac{x + 1}{x}$

28. $f(x) = \dfrac{x - 1}{x}$ **29.** $f(x) = \dfrac{x}{x + 1}$

30. $f(x) = \dfrac{x}{x - 1}$ **31.** $f(x) = x^3 - 12x + 1$

32. $f(x) = x^3 - 27x + 5$

33. (COOLING) Let $T(x) = -2x^2 + 64x + 65$ be the temperature after x seconds of a metal plate undergoing a chemical finishing process. Determine when the metal is cooling.

34. (HEATING) When is the metal of Exercise 33 being heated?

35. (ROCKET LAUNCH) A rocket is launched vertically so that its distance s (in feet) from the ground at any time t (in seconds) is

$$s(t) = -16t^2 + 800t$$

(a) When is the rocket rising?
(b) When does the rocket strike the ground? In other words, when is $s(t)$ equal to zero?
(c) When is the rocket falling?

36. (ROCKET LAUNCH) Redo Exercise 35 assuming that the distance from the ground is given instead by

$$s(t) = -16t^2 + 448t$$

37. (COST) Suppose it costs $C(x) = 36x - .02x^2$ dollars for a company to produce a total of x TV antennas $(0 \le x \le 1200)$. Determine when the total cost is increasing.

38. (REVENUE) Let $R(x) = 8x + .03x^2$ dollars be the revenue from the sale of x calculators. Show that revenue is always increasing.

For each function defined in Exercises 39–44, determine whether it is increasing or decreasing at the given point.

39. $f(x) = x^2 - 7x + 50$ at $(1, 44)$

40. $f(x) = x^3 - 8x$ at $(3, 3)$

41. $f(x) = 10 - \dfrac{1}{x}$ at $(-1, 11)$

42. $f(x) = \sqrt{1 - x}$ at $(-3, 2)$

43. $f(x) = \dfrac{1 - x}{x}$ at $(2, -1/2)$

44. $f(x) = x^{2/3}$ at $(8, 4)$

45. (AVERAGE COST) Since $C(x)$ is the total cost of x units, the average cost per unit is $C(x)$ divided by the number of units x.

$$\bar{C}(x) = \frac{C(x)}{x} = \text{average cost per unit}$$

(a) For the cost function in Exercise 37, determine the average cost function $\bar{C}(x)$.
(b) For what values of x is the average cost decreasing?

46. (PROFIT) Use the approach of Example 5 to show that profit is decreasing when marginal revenue is less than marginal cost.

47. (REVENUE) If revenue from the sale of x units is $80x - .01x^2$ dollars, is revenue increasing or decreasing when 600 units have been sold?

48. (PROFIT) Let $P(x) = 1500 - \dfrac{1200}{x}$ dollars for the sale of x units.
(a) Is $P(x)$ increasing or decreasing for $x \ge 1$?
(b) Make a table showing x and $P(x)$ values for $x = 1, 2, 3, 4, 10, 100,$ and 1200.
(c) Use limits to determine the number of dollars toward which profit tends as the number of units sold gets greater and greater.

W 49. (*BEHAVIOR MODIFICATION*) Psychologists have determined that behavior modification is a better approach to dieting than is crash dieting (see figure).

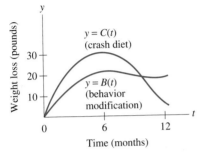

Comparison of crash dieting and behavior modification

(a) Is $B(t) > C(t)$ on [0, 6]? Explain.
(b) Is $B'(t) > C'(t)$ on [0, 6]? Explain.
(c) Based on the graph, explain why behavior modification is considered better than the crash diet approach.

W 50. (*CRIME STUDY*) A sociologist has gathered data showing the number of crimes reported each month of the year in a northern U.S. city. To get a better picture, she plotted the 12 points and then passed a curve $y = c(t)$ through them. Answer the following questions based on her graph, which is shown below.
(a) Is $c'(t)$ positive, negative, or zero on [3, 6]? Interpret your answer from a crime perspective.
(b) Is $c'(t) < 0$ on [9, 12]? What does this mean in terms of crime?

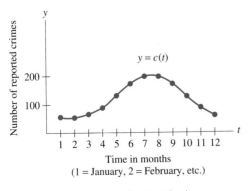

Crime in a northern U.S. city

Determine the critical numbers of each function given in Exercises 51–64.

51. $f(x) = x^2 - 16x$ **52.** $f(x) = x^2 - 7x + 3$

53. $f(x) = 5x^2 + 3x - 2$ **54.** $f(x) = 5x^2$

55. $f(x) = x^3 - 2$ **56.** $f(x) = 1 - x^4$

57. $f(x) = 2x^3 - 27x^2 + 84x + 5$

58. $f(x) = x^3 + x^2 - 8x - 7$

59. $f(x) = x^3 + x^2 - 5x + 2$

60. $f(x) = x^3 - 27x$

61. $f(x) = x^3 - 6x + 1$

62. $f(x) = x^3 - 15x - 7$

63. $f(x) = x^3 + 6x^2 + 3x + 1$

64. $f(x) = x^3 + \frac{1}{2}x^2 - x + 19$

Determine the critical numbers of each function given in Exercises 65–80.

65. $f(x) = x^{1/3} + 6$ **66.** $f(x) = 6x^{2/3}$

67. $f(x) = \dfrac{1}{\sqrt{x}}$ **68.** $f(x) = x^{-1/4}$

69. $f(x) = x^3 + 3x^2 - 15x - 9$

70. $f(x) = x^3 - 3x^2 - 18x + 7$

71. $f(x) = \dfrac{x^2 - 3}{x + 2}$ **72.** $f(x) = \dfrac{x^2 + 3}{x - 4}$

73. $f(x) = \dfrac{1}{x}$ **74.** $f(x) = \dfrac{1}{x^2}$

75. $f(x) = \dfrac{x^2}{x - 1}$ **76.** $f(x) = \dfrac{x + 2}{x^2 + 5}$

77. $f(x) = (3x - 1)(2x + 3)^5$

78. $f(x) = (2x + 1)(x + 5)^3$

79. $g(x) = (4x + 1)(x^2 - 4)^2$

80. $g(x) = (5x - 1)(3x + 2)^3$

Determine where each function given in Exercises 81–86 is increasing and where it is decreasing.

81. $f(x) = \dfrac{1}{x - 2}$ **82.** $f(x) = \dfrac{2}{x^2 + 3}$

83. $f(x) = \sqrt{x^2 + 5}$

84. $f(x) = x^3 - 3x^2 + 3x - 2$

85. $f(x) = x^4 + 4x^3 - 7$

86. $f(x) = x\sqrt{2x + 1}$

Determine the critical numbers of each function in Exercises 87–88.

87. $f(x) = x^4 + 8x^3 + 4x^2 + 15$

88. $f(x) = x\sqrt{4x + 5}$

89. The tangent line to the graph of f (see figure) does not exist at $x = 3$.

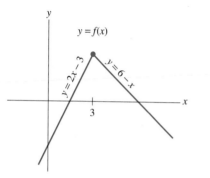

(a) Determine the slope of the tangent line to the left of 3.

(b) Determine the slope of the tangent line to the right of 3.

(c) What is the slope of the tangent line at 3?

90. Consider the function graphed in the figure below. Determine whether the function is (i) continuous and (ii) differentiable at each value of x (a, b, c, d, e).

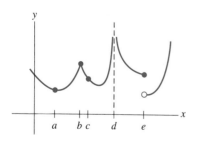

91. Consider the function graphed in the figure below. Determine whether the function is (i) continuous and (ii) differentiable at each value of x (a, b, c, d, e).

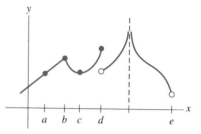

W 92. If $f'(2)$ is undefined, is 2 a critical number of f? Explain.

W 93. If $f'(3) = 0$, is 3 a critical number of f? Explain.

W 94. There are functions that have no critical numbers. If you were testing a function f, how would you conclude that it has no critical numbers? Explain in words.

W 95. Explain in words how you would proceed to determine the critical numbers of a function g.

W 96. Is it possible for $f(x)$ to be negative at some x in an interval on which f is increasing? Explain.

W 97. Explain why critical numbers are the only possible numbers that can lead to relative maximum or relative minimum points. Include in your explanation some consideration of why x values for which $f'(x) > 0$ or $f'(x) < 0$ cannot lead to relative maximum or relative minimum points.

The Exercise Library at the back of the book contains graphing calculator and computer exercises keyed to this section.

4.2 | *RELATIVE EXTREMA AND CURVE SKETCHING*

Some important ideas about relative maximum and relative minimum points were suggested in the previous section. These ideas are organized here and presented as two rules. Together, the rules are known as the **first derivative test**. Briefly, the ideas are:

1. A relative maximum point of f occurs at $(c, f(c))$ if the function changes from increasing to decreasing there.

2. A relative minimum point of f occurs at $(c, f(c))$ if the function changes from decreasing to increasing there.

First Derivative Test

1. Let c be a critical number of f and let f be continuous on an interval containing c. Then $(c, f(c))$ is a *relative maximum point* provided that $f'(x) > 0$ in an interval to the left of c and $f'(x) < 0$ in an interval to the right of c.

2. Let c be a critical number of f and let f be continuous on an interval containing c. Then $(c, f(c))$ is a *relative minimum point* provided that $f'(x) < 0$ in an interval to the left of c and $f'(x) > 0$ in an interval to the right of c.

Note

In practice, a critical number is tested by evaluating the derivative at an x value on each side of the critical number (one x value less than c and one x value greater than c). *Be careful that the interval that includes the x values you select does not include any other critical numbers.*

Note

Not every critical number is associated with a relative maximum or relative minimum. *The derivative can be zero or undefined without there being a relative maximum or relative minimum.* This fact will be illustrated in Example 3.

The expression **relative extremum** of f is often used to refer to either a relative maximum or a relative minimum value of f. The **relative extrema** (plural) of a function include all relative maxima and all relative minima. A point associated with a relative extremum is called a **relative extreme point**.

EXAMPLE 1 Determine all relative extreme points (if any) of the function and sketch a graph.

$$f(x) = x^3 - 3x^2 + 1$$

SOLUTION We seek all relative maximum points and all relative minimum points of the function. Begin by obtaining the derivative,

$$f'(x) = 3x^2 - 6x$$

Since this particular derivative can never be undefined, the only critical numbers will be x for which $f'(x) = 0$. To find them, solve

$$3x^2 - 6x = 0$$

The expression can be factored and the equation solved as

$$(3x)(x - 2) = 0$$
$$x = 0, 2$$

Thus, the critical numbers of f are 0 and 2. Next, apply the first derivative test to the critical numbers. Beginning with 0, evaluate the derivative at an x value on each side of critical number 0. We have decided to evaluate the derivative at -1 and 1. (We were careful not to choose 2 or any number to the right of 2, since 2 is also a critical number.)

$$f'(-1) = +9$$
$$f'(0) = 0$$
$$f'(1) = -3$$

$f' + + + + +$ $f' - - - - -$
f increasing | f decreasing

$x = 0$

Since $f'(x)$ is positive (f increasing) to the left of critical number 0 and negative (f decreasing) to the right, the point $(0, f(0))$ must be a relative maximum point.

Next, apply the first derivative test to critical number 2. Test $f'(x)$ at 1 and 3. (Keep in mind that you cannot use 0 or any number to the left of 0, since 0 is also a critical number.)

$$f'(1) = -3$$
$$f'(2) = 0$$
$$f'(3) = +9$$

$f' - - - - -$ $f' + + + + +$
f decreasing | f increasing

$x = 2$

Since $f'(x)$ is negative (f decreasing) to the left of critical number 2 and positive (f increasing) to the right, the point $(2, f(2))$ is a relative minimum point.

We now return to the original function, $f(x) = x^3 - 3x^2 + 1$, to find the y coordinate of each relative extreme point. Substitute 0 for x to determine that $(0, f(0))$ is in fact $(0, 1)$. Also, substitute 2 for x to determine that $(2, f(2))$ is $(2, -3)$. Thus, $(0, 1)$ is a relative maximum point and $(2, -3)$ is a relative minimum point. The graph will be rounded (rather than pointed) at these relative extreme points because the derivative is equal to zero there. Using this information, the graph thus far appears as shown in Figure 10.

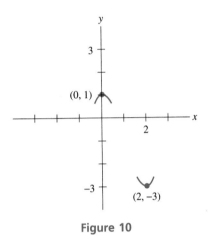

Figure 10

We should now determine a few more points in order to complete the sketch. Good choices for x, after examining the graph thus far, would be 1, 3, and -1. These x values are integers, and they are close to the critical numbers 0 and 2. (If more points are needed after that, you might let x be 4 and -2.) Using 1, 3, and -1 for x yields

x	$f(x)$
1	-1
3	1
-1	-3

After these three points are plotted, the graph appears as shown in Figure 11.

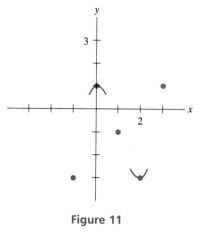

Figure 11

Connecting the points produces a reasonable sketch. See Figure 12.

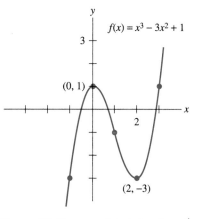

Figure 12 The completed sketch ◆

EXAMPLE 2   Determine all relative maximum and relative minimum points (if any) and sketch a graph.

$$f(x) = x^{2/3} + 1$$

SOLUTION   Begin by obtaining the derivative.

$$f'(x) = \frac{2}{3}x^{-1/3} = \frac{2}{3\sqrt[3]{x}}$$

Clearly $f'(x)$ is never zero (since the numerator, 2, is never zero). But $f'(x)$ is *undefined* when $x = 0$. And since f is defined at 0, this means that 0 is a critical number of the function. The first derivative test, shown next, indicates that f has a relative minimum at $x = 0$.

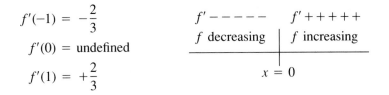

$$f'(-1) = -\frac{2}{3}$$

$$f'(0) = \text{undefined}$$

$$f'(1) = +\frac{2}{3}$$

When $x = 0$, $y = f(0) = 1$. Thus, the point $(0, 1)$ is a relative minimum point. The graph will be pointed (rather than rounded) at the relative minimum point $(0, 1)$ because the derivative is undefined there. See Figure 13.

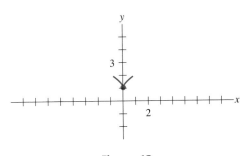

Figure 13

In order to complete the sketch, obtain a few additional points. Considering that the function involves the cube root of x, perfect cubes such as ± 1 and ± 8 would be good choices for x. (Of course, with a calculator at hand, numbers such as ± 3, ± 4, and ± 5 would be all right, too.)

x	y
1	2
−1	2
8	5
−8	5

The graph is shown in Figure 14.

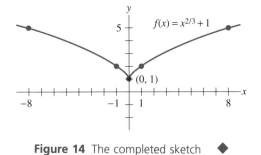

Figure 14 The completed sketch   ◆

EXAMPLE 3   Study the two functions given next for extreme points.

   (a) $f(x) = x^3 + 2$ **(b)** $f(x) = x^{1/3}$

SOLUTION   **(a)** For $f(x) = x^3 + 2$, the derivative is $f'(x) = 3x^2$. Clearly $f'(x) = 3x^2$ is equal to zero when $x = 0$. Since f is defined when $x = 0$, it means that 0 is a critical number (the only critical number of this function). The first derivative test:

$$f'(-1) = +3$$
$$f'(0) = 0$$
$$f'(1) = +3$$

$$\begin{array}{c|c} f' + + + + + & f' + + + + + \\ f \text{ increasing} & f \text{ increasing} \\ \hline & \\ & x = 0 \end{array}$$

Since there is no change in the sign of the derivative, there is no relative maximum or minimum point at $(0, f(0))$. The graph of f is shown in Figure 15.

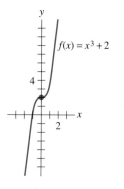

Figure 15

(b) In Example 7 of Section 4.1, we had $f(x) = x^{1/3}$ and $f'(x) = \dfrac{1}{3x^{2/3}}$. It was shown then that the only critical number is 0. We now apply the first derivative test to 0.

$$f'(-1) = +\frac{1}{3}$$

$$f'(0) = \text{undefined}$$

$$f'(1) = +\frac{1}{3}$$

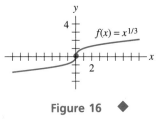

Since there is no change in the sign of the derivative, there is no relative extremum. The graph of f is shown in Figure 16.

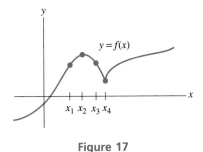

Figure 16 ◆

We end this section with a graphical example showing all four possibilities for the derivative of a function—positive, negative, zero, and undefined.

EXAMPLE 4 Consider the graph of f given next (Figure 17).

Figure 17

Since at x_1 the function is increasing, it follows that $f'(x_1) > 0$. At x_2 the slope of the tangent line is zero, so $f'(x_2) = 0$. At x_3 the function is decreasing, so $f'(x_3) < 0$. The pointed relative minimum at x_4 means that $f'(x_4)$ does not exist. ◆

4.2   Exercises

Determine all relative maximum and all relative minimum points of each function in Exercises 1–16. *Do not sketch graphs.*

1. $f(x) = 15 + 6x - x^2$ **2.** $f(x) = 50 - 8x - x^2$

3. $f(x) = x^2 - 20x$ **4.** $f(x) = x^2 - 9x$

5. $f(x) = x^3 - 3x - 2$ **6.** $f(x) = x^3 - 75x + 14$

7. $f(x) = x^3 - 6x^2$ **8.** $f(x) = x^3 - 27x^2$

9. $f(x) = 12x - x^3$ **10.** $f(x) = 48x - x^3$

11. $f(x) = x^3 + 3x^2 - 9x - 2$

12. $f(x) = x^3 - 9x^2 - 21x + 19$

13. $f(x) = 2x^3 + 21x^2 - 48x + 7$

14. $f(x) = 2x^3 - 3x^2 - 36x + 1$

15. $f(x) = 2x^3 + 3x^2 - 36x + 5$

16. $f(x) = 2x^3 - 12x^2 - 72x + 100$

Obtain all relative extreme points and sketch the graph of each function in Exercises 17–40.

17. $f(x) = x^2 - 8x + 9$ **18.** $f(x) = x^2 - 6x + 10$

19. $f(x) = -x^2 - 6x - 4$ **20.** $f(x) = -x^2 - 8x - 7$

21. $f(x) = 2x^2 + 12x + 11$ **22.** $f(x) = 2x^2 + 12x + 13$

23. $f(x) = -2x^2 + 4x - 1$ **24.** $f(x) = -2x^2 + 8x - 3$

25. $f(x) = x^3 - 3x + 5$ **26.** $f(x) = x^3 - 12x + 2$

27. $f(x) = -x^3 + 12x - 1$ **28.** $f(x) = -x^3 + 3x + 4$

29. $f(x) = x^3 - 6x^2 + 15$ **30.** $f(x) = x^3 - 3x^2 - 1$

31. $f(x) = -x^3 - 3x^2 + 7$ **32.** $f(x) = -x^3 - 6x^2 + 18$

33. $f(x) = 2x^3 - 3x^2 - 12x + 8$

34. $f(x) = 2x^3 + 3x^2 - 12x - 5$

35. $f(x) = 6x^{2/3}$ **36.** $f(x) = 3x^{2/3} + 1$

37. $f(x) = 1 - 3x^{2/3}$ **38.** $f(x) = 5 - 6x^{2/3}$

39. $f(x) = (x - 3)^{2/3}$ **40.** $f(x) = (x + 1)^{2/3}$

In Exercises 41–54, determine all relative maximum and relative minimum points of the function, if any. *Do not sketch the graph of the function.*

41. $f(x) = x^3 + 6x - 19$ **42.** $f(x) = 5x^3 + 2x + 4$

43. $f(x) = 10 - 3x^{2/3}$ **44.** $f(x) = x^{2/3} + 5$

45. $f(x) = x^3$ **46.** $f(x) = 1 - 4x^3$

47. $f(x) = 5x - 11$ **48.** $f(x) = -4x + 1$

49. $f(x) = \dfrac{1}{x}$ **50.** $f(x) = \dfrac{1}{x^2}$

51. $f(x) = (2x + 1)(x + 3)^4$

52. $f(x) = (3x - 7)(x + 1)^3$

53. $f(x) = 3x^{4/3}$ **54.** $f(x) = 4x - 3x^{4/3}$

For each graph in Exercises 55–58, determine whether $f'(x_1)$, $f'(x_2)$, $f'(x_3)$, and $f'(x_4)$ are positive, negative, zero, or undefined. (See Example 4.)

55.

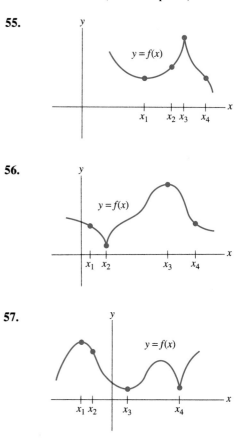

56.

57.

58.

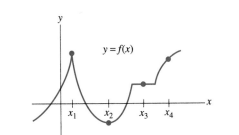

Find all relative extreme points and sketch the graph of each function in Exercises 59–61.

59. $f(x) = 4x^2 - 8x^4$

60. $f(x) = x^{5/3} - \dfrac{5}{2}x^{2/3} - 1$

61. $f(x) = 3x^{2/3} - x$

W 62. The *note* after the statement of the *first derivative test* says "Be careful that the interval . . . does not include any other critical numbers." Explain and justify this note.

W 63. What is the difference between *a relative maximum function value* and *a relative maximum point*?

W 64. Suppose there is a function with exactly one critical number (call it c). Explain how you would use the first derivative test to determine whether the critical number is associated with a relative maximum, a relative minimum, or neither.

65. Refer to Example 3.
   (a) What word describes the tangent line to the graph of $f(x) = x^3 + 2$ at $(0, 2)$?
   (b) What word describes the tangent line to the graph of $f(x) = x^{1/3}$ at $(0, 0)$?

The Exercise Library at the back of the book contains graphing calculator and computer exercises keyed to this section.

| **4.3** | *CONCAVITY, THE SECOND DERIVATIVE TEST, AND CURVE SKETCHING* |

Another useful graphing concept is that of **concavity**. Although concavity can also be applied to rate-of-change situations, for now our consideration will be purely graphical. To begin, consider the graph of f shown in Figure 18.

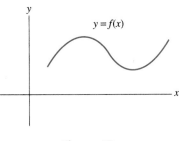

Figure 18

Visually, the graph can be split into two parts.

and

The graph on the left is described as being **concave down**. Casually speaking, if it were filled with water, the water would spill out because it is upside down. The graph on the right is described as **concave up**, and it would hold water.

In some cases, you may feel a need to imagine the graph extended in order to determine the concavity.

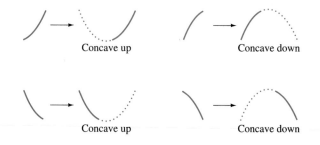

A more technical way of distinguishing concave up from concave down is as follows. A curve is *concave up* on an interval if it lies above its tangent line at every point of the interval. Similarly, a curve is *concave down* on an interval if it lies below its tangent line at every point of the interval. The drawings shown in Figure 19 serve to illustrate this concept.

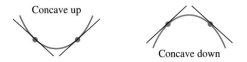

Figure 19 Concavity and the tangent line

EXAMPLE 1   Indicate where the graph in Figure 20 is concave up and where it is concave down.

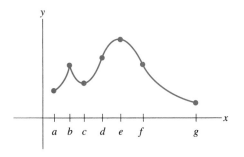

Figure 20

SOLUTION Proceeding from left to right, we see that the graph is concave up over the intervals (a, b) and (b, d). It is then concave down over the interval (d, f) and concave up over the interval (f, g). ◆

When the graph of $y = f(x)$ is concave down, f' is a decreasing function. This can be seen in Figure 21. Look at the left portion of the graph. Notice the slopes of the tangent lines (that is, $f'(x)$) going from large positive values, to small positive values, to zero, and then negative. Accordingly, f' is a decreasing function on the interval being considered, which means that f'' is negative, since f'' is the rate of change of f'.

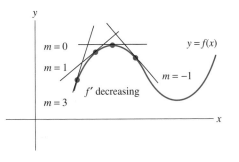

Figure 21 Concave down

A similar argument, made using the right portion of the graph, will show that f'' is positive when the graph is concave up. This study of concavity can be organized into a useful rule.

Test for Concavity

The graph of f is

1. *Concave down* where $f''(x) < 0$.

2. *Concave up* where $f''(x) > 0$.

EXAMPLE 2 Determine where the graph of $f(x) = x^3 - 3x^2 + 1$ is concave down and where it is concave up.

SOLUTION We begin by obtaining $f'(x)$ and $f''(x)$.

$$f'(x) = 3x^2 - 6x$$
$$f''(x) = 6x - 6$$

The graph of f is concave down where $f''(x) < 0$, that is, where $6x - 6 < 0$. We can solve this inequality as

$$6x - 6 < 0$$
$$6x < 6$$
$$x < 1$$

Thus, f is concave down for $x < 1$.

The graph of f is concave up where $f''(x) > 0$, that is, where $6x - 6 > 0$. And $6x - 6 > 0$ when $x > 1$. Thus, f is concave up for $x > 1$.

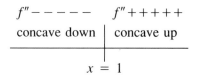

The chart displays our result: f is concave down when $x < 1$ and concave up when $x > 1$.   ◆

Example 2 shows a function whose graph is concave down for $x < 1$ and concave up for $x > 1$. This means that at $x = 1$ the concavity changes from downward to upward. The point $(1, -1)$, where the concavity changes, is given a special name.

> **Point of Inflection**
>
> Any point at which the graph of a continuous function changes concavity is called a **point of inflection**.

Thus, $(1, -1)$ is a point of inflection of the graph of $f(x) = x^3 - 3x^2 + 1$. A graph of this function is shown in Figure 22.

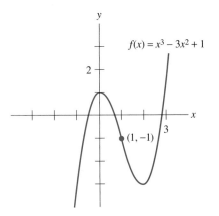

Figure 22 $(1, -1)$ is a point of inflection

In general, *once you have determined the concavity of a graph, it is an easy matter to obtain points of inflection.* For example, with $f(x) = x^3 - 3x^2 + 1$, it was determined that the graph is concave down for $x < 1$ and concave up for $x > 1$. It then follows that there is a point of inflection at $x = 1$, since the function is continuous there.

It is interesting to observe that since $f''(x) < 0$ when the graph is concave down and $f''(x) > 0$ when the graph is concave up, it follows that at a point of inflection $f''(x) = 0$ or else $f''(x)$ does not exist. (After all, if f'' isn't negative or positive, then it must be zero or undefined.) However, this *does not mean* that there is necessarily a point of inflection at $(x, f(x))$ just because $f''(x) = 0$ or just because $f''(x)$ does not exist. There must also be a change of concavity at the point. We see illustrated in Figure 23 that f'' can be zero or fail to exist at a relative minimum or relative maximum point.

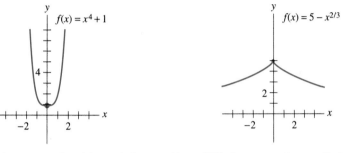

Here $f''(0) = 0$, yet $(0, 1)$ is a relative minimum, not a point of inflection.

Here $f''(0)$ does not exist, yet $(0, 5)$ is a relative maximum, not a point of inflection.

Figure 23

The study of concavity suggests a new test for relative maximum and relative minimum, a test that uses the second derivative. Since $f''(x) < 0$ when the graph is concave down, it follows that if $f''(c) < 0$ for the critical number c, then $(c, f(c))$ is at the "top" of a concave down region and is therefore a relative maximum point. Similarly, since $f''(x) > 0$ where the graph is concave up, it follows that if $f''(c) > 0$ for the critical number c, then $(c, f(c))$ is at the "bottom" of a concave up region and is therefore a relative minimum point. See Figure 24.

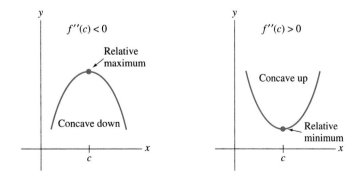

Figure 24 The second derivative and concavity in reference to relative extrema

Thus, we have the **second derivative test** for relative extrema.

Second Derivative Test

Assuming c is a critical number of f,

1. If $f''(c) < 0$, then $(c, f(c))$ is a relative maximum point.

2. If $f''(c) > 0$, then $(c, f(c))$ is a relative minimum point.

Note: If $f''(c) = 0$ or if $f''(c)$ is undefined, you can draw no conclusion from this test. You must use the first derivative test instead.

EXAMPLE 3   Determine the relative maximum and relative minimum points, if any, of the function defined by $f(x) = x^3 - 6x^2 + 9x + 4$. Use the second derivative test.

SOLUTION   From $f(x) = x^3 - 6x^2 + 9x + 4$, we obtain the derivative

$$f'(x) = 3x^2 - 12x + 9$$

The critical numbers are x such that $f'(x) = 0$. We proceed accordingly.

$$3x^2 - 12x + 9 = 0$$
$$3(x^2 - 4x + 3) = 0$$
$$3(x - 3)(x - 1) = 0$$
$$x = 3, 1$$

The critical numbers are 3 and 1. To use the second derivative test, obtain $f''(x) = 6x - 12$. Then evaluate f'' at 3 and at 1.

$$f''(3) = 6(3) - 12 = +6$$

Since $f''(3) > 0$, the point $(3, 4)$ is a relative minimum point. Similarly,

$$f''(1) = 6(1) - 12 = -6$$

Since $f''(1) < 0$, the point $(1, 8)$ is a relative maximum point.   ◆

EXAMPLE 4   Continue Example 3 to find the one point of inflection of the function $f(x) = x^3 - 6x^2 + 9x + 4$.

SOLUTION   We have $f(x) = x^3 - 6x^2 + 9x + 4$ and $f''(x) = 6x - 12$. The graph is concave up wherever $f''(x) > 0$, that is, whenever $6x - 12 > 0$. Thus, the graph is concave up when $x > 2$. The graph is concave down wherever $f''(x) < 0$, that is, whenever $6x - 12 < 0$. Thus, the graph is concave down when $x < 2$.

Since the graph is concave up for $x > 2$ and concave down for $x < 2$, the concavity changes at $x = 2$. Thus, (2, 6) is a point of inflection, the only point of inflection of this function.   ◆

EXAMPLE 5   Continue Examples 3 and 4 to sketch the graph of $f(x) = x^3 - 6x^2 + 9x + 4$.

SOLUTION   From Examples 3 and 4 we know that function f has

1. A relative maximum at (1, 8)

2. A relative minimum at (3, 4)

3. A point of inflection at (2, 6)

In addition, an easy point to obtain for the graph is the y intercept (that is, the point where $x = 0$). Using 0 for x in $f(x) = x^3 - 6x^2 + 9x + 4$ yields the point (0, 4).

Although a few additional points are often needed in order to obtain a good sketch, here we find that once we plot the four points we know, plus the point (4, 8), which can be readily determined, a nice curve can be passed through them. (See Figure 25.)

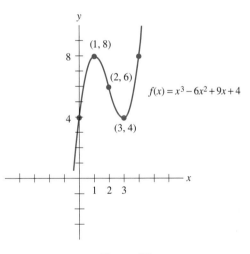

Figure 25

The curve is rounded (rather than pointed) at the relative maximum and relative minimum points because the derivative is zero there (rather than undefined).   ◆

You may wish to organize your approach to sketching graphs of functions, especially since so many aspects of graphs have been presented. Here, then, is a straightforward approach based on the techniques presented in the chapter.

Steps in Sketching the Graph of Function *f*

1. Determine $f'(x)$ and $f''(x)$.

2. Use $f'(x)$ to find all critical numbers.

3. Apply the second derivative test to critical numbers in order to determine relative extrema. If the test fails, use the first derivative test.

4. Determine the y coordinates of all relative maximum and minimum points.

5. Show the graph at relative extreme points as rounded (if $f'(c) = 0$) or pointed (if $f'(c)$ does not exist).

6. Determine where the graph is concave up and where it is concave down.

7. Determine points of inflection.

8. Determine one or more additional points as needed to complete the sketch.

EXAMPLE 6   Sketch the graph of $f(x) = 6x^{2/3} - 4x$.

SOLUTION   Here are the key results and information, plus the graph. You may want to do the step-by-step work to verify these results.

$$f(x) = 6x^{2/3} - 4x$$

$$f'(x) = \frac{4}{x^{1/3}} - 4$$

$$f''(x) = -\frac{4}{3x^{4/3}}$$

To obtain critical numbers, first consider $f'(x) = 0$.

$$\frac{4}{x^{1/3}} - 4 = 0$$

$$\frac{4}{x^{1/3}} = 4$$

$$4 = 4x^{1/3}$$

$$x^{1/3} = 1$$

$$x = 1$$

Also consider when f' is undefined, which happens when $x = 0$. And since f *is* defined at 0, the number 0 is a critical number.

critical numbers: 0, 1

The second derivative test fails to tell anything when $x = 0$ is used. But the first derivative test shows that $(0, f(0))$ is a relative minimum point. Be careful when using the first derivative test here. You can use -1 to test $f'(x)$ to the left of 0, but you cannot use 1 on the right, since 1 is the other critical number. You might want to use 1/8 instead, since 1/8 has a cube root that is easily determined without a calculator.

The second derivative test shows that $x = 1$ is associated with a relative maximum. Thus, $(1, f(1))$ is a relative maximum point.

$$\textit{relative minimum point:} \quad (0, 0)$$
$$\textit{relative maximum point:} \quad (1, 2)$$

The curve is pointed at $(0, 0)$ because $f'(x)$ is undefined at 0. The curve is rounded at $(1, 2)$ because $f'(x) = 0$ at 1.

In testing for concavity, we find that $f''(x)$ is negative whether x is negative or positive, since $x^{4/3}$ is positive for positive or negative values of x. Thus, the curve is always concave down. And since the concavity does not change, there can be no point of inflection.

Two other points on the graph are $(-1, 10)$ and $(8, -8)$. The graph is shown in Figure 26.

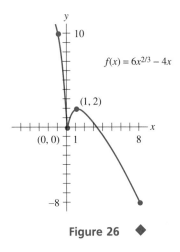

$$f(x) = 6x^{2/3} - 4x$$

Figure 26 ◆

EXAMPLE 7 ◆ *SOCIAL WORK / CELL DIVISION*

Compare the nature of the increases shown in (a) a social worker's caseload (Figure 27) and in (b) cell volume during cell division (Figure 28).

(a) The graph of a social worker's caseload as a function of time shows an increasing function. Furthermore, if you were to draw tangent lines to the curve, you would see that the slopes of those tangent lines are increasing as they go from left to right. This means that there is an increase in the *rate* at which the social worker's caseload is increasing.

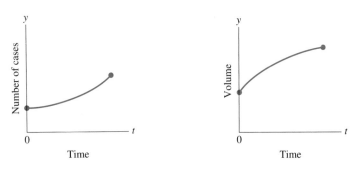

Figure 27 A social worker's caseload as a function of time

Figure 28 Volume of a cell during cell division (before splitting into two cells)

You can also consider that the curve is concave up, which means that $y'' > 0$. Interpreting y'' as the rate of change of y' or the rate of change of the slope of the curve, we see that since $y'' > 0$, *the graph is increasing at an increasing rate.* This means that the social worker's caseload is increasing at an increasing rate.

(b) The graph of cell volume during cell division as a function of time shows an increasing function. If you were to draw tangent lines to the curve, you would see that the slopes of the tangent lines are decreasing as they go from left to right. This means that there is a decrease in the *rate* at which the cell volume is increasing.

You can also consider that the curve is concave down, which means $y'' < 0$. Interpreting y'' as the rate of change of y' or the rate of change of the slope of the curve, we see that since $y'' < 0$, *the graph is increasing at a decreasing rate.* ◆

4.3   Exercises

For each graph in Exercises 1–6, indicate where the graph is concave up and where it is concave down. Use interval notation.

1.

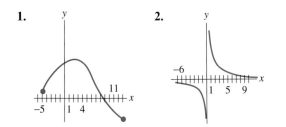

2.

3.

4.

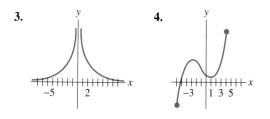

5.

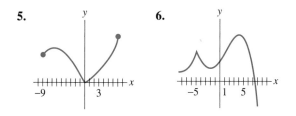

6.

Determine where the graph of each function in Exercises 7–20 is concave up and where it is concave down. Use interval notation.

7. $f(x) = x^2 - 20x + 13$ **8.** $f(x) = x^2 + 8x - 5$

9. $f(x) = -x^2 + 10x - 1$ **10.** $f(x) = -5x^2 + 6x + 2$

11. $f(x) = x^3$ **12.** $f(x) = 7 - x^3$

13. $f(x) = x^3 + 3x^2 + 3x + 1$

14. $f(x) = x^3 - 3x^2 + 3x - 1$

15. $f(x) = -2x^3 + 3x^2 + 12x - 11$

16. $f(x) = -x^3 + 3x^2 + 24x + 5$

17. $f(x) = \dfrac{x+1}{x}$ **18.** $f(x) = \dfrac{x-1}{x}$

19. $f(x) = 3x - \dfrac{1}{x}$ **20.** $f(x) = \dfrac{1}{x} - 2x$

Determine all relative maximum and relative minimum points, if any, for the functions defined in Exercises 21–34. Use the second derivative test when possible.

21. $f(x) = x^3 + 3x^2 + 5$ **22.** $f(x) = x^3 - 6x^2 - 1$

23. $f(x) = x^3 - 48x + 2$ **24.** $f(x) = x^3 - 300x + 50$

25. $f(x) = -x^3 - 3x^2 + 24x + 7$

26. $f(x) = 2 + 3x^2 - x^3$ **27.** $f(x) = 2x^3 - 15x^2$

28. $f(x) = 2x^3 - 3x^2 - 12x + 19$

29. $f(x) = 2x^3 - 7$ **30.** $f(x) = 23 - 6x^3$

31. $f(x) = x^{1/3}$ **32.** $f(x) = 2x^{1/3} + 5$

33. $f(x) = 2x^4 + 5$ **34.** $f(x) = 12x^{2/3} - 4x$

For Exercises 35–40, refer to the graphs given in Exercises 1–6. Estimate to the nearest integer the x coordinate of the point of inflection (if any) of the specified graph.

35. Exercise 1   **36.** Exercise 2

37. Exercise 3   **38.** Exercise 4

39. Exercise 5   **40.** Exercise 6

Each function defined in Exercises 41–48 has *one* point of inflection. Find it.

41. $f(x) = x^3 - 4$ **42.** $f(x) = x^3 + 2$

43. $f(x) = x^3 + 6x^2 + 12x + 12$

44. $f(x) = x^3 - 9x^2 + 27x - 17$

45. $f(x) = x^3 - 3x^2 + 4$

46. $f(x) = x^3 + 3x^2 - 24x + 1$

47. $f(x) = 2x^3 - 3x^2 + 12x - 2$

48. $f(x) = 2x^3 - 39x^2 + 240x + 5$

Sketch the graph of each function defined in Exercises 49–66 by using the methods developed in the chapter.

49. $f(x) = x^3 - 3x^2 + 5$ **50.** $f(x) = x^3 + 3x^2 - 5$

51. $f(x) = x^3 - 3x^2 - 9x + 7$

52. $f(x) = x^3 - 3x + 2$

53. $f(x) = x^3 - 3x^2 + 3x - 1$

54. $f(x) = x^3 + 6x^2 + 12x + 8$

55. $f(x) = 1 - 3x^2 - x^3$ **56.** $f(x) = -x^3 + 12x + 1$

57. $f(x) = -x^3 + 3x^2 + 5$ **58.** $f(x) = -x^3 + 6x^2 - 19$

59. $f(x) = x^{1/3} + 1$ **60.** $f(x) = 2 + 6x^{1/3}$

61. $f(x) = 3 + x^{2/3}$ **62.** $f(x) = 1 - 3x^{2/3}$

63. $f(x) = 2x - 3x^{2/3}$ **64.** $f(x) = 12x^{2/3} - 4x - 7$

65. $f(x) = 3x^{4/3}$ **66.** $f(x) = 5 - 6x^{4/3}$

Sketch the graph of each function defined in Exercises 67–72 by using the methods developed in the chapter.

67. $f(x) = x^3 + 2$ **68.** $f(x) = 1 - x^3$

69. $f(x) = 3 - x^4$ **70.** $f(x) = x^5$

71. $f(x) = x^3 - 6x^2 + 12x - 8$

72. $f(x) = x^3 + 3x^2 + 3x + 1$

73. In Example 2 of Section 4.2 we sketched the graph of $f(x) = x^{2/3} + 1$ without knowing anything about its concavity. Since then we have studied concavity. In view of this, obtain $f''(x)$ and show that the graph is indeed concave down everywhere, as pictured in Example 2.

W 74. In Example 6 the second derivative test fails when $x = 0$. Explain why you really should have known in advance of the test that it would fail.

W 75. You have studied relative extrema and points of inflection. What seems unusual about the points $(a, f(a))$ and $(b, g(b))$ in the graphs shown here?

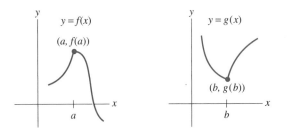

76. Sketch the graph of a continuous function that has a relative maximum and a relative minimum, but no point of inflection.

77. Refer to the graph shown and determine the value of each expression. Answers should be one of the following: positive, negative, or zero.

(a) $f'(6)$ (b) $f'(10)$ (c) $f'(7)$

(d) $f''(7)$ (e) $f'(9)$ (f) $f''(9)$

(g) $f'(5)$ (h) $f''(5)$

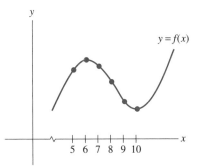

78. Refer to the graph shown and determine the value of each expression. Answers should be one of the following: positive, negative, zero, or undefined.

(a) $f'(7)$ (b) $f''(7)$ (c) $f'(8)$

(d) $f''(8)$ (e) $f'(9)$ (f) $f''(9)$

(g) $f'(10)$ (h) $f'(11)$ (i) $f''(11)$

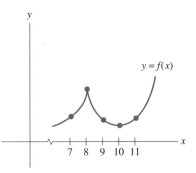

W 79. Is it safe to assume that a graph has a point of inflection wherever $f''(x) = 0$? Explain.

W 80. How do you explain the fact that the function $f(x) = x^2 - 8x + 50$ has no point of inflection?

W 81. The function defined by $f(x) = \dfrac{x + 1}{x}$ is concave up when $x > 0$ and concave down when $x < 0$, yet there is no point of inflection at $x = 0$. Why not?

82. *(Salary Changes)* If faculty salaries S have been rising at a decreasing rate over a period of time t, what is the concavity of the graph of $y = S(t)$?

83. *(Population)* If population P has been rising at an increasing rate over a period of time t, what is the concavity of the graph of $y = P(t)$?

W 84. *(Estrogen Level)* The graph in the figure below shows a woman's estrogen level throughout a 28-day cycle. After reaching a minimum about 3 days into the cycle, the estrogen level climbs toward a maximum that occurs about 14 days into the cycle. Compare

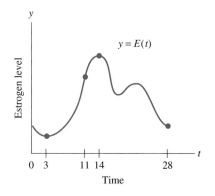

the type of increase in estrogen level during days 3 through 11 (where the curve is concave up) with the type of increase during days 11 through 14 (where the curve is concave down).

W 85. (*REVENUE*) The figure below shows the graph of two revenue functions, R_1 and R_2. For which revenue function is marginal revenue a decreasing function? Explain.

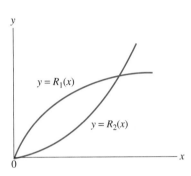

Revenue as a function of the number of units sold

86. In Chapter 1, it was claimed that the x coordinate of the vertex (maximum or minimum point) of the parabola given by $f(x) = ax^2 + bx + c$ is

$$x = -\frac{b}{2a}$$

Use calculus to verify this and show that it is a maximum when $a < 0$ and a minimum when $a > 0$.

In Exercises 87–96, use the methods of this chapter and your knowledge of asymptotes (from Chapter 2) to graph each function.

87. $f(x) = \dfrac{12}{x^2 + 3}$ **88.** $f(x) = \dfrac{6}{x^2 + 6}$

89. $f(x) = \dfrac{2x^2}{x^2 + 1}$ **90.** $f(x) = \dfrac{4x^2}{1 + x^2}$

91. $f(x) = \dfrac{x + 1}{x}$ **92.** $f(x) = \dfrac{1 + 2x}{x}$

93. $f(x) = \dfrac{x^2}{x^2 - 9}$ **94.** $f(x) = \dfrac{x^2}{4 - x^2}$

95. $f(x) = \dfrac{x}{x^2 - 4}$ **96.** $f(x) = \dfrac{x}{x^2 - 9}$

The Exercise Library at the back of the book contains graphing calculator and computer exercises keyed to this section.

4.4 | ABSOLUTE EXTREMA

Before we turn our attention to applied problems involving maxima and minima, it is appropriate to present the idea of absolute extrema. The *absolute maximum value* of a function is the largest possible value of the function. The absolute maximum value of a particular function may or may not be the same as the relative maximum value. Consider the function graphed in Figure 29, defined for x in the interval [2, 8]. The relative maximum value of the function is 5, but the absolute maximum value of the function is 9. Consider next the function graphed in Figure 30, defined for x in the interval [−3, 6]. The relative maximum value of the function is 4, and this is also the absolute maximum value of the function.

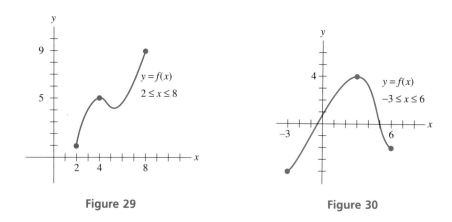

Figure 29   Figure 30

Similar drawings and reasoning can be used to present the *absolute minimum* versus relative minimum. We conclude that for continuous functions defined on a closed interval [*a*, *b*]:

1. The **absolute maximum function** value occurs either where there is a relative maximum or else at an endpoint of the interval.

2. The **absolute minimum function** value occurs either where there is a relative minimum or else at an endpoint of the interval.

Finding Absolute Extrema

Every function f that is continuous on a closed interval [*a*, *b*] will have an absolute maximum and an absolute minimum on the interval. To find the absolute extrema,

1. Find all critical numbers of f that are in the interval (*a*, *b*).

2. Evaluate f at each critical number and also compute $f(a)$ and $f(b)$.

3. The largest value computed in Step 2 is the **absolute maximum**. The smallest value that is computed in Step 2 is the **absolute minimum**.

EXAMPLE 1   Determine the absolute extrema of $f(x) = x^2 + 4$ on the interval [−2, 7].

SOLUTION The only critical number is 0, from $f'(x) = 2x = 0$ when $x = 0$. Considering f at 0 and at the endpoints -2 and 7 of the interval, we have

x	$f(x)$
0	4
-2	8
7	53

From the table it is clear that the absolute maximum value of f is 53, which occurs at the endpoint where $x = 7$. The absolute minimum is 4, which occurs at the critical number 0. ◆

4.4 Exercises

For each function graphed in Exercises 1–6, determine all relative maximum and relative minimum function values, if any. Also determine the absolute maximum and absolute minimum function values, if any.

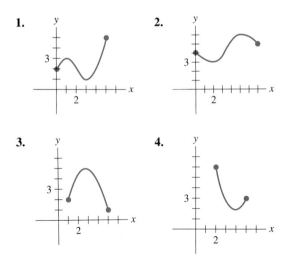

5.

6.

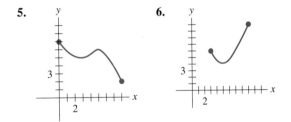

For each function in Exercises 7–20, determine the absolute extrema on the given interval.

7. $f(x) = x^2 + 5$ on $[-1, 8]$

8. $f(x) = 1 - x^2$ on $[-3, 7]$

9. $f(x) = -2x^2 + 8x - 17$ on $[0, 6]$

10. $f(x) = 3x^2 - 6x + 5$ on $[0, 4]$

11. $f(x) = x^3 - 6x^2$ on $[-1, 10]$

12. $f(x) = x^3 + 3x^2 - 9x + 2$ on $[-10, 3]$

13. $f(x) = 2x^3 - 3x^2 - 36x + 4$ on $[-5, 10]$

14. $f(x) = x^3 - 3x - 1$ on $[-5, 5]$

15. $f(x) = -x^3 + 6x^2 - 50$ on $[-1, 2]$

16. $f(x) = x^3 - 12x^2 + 3$ on $[1, 10]$

17. $f(x) = x^3 + 3x^2 + 5$ on $[0, 5]$

18. $f(x) = 2x^3 + 12x^2 - 17$ on $[-1, 1]$

19. $f(x) = x^3 - 15x^2 + 72x + 1$ on $[-4, 3]$

20. $f(x) = x^3 + 9x^2 + 24x + 7$ on $[0, 9]$

W 21. To determine the absolute extrema of a function on a closed interval, we evaluate the function at the endpoints of the interval and at the critical numbers within the interval. Why do we *omit* testing the critical numbers for relative maxima and relative minima?

The Exercise Library at the back of the book contains graphing calculator and computer exercises keyed to this section.

4.5 | ADDITIONAL APPLICATIONS. APPLIED MAXIMUM / MINIMUM

The concepts studied in this chapter can also be used to solve nongraphing problems. A variety of such applications is presented in this section.

The ideas given in Section 4.4 (absolute extrema) should serve to caution you. As we begin now to solve applied maximum and minimum problems using the techniques presented earlier in the chapter, we will be finding *relative* maximum and minimum values. Yet the problems will always ask for *the* largest or smallest value. The application requires us to find the *absolute* maximum or minimum. The only way to be sure that the relative maximum or minimum obtained is indeed the absolute maximum or minimum is to check the endpoints of the interval on which the function is defined. Once you see the problem statements, it will become apparent that determining the interval on which the function is defined is not always a simple matter. Although the applied problems in this section will be such that the relative extrema are also the absolute extrema, some of the examples will explain how to be sure that the relative extremum obtained is indeed the absolute extremum.

EXAMPLE 1 ◆ *MAXIMIZING A PRODUCER'S PROFIT*

A manufacturer of telephones determines that the profit from producing and selling x telephones is $P(x) = -.01x^2 + 60x - 500$ dollars.

(a) How many telephones should be produced in order to maximize the company's profit?

(b) What is the maximum profit possible?

SOLUTION **(a)** To determine x for which $P(x)$ is maximum, begin by obtaining all critical numbers of the function. The derivative is

$$P'(x) = -.02x + 60$$

which is defined for all real numbers. This means that we need only consider $P'(x) = 0$ in order to find the critical numbers. Thus we solve

$$-.02x + 60 = 0$$

which leads to

$$x = 3000$$

The conclusion is that in order to maximize profit, the company should produce 3000 telephones. *However*, before rushing to this conclusion, you really should check to be sure that $x = 3000$ leads to a maximum rather than a minimum. The second derivative test works nicely here. [We have $P''(x) = -.02$, so $P''(3000) = -.02 < 0$. Thus, $P(3000)$ is a maximum.]

(b) The maximum profit is the value of $P(x)$ when $x = 3000$.

$$P(3000) = -.01(3000)^2 + 60(3000) - 500$$
$$= 89,500$$

The maximum profit is \$89,500.

Note that the graph of the (quadratic) profit function is a parabola, which means that the relative maximum profit is also the absolute maximum profit. ◆

EXAMPLE 2 ◆ *FLIGHT OF A ROCKET*

A rocket is launched vertically such that its distance s (in feet) from the ground at any time t (in seconds) is given by

$$s(t) = -16t^2 + 640t$$

How high will the rocket travel before falling back down to the ground?

SOLUTION We seek to maximize s, since s is the height of the rocket. To begin, obtain s' and determine all critical numbers of the function s.

$$s' = -32t + 640$$

Since s' is always defined, consider $s' = 0$ for critical numbers. The equation

$$-32t + 640 = 0$$

leads to

$$t = 20$$

We conclude that 20 is the only critical number of s. The second derivative $(s'' = -32)$ is negative for $t = 20$. (Actually, s'' is negative for all values of t.) Thus, s will have a relative maximum value at 20, and the maximum value of s is $s(20)$, which is computed next.

$$s(20) = -16(20)^2 + 640(20)$$
$$= -16(400) + 12,800$$
$$= -6400 + 12,800$$
$$= 6400$$

The rocket will travel to a maximum height of 6400 feet before falling back to the ground. ◆

> *Note*
>
> In Chapter 6, you will learn how to derive formulas of the type used in Example 2.

In Examples 1 and 2, the function to be maximized was given in the statement of the problem. But when the function is not given directly, you must establish it. In view of this, we now present guidelines to use when solving such applied problems.

Steps for Solving Maximum/Minimum Problems

1. Read the problem carefully.

2. If possible, draw a picture.

3. Determine the constants and variables. Label parts of the drawing.

4. Establish an equation containing the variable to be maximized or minimized.

5. If the equation does not give the variable to be maximized or minimized as a function of *one* variable, then you must make a substitution in order to obtain a function of just one variable.

6. Now that the function is established, proceed to obtain critical numbers and so on—as was done in Examples 1 and 2.

EXAMPLE 3 ◆ *CONSTRUCTING A PATIO*

A family plans to fence in a rectangular patio area behind their house. (See Figure 31.) They have 120 feet of fence to use. What should be the dimensions of the rectangular region if they want to make the patio area enclosed as large as possible?

Patio area

Figure 31

SOLUTION   Let x be the width of the rectangular patio. (See Figure 32.)

Figure 32

Since there is 120 feet of fence, the *length* of the rectangle must be $120 - 2x$. The *area* of the patio is computed as length times width, so we have the following area A:

$$A(x) = (120 - 2x)(x)$$

or

$$A(x) = 120x - 2x^2$$

Differentiating yields

$$A'(x) = 120 - 4x$$

For critical numbers, determine x such that $A'(x) = 0$.

$$120 - 4x = 0$$

when

$$x = 30$$

(It is left for you to verify that $x = 30$ is associated with a maximum.) The length is $120 - 2x = 120 - 2(30) = 60$. Thus, the patio area will be largest when the width is 30 feet and the length is 60 feet.

    We could argue that because the area function is quadratic and its graph is a parabola, the relative maximum obtained is also the absolute maximum. *Another approach*, one that can be used whether or not the function is quadratic, is to consider the interval over which function A is defined. Since x is the width of the rectangle, we insist that $x \geq 0$. Also, since there is only 120 feet of fence and there must be two widths of fence used, it follows that $x \leq 60$. Thus A is defined for x in the interval $[0, 60]$. Testing the value of A at the endpoints of this interval yields $A(0) = 0$ and $A(60) = 0$. Since $A(30) = 1800$, it follows that $x = 30$ leads to an *absolute* maximum value of A. ◆

EXAMPLE 4 ◆ *MAXIMIZING A MANUFACTURER'S PROFIT*

A manufacturer can sell x headphones at a price of $140 - .01x$ dollars *each*. It costs $40x + 15{,}000$ dollars to produce *all x of them*. How many headphones should the manufacturer produce in order to maximize profit?

SOLUTION   The profit function P should be obtained and then maximized. Recall that $P(x) = R(x) - C(x)$; profit equals revenue minus cost. Because the revenue from the sale of *each* headphone is $140 - .01x$ dollars, the revenue $R(x)$ from the sale of x headphones can be obtained by multiplying this revenue per unit by x, the number of units. Thus,

$$R(x) = (140 - .01x)(x) \quad \text{dollars}$$

or

$$R(x) = 140x - .01x^2 \quad \text{dollars}$$

The cost of producing x units is given in the statement of the problem; that is,

$$C(x) = 40x + 15{,}000 \quad \text{dollars}$$

Using $P(x) = R(x) - C(x)$, we can now obtain the profit function.

$$P(x) = (140x - .01x^2) - (40x + 15{,}000)$$

or

$$P(x) = -.01x^2 + 100x - 15{,}000$$

In order to find x for which $P(x)$ will be maximum, obtain $P'(x)$, and set it equal to zero to determine the critical number(s).

$$P'(x) = -.02x + 100$$
$$0 = -.02x + 100$$
$$x = 5000$$

The second derivative is $P''(x) = -.02$, which is always less than zero. This means that $P(5000)$ is indeed a relative *maximum* value. Thus, *5000 units should be produced in order to maximize profit.* (If desired, we could continue and find the maximum profit, $P(5000)$, which is \$235,000.)   ◆

The following result offers an alternative approach to finding maximum profit. We will show that

> Profit is maximized when marginal revenue R' is equal to marginal cost C', provided that $R'' < C''$.

To verify this result, consider that

$$P(x) = R(x) - C(x)$$

Differentiating yields

$$P'(x) = R'(x) - C'(x)$$

For maximum $P(x)$, consider $P'(x) = 0$, which is the same as $R'(x) - C'(x) = 0$.

$$R'(x) - C'(x) = 0$$

or

$$R'(x) = C'(x)$$

Thus, $P'(x) = 0$ when marginal revenue equals marginal cost. According to the second derivative test, $P(x)$ is maximum if $P''(x) < 0$. From

$$P'(x) = R'(x) - C'(x)$$

we can obtain

$$P''(x) = R''(x) - C''(x)$$

Now $P''(x) < 0$ becomes

$$R''(x) - C''(x) < 0$$

or

$$R''(x) < C''(x)$$

We can use this result to solve the problem of Example 4. Consider

$$R'(x) = C'(x)$$

or

$$140 - .02x = 40$$

The solution, as before, is $x = 5000$. Note that $R''(x) = -.02$ and $C''(x) = 0$. Thus, $R''(5000) < C''(5000)$, as required for a maximum.

EXAMPLE 5 ◆ **MINIMIZING THE COST OF MATERIALS**

A manufacturer of storage bins plans to produce some open-top rectangular boxes with square bases. The volume of each box is to be 125 cubic feet. Material for the base costs $6 per square foot, and material for the sides costs $3 per square foot. Determine the dimensions of the box that will minimize the cost of materials.

SOLUTION Suppose we let x be the width of the bin. Since the base is square, the length must also be x. No information is given about the height, so we will use h to represent it. A drawing of the storage bin is shown in Figure 33.

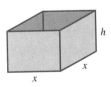

Figure 33 The storage bin (open-top box)

The area of the base is $x \cdot x$, or x^2 square feet. Thus, at \$6 per square foot, the cost of the material for the base is $\$6 \cdot x^2$ or $6x^2$ dollars. The area of each side is xh, so the area of the four sides is $4xh$. Since the cost of material for the sides is \$3 per square foot, the cost of the material for all four sides is $\$3 \cdot 4xh$ or $12xh$ dollars. Thus, the total cost C of all the materials for a bin is

$$C = 6x^2 + 12xh \quad \text{dollars}$$

Unfortunately, C is a function of two variables, x and h. In order to proceed to differentiation, one of the variables should be eliminated. The variable h can be eliminated by obtaining a relationship between x and h. This can be done by considering that the volume of a box is computed as length times width times height. Here the length and width are each x and the height is h. So we have

$$V = x \cdot x \cdot h$$

or

$$V = x^2 h$$

Since the statement of the problem says that the volume is 125 cubic feet, we then have

$$125 = x^2 h$$

This equation can be manipulated into the form

$$h = \frac{125}{x^2}$$

and then $125/x^2$ can be substituted for h in the cost function

$$C = 6x^2 + 12xh$$

to produce

$$C = 6x^2 + 12x \cdot \frac{125}{x^2}$$

or

$$C = 6x^2 + \frac{1500}{x}$$

Now we can differentiate and proceed toward a solution.

$$C' = 12x - \frac{1500}{x^2}$$

In search of critical numbers, consider $C' = 0$.

$$12x - \frac{1500}{x^2} = 0$$

or

$$12x = \frac{1500}{x^2}$$

Multiply both sides of the equation by x^2 in order to eliminate the fraction. The result is

$$12x^3 = 1500$$

After dividing both sides by 12, the result is

$$x^3 = 125$$

or

$$x = 5$$

The value of h can now be found from $h = 125/x^2$.

$$h = \frac{125}{x^2} = \frac{125}{25} = 5$$

We conclude that the base should be made 5 feet by 5 feet and the height should be 5 feet in order to minimize cost.

To verify that C does indeed have a relative minimum at $x = 5$, note that

$$C''(x) = 12 + \frac{3000}{x^3}$$

and $C''(5) > 0$. So by the second derivative test, C has a relative minimum at 5. ◆

Given next are geometric formulas that you will need to use in some of the exercises that follow. The letters used in the formulas are: A (area), P (perimeter), C (circumference), V (volume), S (surface area), r (radius), l (length), w (width), h (height), and π (the irrational number "pi"—approximately 3.14).

1. *Rectangle*

$$A = lw$$
$$P = 2l + 2w$$

2. *Circle*

$$C = 2\pi r$$
$$A = \pi r^2$$

3. *Rectangular solid*

$$V = lwh$$
$$S = 2lw + 2lh + 2wh$$

4. *Cylinder*

$$V = \pi r^2 h$$
$$S = 2\pi rh \quad \text{(curved side \textit{only})}$$
$$S = \pi r^2 \quad \text{(top or bottom \textit{only})}$$

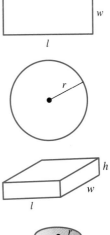

4.5   Exercises

1. (HEATING METAL) In Example 2 of Section 4.1, the function $T(x) = -3x^2 + 60x + 70$ indicated the temperature after x seconds of a metal tray undergoing a chemical finishing process. What is the hottest the metal gets during the process? (The temperature is in degrees Fahrenheit.)

2. (ROCKET LAUNCH) A rocket is launched vertically such that its distance s (in feet) from the ground at any time t (in seconds) is given by

$$s(t) = -16t^2 + 960t$$

   (a) When will the rocket reach its greatest height?
   (b) How high will the rocket travel before falling back down to the ground?

3. (THROWING A BALL) A tennis ball thrown straight up is $-16t^2 + 96t + 7$ feet above the ground after t seconds. How high will the ball go?

4. (LEAPING PORPOISE) The center of gravity of a leaping porpoise describes a parabola. Suppose a particular porpoise follows a path such that its height (in feet) above the water is given by $y = 3x - \frac{1}{4}x^2$. How high does this porpoise leap?

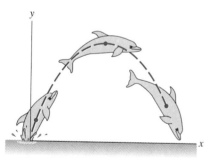

5. (PROFIT) Suppose that a manufacturer's profit on x items is $P(x) = 1300x - x^2$ dollars.
   (a) How many items should be manufactured to maximize profit?
   (b) What is that maximum profit?

6. (REVENUE) The revenue a company obtains from the sale of x items is $R(x) = 160 + 380x - 2x^2$ dollars. How many items must be sold for the company to maximize its revenue? Determine the maximum revenue.

7. (PRICE) A manufacturer can supply x items at a price of $4x^2 - 200x + 2850$ dollars per item. How many items should be ordered from this manufacturer in order to minimize the price per item?

8. (PROFIT) A disgruntled employee with a grudge determines that the company's profit on the production of x *hundred* units seems to be given by the function $P(x) = 100x^2 - 1000x + 3000$ dollars. He plans to "get even with the company" by suggesting that they produce the number of items that he knows will minimize their profit. How many units will he recommend they produce?

9. (MEDICINE CONCENTRATION) When a pill such as aspirin or antihistamine is swallowed, the concentration of the medicine in the bloodstream begins at zero and increases toward a maximum concentration. The concentration then declines until there is none of the medicine present. Suppose the concentration K of a particular medicine t hours after being swallowed is (for $t \geq 0$)

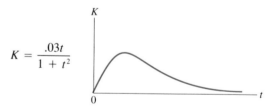

$$K = \frac{.03t}{1 + t^2}$$

When is the concentration highest?

10. (PATIO CONSTRUCTION) Redo Example 3 assuming that the family has 200 feet of fence to use.

11. The sum of two positive numbers is 100. If their product is a maximum, what are the two numbers? (*Hint:* To begin, let one number be x. Then, since their sum is 100, what expression represents the other number?)

12. Determine two positive numbers whose product is 100 and whose sum is a minimum.

13. (FENCING) A farmer has 1600 feet of fencing. He plans to make a rectangular pen for his hogs. What should be the dimensions of the pen if he wants the area to be the largest possible?

14. (*FENCING*) Determine the dimensions of the largest rectangular area that can be enclosed by 2000 meters of fence.

15. (*FENCING*) What is the smallest amount of fencing that can be used to enclose a rectangular garden having an area of 900 square feet?

16. (*FENCING*) The owner of a warehouse decides to fence in an area of 800 square feet behind the warehouse. He plans to use the wall of the building as one of the four sides that will enclose the rectangular area (see figure). He would like to use the least amount of fencing necessary for the other three sides. How many feet of fence will be needed?

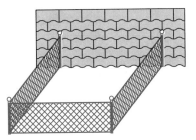

17. (*STORAGE FACILITY*) A storage company wants to create a storage facility by walling in a rectangular region containing 1728 square feet. They will also use walls to subdivide the region into five storage compartments (see figure).

What should be the width (*x*) and length of the storage facility in order to use the least amount of material for the walls?

18. (*BOX SURFACE*) Determine the dimensions of a closed rectangular box with a square base, if the volume must be 1000 cubic centimeters and the area of the outside surface is to be as small as possible.

19. (*BOX SURFACE*) An open rectangular box (that is, a box with no top) with a capacity of 36,000 cubic inches is needed. If the box must be twice as long as it

is wide, what dimensions would require the least material?

20. (*BOX SURFACE*) Redo Exercise 19 assuming that the box will have a top. Approximate each dimension to the nearest tenth of an inch.

21. (*FOLD-UP BOX*) A square piece of cardboard 40 centimeters by 40 centimeters is used to make an open box as shown in the figure. A small square is cut from each corner of the cardboard, and then the sides are folded up. Determine the size of the cut (*x* in the drawing below) that will lead to the box of greatest volume. (*Note*: At some point it may appear that there are *two* answers, but only one of them will make sense given the size of the cardboard.)

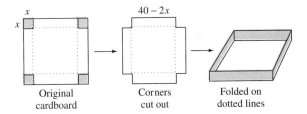

Original cardboard   Corners cut out   Folded on dotted lines

22. (*FOLD-UP BOX*) Redo Exercise 21 using a rectangular piece of cardboard 15 inches by 8 inches.

23. (*GUTTER CONSTRUCTION*) A builder plans to construct a gutter from a long sheet of metal by making two folds of equal size (see figure). The folds are made to create perpendicular sides.

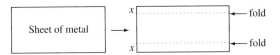

The sheet of metal is 28 centimeters wide and 500 centimeters long. How much (*x*) should be turned up for each side in order for the gutter to hold the greatest amount of water?

24. (*CYLINDRICAL CAN*) A metal can is to be made in the form of a right circular cylinder that will contain 16π cubic inches. What radius of the can will require the least amount of metal (see figure on the next page)? Note that there are three parts—a circular top, a circular bottom, and the curved side.

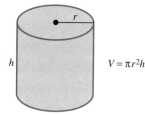

$V = \pi r^2 h$

25. (*PAPER SIZE*) Determine the dimensions of the smallest size rectangular piece of paper that satisfies the following conditions:
   (a) You can print 50 square inches of material on it (the shaded area in the figure).
   (b) There will be 2-inch margins on the top and bottom.
   (c) There will be 1-inch margins on the sides.

26. (*LAND AREA*) A corporation plans to locate its headquarters in a new rectangular one-story building to be constructed. They need 30,000 square feet of floor space. In addition, they would like to have a walkway 15 feet wide at the front and back of the building and a grass area 20 feet wide on the other two sides. Determine the length and width of the smallest area lot that can be used for their headquarters. (Neglect the thickness of the walls.)

27. (*FRUIT PRODUCTION*) Julia wants to have a small but productive orchard. Experienced farmers have told her that if she plants 20 trees, she will get 24 bushels of fruit from each tree. But every additional tree she plants will cause crowding and result in an orchard-wide drop in yield of 1 bushel per tree. How many trees should she plant to get the greatest amount of fruit?

28. (*TICKET SALES*) Concert promoters estimate that they can sell 1200 tickets at $20 each. For every $1 the ticket price is lowered, they can sell 100 additional tickets.

   (a) What should be the ticket price in order to maximize the revenue?
   (b) What is the maximum revenue?

29. (*FENCING*) A gardener wishes to fence in a rectangular area of 1728 square feet. She also wants to insert a fence that will divide the area into two rectangular subareas. The drawing shows that some fencing costs $4 per foot and some costs $2 per foot. Find the dimensions that will minimize the cost of the garden fencing.

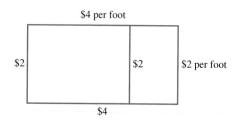

30. (*STORAGE TANK*) A rectangular open storage tank made with a square base is to have a volume of 768 cubic feet. Material for the base costs $6 per square foot, and that for the sides costs $2 per square foot. What dimensions will minimize the cost of materials?

31. (*WINDOW AREA*) Part of the side of a house is made in the shape of a 12-foot-high parabolic arch. The equation $y = 12 - x^2$ describes the arch. (*x* and *y* are in feet.) Determine the dimensions of the rectangular window of largest area that can be placed under the arch (see figure).

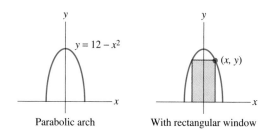

Parabolic arch   With rectangular window

Hint: Express the area of the window in terms of *x* and *y*. Then make a substitution that will leave you with area as a function of *x* alone (that is, without *y*).

32. (*RECTANGULAR AREA*) Determine the dimensions of the rectangle of largest area that can be inscribed in a right triangle with base 10 centimeters and height 20 centimeters. *Hint*: See figure. Use the two unlabeled points (dots) to determine the equation of the line through those points. Then use an approach similar to that suggested in the hint of Exercise 31.

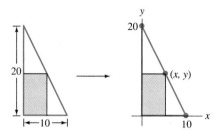

33. (*PROFIT*) Using the fact that profit is maximized when marginal revenue equals marginal cost (provided that $R'' < C''$) determine the number of units x that should be produced in order to maximize profit.
  (a) Revenue: $R(x) = 80x - .02x^2$
    Cost: $C(x) = 130 + 74x - .01x^2$
  (b) Revenue: $R(x) = 7x + .001x^2$
    Cost: $C(x) = 5x + .003x^2$

34. (*PROFIT*) Repeat Exercise 33, using
  Revenue: $R(x) = 120x - .03x^2$
  Cost: $C(x) = 250 + 30x - .02x^2$

35. (*PROFIT*) A manufacturer can sell x calculators at a price of $35 - .01x$ dollars *each*. It costs the manufacturer $14x + 10,000$ dollars to produce *all x of them*.
  (a) How many calculators should the manufacturer produce in order to maximize profit?
  (b) What should be the price of each calculator in order to maximize profit?

36. (*PROFIT*) A manufacturer can sell x chess sets at a price of $42 - .02x$ dollars *each*. It costs $15x + 500$ dollars to produce *all x of them*.
  (a) How many chess sets should the manufacturer produce in order to maximize profit?
  (b) What should be the price of each chess set in order to maximize profit?

37. (*AVERAGE COST*) Given the cost function
$$C(x) = 192 + 7x + .03x^2$$
determine the number of units x that should be produced in order to minimize the average cost per item, namely,
$$\bar{C} = \frac{C(x)}{x}$$

38. (*PROFIT*) Suppose that you work for a corporation and the manager shows you the corporation's profit function $P(x) = 2000x - 8x^2$ dollars, where x is the number of units produced and sold. The manager thinks the company should try to sell 150 units, since the profit from the sale of 150 units is $120,000. What would be your advice to the manager?

39. (*PROFIT*) (See figure below.) As you know, profit is maximum when marginal revenue equals marginal cost. You also know that the derivative can be interpreted as the slope of the tangent line. Combine these two ideas in order to determine the value of x (a, b, c, d, e, or f) for which profit is maximum.

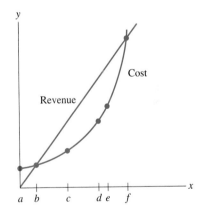

W 40. In Example 5, is 0 a critical number? After all, $C'(0)$ is undefined. Explain.

4.6 ELASTICITY OF DEMAND

The demand for goods and services is usually affected by changes in the price of those goods and services. For example, when the price of automobiles increases, the demand for them decreases. Manufacturers sometimes offer rebates as a way to decrease auto prices and thus increase demand.

The demand for some goods and services is more sensitive to price increases than it is for others. A relatively small price increase of 5% on a candy bar (say from 40¢ to 42¢) will have little effect on the demand for candy bars. On the other hand, a 5% increase in the cost of a home loan (which would create correspondingly higher monthly payments) will have a significant, decreasing effect on the demand for new homes. By contrast, the demand for essential surgery is not sensitive to price increases.

It is natural in business to be concerned about the sensitivity of demand changes to price changes. Calculus provides the **elasticity of demand** *(E)*, which gives a numerical measure of the relative sensitivity of percent changes in demand to percent changes in price. The formula is given next, followed by the derivation of the formula and some examples.

Elasticity of Demand

$$E = -\frac{p}{x} \cdot \frac{dx}{dp}$$

where

E = elasticity of demand

x = demand (quantity)

p = price per unit

Elasticity E is the percent change in demand that will occur for every 1% that the price is changed.

The elasticity formula can be derived by considering a comparison of the percent change in demand, $\Delta x/x$, with the percent change in price, $\Delta p/p$. The ratio is

$$\frac{\dfrac{\Delta x}{x}}{\dfrac{\Delta p}{p}}$$

which simplifies to

$$\frac{p}{x} \cdot \frac{\Delta x}{\Delta p}$$

Assuming that x is a continuous function of p,

$$\lim_{\Delta p \to 0} \frac{\Delta x}{\Delta p} = \frac{dx}{dp}$$

If we let $\Delta p \to 0$ for the entire ratio, we obtain the instantaneous rate of change of the ratio, which we call elasticity E.

$$E = \lim_{\Delta p \to 0} \frac{p}{x} \cdot \frac{\Delta x}{\Delta p} = \frac{p}{x} \cdot \lim_{\Delta p \to 0} \frac{\Delta x}{\Delta p} = \frac{p}{x} \cdot \frac{dx}{dp}$$

Ordinarily, when the price p increases, the quantity demanded x decreases. When the price decreases, the quantity demanded increases. Graphically, this relationship between x and p would appear as a curve with a negative slope. (See Figure 34.) The derivative dx/dp is the slope of the tangent to the curve and is negative. It follows then that the expression derived for the elasticity is always negative. Since economists prefer to use positive values here, a minus is placed in front of the formula in order that the E values will be positive. Thus we have

$$E = -\frac{p}{x} \cdot \frac{dx}{dp}$$

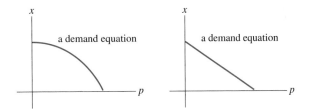

Figure 34 When price p increases, the quantity demanded x decreases.

EXAMPLE 1 ◆ *ELASTICITY OF DEMAND*

Suppose the demand equation is $x = 180 - 2p$.

(a) Determine the elasticity of demand.

(b) Calculate the elasticity for $p = 30$.

(c) Calculate the elasticity for $p = 70$.

SOLUTION **(a)** Elasticity E is computed using

$$E = -\frac{p}{x} \cdot \frac{dx}{dp}$$

From the demand equation $x = 180 - 2p$, it follows that

$$\frac{dx}{dp} = -2$$

Two substitutions can now be made into the formula for E. We can use -2 for dx/dp and $180 - 2p$ for x. The result is

$$E = -\frac{p}{180 - 2p}(-2)$$

or

$$E = \frac{2p}{180 - 2p}$$

(b) When $p = 30$ (that is, the price is \$30),

$$E = \frac{2(30)}{180 - 2(30)} = \frac{60}{120} = .5$$

An elasticity of 1/2 means that a price increase of 1% will result in a 1/2% decrease in demand. This is an example of **inelastic demand**, in which the magnitude of the change in demand is less than that of the change in price.

(c) When $p = 70$ (that is, the price is \$70),

$$E = \frac{2(70)}{180 - 2(70)} = \frac{140}{40} = 3.5$$

An elasticity of 3.5 means that for every 1% the price is increased, there will be a 3.5% decrease in demand. This is an example of **elastic demand**, in which the magnitude of the change in demand is greater than that of the change in price. ◆

The example leads naturally to the three classifications of elasticity.

> If $E < 1$, the demand is called *inelastic*.
> If $E > 1$, the demand is called *elastic*.
> If $E = 1$, the demand has *unit elasticity*.

The case of unit elasticity was not covered in Example 1. If $E = 1$, then a price increase of 1% will cause a 1% change (decrease) in demand.

EXAMPLE 2 ◆ *INELASTIC DEMAND*

Suppose the demand equation is $x = 120 - 5p$. Determine for what prices the demand will be inelastic.

SOLUTION   The demand will be inelastic when $E < 1$, that is, when

$$-\frac{p}{x}\frac{dx}{dp} < 1$$

Here $x = 120 - 5p$, from which

$$\frac{dx}{dp} = -5$$

Upon substituting -5 for dx/dp and $120 - 5p$ for x in the inequality, we obtain

$$-\frac{p}{120 - 5p}(-5) < 1$$

or

$$\frac{5p}{120 - 5p} < 1$$

Multiply both sides of the inequality by $120 - 5p$, which is positive for all p between 0 and 24 (since $120 - 5p > 0$ when $-5p > -120$ or $p < 24$). The result:

$$5p < 120 - 5p$$

or

$$10p < 120$$

Finally,

$$p < 12$$

Thus, the demand will be inelastic when the price is less than $12.   ◆

Some fascinating results can be obtained now by considering the fact that revenue equals quantity x times price p. That is,

$$R = x \cdot p$$

If demand is inelastic ($E < 1$), then raising prices will increase revenue. Why? Because demand will decrease by a smaller percentage than prices increase, when demand is inelastic. As a result, the product xp (that is, revenue) will actually increase.

Similarly, *if demand is elastic (E > 1), then lowering prices will increase revenue*. It is left for you to construct the explanation (see Exercise 18).

It follows that revenue is maximum when demand is of unit elasticity ($E = 1$). The graph in Figure 35 may clarify this point.

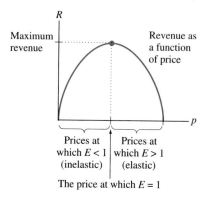

Figure 35 Revenue is maximum when demand is of unit elasticity

In summary,

> **1.** If demand is inelastic, then *raising* prices will increase revenue.
>
> **2.** If demand is elastic, then *lowering* prices will increase revenue.
>
> **3.** Total revenue is *maximum* when demand is of unit elasticity.

4.6   Exercises

In Exercises 1–4, determine the elasticity of demand for the given function at the given price.

1. $x = 50 - 3p, p = 10$ **2.** $x = 78 - 5p, p = 12$

3. $x = \sqrt{200 - p}, p = 40$ **4.** $x = \sqrt{300 - 2p}, p = 85$

In Exercises 5–8, determine the elasticity of demand for the given function.

5. $x = 321 - p^2$ **6.** $x = 400 - .5p^2$

7. $x = \dfrac{200}{p}$ **8.** $x = \dfrac{100}{\sqrt{p}}$

9. (*Lottery ticket sales*) A state lottery believes that the number of tickets it can sell weekly is given by $x = 500{,}000 - 40{,}000p^2$ for $1 \le p \le 3$, where p is the price in dollars.
   (a) How many tickets can be sold at $1.00 each?
   (b) How many tickets can be sold at $3.00 each?
   (c) Find the elasticity of demand when tickets are priced at $2.00 each.

10. (*Demand for lumber*) Suppose that the demand for a certain lumber is given by $x = 3000 - 250p$, for $0 \le p \le 12$, where p is the price in dollars per linear foot and x is the number of linear feet demanded. Determine the elasticity of demand when the price is $2.00 per linear foot.

11. (*Unit elasticity*) Suppose that the demand equation is $x = 800 - 16p$ for $0 \le p \le 50$. Determine the price at which the demand has unit elasticity.

12. (*Elasticity*) Determine the elasticity of demand E for the demand equation $x = a - pb$, where a and b are positive constants and p is between 0 and a/b.

13. (*Price increase*) Suppose that the demand equation is $x = 112 - 8p$ for $0 \le p \le 14$. If the price is $3, what percent decrease in demand will be created by increasing the price 2%?

14. (*Price increase*) Suppose that the demand equation is $x = 360 - 12p$ for $0 \le p \le 30$. If the price is $5, what percent decrease in demand will be created by increasing the price 3%?

15. (*Price increase*) Suppose the demand is given by $x = 280 - 8p$ for $0 \le p \le 35$. At what price (p in dollars) will it be true that for every 1% the price is increased, there will be a 4% decrease in demand?

16. (*Price decrease*) Suppose the demand is given by $x = 300 - 15p$ for $0 \le p \le 20$. At what price (p in dollars) will it be true that for every 1% the price is decreased, there will be a .5% increase in demand?

W 17. (*Public radio*) A local public radio station has priced a year's membership at $30. Their demand equation is $x = 10{,}000 - 80p$ for $0 \le p \le 125$, where x is the number of members and p is the price (in dollars) per membership. Will raising the price to $35 increase or decrease revenue? Explain.

W 18. (*Elastic demand*) Explain why lowering prices will increase revenue when demand is elastic.

W 19. (*Wholesaler's price*) If a brick wholesaler is currently selling bricks at 90¢ each and the demand is given by $x = 10{,}000 - 80p$ for $0 \le p \le 125$ cents, should the wholesaler increase or decrease his price to increase revenue? Explain.

20. (*Maximum revenue*) If the demand is $x = 432 - p^2$ for $0 \le p \le 20$ dollars, determine the price p at which revenue is maximum.

21. (*Maximum revenue*) Suppose $x = 1200 - 20\sqrt{p}$ for $0 \le p \le 3600$ is the demand for a commodity having price p dollars. For what price will the revenue be maximum?

22. (*Mineral cartel*) A mineral cartel is pricing their ore to maximize revenue. If the demand for the ore is $x = 50{,}000 - 40p$ tons when the price is p dollars per ton, what should be the price of a ton of ore?

23. (*Car wash*) A car wash company wants to price a wash at just the right price (p dollars) to maximize its revenue. If the weekly demand is known to be $1000 - 2p^3$, what should be their wash price?

Chapter List   *Important terms and ideas*

increasing function
decreasing function
relative maximum
relative minimum
critical number
first derivative test
concave up

concave down
point of inflection
second derivative test
sketching a graph
absolute maximum
absolute minimum

applied maximum/minimum
   problems
elasticity of demand
inelastic demand
elastic demand
unit elasticity

Review Exercises for Chapter 4

Determine where each function given in Exercises 1–4 is increasing and where it is decreasing.

1. $f(x) = -2x^2$

2. $f(x) = x^3 - 5$

3. $g(x) = \dfrac{2x}{x-1}$

4. $g(x) = \dfrac{100}{x}$

Find the critical numbers of each function in Exercises 5–10.

5. $f(x) = x^3 - 6x^2 - 36x + 5$ **6.** $f(x) = 10 - 2x^3$

7. $f(x) = 8\sqrt{x}$ **8.** $f(x) = 5 - \sqrt[3]{x}$

9. $f(x) = \dfrac{x}{x^2 + 1}$ **10.** $f(x) = 12x - 2x^{3/2}$

Determine all relative extreme points and sketch the graph of each function in Exercises 11–14.

11. $f(x) = x^3 - 3x^2 + 5$ **12.** $f(x) = -x^3 + 3x - 4$

13. $f(x) = 1 - x^{2/3}$ **14.** $f(x) = 1 + 6x^{2/3}$

In Examples 15–18, find all relative extreme points of the function, if any. Do not sketch the graph of the function.

15. $f(x) = x^4 - 8x^3 + 17$ **16.** $f(x) = 75x - x^3$

17. $g(x) = 6x^{1/3} - 20$ **18.** $g(x) = 16x + 6x^{4/3}$

Determine where the graph of each function in Exercises 19–20 is concave up and where it is concave down.

19. $f(x) = 4x^3 - 11x + 1$ **20.** $f(x) = \dfrac{x-2}{x}$

Find the point of inflection of each function in Exercises 21–22.

21. $f(x) = x^3 - 12x^2 + 17x - 4$

22. $f(x) = -x^3 - 9x^2 - 50x + 3$

23. Use the words *positive*, *negative*, and *zero* to complete the entries in the table. Refer to the graph of f shown.

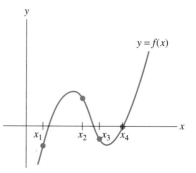

	$f(x)$	$f'(x)$	$f''(x)$
$x = x_1$	(a)	(b)	(c)
$x = x_2$	(d)	(e)	(f)
$x = x_3$	(g)	(h)	(i)
$x = x_4$	(j)	(k)	(l)

24. The questions that follow refer to the functions graphed in the figure below. Match the graph of the function—f or g or h—with the descriptive phrase.
  (a) Increasing at an increasing rate.
  (b) Increasing at a decreasing rate.
  (c) Increasing at a constant rate.
  (d) The slope of the curve is decreasing.
  (e) The curve is concave down on (x_1, x_2).
  (f) The second derivative is 0 on (x_1, x_2).

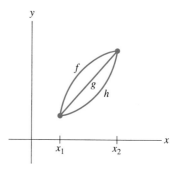

In Exercises 25–26, find the absolute extrema on the given interval.

25. $f(x) = x^3 + 3x^2 - 105x + 20$ on $[-8, 8]$

26. $f(x) = x^3 - 6x^2 + 9x + 2$ on $[1, 3.5]$

27. (*MINIMUM MARGINAL COST*) Suppose that the function $C(x) = x^3 - 42x^2 - 180x + 500$ gives the cost of producing x items. At what production level is the marginal cost minimum?

28. (*MAXIMUM REVENUE*) If the demand equation for your product is $p = 80 - .2x$, how many units should you sell in order to maximize your revenue?

29. (*SWIMMING POOL*) A wealthy family plans to have a rectangular swimming pool built on their estate grounds. If the perimeter of the pool will be 240 feet, what dimensions (length and width) will produce the pool of largest area?

30. (*PLAYGROUND AREA*) The parks department plans to fence in a 2304-square-meter rectangular area for a children's playground. What should be the dimensions in order to use the least amount of fencing?

31. (*ELASTICITY*) Determine the elasticity for $p = 10$, if the demand equation is $x = 200 - 4p$.

32. (*ELASTICITY*) If the demand equation is $x = 150 - 3p$, determine for what price the demand will be inelastic.

W 33. Is it possible for y to be increasing when dy/dt is decreasing? Explain.

5

EXPONENTIAL AND LOGARITHMIC FUNCTIONS

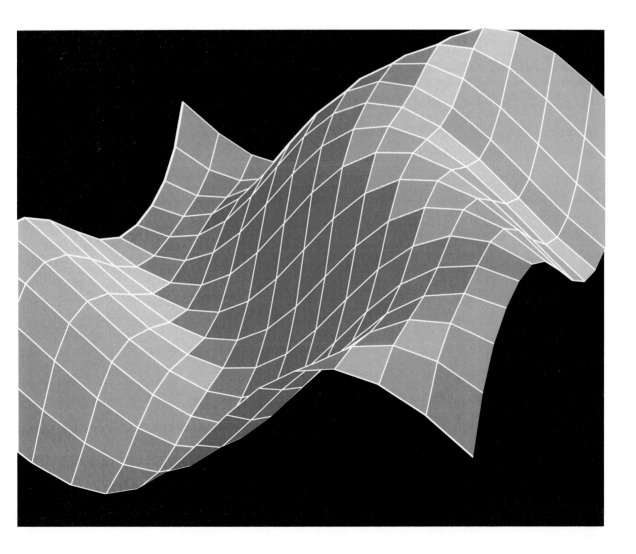

Y̲ou are about to begin the study of some special kinds of functions—exponential functions and logarithmic functions. You will have the opportunity to consider the graphs of these functions and investigate the various limits and derivatives associated with them. An unusual number, called *e*, will be introduced. And, of course, a variety of interesting applications awaits you.

5.1 | *EXPONENTIAL FUNCTIONS*

There are many applications in which the rate of growth of a quantity is proportional to the amount of the quantity present. In such cases, as the amount increases, so does the rate of growth increase. The expression *exponential growth* is used to describe these situations. Bacteria in a laboratory culture provide an example of such exponential growth. When the compounding of interest is done in a special manner called "continuous compounding," the balance in the account will grow exponentially.

Exponential functions provide the means to study such growth situations. The expressions involved have a form we have not seen yet—the base is constant and the exponent is a variable. Here are two examples.

$$f(x) = 3^x \qquad y = .5^x$$

(Notice how 3^x differs from x^3, which has a variable base and a constant exponent.)

In general,

Exponential Functions

$$f(x) = b^x$$

x is any real number

$$b > 0 \quad (b \neq 1)$$

Since the exponent x can be any real number, the domain of f is all the real numbers.

The graph of an exponential function such as $f(x) = 3^x$ provides a picture of these new functions and a chance to visualize exponential growth. We have selected values for x, computed the corresponding y values (3^x), and then passed a smooth curve through the points. See Figure 1.

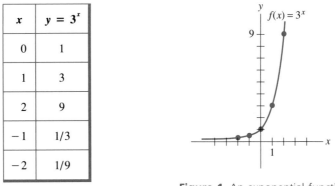

x	$y = 3^x$
0	1
1	3
2	9
−1	1/3
−2	1/9

Figure 1 An exponential function

If desired, points between any two of these points can be obtained by using a calculator. Here are some such points, each of which was determined by using the ⟨y^x⟩ key of a calculator. The 3^x values are approximate, of course.

x	3^x
.5	1.73
1.4	4.66
−.6	.52
−1.2	.27

A graph of $f(x) = .5^x$ can be obtained in a similar manner. Figure 2 shows some points and the graph.

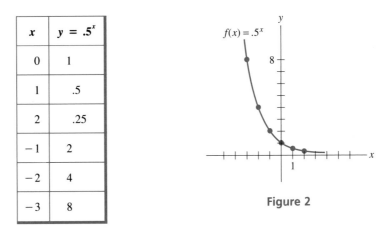

x	$y = .5^x$
0	1
1	.5
2	.25
−1	2
−2	4
−3	8

Figure 2

Note

For graphing and other applications, when evaluating an expression such as $4 \cdot 3^x$ for different values of x, keep in mind the usual order of operations—exponents are applied before multiplication is carried out, unless *parentheses* indicate otherwise. Thus, for $x = 2$, we have $4 \cdot 3^x = 4 \cdot 3^2 = 4 \cdot 9 = 36$.

One particular exponential function stands out because of its calculus properties and its widespread natural applications in a variety of areas. In fact, it is sometimes called "*the* exponential function." The base of this exponential function is the irrational number "e," which is approximately equal to 2.7182818284. We will write

$$e \approx 2.718$$

A study of the expression

$$\left(1 + \frac{1}{n}\right)^n$$

will lead to a formal definition of the number e. Using a calculator to study the value of the expression for various values of n, observe how the value of the expression gets closer and closer to e as n gets larger and larger.

n	$\left(1 + \dfrac{1}{n}\right)^n$
1	2
10	2.59374246
100	2.704813829
1000	2.716923932
10,000	2.718145927
1,000,000	2.718280469
1,000,000,000	2.718281827

As indicated before, the value of e correct to ten decimal places is 2.7182818284. A formal definition of e, based on our work, is given next.

Definition of e

$$e = \lim_{n \to \infty} \left(1 + \frac{1}{n}\right)^n$$

The number e was named by Leonhard Euler (1707–1783), the great Swiss mathematician. Using the definition of e, he calculated its value to 23 decimal places.

Figure 3 Leonhard Euler (1707–1783)

Note

If you find it strange that e has a formal definition of an algebraic nature, recall that the irrational number π (which is approximately 3.14) is defined as the ratio of the circumference of a circle to its diameter.

$$\pi = \frac{C}{d}$$

The exponential function with base e is written

$$f(x) = e^x$$

Figure 4 shows the graph of $f(x) = e^x$, where x can be any real number. Points for the graph were obtained using the rough approximations of 2.7 for e, 7.4 for e^2, and .4 for e^{-1}.

$y = e^x$

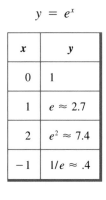

x	y
0	1
1	$e \approx 2.7$
2	$e^2 \approx 7.4$
-1	$1/e \approx .4$

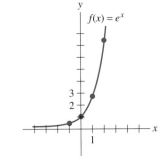

Figure 4 The graph of the exponential function $f(x) = e^x$

From the graph you can see that as x gets larger and larger, so does e^x get larger and larger. In fact,

$$\lim_{x \to \infty} e^x = \infty$$

Also, from this fact it follows that

$$\lim_{x \to \infty} \frac{1}{e^x} = 0 \qquad \text{or, equivalently,} \qquad \lim_{x \to \infty} e^{-x} = 0$$

The graph also shows that

$$\lim_{x \to -\infty} e^x = 0$$

Of course, the usual properties of exponents apply to expressions involving e.

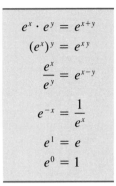

$$e^x \cdot e^y = e^{x+y}$$

$$(e^x)^y = e^{xy}$$

$$\frac{e^x}{e^y} = e^{x-y}$$

$$e^{-x} = \frac{1}{e^x}$$

$$e^1 = e$$

$$e^0 = 1$$

EXAMPLE 1 Simplify the expressions.

  (a) $5 - 2e^0$ **(b)** $\dfrac{e^x}{e}$ **(c)** $(e^{3t})^2$

SOLUTION **(a)** $5 - 2e^0 = 5 - 2 \cdot 1$ since $e^0 = 1$

  $= 5 - 2$

  $= 3$

  (b) $\dfrac{e^x}{e} = \dfrac{e^x}{e^1} = e^{x-1}$

  (c) $(e^{3t})^2 = e^{3t \cdot 2} = e^{6t}$ ◆

Compound Interest and e

Now we present an example that shows how the number e is related to compound interest, a familiar finance topic. Suppose you deposit P dollars into an account that pays annual interest rate r. Furthermore, assume that the interest is compounded m times a year. After t years the balance A in the account would be

$$A = P\left(1 + \frac{r}{m}\right)^{mt}$$

When interest is compounded, the depositor receives interest on the interest as well as on the original investment (principal). The more compoundings per year, the greater the balance will become. The short table below demonstrates this idea for an investment of $1000 at 7% per year.

m compoundings per year	balance after 1 year
1 (annually)	$1070.00
4 (quarterly)	$1071.86
12 (monthly)	$1072.29
365 (daily)	$1072.50

It is natural to wonder what might result if there were thousands or millions of compoundings. The number e enters the picture when an infinite number of compoundings per year, that is, "continuous compounding," is considered. We then have

$$\lim_{m \to \infty} P\left(1 + \frac{r}{m}\right)^{mt}$$

A few manipulations are needed to obtain the desired result.

$$\lim_{m \to \infty} P\left(1 + \frac{r}{m}\right)^{mt} = P \cdot \lim_{m \to \infty} \left(1 + \frac{r}{m}\right)^{mt}$$

$$= P\left[\lim_{m \to \infty} \left(1 + \frac{r}{m}\right)^{\frac{m}{r}}\right]^{rt}$$

Now let m/r be equal to n. This means that r/m is $1/n$ and that $n \to \infty$ as $m \to \infty$. We obtain

$$\lim_{m \to \infty} P\left(1 + \frac{r}{m}\right)^{mt} = P\left[\lim_{n \to \infty} \left(1 + \frac{1}{n}\right)^{n}\right]^{rt}$$

The limit inside the brackets is e, which means

$$\lim_{m \to \infty} P\left(1 + \frac{r}{m}\right)^{mt} = Pe^{rt}$$

In summary, if P dollars is invested at annual interest rate r **compounded continuously** for t years, the total amount accrued will be $A = Pe^{rt}$.

Continuous Compounding

$$A = Pe^{rt}$$

P = amount invested

r = annual interest rate

t = number of years

A = amount accrued

The formula $A = Pe^{rt}$ can be used to show that with continuous compounding, the $1000 at 7% would become $1072.51 in one year. The interest is just slightly more than with daily compounding.

The growth of money is but one example of **exponential growth** applications that involve formulas of this type. When the amount present *decreases* instead, then the application is called **exponential decay**.

> **Note**
>
> In the examples that follow, the value of e^x for specific positive values of x is determined by a *calculator* as follows:
>
> **1.** Enter the value of x.
>
> **2.** Press the $\boxed{e^x}$ key.
>
>   (If your calculator does not have an e^x key, then use $\boxed{\text{inv}}$ and $\boxed{\text{ln}}$ *or* $\boxed{\text{2nd}}$ and $\boxed{\text{ln}}$ *or* $\boxed{\text{shift}}$ and $\boxed{\text{ln}}$.)
>
> If no calculator is available, use Table 1 at the back of the book.

EXAMPLE 2 ◆ **CONTINUOUS COMPOUNDING**

A retiree invests \$10,000 at 8% interest per year compounded continuously. How much will she have after 3 years?

SOLUTION Use the formula for continuous compounding.

$$A = Pe^{rt}$$

with

$$P = \$10,000$$
$$r = .08 \qquad \text{(Express the percent as a decimal.)}$$
$$t = 3$$

The result is

$$
\begin{aligned}
A &= \$10,000 \cdot e^{(.08)(3)} \\
&= \$10,000 \cdot e^{.24} \\
&\approx \$10,000(1.2712) \\
&= \$12,712
\end{aligned}
$$

After 3 years she will have \$12,712. *Also*, note that calculators and tables provide only *approximate* values of $e^{.24}$ and most other e^x values. This is why the approximately equal to symbol was used here. ◆

EXAMPLE 3 ◆ **PRESENT VALUE**

How much should you invest now at 7% interest per year compounded continuously in order to have \$6000 in 5 years?

SOLUTION Use the continuous compounding formula.

$$A = Pe^{rt}$$

with

$$A = \$6000$$
$$r = .07$$
$$t = 5$$

The result is

$$\$6000 = P \cdot e^{(.07)(5)}$$

or

$$\$6000 = Pe^{.35}$$

Thus, the amount that must be invested, P, is

$$P = \frac{\$6000}{e^{.35}} \approx \frac{\$6000}{1.419} \quad \text{or} \quad \$4228$$

You must invest \$4228 now at 7% in order to have \$6000 in 5 years.

This amount invested, \$4228, is known as the **present value**. In general, the present value is the amount that must be invested now in order to have a specified amount at some time in the future. More will be said about present value in Section 5.5. ◆

A variety of other exponential growth applications involve essentially the same formula used for continuous compounding. Here is a short list of some of those applications. In each case, the rate of change is proportional to the amount present.

1. Increase or decrease in the population of a city

2. Increase or decrease in the number of bacteria in a culture

3. Decrease in the amount of radioactive material present

4. Effect of insecticides on an insect population

5. Epidemic spread of a disease

A knowledge of logarithms is needed to solve most of the problems that arise in such settings. Consequently, those applications will be presented in the next section, after logarithms have been introduced.

Logistic Growth

Logistic growth functions describe situations in which the environment inhibits what would otherwise be unrestricted exponential growth. The population begins to grow exponentially, but then levels off because of such problems as overcrowding and lack of food. In many instances it is more realistic to assume logistic growth rather than unlimited growth.

A typical logistic growth curve is shown in Figure 5.

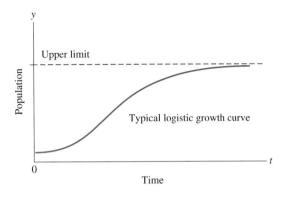

Figure 5 A logistic growth curve

EXAMPLE 4 ◆ *LOGISTIC GROWTH*

The fish population in a breeding lake is growing according to the logistic growth function

$$y = \frac{2000}{1 + 49e^{-.3t}}$$

where y is the number of fish present after t months.

(a) How many fish are present at the beginning?

(b) How many fish are present after 10 months?

(c) What is the maximum number of fish supported by this lake?

SOLUTION **(a)** At the beginning, $t = 0$, and the logistic growth function has the value

$$y = \frac{2000}{1 + 49 \cdot e^0} = \frac{2000}{50} = 40$$

There are 40 fish at the beginning.

(b) After 10 months, $t = 10$, and the logistic growth function has the value

$$y = \frac{2000}{1 + 49e^{-3}}$$

By calculator, $e^{-3} \approx .05$, and

$$y \approx \frac{2000}{1 + 2.45} \approx 580$$

Approximately 580 fish are present after 10 months.

(c) The maximum number of fish would exist, in theory, when $t \to \infty$. Accordingly, we consider the limit

$$\lim_{t \to \infty} e^{-.3t} = \lim_{t \to \infty} \frac{1}{e^{.3t}}$$

As $t \to \infty$, so does $.3t \to \infty$. This means that $e^{.3t} \to \infty$ also. The expression $1/e^{.3t}$ approaches 0 as $e^{.3t}$ approaches ∞. Now we can determine the limit of y as t approaches ∞.

$$\lim_{t \to \infty} \frac{2000}{1 + 49e^{-.3t}} = \frac{2000}{1 + 49 \cdot 0} = \frac{2000}{1} = 2000$$

We conclude that the maximum number of fish this lake will support is 2000. (See Figure 6.) ◆

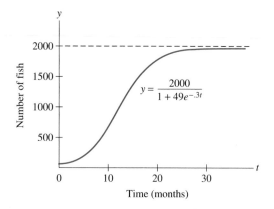

Figure 6 The logistic growth curve for the fish population in a breeding lake

Learning Curve

Psychologists have found that people often learn very quickly at the beginning. Then the rate of learning slows. Eventually, the person reaches a plateau that he or she cannot exceed. For example, someone learning typing will eventually reach a speed (words per minute) that he or she is unlikely to exceed. The general equation of such **learning curves** is

$$y = c(1 - e^{-kt})$$

where t is the time spent learning and y is the amount learned. The plateau, or upper limit of amount learned, is c. Figure 7 shows a graph of a learning curve.

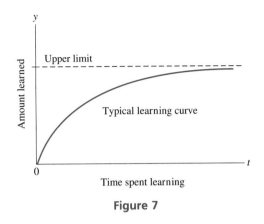

Figure 7

Exercise 53 provides an opportunity to study the learning curve associated with a specific test of learning.

5.1   Exercises

Sketch a graph of each function in Exercises 1–14.

1. $f(x) = 2^x$

2. $f(x) = 4^x$

3. $f(x) = 2^{-x}$

4. $f(x) = 3^{-x}$

5. $f(x) = \left(\dfrac{1}{3}\right)^x$

6. $f(x) = \left(\dfrac{1}{4}\right)^x$

7. $y = .4^x$

8. $y = .6^x$

9. $y = 10^x$

10. $y = 5^x$

11. $y = 3 \cdot 2^x$

12. $y = 2 \cdot 3^x$

13. $f(x) = 2^x + 1$

14. $f(x) = 2^x - 3$

Sketch a graph of each function in Exercises 15–22.

15. $f(x) = e^x$

16. $f(x) = e^{-x}$

17. $f(x) = e^{x+1}$

18. $f(x) = e^{x-1}$

19. $y = 1 + e^x$

20. $y = 2 + e^x$

21. $y = e^x - 2$

22. $y = e^x - 1$

Simplify each expression in Exercises 23–40.

23. $4e^0$

24. $9e^0$

25. $2 - 7e^0$

26. $2e^0 - 10$

27. $\dfrac{e^x}{e^2}$

28. $\dfrac{e^x}{e^3}$

29. $\dfrac{e^{2x}}{e}$

30. $\dfrac{e}{e^x}$

31. $2e^x e^{-x}$

32. $3e^{2x} e^{-2x}$

33. $e^{3x} e$

34. $e^{7x} e$

35. $e^x e^{6x}$

36. $e^{5x} e^x$

37. $(e^{2x})^3$

38. $(e^{3x})^2$

39. $\dfrac{2e^{3x}}{e^{-1}}$

40. $\dfrac{10e^{2x}}{e^{-1}}$

41. (CONTINUOUS COMPOUNDING) If \$1000 is invested at 8% annual interest compounded continuously, how much will accrue in 5 years?

42. (CONTINUOUS COMPOUNDING) If $4000 is invested at 7% annual interest compounded continuously, how much will accrue in 3 years?

43. (CONTINUOUS COMPOUNDING) If Dorys invests $2000 at $6\frac{1}{2}\%$ annual interest compounded continuously, how much will she have at the end of 4 years?

44. (CONTINUOUS COMPOUNDING) If Adolfo invests $3000 at $7\frac{1}{2}\%$ annual interest compounded continuously, how much will he have at the end of 5 years?

45. (PRESENT VALUE) A couple wishes to invest now for their young child's college education. How much should they invest at 9% interest per year compounded continuously in order to have $30,000 in 14 years?

46. (PRESENT VALUE) David is setting aside money for a future vacation trip. How much should he invest at 7% interest per year compounded continuously so that he will have $2500 in 3 years?

47. (MUTUAL FUND) A mutual fund claims that its portfolio value grew at the annual rate of 12% compounded continuously from January 1982 to January 1991. If $10,000 had been invested in the fund in January 1982, what would it have been worth in January 1991?

48. (REAL ESTATE) A corporation has purchased its new headquarters building for $40,000,000. If the building appreciates at the rate of 6% per year compounded continuously, what will be its value 10 years after purchase?

49. (FISH POPULATION) The fish population of a breeding lake is growing according to the logistic growth function

$$y = \frac{1500}{1 + 29e^{-.4t}}$$

where y is the number of fish present after t months.
(a) How many fish are present at the beginning?
(b) How many fish are present after 5 months?
(c) What is the maximum number of fish supported by this lake?

50. (FISH POPULATION) Redo Exercise 49 assuming that the logistic growth function is

$$y = \frac{800}{1 + 39e^{-.35t}}$$

Also, in part (b), use 4 months.

51. (BACTERIA GROWTH) A bacteria culture is growing according to the logistic growth function

$$y = \frac{10,000,000}{1 + 5000e^{-.1t}}$$

where y is the number of bacteria present after t hours.
(a) To the nearest thousand, how many bacteria are present initially?
(b) How large can this culture become?

52. (FRUIT FLY BREEDING) A biology student is breeding fruit flies in a closed container. Suppose the fly population grows according to

$$y = \frac{20}{1 + 9e^{-.2t}}$$

where y is the number of fruit flies present after t days.
(a) How many fruit flies are present after 2 weeks?
(b) How many fruit flies would you expect to have eventually?

53. (LEARNING TEST) To test learning, a psychologist asks people to memorize a long sequence of digits, checking with them every few minutes to see how many digits they have memorized. Assume the learning is described by

$$y = 18(1 - e^{-.3t})$$

where y is the number of digits memorized and t is the time in minutes.
(a) Using the equation given, verify that the number of digits memorized at the very beginning is zero.
(b) Approximately how many digits will people memorize in 1, 2, 3, 4, 5, and 10 minutes? (You will need a calculator.)
(c) According to the equation, what is the upper limit on the number of digits people can memorize in this manner?
(d) Sketch a rough graph of $y = 18(1 - e^{-.3t})$ based on your answers to parts (a), (b), and (c).

54. (PROFIT) If $C(x) = 20xe^{-.03x}$ dollars is the cost of producing x units and $R(x) = 20xe^{.02x}$ dollars is the revenue from the sale of x units, determine the profit from the production and sale of 10 units.

55. (COST) Suppose x units of a product can be supplied at a cost of $30e^{-.02x}$ dollars each.
(a) Determine the cost function $C(x)$.

(b) Determine the cost of 5 units.

56. (*PROFIT*) If x items can be supplied at $20e^{-.03x}$ dollars each and sold at \$25 each, determine the profit function.

57. Graphs of $f(x) = 3^x$ and $f(x) = .5^x$ were presented early in the section. For each of these functions,
  (a) Write the equation of its asymptote.
  (b) Write the limit that shows why that line is an asymptote.

58. The number e can also be defined as

$$e = \lim_{h \to 0}(1 + h)^{1/h}$$

Use a calculator and h values such as 1, .1, .001, .00001, and .0000001 to create a convincing table, as was done for the original definition of e given in the text.

59. Consider $f(x) = b^x$ for $b = 1$ in order to see why b^x is not considered exponential when $b = 1$. Sketch a graph of $f(x) = b^x$ for $b = 1$.

W 60. Compare exponential growth functions and logistic growth functions.
  (a) In what portion of the graph are they similar?
  (b) In what portion of the graph are they different? Explain *how* they are different and *why* each type of function should be expected to behave the way it does.

W 61. Review the graph of the exponential function $f(x) = e^x$ (Figure 4).
  (a) What is the nature of the tangent lines to the graph of f, and how does this relate to the increasing and/or decreasing nature of this function?
  (b) Comment on the concavity of the graph of $f(x) = e^x$.

The Exercise Library at the back of the book contains graphing calculator and computer exercises keyed to this section.

5.2 | LOGARITHMIC FUNCTIONS

The study of **logarithms** is motivated by the need to solve exponential equations such as

$$b^x = a$$

for x. To be able to express x in terms of a and b in this setting requires a manipulation not yet presented. In fact, a new notation must be introduced in order to solve this problem. In view of this, we define $x = \log_b a$ to mean the same as $b^x = a$.

Logarithm Notation

$$x = \log_b a$$

means the same as

$$b^x = a$$

where $b > 0$, $b \neq 1$, and x is any real number.

The equation $x = \log_b a$ is read as "x is the *logarithm* of a to the base b" or as "x is the logarithm to the base b of a." Notice that the logarithm is equal to x, and x is the exponent. Thus, *a logarithm is an exponent*. In this case, x is the exponent to which b must be raised to produce a.

The number a results from raising positive base b to some power x. Consequently, a is always positive in $b^x = a$. In turn, this means that a must be positive in $x = \log_b a$. In other words, $\log_b a$ is defined only for $a > 0$.

> The logarithm of a negative number (or zero) is not defined.

Here are some examples of equations written in logarithmic form along with their equivalent exponential form.

logarithmic equation	exponential equation
$\log_3 9 = 2$	$3^2 = 9$
$\log_2 8 = 3$	$2^3 = 8$
$\log_4 1 = 0$	$4^0 = 1$
$\log_{10} .01 = -2$	$10^{-2} = .01$
$\log_e 20 \approx 3$	$e^3 \approx 20$
$\log_e e = 1$	$e^1 = e$
$\log_e 1 = 0$	$e^0 = 1$

Note

The two most popular bases for logarithms are 10 and e. Logarithms using base 10 are called *common logarithms*, and "$\log_{10}$" is written simply "log." When no base is written, base 10 is assumed. Logarithms using base e are called *natural logarithms*, and "$\log_e$" is written "ln."

> $\log_{10} x$ is written $\log x$
> $\log_e x$ is written $\ln x$

Two of the logarithmic equation examples given above ($\log_e e = 1$ and $\log_e 1 = 0$) include important facts that are often used to make simplifications. They are restated next.

$$\boxed{\begin{aligned} \ln e &= 1 \\ \ln 1 &= 0 \end{aligned}}$$

Here are some examples that show the use of logarithms for manipulation and equation solving.

EXAMPLE 1 Solve for x: $3^x = 11$.

SOLUTION Rewriting the equation in logarithmic form will solve it for x.

$$x = \log_3 11 \quad \blacklozenge$$

EXAMPLE 2 Solve for x: $10^{7x} = 9$.

SOLUTION Rewriting the exponential equation in logarithmic form will isolate the exponent $7x$.

$$7x = \log 9$$

Now, divide both sides by 7.

$$x = \frac{\log 9}{7} \quad \blacklozenge$$

EXAMPLE 3 Solve for x: $4e^{5x} = 12$.

SOLUTION Based on recent experience, it seems natural to want to write this equation in logarithmic form in order to solve for the exponent. But in order to rewrite an equation in logarithmic form, the base and exponent must be alone on one side. Indeed, that was the case in the definition of logarithm notation as well as in Examples 1 and 2. And, of course, it will also be the case here after both sides of the equation $4e^{5x} = 12$ are divided by 4.

$$\frac{4e^{5x}}{4} = \frac{12}{4}$$

or

$$e^{5x} = 3$$

In logarithmic form, this is

$$5x = \ln 3$$

After dividing both sides by 5, the equation is solved.

$$x = \frac{\ln 3}{5} \quad \blacklozenge$$

Related manipulations can be used to solve logarithmic equations.

EXAMPLE 4   Solve for x: $\ln 7x = 50$.

SOLUTION   Just as exponential equations can be solved by changing to logarithmic form, so can logarithmic equations be solved by changing to exponential form. In this example, write

$$\log_e 7x = 50$$

as

$$7x = e^{50}$$

and then

$$x = \frac{e^{50}}{7} \qquad \blacklozenge$$

EXAMPLE 5   Solve for x: $3 \ln 2x = 10$.

SOLUTION   The logarithm must be alone before the equation can be changed to exponential form. Divide both sides by 3 to accomplish this.

$$3 \ln 2x = 10$$

becomes

$$\ln 2x = \frac{10}{3}$$

In exponential form, it is

$$2x = e^{10/3}$$

Finally,

$$x = \frac{e^{10/3}}{2} \qquad \blacklozenge$$

A sketch of the graph of **the natural logarithm function $f(x) = \ln x$** (where $x > 0$) can be made by obtaining several points and then passing a curve through them. Unless you prefer to use a calculator, it is easier to obtain points when the equation $y = \ln x$ is written in exponential form as $x = e^y$. Then you can select y values and obtain corresponding x values.

$$y = \ln x$$

x	y
1	0
$e \approx 2.7$	1
$e^2 \approx 7.4$	2
$1/e \approx .4$	-1

A graph of the natural logarithm function is shown in Figure 8.

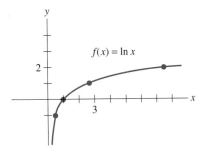

Figure 8 The natural logarithm function

It can be seen by looking at the graph of $y = \ln x$ (or by using a calculator) that as x gets larger and larger, so does $\ln x$ get larger and larger. In fact,

$$\lim_{x \to \infty} \ln x = \infty$$

This result will be needed in Chapter 7. The graph also shows that

$$\lim_{x \to 0^+} \ln x = -\infty$$

Two particularly important results are given next. Each is a restatement of the definition of the natural logarithm: that $y = \ln x$ means the same as $x = e^y$.

> **1.** $e^{\ln x} = x$ for $x > 0$
>
> **2.** $\ln e^x = x$ for all real x

The first result, $e^{\ln x} = x$, is true because when this exponential equation is written in its equivalent logarithmic form, it becomes $\log_e x = \ln x$, which is clearly a true statement. The second result, $\ln e^x = x$, is true because when it is rewritten in exponential form, it becomes $e^x = e^x$, which is a true statement.

EXAMPLE 6 Simplify each expression.   **(a)** $4 \ln e^7$ **(b)** $1 + e^{\ln 3x}$

SOLUTION   **(a)** $4 \ln e^7 = 4 \cdot \ln e^7$

$\qquad\qquad = 4 \cdot 7$ since $\ln e^x = x$

$\qquad\qquad = 28$

(b) $1 + e^{\ln 3x} = 1 + 3x$ since $e^{\ln u} = u$ ◆

There are three important properties of logarithms that are useful in manipulations. These properties are stated next for natural logarithms, although they are also true for other bases. The numbers a and b must be positive; otherwise $\ln a$ and $\ln b$ would not be defined.

Properties of Logarithms

1. $\ln (a \cdot b) = \ln a + \ln b$

2. $\ln \dfrac{a}{b} = \ln a - \ln b$

3. $\ln a^p = p \cdot \ln a$

where $a > 0$, $b > 0$, and p is any real number.

The properties are used to expand or condense algebraic expressions, as shown in the examples that follow. The expanding of algebraic expressions will prove particularly useful with differentiation in Section 5.4. The condensing of algebraic expressions will be useful for simplification of algebraic expressions.

EXAMPLE 7 Use the properties of logarithms to expand each expression.

(a) $\ln \dfrac{x + 2}{x + 5}$ (b) $\ln 3^{2x}$ (c) $\ln e^3 x$

SOLUTION (a) Using logarithm property 2 yields

$$\ln \frac{x + 2}{x + 5} = \ln (x + 2) - \ln (x + 5)$$

(b) Using logarithm property 3 yields

$$\ln 3^{2x} = 2x \ln 3$$

(c) $\ln e^3 x = \ln e^3 + \ln x$ by logarithm property 1

$\qquad\quad = 3 + \ln x$ since $\ln e^x = x$ for any x ◆

EXAMPLE 8 Use properties of logarithms to condense each expression.

(a) $\ln 3 + \ln x$ (b) $\ln x - \ln 2$ (c) $2 \ln x$

SOLUTION (a) Using logarithm property 1,

$$\ln 3 + \ln x = \ln 3x$$

(b) By logarithm property 2,

$$\ln x - \ln 2 = \ln \frac{x}{2}$$

(c) By logarithm property 3,

$$2 \ln x = \ln x^2 \quad \blacklozenge$$

Exponential Growth and Decay

In Section 5.1 it was mentioned that a knowledge of logarithms is needed to solve most of the exponential growth problems described there. Now we can consider those applications, each of which uses basically the same formula.

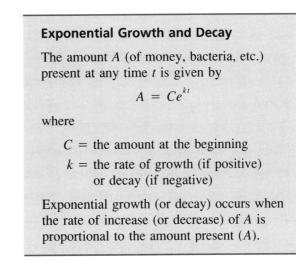

> **Exponential Growth and Decay**
>
> The amount A (of money, bacteria, etc.) present at any time t is given by
>
> $$A = Ce^{kt}$$
>
> where
>
> C = the amount at the beginning
>
> k = the rate of growth (if positive) or decay (if negative)
>
> Exponential growth (or decay) occurs when the rate of increase (or decrease) of A is proportional to the amount present (A).

When we say that the formula $A = Ce^{kt}$ applies to situations in which the rate of growth is proportional to the amount present, we mean that as the amount present increases, so does the rate of growth increase. Exercise 114 of Section 5.3 offers the opportunity to show that when $A = Ce^{kt}$, the rate of growth is indeed proportional to the amount present. Later in the book, the formula will be derived from a differential equation based on this "proportional" idea.

> *Note*
>
> In the examples that follow, the value of $\ln x$ for specific positive values of x is determined by using a *calculator* as follows:
>
> **1.** Enter the value of x.
>
> **2.** Press the ⌊ln⌋ key.
>
> Some calculators may have an exponential key but no logarithm key. In such cases, use the 2nd key or shift key or inverse key *together with* ⌊e^x⌋ to get the effect of the ⌊ln⌋ key.
>
> If a calculator is not available, use Table 2 at the back of the book.

EXAMPLE 9   ◆   CONTINUOUS COMPOUNDING

At what annual interest rate compounded continuously should money be invested in order to double in 8 years?

SOLUTION   In $A = Ce^{kt}$, the interest rate we seek is k, the growth rate. The time is $t = 8$ years. To say that money doubles is to say that the amount C at the beginning becomes $2C$ (that is, *double C*) in 8 years. Thus, $A = Ce^{kt}$ becomes

$$2C = Ce^{k(8)}$$

After we divide both sides by C, the equation is

$$2 = e^{8k}$$

Changing to logarithmic form in order to solve for k yields

$$8k = \ln 2$$

or

$$k = \frac{\ln 2}{8}$$

$$\approx \frac{.6931}{8}$$

$$\approx .087$$

Since the decimal number .087 is equivalent to 8.7%, we conclude that the money should be invested at 8.7% in order to double in 8 years. In this example, 8 years is the **doubling time**. More will be said about doubling times in Section 5.5.   ◆

EXAMPLE 10   ◆   BACTERIA GROWTH

The number of bacteria in a culture increases from 400 to 1000 in 3 hours. Assuming bacteria growth to be exponential, how many bacteria will be present in the culture after 10 hours?

SOLUTION   In order to determine the number of bacteria present after 10 hours, we must have a formula of the form $A = Ce^{kt}$ in which both C and k are known. Then we can substitute 10 for t and obtain the corresponding "amount" A.

To begin, we know that C is 400, since the statement of the problem implies that there are 400 bacteria in the culture at the beginning. Thus we have

$$A = 400e^{kt}$$

We also know that A is 1000 when $t = 3$, since the number of bacteria increases to 1000 in 3 hours.

$$1000 = 400e^{k(3)}$$

This equation can be solved to determine k and thus produce a completed formula of the form $A = Ce^{kt}$. Begin by dividing both sides by 400.

$$2.5 = e^{3k}$$

To solve for k, we obtain logarithmic form.

$$3k = \ln 2.5$$

or

$$k = \frac{\ln 2.5}{3} \approx \frac{.9163}{3} \quad \text{or} \quad .305$$

Thus, the formula is

$$A = 400e^{.305\,t}$$

After 10 hours, when $t = 10$, the number of bacteria A is

$$A = 400e^{.305(10)}$$
$$= 400e^{3.05}$$
$$\approx 400(21.115)$$
$$= 8446$$

After 10 hours there will be approximately 8446 bacteria.
 A graph showing the bacteria growth is given in Figure 9.

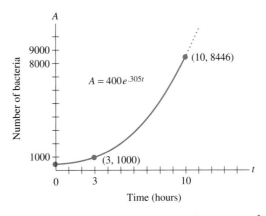

Figure 9 Exponential growth of bacteria ◆

EXAMPLE 11 ◆ *HALF-LIFE OF A RADIOACTIVE ELEMENT*

Radium is a radioactive element that decays exponentially. The half-life of radium is approximately 1600 years. If we begin with an 80-gram mass of radium, how many grams will remain 200 years from now?

SOLUTION The **half-life** of a radioactive element is the amount of time it takes for half of the mass of the element to decay (that is, decompose into some other material). A half-life of 1600 years means that the 80 grams of radium will be 40 grams of radium when $t = 1600$, with the decay occurring exponentially over the 1600-year period. In terms of $A = Ce^{kt}$, we have

$$40 = 80e^{k(1600)}$$

or, upon dividing both sides by 80,

$$.5 = e^{1600k}$$

The value of k can be determined by writing this exponential equation in logarithmic form.

$$1600k = \ln .5$$

or

$$k = \frac{\ln .5}{1600}$$

$$\approx \frac{-.6931}{1600}$$

$$\approx -.0004$$

The decay rate is $-.0004$. The *minus* indicates a decay or decrease.

If the k value $-.0004$ is used in $A = Ce^{kt}$ for this situation in which we have 80 grams of radium, then we will have the formula for the amount of radium at any time t, namely,

$$A = 80e^{-.0004t}$$

In 200 years, $t = 200$, and the amount of radium remaining is

$$A = 80e^{-.0004(200)}$$

$$= 80e^{-.08}$$

$$\approx 80(.9231)$$

$$\approx 73.85$$

Approximately 73.85 grams of radium will remain 200 years from now.

A graph showing the radioactive decay of radium is given in Figure 10.

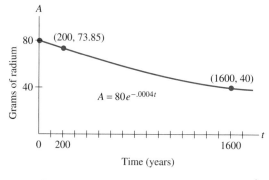

Figure 10 Radioactive decay of radium   ◆

EXAMPLE 12   ◆   *RADIOACTIVE DATING*

In 1948, Dr. Willard Libby devised *radiocarbon dating*, which is used by archaeologists to determine the age of ancient objects they uncover. In life, all plants and animals acquire a certain amount of carbon-14, depending on their weight. Because carbon-14 decays (half-life approximately 5730 years), plants and animals are continually acquiring carbon-14 to replace the amount lost by decay. After death, no new carbon-14 is acquired, and so the amount present is gradually depleted by decay.

Suppose that a leg bone found in 1985 contained 70% of the carbon-14 that would be found in the living animal. How old is the bone?

SOLUTION Using $A = Ce^{kt}$ and the fact that the half-life of carbon-14 is 5730 years, we can determine the decay constant k. Note that whatever the mass C of carbon-14 at the beginning, after 5730 years, it will be reduced to half of C, or $.5C$. Thus, we have

$$.5C = Ce^{k(5730)}$$

or, upon dividing both sides of the equation by C,

$$.5 = e^{5730k}$$

Writing this exponential equation in logarithm form will enable us to solve for k.

$$5730k = \ln .5$$

or

$$k = \frac{\ln .5}{5730}$$

$$\approx \frac{-.6931}{5730}$$

$$\approx -.00012$$

The decay constant is $-.00012$, and the decay formula appears as

$$A = Ce^{-.00012t}$$

Since the leg bone contained 70% of the carbon-14 that a live animal's leg bone would contain, we can replace A by $.70C$ (70% of C).

$$.70C = Ce^{-.00012t}$$

This equation can be solved for t, the time of death of the animal, which is the approximate age of the leg bone. In steps, divide both sides by C and then change to logarithm form to isolate the t term.

$$.70 = e^{-.00012t}$$

$$-.00012t = \ln .70$$

$$t = \frac{\ln .70}{-.00012}$$

$$\approx \frac{-.3567}{-.00012}$$

$$= 2972.5$$

We conclude that the leg bone was approximately 2973 years old when found in 1985. Realistically, the method does not pinpoint the age to the nearest year. Rounding to the nearest hundred years would be reasonable in such cases.   ◆

Note

The carbon-14 dating method can only be used to date objects that are less than 50,000 years old. Older objects contain less than .25% carbon-14 and cannot be dated accurately in this manner. Instead, *potassium-argon dating* is used. All living plants and animals contain potassium-40. Upon death, the potassium-40 decays (to argon-40) with a half-life of 1.3 billion years.

5.2   Exercises

In Exercises 1–6, write each exponential equation in logarithmic form.

1. $2^5 = 32$

2. $3^4 = 81$

3. $10^2 = 100$

4. $10^{-2} = .01$

5. $e^0 = 1$

6. $e^m = u$

In Exercises 7–14, write each logarithmic equation in exponential form.

7. $\log_3 9 = 2$

8. $\log_2 64 = 6$

9. $\log 100 = 2$

10. $\log .01 = -2$

11. $\log .1 = -1$

12. $\ln 1 = 0$

13. $\ln e = 1$

14. $\ln x = t$

In Exercises 15–24, solve each equation for x.

15. $10^x = 2$

16. $10^x = 6$

17. $e^x = 3$

18. $e^x = 4$

19. $e^{3x} = 2$

20. $e^{2x} = 13$

21. $e^{-x} = 50$

22. $e^{-x} = 12$

23. $3e^{5x} = 42$

24. $2e^{4x} = 20$

In Exercises 25–34, solve the equation for x.

25. $\log_3 x = 4$

26. $\log_2 x = 5$

27. $\ln x = -2$

28. $\ln x = 6$

29. $\ln 4x = 30$

30. $\ln 10x = 50$

31. $\ln 3x = 0$

32. $\ln 4x = 1$

33. $5 \ln 3x = 40$

34. $7 \ln 5x = 28$

Simplify each expression in Exercises 35–48.

35. $\ln e^4$

36. $\ln e^{18}$

37. $5 \ln e^2$

38. $4 \ln e$

39. $e \ln 1$

40. $e^{\ln 9}$

41. $3x \ln e^x$

42. $x \ln e^x$

43. $1 + 2e^{\ln 3}$

44. $2 + 5e^{\ln 3}$

45. $\ln \sqrt{e}$

46. $\ln \dfrac{1}{\sqrt{e}}$

47. $\ln \dfrac{1}{e}$

48. $-e^{\ln 7}$

In Exercises 49–60, apply logarithm properties in order to expand each expression. (Simplify afterwards, when possible.)

49. $\ln xy$

50. $\ln 4x$

51. $\ln ex$

52. $\ln e^2 x$

53. $\ln 7x^2$

54. $\ln \dfrac{1}{x}$

55. $\ln \sqrt{x}$

56. $\ln 5\sqrt{x}$

57. $\ln 2^{3x}$

58. $\ln 4^{x+1}$

59. $\ln x^x$

60. $\ln (x + 1)^{5x}$

In Exercises 61–70, apply logarithm properties to condense each expression.

61. $\ln x + \ln 2$

62. $\ln 3 + \ln x$

63. $\ln 4 - \ln x$

64. $\ln (x + 1) - \ln 2$

65. $3 \ln x$

66. $-2 \ln x^{-3}$

67. $\ln 5 + 2 \ln x$

68. $2 \ln 3 + 3 \ln x$

69. $\ln 3 + \ln x + \ln y$

70. $\ln 2 + \ln x - \ln 7$

The use of calculators is recommended for the applied problems that follow. However, the same answers (or nearly the same answers) can be obtained by using the tables at the back of the book.

71. *(COMPOUND INTEREST)* If $4000 is invested at 7% interest per year compounded continuously, how long will it take to double the original investment?

72. *(COMPOUND INTEREST)* If $5000 is invested at 9% interest per year compounded continuously, how long will it take to double the original amount invested?

73. *(COMPOUND INTEREST)* At what annual interest rate, compounded continuously, will money triple in 12 years?

74. *(COMPOUND INTEREST)* At what annual interest rate, compounded continuously, will money double in 9 years?

75. *(POPULATION GROWTH)* The population of Clearwater, Florida was 52,000 in 1970 and 85,000 in 1980. Assuming exponential growth, determine k for this population.

76. *(BACTERIA GROWTH)* The number of bacteria in a lab culture triples every hour. If we start with a culture of 1000 bacteria, how long will it take to have 100,000 bacteria?

77. *(BACTERIA GROWTH)* Assuming that the number of bacteria in a culture doubles every hour, how long will it take to have a million bacteria, if we begin with 2500 bacteria?

78. *(INSECT POPULATION GROWTH)* The insect population in a large field is estimated to be 50,000. An entomologist studies the environment and determines that the population is likely to grow at the rate of 3% a year. How many years will it take for the population to reach 120,000?

79. *(INSECT POPULATION GROWTH)* If the insect population described in Exercise 78 grows at the rate of only 2% per year, how long will it take the population to reach 120,000?

80. *(FRUIT FLY POPULATION GROWTH)* A fruit fly population doubles every 10 days. If there are now 60 fruit flies, how many will there be in two weeks?

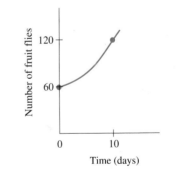

81. *(LAND APPRECIATION)* Suppose that an acre of land purchased in mid 1988 for $20,000 is worth $54,000 in mid 1993.
   (a) What is the annual rate of growth of this investment, assuming continuous compounding?
   (b) If the land continues to appreciate at the same rate, in what year will it be worth $80,000?

82. *(MUTUAL FUND APPRECIATION)* Maureen invested $4000 in a mutual fund in June 1987. In June 1992 her investment was worth $13,500.
   (a) What is the annual rate of growth of this investment, assuming continuous compounding?

(b) If the mutual fund continues to appreciate at the same rate, how much will her investment be worth in June 1995?

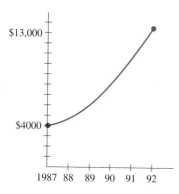

Maureen's investment, from 1987 to 1992

83. *(HALF-LIFE)* If an element has a half-life of 14 years, how much of a 90-gram mass of the element will still remain after 100 years?

84. *(HALF-LIFE)* How much of a 300-gram mass of lead 210 will remain after 200 years? The half-life of lead 210 is 22 years.

85. *(RADIOCARBON DATING)* If a skull found in 1970 contained 35% of the carbon-14 found in a living animal, determine the age of the skull. (Use 5730 years as the half-life of carbon-14. Obtain k to five decimal places.)

86. *(RADIOCARBON DATING)* A wood ax handle found in 1990 contained 40% of the carbon-14 found in a living tree. How old is the ax handle? (Use 5730 years as the half-life of carbon-14. Obtain k to five decimal places.)

87. *(NUCLEAR TESTING)* Atmospheric testing of nuclear weapons introduced radioactive strontium-90 into the environment. Because it is now present in food and water and is chemically similar to calcium, strontium-90 does occur in milk and bones, and it can cause bone cancer. The half-life of strontium-90 is 28 years.
(a) Determine the decay constant k.
(b) A treaty banning the atmospheric testing of nuclear weapons was signed in 1963. Assuming that there has been no such testing since 1963 and that there will be no such testing

in the future, by what year will the level of strontium-90 be reduced to 10% of the amount present in 1963?

88. *(RADIATION IN MEDICINE)* Doctors use the radiation from cobalt-60 to arrest the growth of cancer cells. Cobalt-60 has a half-life of 5.27 years.
(a) Find the decay constant k.
(b) How much of a 2-microgram mass of cobalt-60 remains after 20 years?

89. *(RADIOACTIVE TRACERS)* Iodine-131 is used in medicine as a radioactive tracer. When the radioactive element is combined with other elements, it can be sent through the body and monitored by a radiation detector. Since iodine-131 concentrates in the thyroid gland and the liver, it is used to study thyroid and liver problems. Iodine-131 has a half-life of 8.6 days.
(a) Determine the decay constant k.
(b) What percent of the iodine-131 remains in the body 5 weeks after its use as a tracer?

90. *(RADON DECAY)* The following graph shows the decay of the radioactive element polonium-210, the deadly product of radon decay. Examine the graph to determine (to the nearest 10 days) the half-life of polonium-210.

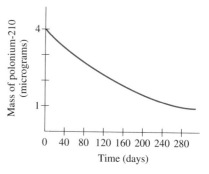

91. *(DRUG CONCENTRATION)* A drug taken intravenously is removed from the blood at a rate that declines exponentially. If the concentration of the drug is 50 units per volume when administered and the decay constant is known to be $-.04$, determine the following, assuming that time is measured in minutes.
(a) The concentration after 30 minutes.
(b) The concentration after 2 hours.

(c) The time it takes the body to *eliminate* 90% of the drug from the bloodstream.

92. (DRUG CONCENTRATION) Redo Exercise 91 assuming that the concentration of the drug is 200 units per volume when administered and that the decay rate is $-.05$.

93. (REVENUE) A manufacturer can sell x units at the price per unit (dollars) given by the following demand function.

$$D(x) = \frac{50}{\ln(x + 3)}$$

(a) Determine the function that gives the manufacturer's revenue from the sale of x units.
(b) What is the revenue from the sale of 30 units?

94. (DEMAND) If the demand function of Exercise 93 were instead

$$D(x) = \frac{50}{\ln x}$$

what restriction would need to be placed on x?

95. (LARVAE DEVELOPMENT) A study of mosquito larvae yielded the following formula for duration D of development as a function of the environment temperature T.

$$\log D = a - b \log T$$

The numbers a and b are constants, a being the number of day-degrees required for completion of a developmental stage and b being the minimum temperature at which development can proceed.
(a) The given equation is solved for $\log D$, rather than D. Solve it for D.
(b) Solve the equation for T.

96. (EARTHQUAKE MEASUREMENT) The *Richter scale* is used to measure and report the magnitude of earthquakes. Specifically, for Southern California,

$$M = \log \frac{I}{I_0}$$

where M is the magnitude of the earthquake, I_0 is the smallest measurable intensity, and I is the intensity of the earthquake being measured.
(a) What magnitude on the Richter scale would correspond to an earthquake of intensity $1000I_0$ (1000 times the intensity of I_0)?

(b) How much greater is the intensity of an earthquake of magnitude 5 than that of an earthquake of magnitude 3?

97. (DOUBLING RATE) The time and interest rate needed for money to double can be studied by considering $A = Pe^{rt}$ with A being equal to $2P$ (that is, double P).
(a) Use $2P$ for A and determine an equation that gives r in terms of t.
(b) Use the result of part (a) to complete the following table of doubling rates and times (in years).

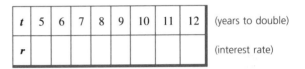

t	5	6	7	8	9	10	11	12	(years to double)
r									(interest rate)

98. (DOUBLING TIME) Redo Exercise 97 by obtaining t in terms of r and completing the table that follows.

r	5	6	7	8	9	10	11	12	(interest rate)
t									(years to double)

99. In Sections 5.1 and 5.2 you were introduced to four exponential-related graphs—exponential growth, exponential decay, logistic growth, and the learning curve.

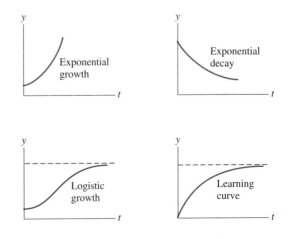

Read each of the situations presented in parts (a) through (h) and decide which of the four types of graphs would best describe it.

(a) (*Lunch business*) A restaurant advertises its lunch specials extensively. As a result, its lunch business begins to increase exponentially. The "lunch hour" is from 11 a.m. to 1:30 p.m. and the restaurant seats 124 people.

(b) (*Pizza making*) Some inexperienced pizza makers have begun work at a pizzeria. Their productivity will improve with experience.

(c) (*Drug elimination*) A nurse gives a patient a drug intravenously. The body acts to remove the drug from the blood at an exponential rate in order to distribute it to organs that will either use it or eliminate it. Our focus is on the amount of the drug remaining in the blood.

(d) (*Population growth*) Although the human population of the world has been growing at a rate of approximately 2% per year, this kind of growth will be restricted by limited space, shortages of food, lack of medical care, and other problems.

(e) (*Investing for retirement*) To accumulate money for retirement needs, Sylvia invests $3000 at 7% annual interest compounded continuously.

(f) (*Use of enzymes*) Increasing the concentration of regular enzymes increases the rate of chemical reaction exponentially. However, the rate of reaction has a bound or limit that cannot be exceeded, regardless of the amount of regular enzyme used.

(g) (*Larva growth*) After a very short period of slow growth, the housefly larva grows quickly. In fact, its weight doubles every 13 hours. Finally, its growth rate slows considerably.

(h) (*Spread of a rumor*) Sociologists studying the spread of a rumor have observed that rumors spread at an exponential rate for a while. However, there is a limited audience (say, at a school), so there is an eventual slowdown in the rate at which the rumor spreads.

100. (*pH*) In chemistry, the pH of a solution is defined as follows:

$$pH = -\log [H^+]$$

where H^+ is the hydrogen concentration (in moles per liter). A "neutral" solution such as water has a pH of 7. Acid solutions have pH values less than 7, and alkaline solutions have pH values greater than 7.

(a) Use the fact that the hydrogen ion concentration of pure water is .0000001 (or 10^{-7}) moles per liter to verify that the pH of water is 7.

(b) What is the pH of acid rain in which the concentration of hydrogen ions is .000031 moles per liter?

(c) What is the concentration of hydrogen ions in the stomach during digestion if the hydrochloric acid solution at that time has a pH of 3.2?

(d) What is the pH of a shampoo in which the concentration of hydrogen ions is .000000009 moles per liter?

101. (*Cooking a roast*) If a cool object (such as meat to be cooked) is placed in a warm environment (such as an oven), the cooler object will be warmed according to a formula that is similar to one we have been using. Specifically, the temperature y of the object placed in a warm environment is given by $y = S - Ce^{kt}$, where S is the temperature of the warmer environment, t is the time the object is in the environment, k is the rate-of-warming constant, and C is a constant that is readily determined from y and S. Assume that a roast at $60°F$ (slightly cooler than room temperature) is placed in a $375°$ oven. After 45 minutes, the roast's temperature is $120°$.

(a) What is the value of the constant C?

(b) Determine the rate-of-warming constant k.

(c) How many minutes will it take for the roast to be $160°$ (cooked)?

102. (*Making ice*) When a warm body is placed in a cooler environment, the body will cool according to $y = Ce^{kt} + S$, where S is the temperature of the cool environment, y is the temperature of the warm body, t is the time, k is the rate-of-cooling constant, and C is a constant. Suppose that $70°$ tap water is used to fill an ice cube tray and then the tray is placed in a freezer that is kept at $15°F$.

(a) Determine C.

(b) Determine the cooling constant k, if the water temperature has been reduced to $45°$ in 15 minutes.

(c) What will be the temperature of the water after 30 minutes?

103. In the formula $A = Ce^{kt}$, C is the amount at the beginning. Prove this statement by showing that $A = C$ when $t = 0$.

W 104. Throughout this section there are statements such as "ln x for $x > 0$" and "ln a for $a > 0$." Why must there be this restriction that we cannot have logarithms of negative numbers or zero?

W 105. We have claimed that a logarithm is an exponent. In $\log_b a = c$, the logarithm ($\log_b a$) is equal to c. Is c an exponent? Explain.

W 106. (*HALF-LIFE*) Explain the concept of half-life by using 64 grams of an element with a half-life of 10 years. Include a time period of 30 years in your explanation.

W 107. Review the graph of the natural logarithm function $f(x) = \ln x$ (Figure 8).
  (a) What is the nature of the tangent lines to the graph of f and how does this relate to the increasing and/or decreasing nature of this function?
  (b) Comment on the concavity of the graph of $f(x) = \ln x$.

108. Using $k = -.00012$, show that the amount of carbon-14 remaining after 50,000 years is less than .25%. This exercise is a reference to the note after Example 12.

5.3 | *DIFFERENTIATION OF EXPONENTIAL FUNCTIONS*

It seems natural to wonder about differentiating the exponential and logarithmic functions. We can now proceed to obtain a rule for differentiating $f(x) = e^x$. In the next section we will consider the derivative of $f(x) = \ln x$.

As suggested in Chapter 3, the *definition* of the derivative is often used when no rule for differentiation appears to be at hand. Accordingly, to determine $f'(x)$ for $f(x) = e^x$, we proceed by using the definition of the derivative.

$$f'(x) = \lim_{\Delta x \to 0} \frac{f(x + \Delta x) - f(x)}{\Delta x}$$

$$= \lim_{\Delta x \to 0} \frac{e^{x+\Delta x} - e^x}{\Delta x}$$

At this point, a property of exponents is used to rewrite $e^{x+\Delta x}$ as $e^x \cdot e^{\Delta x}$.

$$f'(x) = \lim_{\Delta x \to 0} \frac{e^x \cdot e^{\Delta x} - e^x}{\Delta x}$$

Now, e^x can be factored out in the numerator.

$$f'(x) = \lim_{\Delta x \to 0} \frac{e^x(e^{\Delta x} - 1)}{\Delta x}$$

Next, apply the theorem "the limit of a product is the product of the limits."

$$f'(x) = \lim_{\Delta x \to 0} e^x \cdot \lim_{\Delta x \to 0} \frac{e^{\Delta x} - 1}{\Delta x}$$

In the first limit, $\Delta x \to 0$ has no effect on e^x. The limit as $\Delta x \to 0$ of e^x is e^x.

$$f'(x) = e^x \cdot \lim_{\Delta x \to 0} \frac{e^{\Delta x} - 1}{\Delta x} \qquad (*)$$

The limit that remains,

$$\lim_{\Delta x \to 0} \frac{e^{\Delta x} - 1}{\Delta x}$$

is equal to 1.

$$\lim_{\Delta x \to 0} \frac{e^{\Delta x} - 1}{\Delta x} = 1$$

The limit can be considered informally by using a calculator to show that as Δx gets closer and closer to zero, the expression gets closer and closer to 1. The calculated entries in the table are correct to nine decimal places. The table presents a plausible case for the limit being 1.

Δx	$\dfrac{e^{\Delta x} - 1}{\Delta x}$
.1	1.051709181
.01	1.005016708
.001	1.000500160
.0001	1.000050000
.00001	1.000000000

Since that limit is 1, we have at last, from equation ($*$),

$$f'(x) = e^x \cdot 1 = e^x$$

Thus,

$$\frac{d}{dx}(e^x) = e^x$$

That the derivative of e^x is itself e^x is a most unusual result. Furthermore, this property makes e the natural choice of base when applications involve exponential functions. It is a fact that $y = e^x$ and $y = ce^x$ are the only functions (other than $y = 0$) with this property that the derivative of the function is equal to the function.

EXAMPLE 1   Differentiate $f(x) = x^3 e^x$.

SOLUTION   The expression $x^3 e^x$ is a product; $x^3 e^x = x^3 \cdot e^x$. Using the product rule for differentiation, we have the following result.

$$f'(x) = x^3 \cdot \frac{d}{dx}(e^x) + e^x \cdot \frac{d}{dx}(x^3)$$

$$= x^3 \cdot e^x + e^x \cdot 3x^2$$

$$= x^3 e^x + 3x^2 e^x$$

The expression can be simplified algebraically by factoring out $x^2 e^x$ from each term. The result is

$$f'(x) = x^2 e^x (x + 3)$$

This factored form is desirable for determining critical numbers. ◆

EXAMPLE 2   Determine $\dfrac{dy}{dx}$ if $y = \dfrac{e^x + 7}{x^2}$.

SOLUTION   Applying the quotient rule yields

$$\frac{dy}{dx} = \frac{(x^2)(e^x) - (e^x + 7)(2x)}{(x^2)^2}$$

$$= \frac{x^2 e^x - 2x e^x - 14x}{x^4}$$

This expression can be simplified (that is, the fraction reduced) by factoring x out of the numerator and denominator and then eliminating the x's by division. The result is

$$\frac{dy}{dx} = \frac{x e^x - 2 e^x - 14}{x^3} \quad ◆$$

EXAMPLE 3   If $y = \sqrt{e^x - x}$, determine dy/dx.

SOLUTION   Rewrite the expression with a numerical exponent.

$$y = (e^x - x)^{1/2}$$

Differentiation, using the general power rule, yields

$$\frac{dy}{dx} = \frac{1}{2}(e^x - x)^{-1/2} \cdot \frac{d}{dx}(e^x - x)$$

$$= \frac{1}{2}(e^x - x)^{-1/2}(e^x - 1)$$

The factor $(e^x - x)^{-1/2}$ can be written in the denominator as $(e^x - x)^{1/2}$ or as $\sqrt{e^x - x}$. The final result is

$$\frac{dy}{dx} = \frac{e^x - 1}{2\sqrt{e^x - x}} \quad ◆$$

The chain rule can be applied to obtain the derivative of e^u, where u is a function of x. Here

$$\frac{dy}{dx} = \frac{dy}{du} \cdot \frac{du}{dx}$$

becomes

$$\frac{d}{dx}(e^u) = \frac{d}{du}(e^u) \cdot \frac{du}{dx}$$

and then

$$\frac{d}{dx}(e^u) = e^u \cdot \frac{du}{dx}$$

EXAMPLE 4   Obtain the derivative of $y = e^{8x}$.

SOLUTION
$$\frac{dy}{dx} = \frac{d}{dx}(e^{8x})$$

$$= e^{8x} \cdot \frac{d}{dx}(8x)$$

$$= e^{8x} \cdot 8$$

$$= 8e^{8x} \quad \blacklozenge$$

EXAMPLE 5   If $y = (1 + e^{3x})^{12}$, determine dy/dx.

SOLUTION   The general power rule must be applied here.

$$\frac{dy}{dx} = 12(1 + e^{3x})^{11} \cdot \frac{d}{dx}(1 + e^{3x})$$

$$= 12(1 + e^{3x})^{11}(e^{3x} \cdot 3)$$

Upon rearranging and simplifying, we obtain

$$\frac{dy}{dx} = 36e^{3x}(1 + e^{3x})^{11} \quad \blacklozenge$$

EXAMPLE 6   Show that the function $f(x) = e^{2x}$ is always increasing.

SOLUTION   Recall that a function is increasing when its derivative is positive. Since $f(x) = e^{2x}$, it follows that

$$f'(x) = 2e^{2x}$$

Note that $e^{2x} > 0$ for all x, so that $2e^{2x} > 0$ for all x. This means that $f'(x) > 0$ for all x and that the function f is always increasing. $\blacklozenge$

EXAMPLE 7   Determine all critical numbers of $f(x) = 1 + xe^x$.

SOLUTION   Begin by obtaining $f'(x)$.

$$f'(x) = xe^x + e^x = e^x(x + 1)$$

To find the critical numbers of this function, consider the numbers for which the derivative is zero. Since e^x is never zero, $f'(x) = (e^x)(x + 1)$ is zero only when the factor $x + 1$ is zero, that is, when $x = -1$. Thus, -1 is the only critical number. (Keep in mind that the original function f must be defined at -1 for the number -1 to be a critical number. And in this case it is.)   ◆

EXAMPLE 8   Use implicit differentiation to find dy/dx.

$$xe^y + ye^x = x$$

SOLUTION   To begin,

$$\frac{d}{dx}(xe^y) + \frac{d}{dx}(ye^x) = \frac{d}{dx}(x)$$

Observe that since y is (implicitly) a function of x, the term xe^y is a *product* of two functions of x, namely x and e^y. Similarly, ye^x is also a product of two functions of x, namely y and e^x. This means that the product rule must be used in the differentiation. Continuing, we have

$$x \cdot e^y \cdot \frac{dy}{dx} + e^y \cdot 1 + y \cdot e^x + e^x \cdot \frac{dy}{dx} = 1$$

Isolating the dy/dx terms produces

$$xe^y \frac{dy}{dx} + e^x \frac{dy}{dx} = 1 - e^y - ye^x$$

To obtain dy/dx, factor out dy/dx and then divide both sides of the equation by the coefficient of dy/dx.

$$(xe^y + e^x)\frac{dy}{dx} = 1 - e^y - ye^x$$

$$\frac{dy}{dx} = \frac{1 - e^y - ye^x}{xe^y + e^x} \quad ◆$$

Up to this point, the only exponential functions being differentiated have been those involving base e. A knowledge of logarithms can be used to enable us to determine the derivatives of exponential functions having other bases.

To begin, consider $f(x) = a^x$, where a is a positive number. Next, using properties studied earlier in the chapter, we can rewrite a^x as a power of e. This will enable us to differentiate it.

$$a^x = e^{\ln a^x} \qquad \text{since } e^{\ln u} = u$$

$$ = e^{x \ln a} \qquad \text{since } \ln a^p = p \ln a$$

We can now proceed to find the derivative of a^x by differentiating $e^{x \ln a}$.

$$f'(x) = \frac{d}{dx}(a^x)$$

$$= \frac{d}{dx}(e^{x \ln a})$$

$$= e^{x \ln a} \cdot \frac{d}{dx}(x \ln a)$$

Keep in mind that since a is a constant, $\ln a$ is also a constant. This means that the derivative of $x \ln a$ is simply $\ln a$, and

$$f'(x) = e^{x \ln a} \cdot \ln a$$

If $e^{x \ln a}$ is written in its original form, a^x, the result is

$$f'(x) = a^x \ln a$$

Thus,

$$\frac{d}{dx}(a^x) = a^x \ln a$$

EXAMPLE 9   Let $f(x) = 3^x$. Find $f'(x)$.

SOLUTION   The expression 3^x is of the form a^x.

$$f'(x) = 3^x \ln 3 \quad \blacklozenge$$

EXAMPLE 10   Let $y = 10^x$. Determine dy/dx.

SOLUTION   The expression 10^x is of the form a^x.

$$\frac{dy}{dx} = 10^x \ln 10 \quad \blacklozenge$$

If the chain rule is applied to obtain the derivative of a^u, where u is a function of x, the result is

$$\frac{d}{dx}(a^u) = a^u \frac{du}{dx} \ln a$$

EXAMPLE 11   Find the derivative of $f(x) = 4^{x^2+1}$.

SOLUTION   The expression 4^{x^2+1} is of the form a^u, where $u = x^2 + 1$.

$$f'(x) = \frac{d}{dx}(4^{x^2+1})$$

$$= 4^{x^2+1} \cdot 2x \ln 4 \quad \blacklozenge$$

EXAMPLE 12   If $f(x) = \dfrac{3^{2x}}{5x}$, determine $f'(x)$.

SOLUTION   The expression is a quotient, so use the quotient rule with 3^{2x} as numerator and $5x$ as denominator.

$$f'(x) = \frac{d}{dx}\left(\frac{3^{2x}}{5x}\right)$$

$$= \frac{(5x)\dfrac{d}{dx}(3^{2x}) - (3^{2x}) \cdot \dfrac{d}{dx}(5x)}{(5x)^2}$$

$$= \frac{5x \cdot 3^{2x} \cdot 2 \cdot \ln 3 - 3^{2x} \cdot 5}{25x^2}$$

$$= \frac{5 \cdot 3^{2x}(2x \ln 3 - 1)}{5 \cdot 5x^2}$$

$$= \frac{3^{2x}(2x \ln 3 - 1)}{5x^2} \quad \blacklozenge$$

5.3   Exercises

Determine the derivative of each function in Exercises 1–12.

1. $y = x^2 e^x$

2. $y = -e^x$

3. $y = 5e^x$

4. $y = \dfrac{e^x}{x}$

5. $f(x) = (e^x + 2)^4$

6. $f(x) = \sqrt{1 - e^x}$

7. $y = \dfrac{e^x + 1}{x}$

8. $y = \dfrac{x}{e^x + 1}$

9. $f(x) = \dfrac{1}{(e^x + x)^3}$

10. $f(x) = \dfrac{1}{(2x + e^x)^2}$

11. $f(x) = \dfrac{e^x}{e^x + 1}$

12. $f(x) = \dfrac{e^x - 1}{e^x + 1}$

Determine the derivative of each function in Exercises 13–40.

13. $f(x) = e^{6x-1}$

14. $f(x) = 2e^{-3x}$

15. $f(x) = e^{-x^2}$

16. $f(x) = e^{1+x^3}$

17. $f(x) = e^{\sqrt{x}}$

18. $f(x) = e^{1/x}$

19. $y = .4e^{-5x^2}$

20. $y = .5e^{2x}$

21. $y = 1 + \dfrac{e^x}{4}$

22. $y = xe^{1/x}$

23. $f(x) = \dfrac{e^{7x}}{2}$

24. $f(x) = \dfrac{e^{-3x}}{7}$

25. $y = xe^{-x}$

26. $y = x^2 e^{-x}$

27. $y = 5e^{1-x^2}$

28. $f(x) = x^2 e^{-8x}$

29. $f(x) = \dfrac{e^{3x}}{x+1}$ **30.** $f(x) = \dfrac{1+3x}{e^{2x}}$

31. $y = \dfrac{e^{1+2x}}{x^2}$ **32.** $y = \dfrac{e^{2x-1}}{x}$

33. $y = \dfrac{e^x + e^{-x}}{2}$ **34.** $y = \dfrac{e^x - e^{-x}}{2}$

35. $f(x) = \dfrac{5x}{e^x}$ **36.** $f(x) = \dfrac{3}{e^x}$

37. $y = x^2 e^{1+3x}$ **38.** $y = xe^{x^2+1}$

39. $y = (1 + e^{5x})^{10}$ **40.** $y = (2 + e^{4x})^7$

In Exercises 41–44, determine $f''(x)$.

41. $f(x) = e^{3x}$ **42.** $f(x) = 2e^{-x}$

43. $f(x) = xe^x$ **44.** $f(x) = e^{x^2}$

In Exercises 45–52, determine the values of x for which the function is increasing and the values of x for which it is decreasing.

45. $f(x) = 3e^{5x}$ **46.** $f(x) = -2e^{4x}$

47. $f(x) = 1 + e^{-2x}$ **48.** $f(x) = 3 - e^{-x}$

49. $f(x) = 1 + xe^x$ **50.** $f(x) = xe^{2x}$

51. $f(x) = e^{x^2}$ **52.** $f(x) = (1 + x)e^{2x}$

For each function in Exercises 53–62, determine (a) all critical numbers and (b) all relative extrema.

53. $y = x^2 e^x$ **54.** $y = x^3 e^x$

55. $y = xe^{x/2}$ **56.** $y = x^2 e^{3x}$

57. $y = xe^{2x}$ **58.** $y = xe^{3x}$

59. $y = x^2 e^{-x}$ **60.** $y = xe^{-2x}$

61. $y = (1 + x)e^{3x}$ **62.** $y = (x - 1)e^{2x}$

W 63. For $f(x) = e^x$, focus on $f'(x)$ to explain why the exponential function has no relative extrema. Be complete in explaining your search for critical numbers.

W 64. Repeat Exercise 63 using $f(x) = e^{-x}$.

Sketch the graph of each function in Exercises 65–68. Use calculus methods to assist you.

65. $y = x - e^x$ **66.** $y = e^{-x^2}$

67. $y = e^x + e^{-x}$ **68.** $y = e^x - e^{-x}$

69. (*COST/MARGINAL COST*) Suppose that x units of a commodity can be supplied at a cost of $50e^{-.04x}$ dollars *each*.
(a) Construct the cost function $C(x)$.
(b) Determine the marginal cost.
(c) Determine the cost of 10 units and the marginal cost when $x = 10$ units.
(d) Determine when marginal cost is zero.

70. (*MARGINAL REVENUE*) If $R(x) = 1000e^{.02x}$ is the revenue from the sale of x units of a product, determine the marginal revenue.

71. (*MARGINAL PROFIT*) If the cost of producing x units is $C(x) = 40xe^{-.02x}$, and if the revenue from the sale of x units is $R(x) = 40xe^{.03x}$, determine the marginal profit function.

72. (*INCREASING PROFIT*) Suppose that a manufacturer's profit from the production and sale of x units is $P(x) = 25xe^{-.2x}$ dollars. When is the manufacturer's profit increasing?

73. (*MAXIMUM PROFIT*) Let the cost of producing x units be $C(x) = 30xe^{-x/10}$ dollars and the revenue from the sale of x units be $R(x) = 50xe^{-x/10}$ dollars. What should be the production level in order to maximize the profit?

74. (*ATMOSPHERIC PRESSURE*) The atmospheric pressure P in pounds per square inch at a height of x miles above sea level is given by

$$P = 14.7e^{-.21x}$$

(a) Determine the pressure 10 miles above sea level.
(b) What is the rate of change in the atmospheric pressure 10 miles above sea level?

75. (*POLLUTANT CONCENTRATION*) The concentration of pollutants in a river decreases exponentially as the distance downstream from the source of the pollution increases. Suppose that

$$c(x) = .001e^{-2x}$$

gives the concentration of pollutants in grams per cubic centimeter, where x is the number of miles downstream from the source of pollution.
(a) Determine the concentration of pollutants at the source of pollution and one mile downstream.

(b) Determine the rate of change of the concentration of pollutants with respect to the distance downstream from the source of pollution.

76. Determine the slope of the line tangent to the curve $y = e^x$ at any point (x, y).

77. Determine the slope of the line tangent to the curve $y = 1 + e^{4x}$ at the point $(0, 2)$.

78. Determine the slope of the line tangent to the curve $y = 2xe^{-x}$ at the point $(0, 0)$.

79. Determine the equation of the line tangent to the graph of $y = e^x$ at the point where the slope of the tangent line is e.

80. Determine the equation of the line tangent to the graph of $y = e^x$ at the point where the slope of the tangent line is 1.

Use implicit differentiation to find dy/dx in Exercises 81–92.

81. $x + y + xe^y = 10$ **82.** $x + y + ye^x = 2$

83. $e^{xy} - 2x = y + 3$ **84.** $x^2 + 3e^{xy} - y^2 = 1$

85. $xe^y - ye^x = y$ **86.** $x^2 + y^2 - e^{2y} = 0$

87. $e^{x+y} + x^2 + y^2 = 0$ **88.** $e^{y+1} + e^{2x} + 3y = 2x$

89. $e^{-y} + y + e^{2x} = x$ **90.** $3e^x + e^{2y} + 7x = 3$

91. $xe^{y^2} + x^2e^y = y$ **92.** $ye^{x^2} - 3xy + x^2 = y$

Determine the derivative of each function in Exercises 93–106.

93. $y = 2^x$ **94.** $y = 1.5^x$

95. $y = 7^{4x}$ **96.** $y = 2^{3x}$

97. $f(x) = 3^{-x^2}$ **98.** $f(x) = 3^{1/x}$

99. $f(x) = 2 \cdot 9^{1+x^2}$ **100.** $f(x) = 4 \cdot 3^{-x^2}$

101. $y = x \cdot 2^{3x}$ **102.** $y = x^2 10^x$

103. $y = \dfrac{3^x}{2x}$ **104.** $y = \dfrac{x+1}{2^x}$

105. $f(x) = \dfrac{1 - 10^{2x}}{3x}$ **106.** $f(x) = \dfrac{x^2}{1 + 2^x}$

Determine the derivative of each function in Exercises 107–112. Be careful. Think!

107. $y = e^3 + 1$ **108.** $y = \sqrt{e}$

109. $y = e^2x$ **110.** $y = \dfrac{ex}{7}$

111. $y = \dfrac{x}{1 + e}$ **112.** $y = \dfrac{x}{e}$

113. The graph of $f(x) = 2^{-x^2}$ has one relative maximum point. Find it.

114. This exercise is intended to show that when $A = Ce^{kt}$, the rate of growth is proportional to the amount present.
  (a) Differentiate $A = Ce^{kt}$ to obtain dA/dt.
  (b) In $dA/dt = k \cdot Ce^{kt}$, replace Ce^{kt} by A, since $A = Ce^{kt}$. The result is

$$\frac{dA}{dt} = k \cdot A$$

which says that the rate of change of A is a constant times A, that is, proportional to the amount present A.

The Exercise Library at the back of the book contains graphing calculator and computer exercises keyed to this section.

5.4 | DIFFERENTIATION OF LOGARITHMIC FUNCTIONS

To obtain a formula for the derivative of the logarithmic function $f(x) = \ln x$, we begin with a result from Section 5.2.

$$e^{\ln x} = x$$

Differentiating both sides of this equation yields

$$\frac{d}{dx}(e^{\ln x}) = \frac{d}{dx}(x)$$

or

$$e^{\ln x} \cdot \frac{d}{dx}(\ln x) = 1$$

Since $e^{\ln x} = x$, we can substitute x for $e^{\ln x}$ on the left side of the equation above. The result is

$$x \cdot \frac{d}{dx}(\ln x) = 1$$

or

$$\frac{d}{dx}(\ln x) = \frac{1}{x}$$

This is a most important result. We insist that $x > 0$, so that $\ln x$ is defined.

$$\frac{d}{dx}(\ln x) = \frac{1}{x} \qquad x > 0$$

EXAMPLE 1 Find $f'(x)$ when $f(x) = x^2 \ln x$.

SOLUTION The expression $x^2 \ln x$ is the product of x^2 and $\ln x$. Consequently the product rule will be used for the differentiation.

$$f'(x) = x^2 \cdot \frac{d}{dx}(\ln x) + (\ln x) \cdot \frac{d}{dx}(x^2)$$

$$= x^2 \cdot \frac{1}{x} + (\ln x) \cdot 2x$$

$$= x + 2x \ln x$$

If desired (say, for determining critical numbers), the expression $x + 2x \ln x$ can be factored and written as $x(1 + 2 \ln x)$. ◆

EXAMPLE 2 Differentiate $y = 6(\ln x)^3$.

SOLUTION

$$\frac{dy}{dx} = \frac{d}{dx}[6(\ln x)^3]$$

$$= 6 \cdot \frac{d}{dx}(\ln x)^3$$

$$= 6 \cdot 3(\ln x)^2 \cdot \frac{1}{x}$$

$$= \frac{18(\ln x)^2}{x} \quad ◆$$

EXAMPLE 3   Find dy/dx for $y = \dfrac{\ln x}{x^2}$.

SOLUTION   The quotient rule applies here.

$$\frac{dy}{dx} = \frac{d}{dx}\left(\frac{\ln x}{x^2}\right)$$

$$= \frac{x^2 \cdot \dfrac{1}{x} - (\ln x)(2x)}{(x^2)^2}$$

$$= \frac{x - 2x \ln x}{x^4}$$

x can be factored out of the numerator and denominator in order to simplify (reduce) this fraction.

$$\frac{dy}{dx} = \frac{x(1 - 2 \ln x)}{x(x^3)}$$

Finally,

$$\frac{dy}{dx} = \frac{1 - 2 \ln x}{x^3} \qquad \blacklozenge$$

If the chain rule is applied to obtain the derivative of $y = \ln u$, where u is a function of x, then

$$\frac{dy}{dx} = \frac{dy}{du} \cdot \frac{du}{dx} = \frac{d}{du}(\ln u) \cdot \frac{du}{dx} = \frac{1}{u} \cdot \frac{du}{dx}$$

$$\boxed{\frac{d}{dx}(\ln u) = \frac{1}{u} \cdot \frac{du}{dx}}$$

EXAMPLE 4   Differentiate $y = \ln (x^2 + 1)$.

SOLUTION   In this example, $u = x^2 + 1$.

$$\frac{dy}{dx} = \frac{d}{dx} \ln (x^2 + 1)$$

$$= \frac{1}{x^2 + 1} \cdot \frac{d}{dx}(x^2 + 1)$$

$$= \frac{1}{x^2 + 1} \cdot 2x$$

$$= \frac{2x}{x^2 + 1} \qquad \blacklozenge$$

A look at the final form of the derivative obtained in Example 4 suggests that for some situations a different form of the derivative would simplify the differentiation process. Instead of

$$\frac{1}{u} \cdot \frac{du}{dx}$$

we can use

$$\frac{du/dx}{u}$$

$$\frac{d}{dx}(\ln u) = \frac{du/dx}{u}$$

Thus, the differentiation of Example 4 can be done in one step, as

$$\frac{d}{dx} \ln (x^2 + 1) = \frac{2x}{x^2 + 1}$$

This alternative form of the derivative of ln u is often a better choice, but not always. The alternative form will be of great value when techniques of integration are studied in Chapter 7.

EXAMPLE 5 $y = \ln (x^3 + 9)^5$. Find dy/dx.

SOLUTION   Using the formula for the derivative of ln u, we have

$$\frac{dy}{dx} = \frac{\dfrac{d}{dx}(x^3 + 9)^5}{(x^3 + 9)^5}$$

Now, applying the general power rule yields

$$\frac{dy}{dx} = \frac{5(x^3 + 9)^4 \cdot 3x^2}{(x^3 + 9)^5}$$

The factor $(x^3 + 9)^4$ appears in both the numerator and denominator. When it is eliminated by division, the result is

$$\frac{dy}{dx} = \frac{15x^2}{x^3 + 9}$$

This is the final result, but *observe now how property 3 of logarithms could have been used to simplify the procedure.* Using this property,

$$y = \ln (x^3 + 9)^5$$

can be written as

$$y = 5 \ln (x^3 + 9)$$

Now the calculus is much easier.

$$\frac{dy}{dx} = 5 \cdot \frac{d}{dx} \ln (x^3 + 9)$$

$$= 5 \cdot \frac{3x^2}{x^3 + 9}$$

$$= \frac{15x^2}{x^3 + 9} \quad \blacklozenge$$

EXAMPLE 6   Determine the derivative of $f(x) = \ln \sqrt{1 + x^2}$.

SOLUTION   Begin by changing the radical to an exponent.

$$f(x) = \ln (1 + x^2)^{1/2}$$

The calculus will be simpler if property 3 of logarithms is used to rewrite $\ln (1 + x^2)^{1/2}$ as $\frac{1}{2} \ln (1 + x^2)$.

$$f(x) = \frac{1}{2} \ln (1 + x^2)$$

Now,

$$f'(x) = \frac{1}{2} \cdot \frac{2x}{1 + x^2}$$

or

$$f'(x) = \frac{x}{1 + x^2}$$

If you are not convinced of the wisdom of this shortcut approach, try determining the derivative of $f(x) = \ln (1 + x^2)^{1/2}$ without using property 3 of logarithms.   ◆

EXAMPLE 7   Find all relative extrema of $f(x) = 2x - \ln 2x$.

SOLUTION   Begin the search for critical numbers by finding $f'(x)$.

$$f'(x) = 2 - \frac{1}{x}$$

Now determine where $f'(x)$ is zero or undefined.

Clearly $f'(x)$ is undefined when $x = 0$. However, the original function $f(x) = 2x - \ln 2x$ is also undefined for $x = 0$, since $\ln 0$ is not defined. This means that 0 is *not* a critical number.

$f'(x)$ is zero when

$$2 - \frac{1}{x} = 0$$

$$\frac{1}{x} = 2$$

$$2x = 1$$

$$x = \frac{1}{2}$$

And since $f\left(\frac{1}{2}\right)$ is defined, $\frac{1}{2}$ is a critical number.

The second derivative is easy to obtain, so use the second derivative test to see if $f\left(\frac{1}{2}\right)$ is a relative extremum.

$$f''(x) = \frac{1}{x^2}$$

and

$$f''\left(\frac{1}{2}\right) = 4 > 0$$

Since $f''\left(\frac{1}{2}\right) > 0$, function f has a relative minimum at $\frac{1}{2}$, and $f\left(\frac{1}{2}\right)$ is that relative minimum.

$$f\left(\frac{1}{2}\right) = 2 \cdot \frac{1}{2} - \ln\left(2 \cdot \frac{1}{2}\right) = 1 - 0 = 1$$

Thus, 1 is a relative minimum, the only relative extremum of the function $f(x) = 2x - \ln 2x$. See Figure 11. ◆

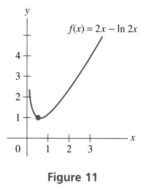

Figure 11

EXAMPLE 8   Is the graph of the function $f(x) = 2x^2 - 5 \ln x$ increasing or decreasing at the point (1, 2)?

SOLUTION   Determine $f'(1)$. If $f'(1) > 0$, then the graph is increasing at (1, 2). If $f'(1) < 0$, then the graph is decreasing at (1, 2).

$$f'(x) = 4x - \frac{5}{x}$$

$$f'(1) = 4 - 5 = -1 < 0$$

Since $f'(1) < 0$, the graph of f is decreasing at $(1, 2)$. ◆

EXAMPLE 9 Use implicit differentiation to find dy/dx.

$$3x + y + \ln(xy) = 0$$

SOLUTION First note that it will be easier to differentiate $\ln(xy)$ if it is written instead as $\ln x + \ln y$, making use of property 1 of logarithms. Thus far we have

$$3x + y + \ln x + \ln y = 0$$

Differentiating term by term yields

$$3 + \frac{dy}{dx} + \frac{1}{x} + \frac{1}{y} \cdot \frac{dy}{dx} = 0$$

Fractions can be eliminated by multiplying each term (both sides of the equation) by the common denominator xy. The result:

$$3xy + xy\frac{dy}{dx} + y + x\frac{dy}{dx} = 0$$

Next, isolate the dy/dx terms.

$$xy\frac{dy}{dx} + x\frac{dy}{dx} = -3xy - y$$

Then

$$(xy + x)\frac{dy}{dx} = -(3xy + y)$$

Finally,

$$\frac{dy}{dx} = -\frac{3xy + y}{xy + x}$$ ◆

EXAMPLE 10 Derive a formula for $\dfrac{d}{dx}\ln|x|$. (You will need this formula in Chapter 6.)

SOLUTION Consider that x can be positive or negative.

(i) *When $x > 0$*

This means that $|x| = x$ and so $\ln|x| = \ln x$. Thus,

$$\frac{d}{dx}\ln|x| = \frac{d}{dx}\ln x = \frac{1}{x}$$

(ii) *When x < 0*

This means that $|x| = -x$ and so $\ln |x| = \ln (-x)$. Thus,

$$\frac{d}{dx} \ln |x| = \frac{d}{dx} \ln (-x)$$

The differentiation can be continued using the formula for the derivative of $\ln u$, with $u = -x$.

$$\frac{d}{dx} \ln (-x) = \frac{1}{-x} \cdot \frac{d}{dx} (-x)$$

$$= \frac{1}{-x} \cdot (-1)$$

$$= \frac{1}{x}$$

Parts (i) and (ii) suggest that whether x is positive or negative, the derivative of $\ln |x|$ is the same, namely, $1/x$. ◆

$$\frac{d}{dx} \ln |x| = \frac{1}{x}$$

There are formulas for differentiating logarithms that have bases other than e, but such logarithms usually do not occur in settings requiring differentiation. In view of this, our coverage will be brief and a formula for such differentiation is presented next without proof. An example follows.

$$\frac{d}{dx} \log_a u = \frac{1}{u} \cdot \frac{du}{dx} \cdot \frac{1}{\ln a}$$

EXAMPLE 11   If $y = \log_{10}(x^2 + 1)$, find dy/dx.

SOLUTION

$$\frac{dy}{dx} = \frac{d}{dx} \log_{10}(x^2 + 1)$$

$$= \frac{1}{x^2 + 1} \cdot 2x \cdot \frac{1}{\ln 10}$$

$$= \frac{2x}{(x^2 + 1) \ln 10} \qquad ◆$$

Logarithmic Differentiation

In Section 5.3 we differentiated functions of the form e^x and a^x, where the base is a constant and the exponent is a variable. In Chapter 3 we differentiated functions of the form x^n—such as x^2, x^3, x^4, x^{-1}—where the base is a variable and the exponent is a constant. Next we consider differentiation of functions in which *both* the base and the exponent are variables. A procedure known as *logarithmic differentiation* will be used. Example 12 demonstrates this procedure.

EXAMPLE 12   Let $y = x^x$ (and $x > 0$). Determine dy/dx.

SOLUTION   The preceding paragraph outlined the nature of this problem. We have no formula for differentiating x^x, in which both the base and the exponent are variables. Logarithms will be introduced in order to "eliminate" the exponent. Since

$$y = x^x$$

it follows that

$$\ln y = \ln x^x$$

By logarithm property 3, $\ln x^x = x \ln x$. Thus,

$$\ln y = x \ln x$$

Note that we no longer have a variable base raised to a variable power. The exponent has been "eliminated" by the use of logarithms. We can now proceed with the calculus by differentiating both sides. Keep in mind that when differentiating $\ln y$, the y is itself a function of x. Also note that $x \ln x$ is a product.

$$\frac{d}{dx}(\ln y) = \frac{d}{dx}(x \cdot \ln x)$$

$$\frac{1}{y} \cdot \frac{d}{dx}(y) = x \cdot \frac{d}{dx}(\ln x) + \ln x \cdot \frac{d}{dx}(x)$$

$$\frac{1}{y}\frac{dy}{dx} = x \cdot \frac{1}{x} + (\ln x) \cdot 1$$

$$\frac{1}{y}\frac{dy}{dx} = 1 + \ln x$$

Now multiply both sides of the equation by y, in order to solve for dy/dx.

$$\frac{dy}{dx} = y(1 + \ln x)$$

Since $y = x^x$, make the substitution of x^x for y. In this way, the derivative will be a function of x explicitly, just as the original function was given explicitly in terms of x. The result:

$$\frac{dy}{dx} = x^x(1 + \ln x)$$

The graph of $y = x^x$ is shown in Figure 12.   ◆

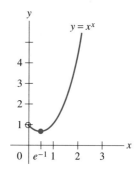

Figure 12 The graph of $y = x^x$. The relative minimum occurs at $x = e^{-1}$. Can you verify that e^{-1} is a critical number?

> *Note*
>
> An alternative to logarithmic differentiation is presented in Exercise 104.

Logarithmic differentiation can also be used to *prove* that the derivative of x^r is rx^{r-1} for any real number r. To begin,

$$y = x^r$$

It follows that

$$\ln y = \ln x^r$$

or

$$\ln y = r \ln x$$

Differentiation yields

$$\frac{1}{y}\frac{dy}{dx} = r \cdot \frac{1}{x}$$

Then

$$\frac{dy}{dx} = y \cdot r \cdot \frac{1}{x}$$

Upon substituting x^r for y,

$$\frac{dy}{dx} = x^r \cdot r \cdot \frac{1}{x}$$

$$= r \cdot x^r \cdot x^{-1} \qquad \text{after rearranging}$$

$$= rx^{r-1} \qquad\qquad \text{a property of exponents}$$

5.4   Exercises

Determine the derivative of each function in Exercises 1–12.

1. $y = \ln x$

2. $y = 3 \ln x$

3. $y = x \ln x$

4. $y = (\ln x)^3$

5. $y = \dfrac{\ln x}{x}$

6. $y = \dfrac{2x}{\ln x}$

7. $y = \dfrac{x + 1}{\ln x}$

8. $y = \dfrac{1}{(\ln x)^2}$

9. $f(x) = \sqrt{\ln x}$

10. $f(x) = e^x \ln x$

11. $f(x) = \dfrac{x}{1 + \ln x}$

12. $f(x) = x(\ln x)^2$

Find the derivative of each function in Exercises 13–28.

13. $y = \ln (x^2 + 7)$

14. $y = \ln (1 + 2x)$

15. $y = \ln (2x + 1)^3$

16. $y = \ln (x^2 - 3)^4$

17. $f(x) = \ln \dfrac{1}{x}$

18. $f(x) = \dfrac{\ln x}{e^x}$

19. $f(x) = 30 \ln (1 + e^x)$

20. $f(x) = \ln (x^2 - e^x)$

21. $f(x) = \ln \dfrac{x}{x + 1}$

22. $f(x) = \ln \dfrac{x + 1}{x}$

23. $f(x) = \ln (\ln x)$

24. $f(x) = \ln (\ln x^2)$

25. $y = \ln (xe^x)$

26. $y = x^2 \ln \sqrt{x}$

27. $y = e^{-x} \ln x^2$

28. $y = \dfrac{1}{\sqrt{\ln x}}$

29. Determine the equation of the line tangent to the graph of $f(x) = 3 \ln x$ at the point $(1, 0)$.

30. Determine the equation of the line tangent to the graph of $f(x) = 6x - \ln x^2$ at the point $(1, 6)$.

Find the second derivative of each function in Exercises 31–36.

31. $f(x) = x \ln x$

32. $f(x) = x^2 \ln x$

33. $f(x) = (\ln x)^2$

34. $f(x) = (\ln x)^3$

35. $f(x) = \dfrac{\ln x}{x}$

36. $f(x) = \dfrac{\ln x^2}{x}$

37. (*MARGINAL REVENUE*) Suppose $R(x) = 70x + 100 \ln x$ is the revenue in dollars when x units are produced and sold.
   (a) Determine the marginal revenue function.
   (b) What is the marginal revenue when the production level is 20 units?

38. (*REVENUE/MARGINAL REVENUE*) Suppose that the price p at which x units $(x > 1)$ can be sold is given by the demand equation

$$p = 60 + \dfrac{10}{\ln x}$$

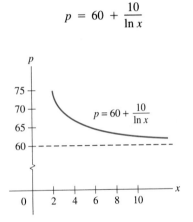

   (a) Determine the revenue function.
   (b) Determine the marginal revenue function.

39. (*COST/MARGINAL COST*) Suppose that the price p at which x units can be produced is given by the supply equation

$$p = 3 \ln (x + 1)$$

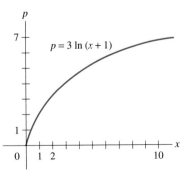

(a) Determine the cost function.

(b) Determine the marginal cost function.

40. (*RACING CAR VELOCITY*) Suppose that a racing car can travel according to $s = 4t^2 - 4 \ln t$ for $1 \le t \le 4$. Distance s is measured in meters and time t is in seconds.

(a) Determine the function that gives the car's velocity at any time t, where $1 \le t \le 4$.

(b) What is the car's velocity after 1 second, after 2 seconds, and after 4 seconds?

For each function in Exercises 41–48, determine whether its graph is increasing or decreasing at the given point.

41. $f(x) = \ln x^2$ at $(1, 0)$

42. $f(x) = x \ln x$ at $(1, 0)$

43. $f(x) = \dfrac{\ln x}{x}$ at $(1, 0)$

44. $f(x) = \ln \dfrac{1}{x}$ at $\left(\dfrac{1}{e}, 1\right)$

45. $f(x) = e^x \ln (x + 1)$ at $(0, 0)$

46. $f(x) = e \ln ex$ at $(1, e)$

47. $f(x) = \dfrac{x + 1}{\ln x}$ at $(e, e + 1)$

48. $f(x) = \dfrac{x + 2}{1 + \ln x}$ at $(1, 3)$

Find all critical numbers and all relative extrema, if any, for each function in Exercises 49–58.

49. $f(x) = x - \ln x$ **50.** $f(x) = x^2 - 18 \ln x$

51. $f(x) = 2x - \ln x^2$ **52.** $f(x) = x \ln x$

53. $f(x) = \dfrac{\ln x^2}{x}$ **54.** $f(x) = \dfrac{\ln x}{x^2}$

55. $f(x) = x^2 \ln x$ **56.** $f(x) = x^3 \ln x$

57. $f(x) = \ln (2x - x^2)$ **58.** $f(x) = \ln (3x - x^3)$

Sketch the graph of each function in Exercises 59–62.

59. $y = x - \ln x$ **60.** $y = x \ln x$

61. $y = \ln (x^2 + 1)$ **62.** $y = x^2 \ln x$

Use implicit differentiation to find dy/dx in Exercises 63–72.

63. $3 \ln x + xy + y = 1$ **64.** $y^2 - \ln y + 2x = 6$

65. $x^2 + 2y + \ln (xy) = 0$ **66.** $xy^3 + \ln (xy) = 5x$

67. $x \ln y - y \ln x = 8$ **68.** $x^2 \ln x + y^2 \ln y = 2$

69. $e^x \ln y - 3xy + x = 10$ **70.** $e^y \ln x + xy^2 = y$

71. $5 \ln y + e^{3y} = 2x$ **72.** $4 \ln x + e^{2y} = 3y$

Determine the derivative of each function in Exercises 73–80.

73. $y = \log_{10} x$ **74.** $y = \log_3 x$

75. $y = \log_2 9x$ **76.** $y = \log_2 3x^2$

77. $y = x \log_{10}(1 - 3x)$ **78.** $y = x^2 \log_2 5x$

79. $y = \dfrac{\log_{10} x}{x}$ **80.** $y = \dfrac{\log_2 x}{x + 1}$

Use logarithmic differentiation (as explained in Example 12) to obtain dy/dx for each function in Exercises 81–90.

81. $y = x^{2x}$ **82.** $y = (2x)^x$

83. $y = (x + 1)^x$ **84.** $y = (x + 1)^{2x}$

85. $y = x^{x^2}$ **86.** $y = x^{x+1}$

87. $y = (3x)^{x+1}$ **88.** $y = (x - 1)^{x+1}$

89. $y = x^{1/x}$ **90.** $y = x^{\ln x}$

W 91. Consider the slope of the tangent line to the graph of $y = \ln x$, where $x = 1, 10, 100$, and 1000. Where will the slope of the tangent line be zero, or is that not possible? Explain.

92. Determine the rate of change of y with respect to x for the function $y = \ln x$, when $x = 1/2$.

93. Use a calculator to investigate the limit

$$\lim_{x \to \infty} \frac{\ln x}{x}$$

Determine dy/dx in Exercises 94–97.

94. $y = \ln \sqrt{1 + \sqrt{x}}$ **95.** $e^{xy} + \ln \sqrt{xy} = 1$

96. $e^{x+y} + \ln(x+y) = 0$

97. $y = x \ln \dfrac{1}{\sqrt{x}}$

W 98. Explain why -1 is not a critical number of the function $y = \ln(3x - x^3)$.

99. Is the graph of $y - x + y \ln x = 1 - e$ increasing or decreasing at the point $\left(e, \frac{1}{2}\right)$?

W 100. For $f(x) = \ln x$, we have $f'(x) = 1/x$. It would appear that $f'(x) > 0$ for $x > 0$ and $f'(x) < 0$ for $x < 0$. Is it correct to conclude that f is increasing for $x > 0$ and decreasing for $x < 0$? Explain.

W 101. For $f(x) = \ln x$, use $f'(x)$ to explain why the natural logarithm function has no relative extrema. Be complete in the explanation of your search for critical numbers.

102. Let $y = u^v$, where u and v are functions of x. Use logarithmic differentiation to obtain the result

$$\frac{dy}{dx} = u^v \left[\frac{v}{u} \cdot \frac{du}{dx} + (\ln u) \frac{dv}{dx} \right]$$

103. If $f(x) = \ln x$, determine $f^{(40)}(x)$, the 40th derivative. (*Hint*: Obtain a few derivatives and determine a pattern.)

104. *Alternative to logarithmic differentiation* (which was demonstrated in Example 12).

The expression x^x can be written as follows:

$$x^x = e^{\ln x^x}$$

or

$$x^x = e^{x \ln x}$$

Now x^x can be differentiated by differentiating $e^{x \ln x}$.

$$\frac{d}{dx} x^x = \frac{d}{dx} e^{x \ln x}$$

$$= e^{x \ln x} \cdot \frac{d}{dx} (x \ln x)$$

$$= e^{x \ln x} \left[x \cdot \frac{1}{x} + (\ln x) \cdot 1 \right]$$

$$= e^{x \ln x} (1 + \ln x)$$

$$= x^x (1 + \ln x)$$

Redo Exercises 81 and 83 by using this alternative method.

The Exercise Library at the back of the book contains graphing calculator and computer exercises keyed to this section.

5.5 | *SOME ADDITIONAL BUSINESS APPLICATIONS*

Effective Rate of Interest

When a bank offers you an annual interest rate of 6% compounded continuously, they are really paying you more than 6%. Because of compounding, the 6% is in fact a yield of 6.18% for the year. To see this, consider investing \$1 at 6% per year compounded continuously for 1 year. The total return is

$$A = Pe^{rt} = 1 \cdot e^{.06(1)} = e^{.06} = \$1.0618$$

If we subtract from \$1.0618 the \$1 we invested, the return is \$.0618, which is 6.18% of the amount invested. The 6% annual interest rate of this example is called the **nominal rate** and the 6.18% is called the **effective rate**. In practice, the effective rate is calculated using a formula based on the reasoning we used.

> **Effective Rate**
>
> If r is the annual interest rate
> (nominal rate) and the interest is
> compounded continuously, then the
> **effective rate** is
>
> $$e^r - 1$$

EXAMPLE 1 ◆ *EFFECTIVE RATE*

A savings and loan pays an annual interest rate (nominal rate) of 7.5%. What is
the effective rate?

SOLUTION   The effective rate is

$$e^r - 1$$

Here $r = 7.5\%$, or .075. Thus,

$$e^r - 1 = e^{.075} - 1$$
$$\approx 1.0779 - 1$$
$$= .0779$$

The effective rate is 7.79%. Notice that we rounded the value of $e^{.075}$ to four
decimal places. As a result, the effective rate came out to the nearest hundredth
of a percent.   ◆

EXAMPLE 2 ◆ *EFFECTIVE RATE*

A bank offers an effective rate of 5.41%. What is the nominal rate?

SOLUTION   The effective rate is given as 5.41%, which is .0541 in decimal form. The
effective rate is also known to be $e^r - 1$, *where r is the nominal rate*. Thus,
we have the equation

$$e^r - 1 = .0541$$

or

$$e^r = 1.0541$$

Changing to logarithm notation, we have

$$r = \ln 1.0541$$

or

$$r \approx .0527$$

This means the nominal rate is 5.27%.   ◆

Rule of 70

The rule of 70 provides a quick and easy method for approximating the time needed for the amount of an investment to double in value. To obtain the rule, consider

$$A = Pe^{rt}$$

When amount A becomes $2P$, it will be double the value of the initial investment P.

$$2P = Pe^{rt}$$

After dividing both sides of this equation by P, we have

$$2 = e^{rt}$$

or

$$rt = \ln 2$$

Finally,

$$t = \frac{\ln 2}{r} \approx \frac{.6931}{r}$$

Since $.6931 \approx .70$, it follows that

$$t \approx \frac{.70}{r}$$

If we multiply both the numerator and denominator of $.70/r$ by 100, we obtain

$$t \approx \frac{70}{100r}$$

Consider now that r is the decimal form of the interest rate. This means that $100r$ is the percent form. (Compare, for example, .15 and 15%.) If we use R to represent the interest in percent form, then we have

$$t \approx \frac{70}{R}$$

This result shows that *if the interest rate is divided into 70, the result is the time needed to double the value of the original investment.*

Rule of 70

The number of years t needed to double the value of an investment that yields $R\%$ annual interest is given by

$$t = \frac{70}{R}$$

EXAMPLE 3   ◆   *RULE OF 70*

Use the rule of 70 to approximate the time needed to double the value of each investment.

(a) $5000 invested at 5%

(b) $5000 invested at 6%

(c) $5000 invested at 7%

SOLUTION   **(a)** The doubling time t when $R = 5\%$ is approximately 14 years, as shown next.

$$t \approx \frac{70}{5} = 14 \text{ years}$$

(b) The doubling time t when $R = 6\%$ is approximately 11.7 years.

$$t \approx \frac{70}{6} \approx 11.7 \text{ years}$$

(c) The doubling time t when $R = 7\%$ is approximately 10 years.

$$t \approx \frac{70}{7} = 10 \text{ years}$$ ◆

Figure 13 shows a graph of doubling times as a function of interest rate for rates between 5% and 14%.

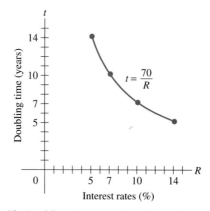

Figure 13 Doubling time as a function of interest rate

Present Value

The **present value** is the amount that must be invested now in order to have a specified amount at some time in the future. In

$$A = Pe^{rt}$$

P is the present value. A formula that gives present value explicitly can be obtained from $A = Pe^{rt}$ by dividing both sides of the equation by e^{rt}.

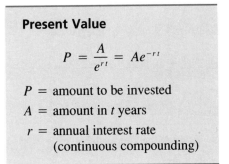

Present Value

$$P = \frac{A}{e^{rt}} = Ae^{-rt}$$

P = amount to be invested

A = amount in t years

r = annual interest rate
(continuous compounding)

EXAMPLE 4 ◆ *PRESENT VALUE*

The Kims want to have $60,000 available when their child begins college in 15 years. How much should they invest now at 7% interest per year compounded continuously in order to meet their objective?

SOLUTION The amount to invest now is the present value P in

$$P = Ae^{-rt}$$

The $60,000 is A, t is 15 years, and r is .07. Thus,

$$P = \$60,000e^{-(.07)(15)}$$
$$= \$60,000e^{-1.05}$$
$$\approx \$60,000(.3499)$$
$$= \$20,994$$

The Kims should invest $20,994 now at 7% in order to have $60,000 available in 15 years. ◆

5.5 Exercises

Assume continuous compounding unless otherwise indicated.

1. (*EFFECTIVE RATE*) A savings and loan offers an annual interest rate (nominal rate) of 6.3%. What is the effective rate?

2. (*EFFECTIVE RATE*) Your bank offers an annual interest rate (nominal rate) of 5.7%. What is the effective rate?

3. (*EFFECTIVE RATE*) If the nominal rate is 6.1%, what is the effective rate?

4. (*EFFECTIVE RATE*) If the nominal rate is 5.6%, what is the effective rate?

5. (*EFFECTIVE AND NOMINAL RATES*) How much interest (in dollars) will you get in a year if you deposit $2000 into an account that pays as follows?
(a) An effective rate of 5.82%
(b) A nominal rate of 6.31%

6. (*EFFECTIVE RATE*) If you deposit $1000 into a bank account paying an effective rate of 6.13%, how much will your deposit be worth at the end of the year?

7. (*NOMINAL RATE*) A credit union offers an effective rate of 6.12%. What is the nominal rate?

8. (*NOMINAL RATE*) A savings bank offers an effective rate of 5.94%. What is the nominal rate?

9. (*EFFECTIVE AND NOMINAL RATES*) The following graph shows the difference D between the effective rate and the nominal rate as a function of the nominal rate n. Answer the following questions based on the graph.
(a) As the nominal rate increases, the difference between the effective rate and the nominal rate _____.
(b) Is D' positive, negative, or zero?
(c) Is D'' positive, negative, or zero?
(d) Use the words *increasing* and/or *decreasing* to fill in the blanks. "The function D is _____ at a(n) _____ rate."

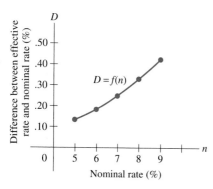

W 10. (*INTEREST COMPARISON*) County Bank offers a nominal rate of 5.9% with continuous compounding. Town

Bank offers a simple interest payment at the end of the year of 6% (no compounding). If interest is your only concern, which is the better bank for you for a 1-year investment? Explain.

11. (*RULE OF 70*) If money is invested at an annual rate of 7%, how long will it take to double the value of the investment?

12. (*RULE OF 70*) If money is invested at an annual rate of 10%, how long will it take to double the value of the investment?

13. (*DOUBLING TIME*) If your investment earns 9% annual interest, how long (to the nearest year) will it take to double its value?

14. (*DOUBLING TIME*) If your investment earns 8% annual interest, how long (to the nearest year) will it take to double its value?

15. (*RULE OF 70*) Sylvia hopes to double her money in a 6-year investment. What annual rate of interest (to the nearest tenth of a percent) will she need in order to accomplish this?

16. (*RULE OF 70*) In order to double your money in an investment lasting 8.2 years, what annual rate of interest (to the nearest tenth of a percent) will you need to obtain?

17. (*DOUBLING TIME*)
(a) At an annual inflation rate of 5%, how long will it take the price of consumer goods to double?
(b) If the inflation rate is 10%, how long is the doubling time?

18. (*RULE OF 70*) At what annual rate of inflation will the price of consumer goods double in 8 years?

19. (*DOUBLING TIME*) How long will it take to double the value of a $5000 investment that earns 6.5% annual interest?
(a) Solve the problem first by using $A = Pe^{rt}$.
(b) Solve the problem by using the approximation offered by the rule of 70.
(c) Compare the results of (a) and (b).

20. (*RULE OF 70*) Use the rule of 70 to estimate how long it will take to quadruple the value of an investment made at an annual rate of 7%.

21. (*PRESENT VALUE*) How much should the Martinez family invest now at 7.3% annual interest in order to have $80,000 available when their child begins college in 10 years?

22. (PRESENT VALUE) Ron and Doris plan to retire in 10 years. They want to establish an account that will be worth $150,000 then. How much must they invest now at 7.5% annual interest to reach their goal?

23. (PRESENT VALUE) The Johnsons have been married for 20 years. They plan to go to Hawaii for their 25th anniversary. How much should they invest now at 6.8% annual interest in order to have $2500 for their Hawaiian vacation?

24. (MUTUAL FUND GROWTH) A mutual fund claims that its portfolio value grew at an annual rate of 14% from July 1984 to July 1992. If you had invested $5000 in the fund in July 1984, what would it have been worth in July 1992?

Chapter List *Important terms and ideas*

exponential functions	logarithm	radiocarbon dating
e	common logarithms	logarithmic differentiation
continuous compounding	natural logarithms	nominal rate
exponential growth	natural logarithm function	effective rate
exponential decay	properties of logarithms	rule of 70
logistic growth	doubling time	present value
learning curve	half-life	

Review Exercises for Chapter 5

In Exercises 1–8, solve each equation for x.

1. $5e^{7x} = 30$

2. $10^{3x} = 4$

3. $3 \ln 4x = 18$

4. $\log 4x = 2$

5. $4e^{.03x} = 2$

6. $e^{-.004x} = 7$

7. $5 - \ln x = 0$

8. $1 - 2 \ln x = 0$

Simplify each expression in Exercises 9–12.

9. $-e^{\ln 7}$

10. $xe^{\ln x}$

11. $\ln \dfrac{1}{e^2}$

12. $-e \ln e$

Determine the derivative of each function in Exercises 13–22.

13. $y = x^4 e^4$

14. $y = \dfrac{e^x}{x^2}$

15. $f(x) = (3 - e^x)^7$

16. $f(x) = \sqrt{1 + e^x}$

17. $y = xe^{1/x^2}$

18. $y = \dfrac{e^x + 1}{e^x}$

19. $f(x) = x^3 e^{-3x}$

20. $f(x) = \sqrt{x + e^{4x}}$

21. $y = 5^x$

22. $y = x \cdot 3^x$

Determine the derivative of each function in Exercises 23–32.

23. $f(x) = e^{3x} \ln x$

24. $f(x) = \sqrt{1 + \ln x}$

25. $y = x^2 (\ln x)^2$

26. $y = x^2 \ln x^2$

27. $f(x) = \ln \dfrac{1}{x^2}$

28. $f(x) = \dfrac{\ln x}{x^2}$

29. $y = x \ln x - x$

30. $y = x^2 \ln \sqrt{x}$

31. $y = \ln (x^2 e^x)$

32. $y = x \ln 5x$

In Exercises 33–36, find all critical numbers of the given function.

33. $f(x) = 4x - 3e^x$ **34.** $f(x) = .5e^x - x$

35. $y = \ln(x - 3) - \dfrac{x}{7}$ **36.** $y = x - \ln 2x$

In Exercises 37–38, use the properties of logarithms to show that each result is true.

37. $\ln \dfrac{1}{x}$ can be simplified to $-\ln x$.

38. $\dfrac{\ln x^3}{\ln x^5}$ can be simplified to 3/5.

39. (*PRESENT VALUE*) How much should be invested now at 7.5% interest per year compounded continuously in order to have $10,000 in 4 years?

40. (*DECLINING REAL ESTATE VALUE*) If the value of an old industrial property declines at the rate of 3% per year, what percent of its current value will it have 10 years from now?

41. (*FUNGUS GROWTH*) If a fungus doubles in size every 7 hours, how many times its present size will it be in 32 hours?

42. (*NUMBER OF SHEEP*) The number of sheep on a ranch increases from 300 to 520 in 3 years. Assuming exponential growth, determine the growth constant k and express it as a percent.

43. (*POPULATION GROWTH*) The population of a small sunbelt town has grown from 5000 to 8400 in 4 years. If the exponential growth continues at the same rate, what will be the town's population in another 4 years?

44. (*NUMBER OF ZEBRAS*) As a result of 6 months of severe drought, the number of zebras in a particular region has been reduced from 400 to 320. Assuming exponential decline and assuming the drought continues, what will be the zebra population in another 4 months?

45. (*SPREAD OF AN EPIDEMIC*) A logistic growth function can often be used to describe the spread of an epidemic through a city. For example, if y is the number of cases of the flu after t weeks, then we may have

$$y = \frac{200,000}{1 + 1999e^{-.8t}}$$

(a) How many cases of flu were there at the beginning, when the epidemic was first declared?
(b) How many cases of flu were there after 5 weeks?
(c) What is the maximum number of people in this city who can get the flu?

46. (*RADIOACTIVE DECAY*) If a 3-microgram mass of a radioactive element decays according to

$$A(t) = 3e^{-.001t}$$

determine the limit

$$\lim_{t \to \infty} A(t)$$

47. (*PHOTOSYNTHESIS*) Studies of leaves and photosynthesis led to the formula

$$\log P = a - b \log A$$

in which P is the rate at which photosynthesis occurs and A is the age of the leaf.
(a) Solve the equation for P.
(b) Solve the equation for A.

48. (*PROFIT*) Profit on the production and sale of x items is given by

$$P(x) = 50e^{.01x} - 80$$

Use calculus to determine if profit is increasing or decreasing when $x = 10$ units.

W 49. What happens when you use a calculator to determine $\ln 0$ or $\ln(-1)$? Explain why it happens.

6

INTEGRATION

The previous three chapters presented a thorough study of the calculus operation called differentiation. The definition of derivative was motivated by our desire to define the slope of the tangent line. Many other applications followed.

In this chapter, we begin the study of another calculus operation. Once again, there will be a geometric interpretation and many non-geometric applications. But this time you have a head start, because this calculus operation is related to differentiation. One name for this operation is *integration*. However, it is its other name, *antidifferentiation*, that provides a hint of what is about to unfold.

6.1 ANTIDIFFERENTIATION

The preceding chapters focused on the calculus operation called differentiation, in which we begin with a function f and obtain the derivative f'. Interpreting f' as a rate of change of f led to a variety of applications. By contrast, there are situations in which we know the rate of change and seek the function. We need to be able to reverse the differentiation process in such cases.

One view of the calculus operation called **antidifferentiation** is that of reversing differentiation, that is, beginning with the derivative f' and obtaining an **antiderivative** f.

$$f \xrightarrow{\quad \text{differentiation} \quad} f'$$
$$f' \xrightarrow{\quad \text{antidifferentiation} \quad} f$$

Another view of antidifferentiation suggests that we may seek an antiderivative even though no differentiation has taken place; we begin with a function f and obtain an antiderivative denoted by F.

The rules for antidifferentiation are easy to obtain from known differentiation rules, especially since the process essentially reverses the effect of differentiation. Before obtaining rules for antidifferentiation, let us consider an example and some basic notation.

EXAMPLE 1 Antidifferentiate $f(x) = 3x^2$.

SOLUTION We need to determine the function F that yields $f(x) = 3x^2$ when differentiated. It does not take long to realize that the derivative of x^3 is $3x^2$.

We might conclude that the antiderivative of $f(x) = 3x^2$ is $F(x) = x^3$, since

$$\frac{d}{dx}(x^3) = 3x^2$$

But it is also true that

$$\frac{d}{dx}(x^3 + 1) = 3x^2$$

and

$$\frac{d}{dx}(x^3 + 7) = 3x^2$$

as well as

$$\frac{d}{dx}(x^3 - 5) = 3x^2$$

In fact,

$$\frac{d}{dx}(x^3 + C) = 3x^2$$

for any constant C. In view of this, we say that the antiderivative of $f(x) = 3x^2$ is

$$F(x) = x^3 + C. \quad \blacklozenge$$

The example has demonstrated why antiderivatives are written in this most general "plus C" form. Here are a few more examples of functions and their antiderivatives. Also, notice that *differentiation can be used to check antiderivatives*. The derivative of the antiderivative is equal to the function.

function	antiderivative	check
$f(x) = 2x$	$F(x) = x^2 + C$	$\frac{d}{dx}(x^2 + C) = 2x$
$f(x) = 6$	$F(x) = 6x + C$	$\frac{d}{dx}(6x + C) = 6$
$f(x) = e^x$	$F(x) = e^x + C$	$\frac{d}{dx}(e^x + C) = e^x$

Note

We now state without proof a result that is consistent with the result of Example 1. *If functions F and G are both antiderivatives of the same function f, then F(x) and G(x) can differ only by an added constant.*

Just as there are special notations for the derivative, there is also a special notation for antiderivatives.

The antiderivative of f is denoted by

$$\int f(x)\, dx$$

The symbol $\int$ is called an **integral sign** and was introduced by Leibniz. $f(x)$ is the **integrand**. The dx specifies that this is the integral of $f(x)$ with respect to x. (Later, we will say more about dx and how it relates to Δx and to differentials.) $\int f(x)\, dx$ is called an **indefinite integral**. The indefinite integral is the same as the most general antiderivative. The process of antidifferentiation can also be called **integration**. To integrate is to antidifferentiate.

Even though notation such as $\int f(x)\, dx$ clearly indicates integration, the directions accompanying examples and exercises may say perform the antidifferentiation, evaluate, calculate, or perform the integration.

EXAMPLE 2 Perform the integration. $\int 3x^2\, dx$

SOLUTION Here we are asked to antidifferentiate (integrate) $3x^2$. In fact, this is the problem of Example 1. The only thing new or different here is the notation. Thus, from previous experience we know that

$$\int 3x^2\, dx = x^3 + C$$

The number C is called the **constant of integration**.   ◆

Recall the differentiation formula

$$\frac{d}{dx}(x^n) = nx^{n-1}$$

The corresponding antidifferentiation formula is given next.

Power Rule

$$\int x^n\, dx = \frac{x^{n+1}}{n+1} + C \qquad n \neq -1$$

To antidifferentiate (integrate) a power of x, increase the exponent by 1 and then divide by the new exponent.

The formula is true because the derivative of $\frac{x^{n+1}}{n+1} + C$ is in fact x^n, as we now show. (Note that when $n = -1$ the expression is not defined.)

$$\frac{d}{dx}\left(\frac{x^{n+1}}{n+1} + C\right) = \frac{d}{dx}\left(\frac{x^{n+1}}{n+1}\right) + \frac{d}{dx}(C)$$

$$= \frac{(n+1)x^n}{(n+1)} + 0$$

$$= x^n$$

EXAMPLE 3   Perform the indicated antidifferentiation. $\int x^4\, dx$

SOLUTION
$$\int x^4\, dx = \frac{x^{4+1}}{4+1} + C$$

$$= \frac{x^5}{5} + C$$

Note that in using the formula for $\int x^n\, dx$, we added 1 to the exponent and then divided by the new exponent. As a check, the derivative of $x^5/5 + C$ is x^4.

The check:

$$\frac{d}{dx}\left(\frac{x^5}{5} + C\right) = \frac{5x^4}{5} + 0 = x^4 \quad \blacklozenge$$

EXAMPLE 4   Determine $\int t^{-3}\, dt.$

SOLUTION
$$\int t^{-3}\, dt = \frac{t^{-3+1}}{-3+1} = \frac{t^{-2}}{-2} + C = -\frac{1}{2t^2} + C \quad \blacklozenge$$

EXAMPLE 5   Perform the integration. $\int \frac{1}{\sqrt{x}}\, dx$

SOLUTION   To begin, write $\sqrt{x}$ as $x^{1/2}$.

$$\int \frac{1}{\sqrt{x}}\, dx = \int \frac{1}{x^{1/2}}\, dx$$

Rewriting the integrand as a power of x, we obtain

$$\int x^{-1/2}\, dx$$

We can now apply the antidifferentiation formula. Add 1 to the exponent and divide by the new exponent. The result is

$$\frac{x^{1/2}}{1/2} + C$$

which simplifies to

$$2x^{1/2} + C \quad or \quad 2\sqrt{x} + C \quad \blacklozenge$$

The three rules given next are based on differentiation rules. Each one can be proved by differentiating the right-hand side to obtain the integrand given on the left-hand side. (Keep in mind that the derivative of the antiderivative of a function is the function.)

Properties of Indefinite Integrals

$$\int [f(x) + g(x)] \, dx = \int f(x) \, dx + \int g(x) \, dx$$

$$\int [f(x) - g(x)] \, dx = \int f(x) \, dx - \int g(x) \, dx$$

$$\int kf(x) \, dx = k \int f(x) \, dx \qquad k = \text{constant}$$

In words, *the integral of a sum is the sum of the integrals, the integral of a difference is the difference of the integrals*, and *the integral of a constant times a function is that constant times the integral of the function.*

EXAMPLE 6 Evaluate $\int (x^2 + x^{3/2}) \, dx$.

SOLUTION
$$\int (x^2 + x^{3/2}) \, dx = \int x^2 \, dx + \int x^{3/2} \, dx$$

$$= \frac{x^3}{3} + \frac{x^{5/2}}{5/2} + C$$

$$= \frac{x^3}{3} + \frac{2}{5}x^{5/2} + C$$

Notice that although there were two separate integrals along the way, there is only one constant of integration (C) shown in the final result. There is no need to list two constants, since the sum of two constants is simply one constant. Using the one constant C is sufficient. ◆

EXAMPLE 7 Perform the antidifferentiation. $\int (4x^2 - 3x) \, dx$

SOLUTION
$$\int (4x^2 - 3x) \, dx = \int 4x^2 \, dx - \int 3x \, dx$$

$$= 4 \int x^2 \, dx - 3 \int x \, dx$$

$$= 4 \cdot \frac{x^3}{3} - 3 \cdot \frac{x^2}{2} + C$$

$$= \frac{4x^3}{3} - \frac{3x^2}{2} + C \quad ◆$$

As you know, if k is a constant, then the derivative of kx is k. Reversing the direction of this process yields the following antidifferentiation rule.

Integral of a Constant

$$\int k \, dx = kx + C$$

where k is a constant.

EXAMPLE 8   Perform the integration. $\int 7 \, dx$

SOLUTION $$\int 7 \, dx = 7x + C \quad \blacklozenge$$

EXAMPLE 9   Perform the integration. $\int dx$

SOLUTION $$\int dx = \int 1 \, dx$$
$$= x + C \quad \blacklozenge$$

EXAMPLE 10   Perform the integration. $\int \left(5x^4 - \dfrac{3}{x^2} + 6 \right) dx$

SOLUTION $$\int \left(5x^4 - \frac{3}{x^2} + 6 \right) dx = \int 5x^4 \, dx - \int \frac{3}{x^2} \, dx + \int 6 \, dx$$
$$= 5 \int x^4 \, dx - 3 \int x^{-2} \, dx + \int 6 \, dx$$
$$= \frac{5x^5}{5} - \frac{3x^{-1}}{-1} + 6x + C$$
$$= x^5 + \frac{3}{x} + 6x + C \quad \blacklozenge$$

Two other integration formulas follow from differentiation formulas presented in Chapter 5.

$$\int e^x \, dx = e^x + C$$

$$\int \frac{1}{x} \, dx = \ln |x| + C$$

If $x > 0$, then $\ln |x|$ can be written $\ln x$. Note too that $1/x$ is the same as x^{-1}. This means we now have a formula for $\int x^{-1} \, dx$, the case of $\int x^n \, dx$ when $n = -1$.

EXAMPLE 11   Determine $\int (e^x + 3x)\, dx$.

SOLUTION

$$\int (e^x + 3x)\, dx = \int e^x\, dx + \int 3x\, dx$$

$$= e^x + 3\int x\, dx$$

$$= e^x + \frac{3x^2}{2} + C \quad \blacklozenge$$

EXAMPLE 12   Determine $\int \left(t - \dfrac{1}{t} \right) dt$.

SOLUTION

$$\int \left(t - \frac{1}{t} \right) dt = \int t\, dt - \int \frac{1}{t}\, dt$$

$$= \frac{t^2}{2} - \ln |t| + C \quad \blacklozenge$$

EXAMPLE 13   Calculate $\int \dfrac{1 + 3x^3}{x}\, dx$.

SOLUTION   The integrand can be separated into two fractions. Then, the integration can be performed.

$$\int \frac{1 + 3x^3}{x}\, dx = \int \left(\frac{1}{x} + \frac{3x^3}{x} \right) dx$$

$$= \int \frac{1}{x}\, dx + \int 3x^2\, dx$$

$$= \ln |x| + x^3 + C \quad \blacklozenge$$

The formula shown next is needed for some of the applications presented later in the chapter. Verification of the formula follows.

$$\int e^{kx}\, dx = \frac{1}{k} e^{kx} + C \qquad k \neq 0$$

The formula can be verified by showing that the derivative of $\frac{1}{k} e^{kx} + C$ is e^{kx}.

$$\frac{d}{dx} \left(\frac{1}{k} e^{kx} + C \right) = \frac{d}{dx} \left(\frac{1}{k} e^{kx} \right) + \frac{d}{dx}(C)$$

$$= \frac{1}{k} \cdot \frac{d}{dx}(e^{kx}) + 0$$

$$= \frac{1}{k} \cdot e^{kx} \cdot k$$

$$= e^{kx}$$

EXAMPLE 14 Evaluate the integrals.

(a) $\int e^{3x} \, dx$ (b) $\int e^{-2x} \, dx$ (c) $\int e^{.02x} \, dx$ (d) $\int 4e^{.1x} \, dx$

SOLUTION (a) $\int e^{3x} \, dx = \dfrac{1}{3} e^{3x} + C$ (b) $\int e^{-2x} \, dx = \dfrac{1}{-2} e^{-2x} + C$

$$= -\dfrac{1}{2} e^{-2x} + C$$

(c) $\int e^{.02x} \, dx = \dfrac{1}{.02} e^{.02x} + C$ (d) $\int 4e^{.1x} \, dx = 4 \int e^{.1x} \, dx$

$$= 50 e^{.02x} + C$$

$$= 4 \cdot \dfrac{1}{.1} e^{.1x} + C$$

$$= 4(10) e^{.1x} + C$$

$$= 40 e^{.1x} + C \quad \blacklozenge$$

6.1 Exercises

Perform each antidifferentiation in Exercises 1–8.

1. $\int 8x \, dx$

2. $\int 4x \, dx$

3. $\int 6x^2 \, dx$

4. $\int 5x^2 \, dx$

5. $\int t^3 \, dt$

6. $\int 4x^3 \, dx$

7. $\int 10x^5 \, dx$

8. $\int 5x^5 \, dx$

Evaluate each indefinite integral in Exercises 9–24.

9. $\int x^{-2} \, dx$

10. $\int t^{-5} \, dt$

11. $\int 3x^{-4} \, dx$

12. $\int 10x^{-6} \, dx$

13. $\int \dfrac{1}{x^5} \, dx$

14. $\int \dfrac{1}{x^3} \, dx$

15. $\int \dfrac{20}{z^6} \, dz$

16. $\int \dfrac{6}{t^4} \, dt$

17. $\int \sqrt{x} \, dx$

18. $\int \sqrt[3]{x} \, dx$

19. $\int x^{3/4} \, dx$

20. $\int t^{3/2} \, dt$

21. $\int x^{-2/3} \, dx$

22. $\int x^{-3/4} \, dx$

23. $\int 7\sqrt[3]{t} \, dt$

24. $\int 5\sqrt{z} \, dz$

Perform each integration in Exercises 25–34.

25. $\int 3 \, dx$

26. $\int 10 \, dx$

27. $\int (x^2 + 6x) \, dx$

28. $\int (4x^3 - 7) \, dx$

29. $\int (\sqrt{x} - 3x^2) \, dx$

30. $\int (1 - \sqrt{t}) \, dt$

31. $\int (x^{-1/2} + 9) \, dx$

32. $\int (1 - x^{-1/2}) \, dx$

33. $\int (t^2 - 8t + 1) \, dt$

34. $\int (x^2 + 7x - 2) \, dx$

Determine the value of each integral in Exercises 35–62.

35. $\int e^x \, dx$

36. $\int e^t \, dt$

37. $\int (2x - e^x) \, dx$

38. $\int (e^x - x^4) \, dx$

39. $\int (e^x + 1) \, dx$

40. $\int (x^2 - e^x) \, dx$

41. $\int \frac{1}{z}\, dz$

42. $\int \frac{7}{t}\, dt$

43. $\int \left(\frac{4}{x} + 6x\right) dx$

44. $\int \left(\frac{2}{x} - 4x\right) dx$

45. $\int \frac{1}{5x}\, dx$

46. $\int \frac{1}{7x}\, dx$

47. $\int (3 + x^{-1})\, dx$

48. $\int (x^{-1} + 3x^2)\, dx$

49. $\int e^{7x}\, dx$

50. $\int e^{4x}\, dx$

51. $\int 5e^x\, dx$

52. $\int 3e^x\, dx$

53. $\int (4 - e^{.1x})\, dx$

54. $\int (1 - e^{.02x})\, dx$

55. $\int (e^{.05t} + 1)\, dt$

56. $\int (e^{.01t} + 2)\, dt$

57. $\int e^{-6x}\, dx$

58. $\int e^{-2x}\, dx$

59. $\int 8e^{.01x}\, dx$

60. $\int 4e^{.2x}\, dx$

61. $\int e^{-.01x}\, dx$

62. $\int e^{-.02x}\, dx$

Evaluate each indefinite integral in Exercises 63–70.

63. $\int \frac{1 + 2x^2}{x}\, dx$

64. $\int \frac{x^3 - 4}{x}\, dx$

65. $\int \frac{x^5 + 2}{x}\, dx$

66. $\int \frac{3 + x^4}{x}\, dx$

67. $\int \frac{t + 1}{t^{1/2}}\, dt$

68. $\int \frac{1 - t^2}{t^{1/2}}\, dt$

69. $\int \frac{x^2 + x}{\sqrt{x}}\, dx$

70. $\int \frac{1 + \sqrt{x}}{x}\, dx$

Evaluate each indefinite integral in Exercises 71–78.

71. $\int \frac{e^x + 1}{e^x}\, dx$

72. $\int \frac{1 - e^{2x}}{e^x}\, dx$

73. $\int x^2(1 + x)\, dx$

74. $\int \frac{1}{x}\left[x + \frac{1}{x}\right] dx$

75. $\int \frac{\sqrt{x} + \sqrt[3]{x}}{x}\, dx$

76. $\int \frac{x^2 + x + 1}{x}\, dx$

77. $\int (1 - e^{-x})^2\, dx$

78. $\int \frac{x + e^x}{xe^x}\, dx$

W 79. Consider a function such as $f(x) = x^2 + 5$ or $f(x) = \sqrt{x}$ or some other function f.
  (a) If you integrate the function f and then differentiate the result, will you then have the function again?
  (b) If you differentiate the function f and then integrate the result, will you then have the function f again? Explain.

W 80. In the formula for the integral of e^{kx}, it is stated that k cannot be 0. Explain the problem that arises when $k = 0$.

6.2 | SOME APPLICATIONS OF ANTIDIFFERENTIATION

This section includes a variety of applications of antidifferentiation. In each instance, the value of the constant C will need to be determined for the specific situation. When this is done, the resulting antiderivative is called a **particular antiderivative** or may be thought of as a *particular solution*. To begin, here is an example that demonstrates how the value of C can be determined when specific information is given.

EXAMPLE 1   Find $f(x)$ if $f'(x) = 3x^2 + 2x - 1$ and $f(2) = 14$.

SOLUTION $f(x)$ is determined from its derivative $f'(x)$ by antidifferentiation. Since $f'(x) = 3x^2 + 2x - 1$, we have

$$f(x) = \int (3x^2 + 2x - 1)\, dx$$

or

$$f(x) = x^3 + x^2 - x + C$$

The value of C can be determined by using the fact that $f(2) = 14$. Recall that $f(2) = 14$ means that when $x = 2$, $f(x) = 14$. Upon substituting 2 for x and 14 for $f(x)$ in $f(x) = x^3 + x^2 - x + C$, we have

$$14 = 2^3 + 2^2 - 2 + C$$
$$14 = 10 + C$$
$$C = 4$$

Thus C is 4 and the function is

$$f(x) = x^3 + x^2 - x + 4$$

Note that because we have determined the value of C, this function can be called a "particular" antiderivative of $f'(x)$. ◆

EXAMPLE 2 Determine the equation of the curve that passes through the point (3, 11) and has slope $2x$.

SOLUTION Since the slope is $2x$, we can write

$$\frac{dy}{dx} = 2x$$

It follows that

$$y = \int 2x\, dx$$

or

$$y = x^2 + C$$

The curve passes through (3, 11), which means that $y = 11$ when $x = 3$. So let $x = 3$ and $y = 11$ in the equation $y = x^2 + C$.

$$11 = 3^2 + C$$

or

$$C = 2$$

We can now replace C by 2 in the equation $y = x^2 + C$. The result is

$$y = x^2 + 2 \quad ◆$$

EXAMPLE 3 ◆ Oʙᴛᴀɪɴɪɴɢ ᴄᴏꜱᴛ ꜰʀᴏᴍ ᴍᴀʀɢɪɴᴀʟ ᴄᴏꜱᴛ

For a company, the marginal cost when x units of merchandise are produced is $50 - .08x$ dollars. If the fixed cost (overhead) is $700, determine

(a) the cost of producing x units

(b) the cost of producing 10 units.

SOLUTION Recall that marginal cost is the derivative of cost (see **Chapter 3**). Using $C(x)$ for the cost function, we have

$$C'(x) = 50 - .08x$$

Then

$$C(x) = \int (50 - .08x)\, dx$$

or

$$C(x) = 50x - .04x^2 + C$$

Since the fixed cost (overhead) is $700, we know that the cost of producing zero units is $700. In other words, $C(0) = 700$. This information can be used to determine the constant C.

$$C(0) = 50(0) - .04(0)^2 + C$$
$$700 = 0 - 0 + C$$
$$C = 700$$

Thus,

(a) $C(x) = 50x - .04x^2 + 700$

(b) $C(10) = 50(10) - .04(10)^2 + 700 = 1196$

The cost of producing 10 units is $1196. Figure 1 shows the cost function we have determined.

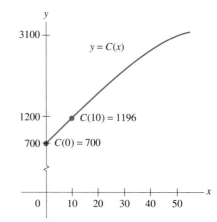

Figure 1 The cost function determined in Example 3 ◆

EXAMPLE 4 ◆ ROCKET FLIGHT

A toy rocket is shot vertically upward from the ground with an initial velocity of 300 feet per second. The acceleration due to gravity is −32 feet per second per second—negative because it is downward. (No other acceleration is applied to the rocket.)

(a) Determine a formula for the rocket's velocity t seconds after the launch.

(b) Determine a formula for the rocket's distance above the ground at any time t.

SOLUTION **(a)** Recall from Chapter 3 that acceleration is the derivative of velocity. That is,

$$a = \frac{dv}{dt}$$

Since the acceleration is given as −32, we have

$$\frac{dv}{dt} = -32$$

Antidifferentiation yields

$$v = \int (-32)\, dt$$

or

$$v = -32t + C$$

To determine C, use the fact that the initial velocity is 300 feet per second. This means $v = 300$ when $t = 0$. Substituting these two numbers into the equation $v = -32t + C$ yields

$$300 = -32(0) + C$$

or

$$300 = C$$

Thus we have

$$v = -32t + 300$$

(b) Recall from Chapter 3 that velocity is the derivative of distance. That is,

$$v = \frac{ds}{dt}$$

And since $v = -32t + 300$, we have

$$\frac{ds}{dt} = -32t + 300$$

Antidifferentiation yields

$$s = \int (-32t + 300)\, dt$$

or

$$s = -16t^2 + 300t + C$$

To determine C, note that at the beginning (when $t = 0$), the rocket's distance s above the ground is zero, since it is shot upward from the ground. Substituting 0 for t and 0 for s into $s = -16t^2 + 300t + C$ yields

$$0 = -16(0)^2 + 300(0) + C$$

or

$$0 = C$$

Thus,

$$s = -16t^2 + 300t$$

is the formula for distance. ◆

EXAMPLE 5 ◆ *A LEARNING EXPERIMENT*

To test learning, a psychologist asks people to memorize a long sequence of digits. Assume that the rate at which digits are being memorized is

$$\frac{dy}{dt} = 5.4e^{-.3t} \quad \text{words per minute}$$

where y is the number of digits memorized and t is the time in minutes.

(a) Determine y as a function of t, which will tell us the number of digits memorized after t minutes.

(b) How many digits will be memorized after 5 minutes?

SOLUTION **(a)** From the given equation

$$\frac{dy}{dt} = 5.4e^{-.3t}$$

we can obtain y by antidifferentiation.

$$y = \int 5.4e^{-.3t}\, dt$$

$$= 5.4\int e^{-.3t}\, dt$$

$$= \frac{5.4}{-.3} e^{-.3t} + C$$

$$= -18e^{-.3t} + C$$

To determine C, we use the fact that in the beginning (when $t = 0$), the number of digits memorized is zero (that is, $y = 0$). Thus, $y = 0$ when $t = 0$ and

$$0 = -18e^{-.3(0)} + C$$

$$0 = -18(1) + C$$

$$C = 18$$

Thus,

$$y = 18 - 18e^{-.3t}$$

gives the number of digits memorized (y) as a function of time (t). A graph of the function is given in Figure 2.

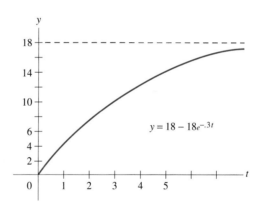

Figure 2 The learning function of Example 5: The number of digits memorized as a function of time

(b) After 5 minutes ($t = 5$), the number of digits memorized is

$$y = 18 - 18e^{-.3(5)}$$

$$= 18 - 18e^{-1.5}$$

$$\approx 18 - 18(.2231)$$

$$= 18 - 4.0158$$

$$\approx 14$$

Approximately 14 digits will be memorized after 5 minutes. ◆

6.2   Exercises

In Exercises 1–14, determine the particular antiderivative $f(x)$. See Example 1.

1. $f'(x) = 3x^2 - 2x + 5$, $f(1) = 8$

2. $f'(x) = 12x^2 + 6x$, $f(2) = 27$

3. $f'(x) = -3x^{-4}$, $f(1) = 3$

4. $f'(x) = 1 - \dfrac{20}{x^3}$, $f(1) = 20$

5. $f'(x) = 1 + 3\sqrt{x}$, $f(4) = 16$

6. $f'(x) = x^{3/2} + 6x$, $f(1) = 22/5$

7. $f'(x) = -\dfrac{1}{2}x^{-1/2}$, $f(9) = 2$

8. $f'(x) = 8x^{1/3} + 3x^{1/2}$, $f(1) = 7$

9. $f'(x) = e^x - 2x$, $f(0) = 6$

10. $f'(x) = e^x + 6x - 3$, $f(1) = e$

11. $f'(x) = \dfrac{1}{x}$, $f(1) = 5$

12. $f'(x) = 3 - \dfrac{1}{x}$, $f(1) = 8$

13. $f'(x) = e^{2x} + 8x$, $f(0) = 2$

14. $f'(x) = 1 + 10e^{5x}$, $f(0) = 2$

In Exercises 15–22, determine the equation of the curve that passes through the given point and has the given slope.

15. Through $(3, 14)$ and having slope $2x$

16. Through $(0, 3)$ and having slope $4x - 2$

17. Through $(2, 14)$ and having slope $3x^2 + 5$

18. Through $(1, 0)$ and having slope $4x^3 - 1$

19. Through $(9, 19)$ and having slope $\sqrt{x}$

20. Through $(1, 3)$ and having slope $3/x$

21. Through $(0, 1)$ and having slope e^x

22. Through $(0, 3)$ and having slope $2e^x$

(*Cost*) In Exercises 23–26, determine the cost function $C(x)$ that corresponds to the marginal cost given.

23. Marginal cost $= 40 - .06x$,   fixed cost $= \$200$

24. Marginal cost $= 10 - .02x$,   fixed cost $= \$150$

25. Marginal cost $= \dfrac{10}{\sqrt{x}}$,   fixed cost $= \$50$

26. Marginal cost $= \dfrac{2}{\sqrt{x}}$,   fixed cost $= \$10$

27. (*Cost*) If the marginal cost when x units are produced is $100 - .50x$ dollars and the overhead is $\$40$, what is the cost of producing 10 units?

28. (*Revenue*) Let $R(x)$ be the revenue a company receives from the sale of x units of its product. If their marginal revenue $R'(x)$ is $100 - .2x$ dollars, determine
(**a**) $R(x)$
(**b**) The company's revenue from the sale of 20 units.
Assume there is no revenue when zero units are sold.

(*Revenue*) In Exercises 29–33, determine the revenue function $R(x)$ that corresponds to the marginal revenue given. Assume there is no revenue when zero units are sold.

29. Marginal revenue $= 50 - .4x$

30. Marginal revenue $= 100 - .03x^2$

31. Marginal revenue $= \dfrac{1}{\sqrt{x}} - \dfrac{1}{10}$

32. Marginal revenue $= 1 + .0002x$

33. Marginal revenue $= 10 - e^{.05x}$

34. (*Demand*) Determine the demand function corresponding to the revenue function that was obtained in Exercise 28.

(*Profit*) In Exercises 35–41, determine the profit function $P(x)$ that corresponds to the given marginal profit.

35. Marginal profit $= 40 - .8x$, $P(0) = -\$30$

36. Marginal profit $= 35 - .6x^2$, $P(0) = -\$50$

37. Marginal profit $= 100 + .4x - .06x^2$, $P(0) = 0$

38. Marginal profit $= 25 + .02x$, $P(0) = 0$

39. Marginal profit $= 50 - .3\sqrt{x}$, $P(0) = -\$130$

40. Marginal profit $= 100 - e^{.02x}$, $P(0) = -\$50$

41. Marginal profit $= 70 - e^{.01x}$, $P(0) = -\$30$

W 42. (*PROFIT*) In Exercises 35–41, the value of $P(0)$ was given.
(a) What is the meaning of $P(0)$?
(b) What does it mean when $P(0)$ is negative?

43. (*VELOCITY/DISTANCE*) A ball is thrown vertically upward from the ground with an initial velocity of 105 feet per second. The acceleration due to gravity is -32 feet per second per second.
(a) Determine a formula for the velocity of the ball t seconds after being thrown.
(b) How far is the ball from the ground t seconds after being thrown?

44. (*VELOCITY/DISTANCE*) A ball is shot vertically upward from the edge of a building with initial velocity 352 feet per second. The building is 768 feet tall. Acceleration due to gravity is -32 feet per second per second.
(a) Determine the equations that describe the velocity of the ball and its distance from the ground.
(b) How far above the ground is the ball after 6 seconds, and how fast is it going then?

45. (*VELOCITY/DISTANCE*) A tourist accidentally drops his camera from the top of a cliff that is 576 feet above the water below. Assume the acceleration due to gravity to be -32 feet per second per second.
(a) Determine the velocity $v(t)$ of the camera at any time t during its fall.

(b) Determine $s(t)$, the height of the camera above the water at any time t during its fall.
(c) How fast is the camera falling 4 seconds after it is dropped?
(d) How long will it take the camera to hit the water? (*Hint*: What is the value of s when the camera hits the water?)

46. (*VELOCITY/DISTANCE*) A woman gets into her car and then drives it with a constant acceleration of 22 feet per second per second.
(a) Determine the velocity function.
(b) Determine the distance function.
(c) How far does the car go in 6 seconds?

47. (*VELOCITY/DISTANCE*) On the *moon* the magnitude of the acceleration due to gravity is less than that on the earth; it is approximately -5.3 feet per second per second. Consider a ball thrown upward from the surface of the moon with an initial velocity of 120 feet per second.
(a) Obtain a function that gives the velocity of the ball at any time t.
(b) Determine a function that shows the distance of the ball from the moon's surface at any time t.

48. (*TREE HEIGHT*) The height h (in feet) of a tree is a function of time t (in years). Suppose you begin ($t = 0$) by planting a 5-foot tree in your yard. Assume the tree will grow to maturity according to the formula

$$\frac{dh}{dt} = 1.5 + \frac{.25}{\sqrt{t}} \qquad t > 0$$

(a) Determine a formula for the height of the tree at any time t.
(b) Find the height of the tree after 1 year, 4 years, 9 years, and 16 years.

49. (*FLU OUTBREAK*) From data collected by the county health office, it is estimated that a flu virus is spreading through the county at the rate of $5t^{2/3} + 22$ people per day.
(a) If n is the number of people who have the flu at any time t, where t is the time in days, complete the equation. $dn/dt = $ _____
(b) If 50 people had the flu at the beginning of the outbreak, determine an equation that expresses n as a function of t.
(c) How many people have the flu after 8 days?

50. **(FLU EPIDEMIC)** A flu epidemic is spreading at the rate

$$\frac{dn}{dt} = 180t - 6t^2$$

where n is the number of people who are sick with flu on any particular day t after the start of the outbreak (see figure).

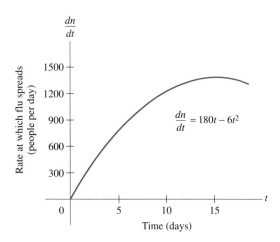

(a) Determine an equation for n as a function of t. Assume no one has the flu at the beginning (when $t = 0$).
(b) How many people have the flu the tenth day after the outbreak begins?

51. **(INHIBITING GROWTH)** A colony of 2000 bacteria is introduced to a growth-inhibiting environment and grows at the rate

$$\frac{dn}{dt} = 30 + 2t$$

where n is the number of bacteria present at any time t (t is measured in hours).
(a) Determine a function that gives the number of bacteria present at any time t.
(b) How many bacteria are present after 3 hours?

52. **(INHIBITING GROWTH)** Redo Exercise 51 assuming that a colony of 1000 bacteria is introduced to the growth-inhibiting environment and that

$$\frac{dn}{dt} = 50 + 2t$$

53. **(MOLD GROWTH)** The weight of a mold is growing exponentially at the rate of

$$\frac{dw}{dt} = e^{.2t} \quad \text{milligrams per hour}$$

How much will the mold weigh in 10 hours, if it weighs 70 milligrams now?

54. **(TEMPERATURE)** The rate of change of the temperature T inside a furnace after x minutes ($0 \le x \le 20$) is

$$\frac{dT}{dx} = 2x + 15 \quad \text{degrees per minute}$$

Assume the furnace is 200°F at the beginning.
(a) Find the formula for the temperature at any time x.
(b) What is the temperature inside the furnace after 14 minutes?

55. **(ATMOSPHERIC PRESSURE)** The rate at which atmospheric pressure P changes as the height x above sea level changes is

$$\frac{dP}{dx} = -3.087e^{-.21x}$$

(P is measured in pounds per square inch and x is in miles, which means that dP/dx is in pounds per square inch per mile.)
Determine P as a function of x. At sea level, P is 14.7 pounds per square inch.

56. **(GEOMETRY)** The rate of change of the area of a circular region with respect to its radius is

$$\frac{dA}{dr} = 2\pi r$$

Use this fact, and the fact that $A = 0$ when $r = 0$, to determine the area of a circular region when the radius is 4 centimeters.

57. **(GEOMETRY)** The rate of change of the volume of a spherical balloon with respect to its radius is

$$\frac{dV}{dr} = 4\pi r^2$$

Use this fact, and the fact that $V = 0$ when $r = 0$, to determine the volume of the balloon when its radius is 6 centimeters.

W 58. **(COST)** Explain in words how you can determine the cost function when you are given the marginal cost function and the fixed cost (or overhead).

6.3 | THE DEFINITE INTEGRAL AS THE AREA UNDER A CURVE

As we prepare to introduce the definite integral, the need will arise for a compact way of writing sums. Consider the sum of the integers from 1 through 50.

$$1 + 2 + 3 + \cdots + 50$$

We shall use the capital Greek letter sigma (Σ) to specify a sum. Along with sigma, a letter such as i, j, or k is used as the *index*, or counter. The first and last values of the index are written on the sigma as shown next.

$$1 + 2 + 3 + \cdots + 50 = \sum_{i=1}^{50} i$$

This particular example of **sigma notation**, or **summation notation**, specifies a sum of numbers of the form "i," where i begins at 1 and counts up to 50. Thus, it specifies the sum of the integers from 1 through 50.

Here is another sum written in sigma notation.

$$\sum_{j=2}^{5} j(j-2)$$

This one represents the sum of terms of the form $j(j-2)$. The j values begin at 2 and count up to 5; that is, j is 2, 3, 4, 5.

$$\begin{aligned}
\sum_{j=2}^{5} j(j-2) &= 2(2-2) + 3(3-2) + 4(4-2) + 5(5-2) \\
&= 2(0) + 3(1) + 4(2) + 5(3) \\
&= 0 + 3 + 8 + 15 \\
&= 26
\end{aligned}$$

The index can also be used for subscripts. Consider the sum

$$x_1 + x_2 + x_3 + x_4 + \cdots + x_n$$

Using summation notation, this sum can be written as

$$\sum_{i=1}^{n} x_i$$

EXAMPLE 1 Evaluate the sum

$$\sum_{i=1}^{3} f(x_i)$$

assuming

$$f(x) = x^2$$
$$x_1 = 2$$
$$x_2 = 3$$
$$x_3 = 7$$

SOLUTION

$$\sum_{i=1}^{3} f(x_i) = f(x_1) + f(x_2) + f(x_3)$$
$$= f(2) + f(3) + f(7)$$
$$= 2^2 + 3^2 + 7^2$$
$$= 4 + 9 + 49$$
$$= 62 \quad \blacklozenge$$

Area and the Definite Integral

The study of geometry includes formulas for determining the area bounded by such geometric figures as circles, triangles and rectangles. (See Figure 3.) In this section we will develop the calculus necessary to determine the area of other types of regions—regions bounded by various curves. See Figure 4. We will also pursue a variety of applications and the role of antidifferentiation in this setting.

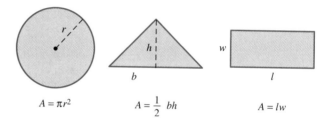

Figure 3 Areas enclosed by geometric figures

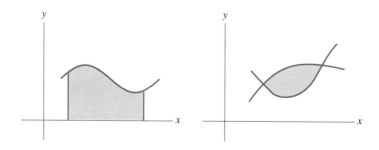

Figure 4 Regions bounded by curves

To begin, let us consider the area bounded by the graph of $y = f(x)$, the x axis, and the vertical lines $x = a$ and $x = b$. This is usually called simply "the area under the curve." We will be considering the interval from $x = a$ to $x = b$ and we will assume that the graph of $y = f(x)$ is continuous, that is, has no breaks or gaps. We will also assume that $f(x) \geq 0$ for all x between a and b. The desired area is shown shaded in Figure 5.

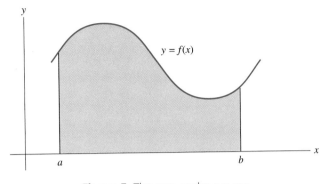

Figure 5 The area under a curve

At this time we have no means of determining the *exact* value of the shaded area. However, we can approximate the area by using rectangles, since it is an easy matter to determine the area of any rectangle.

We begin by dividing the interval from a to b into n equal subintervals. Since the whole interval has width $b - a$, it follows that the width of each of the n subintervals is

$$\frac{b - a}{n}$$

We shall call this width Δx. That is,

$$\Delta x = \frac{b - a}{n}$$

One rectangle will be constructed for each subinterval, and the width of each rectangle will be Δx. The length of each rectangle will be the distance between the x axis and the graph, measured vertically at the right end of each subinterval. This is illustrated in Figure 6. Here $n = 4$, so that there are 4 subintervals, each of width Δx. If the area of each rectangle is computed, and then all four areas are added, the result will be an approximation to the area under the curve.

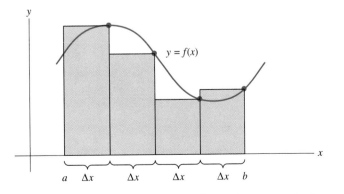

Figure 6 Rectangles constructed to approximate the area under the curve

The area of a rectangle is computed as length times width. Here the width is Δx in each instance. The length is the distance from the x axis to the graph. That distance is the value of $f(x)$ for the particular x. If we call the x values x_1, x_2, x_3, and x_4, then the lengths of the rectangles are $f(x_1)$, $f(x_2)$, $f(x_3)$, and $f(x_4)$. The area of the first rectangle is $f(x_1) \cdot \Delta x$. (See Figure 7.) For the entire region,

$$\text{Area} = f(x_1)\Delta x + f(x_2)\Delta x + f(x_3)\Delta x + f(x_4)\Delta x$$

See Figure 8.

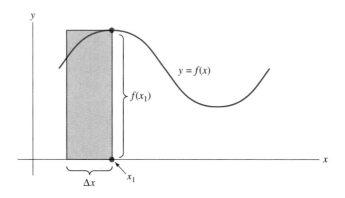

Figure 7 The area of this (shaded) rectangle is $f(x_1) \cdot \Delta x$

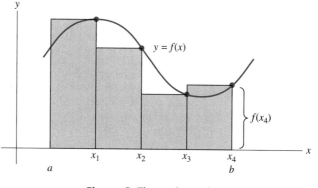

Figure 8 The entire region

Using summation notation, the total area of the rectangles is

$$\text{Area} = \sum_{i=1}^{4} f(x_i)\,\Delta x$$

This is a rough approximation to the area under the curve. The approximation can be improved by using more rectangles. Compare now the approximation shown by using four rectangles (Figure 9) and the better approximation obtained by using eight rectangles (Figure 10).

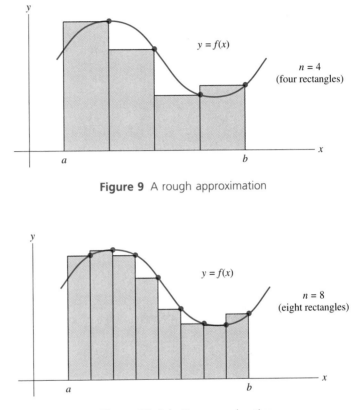

Figure 9 A rough approximation

Figure 10 A better approximation

EXAMPLE 2   Approximate the area bounded by the graph of $f(x) = 4 - x^2$, the x axis, and the lines $x = -1$ and $x = 1$.

(a) Use $n = 2$ subintervals.

(b) Use $n = 4$ subintervals.

SOLUTION   **(a)** If $n = 2$ subintervals are used, then

$$\Delta x = \frac{b - a}{n} = \frac{1 - (-1)}{2} = 1$$

The rectangles are shown in Figure 11.

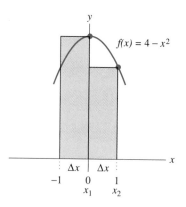

Figure 11 The rectangles of Example 2(a)

The area of the rectangles is

$$
\begin{aligned}
A &= f(x_1)\,\Delta x + f(x_2)\,\Delta x \\
&= f(0) \cdot 1 + f(1) \cdot 1 \\
&= (4 - 0^2) \cdot 1 + (4 - 1^2) \cdot 1 \\
&= 7
\end{aligned}
$$

(b) If $n = 4$ subintervals are used, then

$$
\Delta x = \frac{b - a}{n} = \frac{1 - (-1)}{4} = .5
$$

The rectangles are shown in Figure 12.

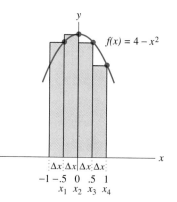

Figure 12 The rectangles of Example 2(b)

The area of the approximating rectangles is

$$A = f(x_1) \, \Delta x + f(x_2) \, \Delta x + f(x_3) \, \Delta x + f(x_4) \, \Delta x$$
$$= f(-.5)(.5) + f(0)(.5) + f(.5)(.5) + f(1)(.5)$$
$$= (3.75)(.5) + (4)(.5) + (3.75)(.5) + (3)(.5)$$
$$= 7.25$$

The approximation to the area under the curve improves as the number of rectangles n increases. In the next section, you will be able to determine that the exact area is $7\frac{1}{3}$ square units. ◆

Returning now to the development of the area under the graph of $y = f(x)$, we find the area of n rectangles to be

$$A = f(x_1) \, \Delta x + f(x_2)\Delta x + f(x_3) \, \Delta x + \cdots + f(x_n) \, \Delta x$$
$$= \sum_{i=1}^{n} f(x_i) \, \Delta x$$

The *exact* area between the graph of $y = f(x)$ and the x axis (on the interval from a to b) is the limit of this sum as the number of rectangles n approaches infinity. Thus, assuming the limit exists,

$$A = \lim_{n \to \infty} \sum_{i=1}^{n} f(x_i) \, \Delta x$$

Saying that $n \to \infty$ is the same as saying that $\Delta x \to 0$, since the width of each rectangle gets smaller and smaller as the number of rectangles increases. Thus, we can also write

$$A = \lim_{\Delta x \to 0} \sum_{i=1}^{n} f(x_i) \, \Delta x$$

These limits are given a special name and notation. They are known as the **definite integral** of f from a to b.

The Definite Integral of f from a to b

If f is continuous on $[a,b]$, then the **definite integral** of f from a to b is given by

$$\int_{a}^{b} f(x) \, dx = \lim_{\Delta x \to 0} \sum_{i=1}^{n} f(x_i) \, \Delta x$$

assuming the limit exists. The interval $[a,b]$ is divided into n equal subintervals of width Δx, where Δx is $(b - a)/n$, and x_i is the rightmost point in the ith interval.

Recall that we began with the area under a curve and insisted that $f(x) \geq 0$ on the interval from $x = a$ to $x = b$. This means that the definite integral represents the area under the curve only when $f(x) \geq 0$. Nevertheless, many other applications do not require that $f(x) \geq 0$.

6.3   Exercises

In Exercises 1–8, compute the value of each expression and simplify it completely.

1. $\displaystyle\sum_{i=1}^{5} i^2$

2. $\displaystyle\sum_{i=0}^{10} (i + 2)$

3. $\displaystyle\sum_{k=1}^{6} (2k + 1)$

4. $\displaystyle\sum_{j=-1}^{5} 2j$

5. $\displaystyle\sum_{j=0}^{5} j(j + 3)$

6. $\displaystyle\sum_{k=1}^{4} 3k$

7. $\displaystyle\sum_{n=1}^{3} \frac{n + 1}{2n}$

8. $\displaystyle\sum_{n=0}^{3} \frac{n}{n + 1}$

In Exercises 9–13, use summation notation to write each expression in condensed form.

9. $1 + 2 + 3 + 4 + 5 + 6 + 7 + 8 + 9$

10. $1 + \dfrac{1}{2} + \dfrac{1}{3} + \dfrac{1}{4} + \cdots + \dfrac{1}{100}$

11. $4 + 5 + 6 + \cdots + n$

12. $\dfrac{3}{7} + \dfrac{4}{7} + \dfrac{5}{7} + \cdots + \dfrac{20}{7}$

13. $\dfrac{1}{2} + \dfrac{2}{3} + \dfrac{3}{4} + \dfrac{4}{5} + \cdots + \dfrac{49}{50}$

Use summation notation to write each expression in Exercises 14–18 in a condensed form.

14. $x_1 + x_2 + x_3 + x_4 + x_5 + x_6$

15. $a_1 x_1 + a_2 x_2 + a_3 x_3 + \cdots + a_{10} x_{10}$

16. $x_1^2 + x_2^2 + x_3^2 + \cdots + x_n^2$

17. $f(x_0) + f(x_1) + f(x_2) + f(x_3) + \cdots + f(x_n)$

18. $a_0 x^0 + a_1 x^1 + a_2 x^2 + \cdots + a_{n-1} x^{n-1}$

Evaluate each sum in Exercises 19–22.

19. $\displaystyle\sum_{i=1}^{4} f(x_i)$ assuming that $f(x) = x^3$, $x_1 = 0$, $x_2 = 1$, $x_3 = 2$, and $x_4 = 3$

20. $\displaystyle\sum_{i=1}^{3} x_i f(x_i)$ assuming $f(x) = 3x$, $x_1 = 1$, $x_2 = 2$, and $x_3 = 3$

21. $\displaystyle\sum_{i=1}^{3} f(x_i)\, \Delta x$ assuming $f(x) = x^2$, $x_1 = 1$, $x_2 = 3$, $x_3 = 5$, and $\Delta x = 2$

22. $\displaystyle\sum_{i=1}^{5} f(x_i)\, \Delta x$ assuming $f(x) = 4x$, $x_1 = 0$, $x_2 = .5$, $x_3 = 1$, $x_4 = 1.5$, $x_5 = 2$, and $\Delta x = .5$

In Exercises 23–30, use rectangles to approximate the area bounded by the graph of function f, the x axis, and the two vertical lines given. Use n subintervals. Refer to Example 2.

23. $f(x) = x^2 + 2$, $x = 0$, $x = 2$, $n = 2$

24. $f(x) = x^2 + 2$, $x = 0$, $x = 2$, $n = 4$

25. $f(x) = 6 - x^2$, $x = -1$, $x = 1$, $n = 4$

26. $f(x) = 10 - x^3$, $x = 0$, $x = 2$, $n = 4$

27. $f(x) = 1 + x^3$, $x = 0$, $x = 2$, $n = 4$

28. $f(x) = 5 + \sqrt{x}$, $x = 0$, $x = 4$, $n = 4$

29. $f(x) = e^x$, $x = 0$, $x = 2$, $n = 4$

30. $f(x) = \dfrac{1}{x}$, $x = 1$, $x = 4$, $n = 4$

Perhaps you were surprised to see that the notation used for the definite integral looks so much like the notation used for antidifferentiation.

$$\text{The definite integral:} \quad \int_a^b f(x)\, dx$$

$$\text{Antidifferentiation:} \quad \int f(x)\, dx$$

As it happens, determining the area under a curve is indeed related to antidifferentiation. To see this, consider the following.

1. (a) Distance can be determined from velocity by antidifferentiation. Since $v = ds/dt$, it follows that

$$s = \int v\, dt$$

(b) Distance can also be determined as the area under the graph of a velocity function. Consider a train traveling at $v = 60$ miles per hour for 3 hours. As shown in Figure 13, the area is 180, which is in fact the distance traveled in 3 hours at 60 miles per hour. If the velocity were to vary, then we would have the situation illustrated in Figure 14.

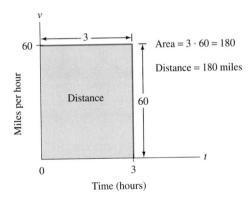

Figure 13 Distance as the area under the graph of a velocity function

2. (a) Total cost can be determined from marginal cost by antidifferentiation.

$$C(x) = \int C'(x)\, dx$$

(b) In a manner similar to the distance example in 1(b), total cost $C(x)$ can be determined as the area under the graph of the marginal cost function $y = C'(x)$. See Figure 15.

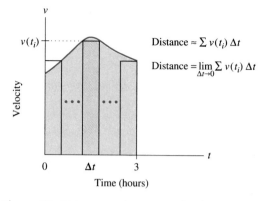

Figure 14 Distance as the area under the graph of a (variable) velocity function

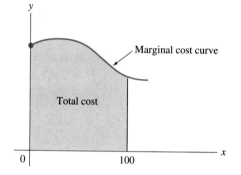

Figure 15 The shaded area represents the total cost of producing 100 units

Assuming that f is nonnegative and continuous at all x in the interval from a to b, the area $A(x)$ under the graph of f from a to x (as shown in Figure 16) is given by

$$A(x) = \int_a^x f(x)\, dx$$

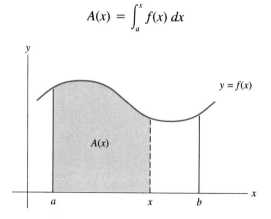

Figure 16 The area under the graph of f from a to x

Next we will show that A is an antiderivative of f. The shaded area in Figure 17 is $A(x + \Delta x) - A(x)$. This shaded area can be approximated by the rectangle having width Δx and length $f(x)$. The area of this approximating rectangle is $f(x) \cdot \Delta x$. Thus,

$$A(x + \Delta x) - A(x) \approx f(x)\,\Delta x$$

or

$$\frac{A(x + \Delta x) - A(x)}{\Delta x} \approx f(x)$$

The smaller Δx becomes, the better this approximation. In the limit, we have

$$\lim_{\Delta x \to 0} \frac{A(x + \Delta x) - A(x)}{\Delta x} = f(x)$$

or

$$A'(x) = f(x)$$

Thus, A is an antiderivative of f.

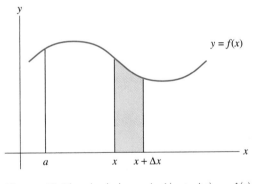

Figure 17 The shaded area is $A(x + \Delta x) - A(x)$

Knowing that A is an antiderivative of f, and letting F represent any antiderivative of f, we have the following result (which can be proved formally):

$$A(x) = F(x) + C \qquad (C = \text{some constant})$$

Considering that

$$A(b) = \int_a^b f(x)\,dx$$

and

$$A(a) = \int_a^a f(x)\,dx = 0$$

it follows that

$$\int_a^b f(x)\, dx = A(b)$$

$$= A(b) - A(a)$$
$$= (F(b) + C) - (F(a) + C)$$
$$= F(b) - F(a)$$

This result,

$$\int_a^b f(x)\, dx = F(b) - F(a)$$

is so important to the study of calculus that it is known as the **Fundamental Theorem of Calculus**. It provides a key link between differential calculus and integral calculus.

The Fundamental Theorem of Calculus

If f is continuous on the closed interval $[a, b]$, then

$$\int_a^b f(x)\, dx = F(b) - F(a)$$

where F is any antiderivative of f.

Note

1. As a convenience in applications, we will use the notation

$$[F(x)]_a^b$$

for $F(b) - F(a)$. This use is demonstrated in the examples that follow, where the Fundamental Theorem of Calculus is used to evaluate definite integrals.

2. The numbers a and b are called **limits of integration**. Specifically, b is the **upper limit** and a is the **lower limit**.

Using the Fundamental Theorem

To evaluate the definite integral of f from a to b by using the Fundamental Theorem of Calculus, follow these steps.

1. Determine an antiderivative of f.

2. Evaluate the antiderivative F at the upper limit; that is, obtain $F(b)$.

3. Evaluate the antiderivative F at the lower limit; that is, obtain $F(a)$.

4. Subtract as follows: $F(b) - F(a)$.

EXAMPLE 1 Evaluate $\displaystyle\int_1^2 x^2 \, dx$ using the Fundamental Theorem of Calculus.

SOLUTION For this integral, $f(x) = x^2$. Thus, $F(x) = x^3/3$, and so

$$\int_1^2 x^2 \, dx = \left[\frac{x^3}{3}\right]_1^2 \qquad \text{This is } [F(x)]_a^b.$$

$$= \frac{(2)^3}{3} - \frac{(1)^3}{3} \qquad \text{This is } F(b) - F(a).$$

$$= \frac{8}{3} - \frac{1}{3}$$

$$= \frac{7}{3} \quad \blacklozenge$$

Note

Since the Fundamental Theorem of Calculus says we can use *any* antiderivative, *we will always use the antiderivative with $C = 0$*. However, if you did use some other value of C, it would be eliminated in the process anyway. If $x^3/3 + C$ (rather than just x^3) had been used in Example 1, we would have had

$$\left[\frac{x^3}{3} + C\right]_1^2 = \left(\frac{8}{3} + C\right) - \left(\frac{1}{3} + C\right) = \frac{7}{3} + C - C = \frac{7}{3}$$

EXAMPLE 2 Evaluate the definite integral. $\int_0^4 (5x + 3)\, dx$

SOLUTION
$$\int_0^4 (5x + 3)\, dx = \left[\frac{5x^2}{2} + 3x \right]_0^4$$

$$= \left(\frac{5(4)^2}{2} + 3(4) \right) - \left(\frac{5(0)^2}{2} + 3(0) \right)$$

$$= (40 + 12) - (0 + 0)$$

$$= 52 \quad \blacklozenge$$

EXAMPLE 3 Integrate. $\int_0^1 (1 - e^t)\, dt$

SOLUTION
$$\int_0^1 (1 - e^t)\, dt = [t - e^t]_0^1$$

$$= (1 - e^1) - (0 - e^0)$$

$$= 1 - e - 0 + 1$$

$$= 2 - e \quad \blacklozenge$$

EXAMPLE 4 Determine the (exact) area under the curve $y = \sqrt{x}$ from $x = 1$ to $x = 4$.

SOLUTION The area is shown shaded in Figure 18.

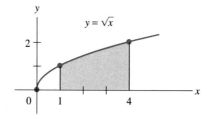

Figure 18 The area under $y = \sqrt{x}$ from $x = 1$ to $x = 4$

The area under the curve $y = f(x)$ from $x = a$ to $x = b$ is given by the definite integral

$$\int_a^b f(x)\, dx$$

In this instance, $f(x) = \sqrt{x}$, $a = 1$, and $b = 4$. Thus,

$$\text{Area} = \int_1^4 \sqrt{x}\,dx$$

$$= \int_1^4 x^{1/2}\,dx$$

$$= \left[\frac{x^{3/2}}{3/2}\right]_1^4$$

$$= \left[\frac{2}{3}x^{3/2}\right]_1^4$$

$$= \frac{2}{3} \cdot 4^{3/2} - \frac{2}{3} \cdot 1^{3/2}$$

$$= \frac{2}{3} \cdot 8 - \frac{2}{3} \cdot 1$$

$$= \frac{16}{3} - \frac{2}{3}$$

$$= \frac{14}{3}$$

The area between the graph of $y = \sqrt{x}$ and the x axis on the interval $[1, 4]$ has been shown to be 14/3, or 4 2/3 square units. ◆

EXAMPLE 5   Determine the area under the curve $y = 1/x$ from $x = 1$ to $x = 7$.

SOLUTION   The area is

$$\int_1^7 \frac{1}{x}\,dx = \Big[\ln|x|\Big]_1^7$$

$$= \ln 7 - \ln 1$$

$$= \ln 7 - 0$$

$$= \ln 7 \quad ◆$$

Note

Example 5 showed that ln 7 is the area under the graph of $y = 1/x$ from $x = 1$ to $x = 7$. Can you see that for any $t > 0$,

$$\ln t = \int_1^t \frac{1}{x}\,dx$$

This is a calculus interpretation of the natural logarithm. See Figure 19.

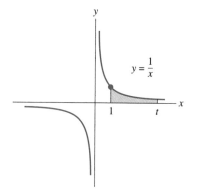

Figure 19 The shaded area is ln t

The three properties given next resemble familiar properties of indefinite integrals (from Section 6.1). You may need them for some applications later in the chapter.

Properties of Definite Integrals

$$\int_a^b [f(x) + g(x)]\, dx = \int_a^b f(x)\, dx + \int_a^b g(x)\, dx$$

$$\int_a^b [f(x) - g(x)]\, dx = \int_a^b f(x)\, dx - \int_a^b g(x)\, dx$$

$$\int_a^b kf(x)\, dx = k\int_a^b f(x)\, dx \qquad k = \text{constant}$$

6.4   Exercises

It has been shown that the area under the graph of a velocity function is the (total) distance traveled and that the area under the marginal cost curve is the total cost. In Exercises 1–8, interpret the area under the graph of each function.

1. (*MARGINAL REVENUE*) The function gives the marginal revenue when x units have been sold.

2. (*MARGINAL PROFIT*) The function gives the marginal profit when x units have been produced and sold.

3. (*GASOLINE CONSUMPTION*) The function gives the rates at which gasoline was consumed in the United States from 1964 to 1976.

4. (*SALES INCREASE*) The function gives the rate at which a corporation's sales have grown from 1988 to 1993.

5. (*INTEREST RATES*) The function gives the interest rates paid by a bank on its money market account from 1984 to 1992.

6. (*FLU EPIDEMIC*) The function gives the rate (in people per day) at which a flu virus is spreading from the fourth day through the tenth day of the epidemic.

7. (*YEAST CULTURE*) The function gives the rate (in milligrams per hour) at which a yeast culture has been growing this week.

8. (*POPULATION GROWTH*) The function gives the rate at which a city's population has grown from 1990 to 1993.

Evaluate each definite integral in Exercises 9–34.

9. $\displaystyle\int_0^3 8x \, dx$

10. $\displaystyle\int_1^2 3x^2 \, dx$

11. $\displaystyle\int_1^4 x^2 \, dx$

12. $\displaystyle\int_2^6 (3t - 8) \, dt$

13. $\displaystyle\int_0^2 (1 + x^3) \, dx$

14. $\displaystyle\int_0^2 (y^3 - y) \, dy$

15. $\displaystyle\int_1^6 (5t - t^2) \, dt$

16. $\displaystyle\int_3^4 (1 - 2x + 3x^2) \, dx$

17. $\displaystyle\int_0^3 (x^2 + 2x - 5) \, dx$

18. $\displaystyle\int_0^6 (x^2 - 8x + 17) \, dx$

19. $\displaystyle\int_0^1 (e^x - 2x) \, dx$

20. $\displaystyle\int_0^1 (x^2 + e^x) \, dx$

21. $\displaystyle\int_1^4 x^{-3} \, dx$

22. $\displaystyle\int_1^{10} 3t^{-2} \, dt$

23. $\displaystyle\int_2^5 \frac{1}{t^2} \, dt$

24. $\displaystyle\int_3^4 \frac{2}{x^2} \, dx$

25. $\displaystyle\int_4^9 6\sqrt{x} \, dx$

26. $\displaystyle\int_0^8 \sqrt[3]{x} \, dx$

27. $\displaystyle\int_1^2 \frac{3}{x^4} \, dx$

28. $\displaystyle\int_1^3 \frac{8}{t^3} \, dt$

29. $\displaystyle\int_1^3 \frac{1}{x} \, dx$

30. $\displaystyle\int_2^e \frac{1}{x} \, dx$

31. $\displaystyle\int_0^2 (1 - 4e^t) \, dt$

32. $\displaystyle\int_0^3 (5 - 2e^x) \, dx$

33. $\displaystyle\int_0^{100} e^{.05x} \, dx$

34. $\displaystyle\int_0^{50} e^{.04x} \, dx$

Evaluate each definite integral in Exercises 35–42.

35. $\displaystyle\int_1^2 \frac{3x^2 + 4}{x} \, dx$

36. $\displaystyle\int_1^e \frac{1 + x^3}{x} \, dx$

37. $\displaystyle\int_e^4 \frac{t + 1}{t^2} \, dt$

38. $\displaystyle\int_2^e \frac{1 - t^2}{t} \, dt$

39. $\displaystyle\int_1^9 \frac{1 + x}{\sqrt{x}} \, dx$

40. $\displaystyle\int_4^9 \frac{x + \sqrt{x}}{x} \, dx$

41. $\displaystyle\int_1^4 \frac{\sqrt{x} + x}{\sqrt{x}} \, dx$

42. $\displaystyle\int_1^4 \frac{1 - x^2}{x^{1/2}} \, dx$

Determine each area in Exercises 43–56.

43. The area under $y = x^2$ from $x = 0$ to $x = 3$.

44. The area under $y = x^2 + 3$ from $x = 1$ to $x = 4$.

45. The area under $y = 4 - x^2$ from $x = 0$ to $x = 2$.

46. The area under $y = 9 - x^2$ from $x = 1$ to $x = 2$.

47. The area under $y = 3x - x^2$ from $x = 0$ to $x = 2$.

48. The area under $y = -5x - x^2$ from $x = -3$ to $x = 0$.

49. The area under $y = \sqrt[3]{x}$ from $x = 1$ to $x = 8$.

50. The area under $y = \sqrt{x}$ from $x = 9$ to $x = 16$.

51. The area under $y = e^x$ from $x = 0$ to $x = 1$.

52. The area under $y = e^x$ from $x = 0$ to $x = 2$.

53. The area under $y = 1/x$ from $x = 1$ to $x = 6$.

54. The area under $y = 1/x$ from $x = 1$ to $x = e$.

55. The area under $y = e^{.5x}$ from $x = 0$ to $x = 2$.

56. The area under $y = e^{.2x}$ from $x = 0$ to $x = 10$.

57. A property of definite integrals is

$$\int_a^a f(x) \, dx = 0$$

Verify this property for the three integrals shown below by evaluating the integrals and showing that in each instance the integral is zero.

$$\int_2^2 6x \, dx \qquad \int_1^1 x^2 \, dx \qquad \int_{-2}^{-2} (x^3 + 1) \, dx$$

58. Another property of definite integrals allows the splitting of an integral into two integrals by changing the limits of integration. Specifically,

$$\int_a^b f(x)\,dx = \int_a^c f(x)\,dx + \int_c^b f(x)\,dx$$

for any number c in the interval from a to b.

Verify this property for the two examples shown below by evaluating all integrals and showing that in each case the integral on the left is indeed equal to the sum of the two integrals on the right.

(a) $\displaystyle\int_0^5 (4x + 3)\,dx = \int_0^2 (4x + 3)\,dx + \int_2^5 (4x + 3)\,dx$

(b) $\displaystyle\int_1^8 3x^2\,dx = \int_1^4 3x^2\,dx + \int_4^8 3x^2\,dx$

W 59. Give an intuitive geometric interpretation of why you

would expect the integral property of Exercise 57 to be true.

W 60. Give an intuitive geometric interpretation of why you would expect the integral property of Exercise 58 to be true. It may help to draw the graph of $y = f(x)$ defined on an interval $[a, b]$ with c in the interval.

Evaluate each definite integral in Exercises 61–66.

61. $\displaystyle\int_{-1}^0 \frac{e^x - 1}{e^x}\,dx$

62. $\displaystyle\int_1^2 \frac{2x + e^x}{xe^x}\,dx$

63. $\displaystyle\int_0^3 |x|\,dx$

64. $\displaystyle\int_{-2}^0 |x|\,dx$

65. $\displaystyle\int_0^2 \frac{x^2 - 9}{x - 3}\,dx$

66. $\displaystyle\int_0^2 \frac{x^3 + 8}{x + 2}\,dx$

6.5 | SOME APPLICATIONS OF THE DEFINITE INTEGRAL

This section presents an introduction to applications of the definite integral. The first three examples here can be compared with those of Section 6.2 to see the similarities and differences between applications of the antiderivative (indefinite integral) and the definite integral.

EXAMPLE 1 ◆ OBTAINING COST FROM MARGINAL COST

For a company, the marginal cost when x units of merchandise have been produced is $50 - .08x$ dollars. The fixed cost is $700.

(a) Determine the cost of producing x units.

(b) Find the cost of producing 10 units.

(c) Find the cost of producing 15 units.

(d) What is the total cost of raising production from 10 units to 15 units?

SOLUTION Parts (a) and (b) are the same as Example 3 from Section 6.2, where we found by antidifferentiating $C'(x) = 50 - .08x$ that $C(x) = 50x - .04x^2 + 700$ dollars and $C(10) = \$1196$.

(c) Let $x = 15$ in $C(x) = 50x - .04x^2 + 700$ to obtain $C(15) = \$1441$.

(d) The total cost of raising production from 10 units to 15 units is

$$C(15) - C(10) = \$1441 - \$1196$$
$$= \$245. \quad ◆$$

Note

The answer to part (d) of Example 1 is the value of $C(15) - C(10)$. We began with $C'(x)$. That should look like a familiar form. It is an example of the Fundamental Theorem of Calculus, and could have been obtained *directly* from a definite integral. Specifically,

$$\int_{10}^{15} C'(x) \, dx = C(15) - C(10)$$

This means that the cost of raising production from 10 units to 15 units could have been determined simply as

$$\int_{10}^{15} (50 - .08x) \, dx = [50x - .04x^2]_{10}^{15}$$

$$= (750 - 9) - (500 - 4)$$

$$= 245 \text{ dollars}$$

Note that there was no need to determine the constant 700. In fact, part (d) could have been done without first doing parts (a), (b), and (c).

In the next two examples you will see how the concept explained in the note (above) can be applied to completely different settings.

EXAMPLE 2 ◆ *FALLING OBJECT*

A ball is dropped from a high-altitude balloon. If the ball falls with velocity $v = 32t$ feet per second, how far does the ball travel during the first 4 seconds?

SOLUTION We are given $v = 32t$. Since v is the same as ds/dt or $s'(t)$, we can use the equation

$$s'(t) = 32t$$

We seek the total distance traveled from $t = 0$ to $t = 4$. That distance is $s(4) - s(0)$, or

$$\int_0^4 s'(t) \, dt = \int_0^4 32t \, dt$$

$$= [16t^2]_0^4$$

$$= 256 \text{ feet}$$

The ball falls 256 feet in the first 4 seconds. (Notice that the height of the balloon does not matter, unless it is lower than 256 feet, in which case the ball hits the ground within the first 4 seconds.) ◆

EXAMPLE 3 ◆ GASOLINE CONSUMPTION

The rate at which gasoline was consumed in the United States was approximately $c'(t) = .075t + 1.7$ billion gallons per year from 1964 ($t = 0$) to 1976 ($t = 12$). Determine the total amount of gasoline consumed from 1964 to 1976.

SOLUTION The function $c'(t) = .075t + 1.7$ gives the *rate* of consumption, dc/dt. So the consumption itself $c(t)$ is an antiderivative of the given function $c'(t)$. Specifically, the total amount of gasoline consumed from 1964 ($t = 0$) to 1976 ($t = 12$) is given by the following definite integral:

$$\int_0^{12} c'(t)\, dt = \int_0^{12} (.075t + 1.7)\, dt$$

$$= [.0375t^2 + 1.7t]_0^{12}$$

$$= 5.4 + 20.4$$

$$= 25.8$$

The total consumption of gasoline was 25.8 billion gallons.   ◆

Average Value of a Function

You probably know that the average value of n numbers $x_1, x_2, x_3, \ldots, x_n$ is

$$\frac{x_1 + x_2 + x_3 + \cdots + x_n}{n}$$

To obtain the average value of a *function* over an interval, consider n functional values spread out evenly over the interval. The average of the n functional values $f(x_1), f(x_2), f(x_3), \ldots, f(x_n)$ is

$$\frac{f(x_1) + f(x_2) + f(x_3) + \cdots + f(x_n)}{n}$$

or

$$f(x_1) \cdot \frac{1}{n} + f(x_2) \cdot \frac{1}{n} + f(x_3) \cdot \frac{1}{n} + \cdots + f(x_n) \cdot \frac{1}{n}$$

If each term is multiplied by 1 in the form

$$\frac{b - a}{b - a}$$

the average will appear as

$$\frac{1}{b - a} \cdot f(x_1) \frac{b - a}{n} + \frac{1}{b - a} \cdot f(x_2) \frac{b - a}{n} + \cdots + \frac{1}{b - a} \cdot f(x_n) \cdot \frac{b - a}{n}$$

The fraction $(b - a)/n$ should look familiar; it's the Δx from Section 6.3, where the interval $[a, b]$ was divided into n equal subintervals. If we substitute Δx for each $(b - a)/n$, the result is

$$\frac{1}{b - a} f(x_1) \Delta x + \frac{1}{b - a} f(x_2) \Delta x + \cdots + \frac{1}{b - a} f(x_n) \Delta x$$

Factoring out $1/(b - a)$ from each term yields

$$\frac{1}{b - a} [f(x_1)\, \Delta x + f(x_2)\, \Delta x + \cdots + f(x_n)\, \Delta x]$$

or

$$\frac{1}{b - a} \sum_{i=1}^{n} f(x_i)\, \Delta x$$

Letting the number of functional values approach infinity will give the average value of the function.

$$\lim_{n \to \infty} \frac{1}{b - a} \sum_{i=1}^{n} f(x_i)\, \Delta x = \frac{1}{b - a} \int_{a}^{b} f(x)\, dx$$

Thus,

Average Value of a Function

The **average value of $f(x)$** over the interval $[a, b]$ is

$$\frac{1}{b - a} \int_{a}^{b} f(x)\, dx$$

EXAMPLE 4   Find the average value of $f(x) = 3x^2 + 4x - 5$ over the interval $[1, 3]$.

SOLUTION   The average value is

$$\frac{1}{3 - 1} \int_{1}^{3} (3x^2 + 4x - 5)\, dx = \frac{1}{2} \int_{1}^{3} (3x^2 + 4x - 5)\, dx$$

$$= \frac{1}{2} [x^3 + 2x^2 - 5x]_{1}^{3}$$

$$= \frac{1}{2} [(27 + 18 - 15) - (1 + 2 - 5)]$$

$$= 16 \quad \blacklozenge$$

EXAMPLE 5   ◆   *BLOOD FLOW THROUGH AN ARTERY*

Blood does not flow with a constant velocity. It flows fastest in the center of an artery and slowest next to the wall of the artery. In fact, the velocity v at any distance x from the center can be expressed as a function of x. For example, for an artery of radius .2 centimeter, the velocity is

$$v = 40 - 990x^2$$

The distance x is measured in centimeters and the velocity is in centimeters per second. See Figure 20.

Figure 20 Blood flow through an artery

It is natural to wonder what is the *average* velocity of the blood flowing through the artery. Using

$$\text{Average value of } f(x) = \frac{1}{b-a} \int_a^b f(x)\, dx$$

the average velocity $\bar{v}$ is

$$\bar{v} = \frac{1}{.2 - 0} \int_0^{.2} (40 - 990x^2)\, dx$$

Keep in mind that x is in the interval $[0, .2]$ because it is the distance from the center. That distance could be as small as zero (at the center) or as large as .2 (at the artery wall).

Continuing,

$$\bar{v} = \frac{1}{.2} \int_0^{.2} (40 - 990x^2)\, dx$$

$$= 5[40x - 330x^3]_0^{.2}$$

$$= 5\{[(40)(.2) - (330)(.2)^3] - [(40)(0) - (330)(0)^3]\}$$

$$= 26.8 \text{ centimeters per second} \qquad .$$

We have used an equation developed by French physician J. L. M. Poiseuille in 1842. The equation will be different if an artery of different radius is used. The length of the artery and the person's blood pressure also affect the equation. Some assumptions were made in order to simplify the example. ◆

Volume

In this presentation you will see how a definite integral can be used to find the volume of a *solid of revolution*. Such a solid is produced by revolving a plane region (an area) about a line (such as the x axis). Consider the plane region shown shaded in Figure 21.

Now, imagine revolving this region about the x axis. As it spins around the axis, it sweeps out a three-dimensional figure—a solid of revolution. See Figure 22.

This particular solid is cone-shaped. A formula for finding the volume of such solids of revolution can be obtained in a manner similar to the way the basic area formula was obtained. Consider a function $y = f(x)$ that is nonnegative

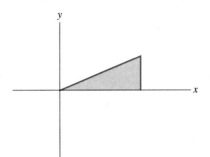

Figure 21 A plane region—to be revolved about the *x* axis

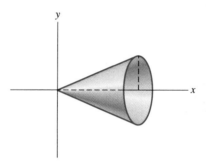

Figure 22 A solid of revolution

and continuous on an interval $[a, b]$. Divide the interval of width $b - a$ into n equal subintervals of width Δx, where

$$\Delta x = \frac{b - a}{n}$$

Let us draw the rectangles as we did in Section 6.3. The heights of the rectangles are $f(x_1)$, $f(x_2)$, and so on. See Figure 23.

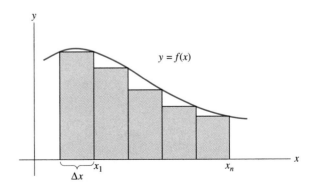

Figure 23 Rectangles in the region under the curve

As the region is spun around the x axis, each rectangle generates a cylinder. For example, the first rectangle with width Δx and height $f(x_1)$ generates a cylinder having radius $f(x_1)$ and height Δx. This is illustrated in Figures 24 and 25.

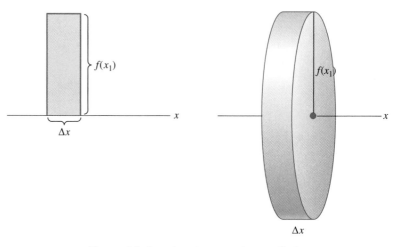

Figure 24 A rectangle generates a cylinder

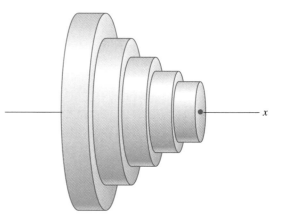

Figure 25 The rectangles generate cylinders

The volume of a cylinder is $\pi r^2 h$, where r is the radius and h is the height. Here $r = f(x_1)$ and $h = \Delta x$. Thus, the volume of the first cylinder generated is $\pi[f(x_1)]^2 \Delta x$. The volume of the second cylinder is $\pi[f(x_2)]^2 \Delta x$. The total volume generated by revolving all n rectangles is

$$\pi[f(x_1)]^2 \Delta x + \pi[f(x_2)]^2 \Delta x + \cdots + \pi[f(x_n)]^2 \Delta x$$

In summation notation, we have

$$V = \sum_{i=1}^{n} \pi[f(x_i)]^2 \, \Delta x$$

This volume is approximately the volume of the solid of revolution. The larger n becomes, the smaller Δx becomes, and the better is the approximation. The exact volume of the solid is the limit of this sum as $n \to \infty$ or as $\Delta x \to 0$.

$$V = \lim_{\Delta x \to 0} \sum_{i=1}^{n} \pi[f(x_i)]^2 \, \Delta x$$

The limit on the right is the definite integral shown next in the box.

Volume

The **volume** V of the solid produced by revolving the region bounded by $y = f(x)$ and the x axis (between $x = a$ and $x = b$) about the x axis is

$$V = \int_a^b \pi[f(x)]^2 \, dx$$

provided that f is continuous and nonnegative on $[a, b]$.

EXAMPLE 6 ◆ *VOLUME*

Find the volume of the solid produced by revolving about the x axis the region bounded by $y = \sqrt{x}$, the x axis, $x = 1$, and $x = 3$.

SOLUTION   A graph of the region being revolved is shown in Figure 26.

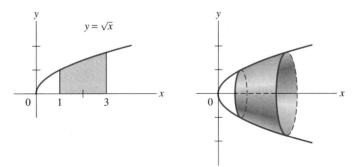

Figure 26 The region and the volume of revolution

$$V = \int_1^3 \pi \left[\sqrt{x}\right]^2 dx$$

$$= \pi \int_1^3 x \, dx$$

$$= \pi \left[\frac{x^2}{2}\right]_1^3$$

$$= \pi \left[\frac{9}{2} - \frac{1}{2}\right]$$

$$= 4\pi$$

The volume is 4π cubic units. ◆

6.5 Exercises

1. *(COST)* Suppose the marginal cost when x units of merchandise have been produced is $40 - .06x$ dollars. What is the cost of increasing the number of units produced from 5 to 10?

2. *(OIL CONSUMPTION)* Assume the rate of oil consumption over a 4-year period was $c'(t) = 10e^{.05t}$ billions of barrels per year. Determine the total amount of oil consumed during the period.

3. *(SALES)* A corporation's sales have grown at the rate $ds/dt = 1800 - 200e^{.01t}$, where s is the amount of sales in dollars and t is the time in days. Determine the total amount of sales for the first 100 days.

4. *(SALES)* A corporation's sales are declining at the rate $s' = 2000e^{-.01t}$, where s is the amount of sales in dollars and t is the time in days. (Assume $t = 0$ today.)
 (a) Determine the total amount of sales (to the nearest hundred dollars) for the next 100 days.
 (b) Determine the total amount of sales for the hundred days after that (that is, for the second hundred days).

W 5. *(PROFIT)* A computer manufacturer has a marginal profit function $P'(x) = 300 - .2x$ dollars, where x is the number of personal computers produced. The company is currently producing 1400 computers. Would it be wise for them to raise the production level to 1700 computers? Explain.

6. *(FALLING STONE)* A stone is dropped from a high cliff and falls with velocity $v = 32t$ feet per second. How far does the stone travel during the first 3 seconds?

7. *(FALLING STONE)* On the moon a stone is dropped from a high cliff and falls with velocity $v = 5.3t$ feet per second. How far does the stone travel during the first 3 seconds?

8. *(CAR TRAVEL)* A car is traveling at the rate of $v = 12t$ feet per second.
 (a) How far does it travel in the first 5 seconds?
 (b) How far does it travel during the 10th second?

9. *(SPREAD OF FLU VIRUS)* A flu virus is spreading at the rate of $dn/dt = 5 + 3\sqrt{t}$ people per day. Here n is the number of people and t is the time in days. How many people will get the flu between the ninth and sixteenth days?

10. *(TREE GROWTH)* Assume that a small tree will grow to maturity according to

$$\frac{dh}{dt} = 2 + \frac{1}{2\sqrt{t}} \qquad t > 0$$

where h is the height of the tree in feet and t is the time in years.
 (a) How much does the tree grow between the first and fourth years?
 (b) How much does it grow between the fourth and ninth years?

11. (AIR POLLUTION) In response to pressure from the mayor and the city council, the owners of a local factory have agreed to reduce the volume of polluted gas spewed into the air at their plant. Accordingly, over the next 6 months ($0 \leq t \leq 180$ days), the rate of pollution dp/dt will be given by

$$\frac{dp}{dt} = 3000 - 10t \quad \text{cubic meters per day}$$

Find the total volume of polluted gas that will be vented into the atmosphere over the next 6 months.

12. (DEPRECIATION) A business anticipates that $10,000 worth of newly purchased equipment will depreciate at the rate of $3000 - 400t$ dollars per year for 7 years. Use a definite integral to determine the total amount of depreciation that will occur during the first four years.

13. (POPULATION INCREASE) It is anticipated that the county's population will be increasing at the rate of $10,000e^{.04t}$ people per year for the next t years. What will be the county's total increase in population in the next 5 years, based on this rate?

14. (WATER FLOW) Water will be added to the city's reservoir tonight for a 4-hour period. The rate at which water is added depends on the time and is given by

$$\frac{dw}{dt} = 1000 + 100t \qquad 0 \leq t \leq 4$$

where w is the volume in gallons and t is the time in hours. Determine the total amount of water added during the 4-hour period.

15. (WATER USAGE) The rate at which water is used in a small Maryland town depends on the time of day. From 2 p.m. ($t = 2$) to 10 p.m. ($t = 10$), water is used at the rate of $36t - 3t^2$ gallons per hour. Find the total number of gallons of water used over this 8-hour period.

In Exercises 16–24, determine the average value of each function over the given interval.

16. $f(x) = x^2$ over [0, 1]

17. $f(x) = x^2 + 4$ over [0, 1]

18. $f(x) = 3x^2 - 2x + 10$ over [1, 4]

19. $f(x) = x^2 + 6x - 2$ over [0, 3]

20. $f(x) = \sqrt{x}$ over [4, 9]

21. $f(x) = \dfrac{1}{\sqrt{x}}$ over [9, 16]

22. $f(x) = e^{2x}$ over [0, 1]

23. $f(x) = e^{.01x}$ over [0, 100]

24. $f(x) = \dfrac{1}{x}$ over [1, e]

25. (BLOOD FLOW) Determine the average velocity of the blood flowing through an artery having radius .25 centimeter. Assume that $v = 63 - 960x^2$. (See Example 5.)

26. (BLOOD FLOW) Determine the average velocity of blood flowing through an artery having radius .18 centimeter. Assume that $v = 32 - 1020x^2$. (See Example 5.)

27. (FALLING STONE) A stone is dropped from a high cliff and falls with velocity $v = 32t$ feet per second. What is the average velocity of the stone during the first 5 seconds?

28. (HEATED METAL) As heat is applied to a metal bar, its temperature T increases. After t seconds the temperature of the bar is

$$T(t) = 70e^{.02t}$$

(a) What is the temperature of the bar at the beginning?

(b) What is the temperature of the bar after 20 seconds?

(c) What is the average temperature of the bar during the first 20 seconds?

29. (TREE GROWTH) Suppose a small tree grows so that its height h after t years is

$$h = 1.4t + .6\sqrt{t} + 7 \quad \text{feet}$$

What is the average height of the tree from the fourth year to the ninth year?

30. (BACTERIA COUNT) The number of bacteria present in a particular culture is

$$A = 200e^{.3t}$$

where A is the number of bacteria present after t hours. Determine the average number of bacteria present during the first 10 hours.

31. (COMPOUND INTEREST) If you deposit $3000 into a savings account that pays 8% interest per year compounded continuously for 10 years, what will be the average amount of money in your account during the 10-year period?

32. (BUYER POWER) If an economist estimates that the buying power b of the dollar will be

$$b = .98^t$$

in t years, what is the average buying power of the dollar during the next 2 years?

W 33. Use the calculus formula for average value to determine the average value of $f(x) = 5$ over the interval $[1, 8]$. Comment on the answer and suggest an easy, noncalculus method of determining the same result.

W 34. Use calculus to determine the average value of $f(x) = x^3$ over the interval $[-2, 2]$. Then sketch a graph of the function and use the graph to explain the result of the integration.

(VOLUME) In Exercises 35–52, find the volume of the solid produced by revolving about the x axis the region whose boundary is given. (*Note:* $y = 0$ is the equation of the x axis.)

35. $y = \sqrt{x}$, the x axis, $x = 2$, $x = 4$

36. $y = 2x$, $y = 0$, $x = 1$, $x = 3$

37. $y = x + 1$, the x axis, $x = 0$, $x = 4$

38. $y = 2x + 3$, the x axis, $x = 0$, $x = 5$

39. $y = x^2 + 1$, $y = 0$, $x = 0$, $x = 1$

40. $y = x^2 + 2$, $y = 0$, $x = 1$, $x = 3$

41. $y = x^3$, the x axis, $x = 1$, $x = 2$

42. $y = \sqrt{x - 1}$, $y = 0$, $x = 3$, $x = 7$

43. $y = \sqrt{x + 2}$, $y = 0$, $x = 0$, $x = 8$

44. $y = e^x$, the x axis, $x = 0$, $x = 1$

45. $y = e^{2x}$, the x axis, $x = 0$, $x = 3$

46. $y = \dfrac{1}{x}$, the x axis, $x = 1/2$, $x = 1$

47. $y = \dfrac{1}{x}$, the x axis, $x = 1/4$, $x = 1$

48. $y = \dfrac{1}{\sqrt{x}}$, $y = 0$, $x = 1$, $x = e$

49. $y = \dfrac{1}{\sqrt{x}}$, the x axis, $x = e$, $x = 10$

50. $y = \sqrt[3]{x}$, $y = 0$, $x = 1$, $x = 8$

51. $y = 1 + \dfrac{1}{x}$, $y = 0$, $x = 1$, $x = 2$

52. $y = 1 + \sqrt{x}$, $y = 0$, $x = 1$, $x = 4$

53. (VOLUME) If the region bounded by the line $y = \frac{r}{h}x$, the x axis, and $x = h$ is revolved about the x axis, the result will be a cone. Use calculus to determine the volume of the cone. (From geometry, a cone with radius r and height h has volume $V = \frac{1}{3}\pi r^2 h$, which is the answer.)

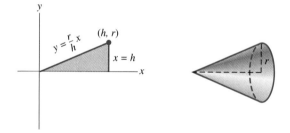

6.6 | SURPLUS

Now we will combine our knowledge of supply and demand with what we know about the area under a curve, to study the idea of surplus. To begin, consider a graph (Figure 27) showing a typical demand curve $p = D(x)$ and supply curve $p = S(x)$. Recall that x is the number of units and p is the price per unit.

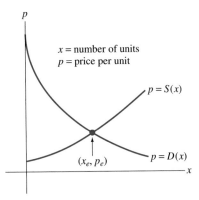

p

x = number of units
p = price per unit

p = S(x)

(x_e, p_e)

p = D(x)

x

Figure 27 Supply and demand curves

Recall from Section 1.7 that the point (x_e, p_e) where the two curves meet is called the **equilibrium point**; it is the point where supply equals demand. To buy x_e units at price p_e dollars each, a consumer would spend $x_e \cdot p_e$ dollars. The amount $x_e p_e$ happens to be the area of a rectangle—the rectangle shown shaded in Figure 28. The length of the rectangle is x_e and the width is p_e.

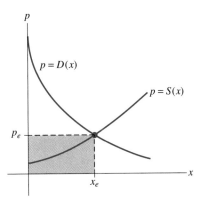

p

p = D(x)

p = S(x)

p_e

x_e

x

Figure 28 The area of the rectangle is $x_e p_e$

However, if we examine the demand curve alone (see Figure 29), we see that there are consumers willing to pay more than p_e dollars. For example, they will pay approximately p_1 dollars for each of the first Δx units, p_2 dollars for each of the next Δx units, and so on. The sum $p_1 \Delta x + p_2 \Delta x + \cdots + p_e \Delta x$ is the approximate cost of buying x_e units at the higher prices.

The exact cost of buying x_e units this more expensive way is the area under the graph of $p = D(x)$ between $x = 0$ and $x = x_e$, namely,

$$\int_0^{x_e} D(x)\, dx$$

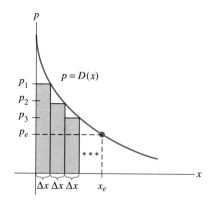

Figure 29 Some consumers are willing to pay more than p_e dollars

So if the consumer can buy x_e units at the equilibrium price p_e, he will save an amount equal to

$$\int_0^{x_e} D(x)\, dx - x_e p_e$$

This amount of savings is called the **consumer's surplus**.

Consumer's Surplus

$$\int_0^{x_e} D(x)\, dx - x_e p_e$$

EXAMPLE 1 ◆ CONSUMER'S SURPLUS

Let $p = D(x) = 15 - \frac{1}{2}x$ dollars and $p = S(x) = \frac{3}{2}x + 1$ dollars. Determine the equilibrium point and the consumer's surplus.

SOLUTION The equilibrium point is the point where the graphs of $p = D(x)$ and $p = S(x)$ meet. The point is found in this example by setting $15 - \frac{1}{2}x$ equal to $\frac{3}{2}x + 1$.

$$15 - \frac{1}{2}x = \frac{3}{2}x + 1$$

leads to

$$2x = 14$$

or

$$x = 7$$

And when $x = 7$, $p = 23/2$ or 11.5. Thus, the equilibrium point (x_e, p_e) is (7, 11.5).

The consumer's surplus is given by

$$\int_0^{x_e} D(x)\, dx - x_e p_e = \int_0^7 \left(15 - \frac{1}{2}x\right) dx - (7)(11.5)$$

$$= \left[15x - \frac{1}{4}x^2\right]_0^7 - 80.5$$

$$= 105 - 12.25 - 80.5$$

$$= 12.25$$

The consumer's surplus is \$12.25.   ◆

From the producer's point of view there is also a saving made by selling at the equilibrium price. The supply curve shows that some suppliers are willing to sell units at a lower price, in which case the amount they receive would be the area under the supply curve from $x = 0$ to $x = x_e$. See Figure 30.

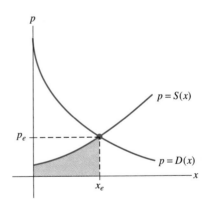

Figure 30 The amount received (area under the curve) is $\int_0^{x_e} S(x)\, dx$

By selling at the equilibrium price, the producer receives $x_e p_e$. Thus, the producer's gain by selling at the equilibrium price is

$$x_e p_e - \int_0^{x_e} S(x)\, dx$$

This amount is called the **producer's surplus**.

Producer's Surplus

$$x_e p_e - \int_0^{x_e} S(x)\, dx$$

EXAMPLE 2 ◆ **PRODUCER'S SURPLUS**

Find the producer's surplus for the conditions of Example 1.

SOLUTION   In Example 1 we determined the equilibrium point:

$$(x_e, p_e) = (7, 11.5)$$

Using this information and the given supply function $S(x) = \frac{3}{2}x + 1$, the producer's surplus is

$$x_e p_e - \int_0^{x_e} S(x)\, dx = (7)(11.5) - \int_0^7 \left(\frac{3}{2}x + 1\right) dx$$

$$= 80.5 - \left[\frac{3x^2}{4} + x\right]_0^7$$

$$= 80.5 - (36.75 + 7)$$

$$= 36.75$$

The producer's surplus is $36.75.   ◆

6.6   Exercises

(SURPLUS) In each exercise, both a supply function and a demand function are given. Determine (a) the equilibrium point; (b) the consumer's surplus; (c) the producer's surplus. Assume the monetary unit is dollars.

1. $S(x) = 2x + 7$, $D(x) = 40 - x$

2. $S(x) = 3x + 16$, $D(x) = 100 - x$

3. $S(x) = \frac{3}{2}x + 4$, $D(x) = 30 - \frac{1}{2}x$

4. $S(x) = \frac{4}{3}x + 2$, $D(x) = 20 - \frac{2}{3}x$

5. $S(x) = .25x + 1$, $D(x) = 15 - .25x$

6. $S(x) = .3x + 2$, $D(x) = 10 - .2x$

7. $S(x) = \frac{1}{10}x^2 + 1$, $D(x) = 11 - \frac{3}{10}x^2$

8. $S(x) = .2x^2 + 5$, $D(x) = 77 - .3x^2$

9. $S(x) = 2\sqrt{x} + 1$, $D(x) = 13 - \sqrt{x}$

10. $S(x) = \frac{3}{2}\sqrt{x} + 3$, $D(x) = 15 - \frac{1}{2}\sqrt{x}$

11. $S(x) = x^2 + 1$, $D(x) = 13 - x$

12. $S(x) = x^2 + 5$, $D(x) = 29 - 2x$

6.7 │ AREA IN THE PLANE

The development of the definite integral produced a natural application—area under a curve. By area under a curve we mean the area of the region between the graph of $y = f(x)$ and the x axis over some interval (from $x = a$ to $x = b$) for which $f(x) \geq 0$. See Figure 31.

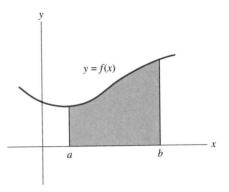

Figure 31 The area under a curve

We can easily extend the area concept to include the area *between* two curves. Let the curves be graphs of the functions $y = f(x)$ and $y = g(x)$ and consider the region between them from $x = a$ to $x = b$. (See Figure 32.)

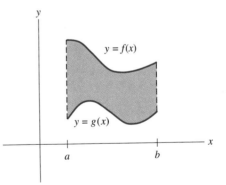

Figure 32 The region between two curves

The area of this region can be computed as the area under the curve $y = f(x)$ *minus* the area under the curve $y = g(x)$, as can be seen from drawings in Figure 33. The *unshaded* region in the drawing in Figure 33(b) is the area between the two curves, and it is, in fact, the area under $y = f(x)$ minus the area under $y = g(x)$. Thus, assuming that both functions are continuous and that the graph of $y = f(x)$ is above the graph of $y = g(x)$ everywhere on the interval from a to b, the area between the curves from $x = a$ to $x = b$ is

$$\int_a^b f(x)\,dx - \int_a^b g(x)\,dx \qquad \text{("top curve } minus \text{ bottom curve")}$$

The two integrals can be combined as

$$\int_a^b [f(x) - g(x)] \, dx$$

Area Between Two Curves

The area between the graphs of $y = f(x)$ and $y = g(x)$, from $x = a$ to $x = b$, is

$$\int_a^b [f(x) - g(x)] \, dx$$

provided that f and g are continuous on $[a, b]$ and $f(x) \geq g(x)$.

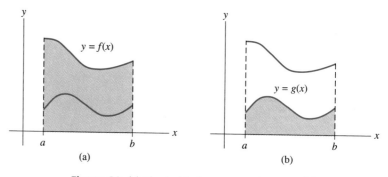

(a) (b)

Figure 33 (a) Shaded is the area under $y = f(x)$
(b) Shaded is the area under $y = g(x)$

For area, we have insisted that the integrand be nonnegative. As long as $f(x) \geq g(x)$, the integrand $f(x) - g(x)$ will indeed be nonnegative. (Add $-g(x)$ to both sides of the inequality $f(x) \geq g(x)$ and the result is $f(x) - g(x) \geq 0$.) Note that this is true even if $g(x)$ is negative or if both $f(x)$ and $g(x)$ are negative.

EXAMPLE 1 Find the area between the curves $y = x^2 + 1$ and $y = x - 2$ from $x = -1$ to $x = 2$.

SOLUTION Begin with a graph of the two functions and shade in the desired region, as shown in Figure 34.

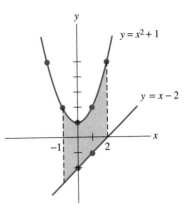

Figure 34 The area between the curves

The function $y = x^2 + 1$ is the greater (top) function, so call if f. This means $f(x) = x^2 + 1$ and $g(x) = x - 2$, and

$$
\begin{aligned}
\text{Area} &= \int_a^b [f(x) - g(x)]\, dx \\
&= \int_{-1}^2 [(x^2 + 1) - (x - 2)]\, dx \\
&= \int_{-1}^2 (x^2 - x + 3)\, dx \\
&= \left[\frac{x^3}{3} - \frac{x^2}{2} + 3x\right]_{-1}^2 \\
&= \left(\frac{8}{3} - 2 + 6\right) - \left(-\frac{1}{3} - \frac{1}{2} - 3\right) \\
&= 10\frac{1}{2} \quad \blacklozenge
\end{aligned}
$$

EXAMPLE 2 Find the area of the region enclosed by the curves $y = x^2$ and $y = 2x$.

SOLUTION The statement of the example gives no values for a and b. However, the curves intersect and create an enclosed region. Thus, the x coordinates of the points of intersection will be the a and b values. A graph makes this more apparent. The desired region is shown shaded in Figure 35.

Several observations need to be made.

1. From Figure 35, we see that $y = 2x$ is the greater ("top") function, so let $f(x) = 2x$ and $g(x) = x^2$.

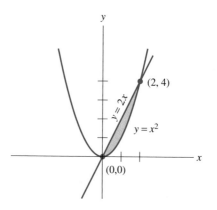

Figure 35 The area between the curves

2. The graphs intersect at the points $(0, 0)$ and $(2, 4)$, that is, where $x = 0$ and where $x = 2$. So $a = 0$ and $b = 2$ for the limits of integration. The area will be determined from $x = 0$ to $x = 2$.

3. The area is

$$\text{Area} = \int_0^2 [(2x) - (x^2)] \, dx$$

$$= \int_0^2 (2x - x^2) \, dx$$

$$= \left[x^2 - \frac{x^3}{3} \right]_0^2$$

$$= \frac{4}{3}$$

4. In *this* example, the points of intersection $(0, 0)$ and $(2, 4)$ were likely to have been chosen as points to plot when obtaining points. So by luck you might have just happened upon the points of intersection. But you won't always be so fortunate. Often the points of intersection must be found algebraically by solving a system of two equations in the unknowns x and y. In this example, we would solve the system

$$\begin{cases} y = x^2 \\ y = 2x \end{cases}$$

The method of substitution can be used. Since $y = x^2$ and $y = 2x$, it follows that $x^2 = 2x$. Solving:

$$x^2 = 2x$$
$$x^2 - 2x = 0$$
$$(x)(x - 2) = 0$$
$$x = 0, x = 2 \qquad \text{(as we expected!)}$$

The corresponding y values can be found by substituting these x values into either $y = x^2$ or $y = 2x$. *However*, we have no need for the y values in this area application. ◆

EXAMPLE 3   Find the area enclosed by $y = x^2 - 4x + 3$ and the x axis.

SOLUTION   This example is different because the selected portion of the graph of the function $y = x^2 - 4x + 3$ is *below* the x axis. The desired region is shown shaded in Figure 36.

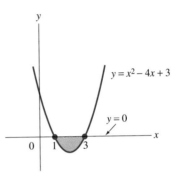

Figure 36 The region is below the x axis

From the graph you can see that the area we seek is from $x = 1$ to $x = 3$. Notice how the x axis (the line $y = 0$) serves as the greater (top) function by helping to form the enclosed region. Thus, $f(x) = 0$ and $g(x) = x^2 - 4x + 3$.

$$\text{Area} = \int_1^3 [(0) - (x^2 - 4x + 3)]\, dx$$

$$= \int_1^3 (-x^2 + 4x - 3)\, dx$$

$$= \left[-\frac{x^3}{3} + 2x^2 - 3x \right]_1^3$$

$$= \frac{4}{3}$$

By the way, the limits of integration ($x = 1$ and $x = 3$) *can* be determined algebraically. They are the numbers for which $g(x) = x^2 - 4x + 3$ equals zero. So solve the equation $x^2 - 4x + 3 = 0$ and you will obtain $x = 1$ and $x = 3$ algebraically. ◆

Note

To find the area of the region between a curve $y = f(x)$ and the x axis, *when the region is entirely below the x axis*, use either one of these equivalent methods.

$$\int_a^b [0 - f(x)]\, dx \qquad \text{as done in Example 3}$$

$$\int_a^b [-f(x)]\, dx \qquad \text{simplified form of the above}$$

If you are careless and compute this type of area as simply

$$\int_a^b f(x)\, dx$$

your result will be negative.

EXAMPLE 4   Determine the area between $y = x^2 - 4$ and the x axis from $x = 1$ to $x = 4$.

SOLUTION   You will get the *wrong* result if you do not graph the function and instead simply write down the integral

$$\int_1^4 (x^2 - 4)\, dx = 9 \qquad \text{(wrong!)}$$

A graph will show why this approach is wrong. The desired area is shown shaded in Figure 37.

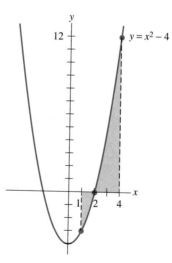

Figure 37 A region that consists of two areas—one above the x axis and one below

The area is in fact the sum of two areas, one below the x axis and one above it. As shown in Figure 37, the curve crosses the x axis at 2. Thus,

$$\text{Area} = \int_1^2 [0 - (x^2 - 4)]\, dx + \int_2^4 (x^2 - 4)\, dx$$

$$= \int_1^2 (4 - x^2)\, dx + \int_2^4 (x^2 - 4)\, dx$$

$$= \left[4x - \frac{x^3}{3}\right]_1^2 + \left[\frac{x^3}{3} - 4x\right]_2^4$$

$$= \left[\left(8 - \frac{8}{3}\right) - \left(4 - \frac{1}{3}\right)\right] + \left[\left(\frac{64}{3} - 16\right) - \left(\frac{8}{3} - 8\right)\right]$$

$$= \frac{5}{3} + \frac{32}{3}$$

$$= \frac{37}{3} \quad \blacklozenge$$

EXAMPLE 5 Determine the area of the region enclosed by $y = x^3$ and $y = x$.

SOLUTION The graph is shown in Figure 38, where the desired region is shaded.

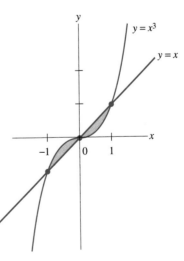

Figure 38 The area enclosed by the curves

The region is in fact two separate regions. Notice that for the region on the left, x goes from -1 to 0, and $y = x^3$ is the greater function. However, for the region on the right, x goes from 0 to 1, and $y = x$ is the greater function. Thus,

$$\text{Area} = \int_{-1}^{0} [(x^3) - (x)]\, dx + \int_{0}^{1} [(x) - (x^3)]\, dx$$

$$= \int_{-1}^{0} (x^3 - x)\, dx + \int_{0}^{1} (x - x^3)\, dx$$

$$= \left[\frac{x^4}{4} - \frac{x^2}{2} \right]_{-1}^{0} + \left[\frac{x^2}{2} - \frac{x^4}{4} \right]_{0}^{1}$$

$$= (0 - 0) - \left(\frac{1}{4} - \frac{1}{2} \right) + \left(\frac{1}{2} - \frac{1}{4} \right) - (0 - 0)$$

$$= \frac{1}{2}$$

Incidentally, the points of intersection of $y = x^3$ and $y = x$ can be found algebraically by solving the system

$$\begin{cases} y = x^3 \\ y = x \end{cases}$$

By substitution, $x^3 = x$. Then $x^3 - x = 0$, or $(x)(x^2 - 1) = 0$. Considering the factors equal to zero separately leads to $x = 0$, 1, and -1. ◆

6.7   Exercises

In Exercises 1–8, determine the area between the two curves. Graphing is recommended.

1. $y = x^2 + 2$ and $y = x$ from $x = 0$ to $x = 6$

2. $y = x^2 + 4$ and $y = x - 1$ from $x = 0$ to $x = 3$

3. $y = x + 5$ and $y = \sqrt{x}$ from $x = 0$ to $x = 4$

4. $y = 2x + 4$ and $y = \sqrt{x}$ from $x = 1$ to $x = 9$

5. $y = \sqrt{x}$ and $y = 9 - x$ from $x = 0$ to $x = 1$

6. $y = \sqrt{x}$ and $y = 7 - x$ from $x = 1$ to $x = 4$

7. $y = 1 - x^2$ and $y = x + 4$ from $x = -3$ to $x = 1$

8. $y = 2 - x^2$ and $y = 5 - x$ from $x = -1$ to $x = 1$

In Exercises 9–19, determine the area enclosed by the curves. Graphing is recommended.

9. $y = x^2$ and $y = 3x$

10. $y = x^2$ and $y = 4x$

11. $y = x^2 + 1$ and $y = 2x + 1$

12. $y = x^2 - 2$ and $y = 3x - 2$

13. $y = x^3$ and $y = x^2$ (Graph very carefully.)

14. $y = 4 - x^2$ and $y = x^2 - 4$

15. $y = \sqrt{x}$ and $y = x^2$

16. $y = x^2 + 3x + 2$ and $y = 3x + 6$

17. $y = x^2 - 2x + 1$ and $y = -2x + 2$

18. $y = e^x$, $y = e$, and $x = 0$

19. $y = e^x$, $y = 1$, and $x = 1$

31. $y = x^3$ and $y = 4x$ **32.** $y = x^3$ and $y = 9x$

33. $y = -x^3$ and $y = -x$ **34.** $y = x^3 + 1$ and $y = x + 1$

35. $y = x^3 - 4$ and $y = x - 4$

36. $y = x^3 - 12x$ and $y = 0$

37. $y = x^3 - 3x$ and $y = 0$

38. $y = x^4$ and $y = 4x^2$

39. $y = x^4$ and $y = x^2$

In Exercises 20–30, determine the area between the curve and the x axis. For some curves, the desired area is entirely below the x axis. For other curves, part of the desired area is below the x axis and part is above the x axis. See Examples 3 and 4. It is recommended that you graph before computing the area.

20. $y = x^2 - 5x + 4$ from $x = 1$ to $x = 4$

21. $y = x^2 - 2x - 3$ from $x = -2$ to $x = 2$

22. $y = x^2 - 7x + 10$ from $x = 3$ to $x = 5$

23. $y = x^2 - 3x - 4$ from $x = 2$ to $x = 6$

24. $y = x^2 - 1$ from $x = 0$ to $x = 4$

25. $y = x^2 - 4$ from $x = -1$ to $x = 4$

26. $y = x^2 - 3x$ from $x = 0$ to $x = 3$

27. $y = x^2 - 2x$ from $x = 0$ to $x = 2$

28. $y = x^2 - 3x$ from $x = 1$ to $x = 4$

29. $y = x^2 - 2x$ from $x = 1$ to $x = 3$

30 $y = -e^{-x}$ from $x = 0$ to $x = 1$

W 40. Explain how to select the *greater* function when finding the area between the graphs of two functions.

41. (*CONSUMER'S SURPLUS*) Consider the consumer's surplus as the area between the curves $y = D(x)$ and $y = p_e$ and obtain a definite integral that represents consumer's surplus.

42. (*PRODUCER'S SURPLUS*) Use the type of approach suggested in Exercise 41 to obtain a definite integral that represents producer's surplus.

In Exercises 43–48, determine the area enclosed by the curves.

43. $y = x^3 - 3x + 3$ and $y = 2x^2 + 3$

44. $y = x^3 - 2x + 1$ and $y = x^2 + 1$

45. $y = \dfrac{1}{x}$, $y = \dfrac{1}{e}$, and $x = 2$

46. $y = \sqrt{x}$, $y = \dfrac{1}{x}$, and $x = 4$

47. $y = |x|$ and $y = -x^2 + 2$

48. $y = |2x|$ and $y = -x^2 + 3$

In Exercises 31–39, determine the area enclosed by the curves. In each case the region enclosed consists of two regions. A careful study of the graphs is essential.

Chapter List   *Important terms and ideas*

antidifferentiation

antiderivative

integral sign

integrand

indefinite integral

constant of integration

power rule

particular antiderivative

summation notation

definite integral

Fundamental Theorem of Calculus

limits of integration

area under a curve

average value of $f(x)$

volume of a solid of revolution

consumer's surplus

producer's surplus

area between two curves

Review Exercises for Chapter 6

Evaluate each indefinite integral in Exercises 1–12.

1. $\int (8x + 2)\, dx$

2. $\int \frac{x}{2}\, dx$

3. $\int (10x^4 - x^2)\, dx$

4. $\int (3 - x^7)\, dx$

5. $\int (1 + \sqrt{x})\, dx$

6. $\int x^{-2/3}\, dx$

7. $\int \frac{3}{t^2}\, dt$

8. $\int 4t^{-3}\, dt$

9. $\int \frac{2}{x}\, dx$

10. $\int \frac{5}{\sqrt{x}}\, dx$

11. $\int e^{-.02x}\, dx$

12. $\int 4e^{4x}\, dx$

Evaluate each definite integral in Exercises 13–20.

13. $\int_0^5 (2x - 3)\, dx$

14. $\int_0^2 (x^3 + x)\, dx$

15. $\int_1^2 (3 - t^{-2})\, dt$

16. $\int_0^1 e^{-5x}\, dx$

17. $\int_1^6 \frac{1}{t}\, dt$

18. $\int_2^3 \frac{1}{t^2}\, dt$

19. $\int_1^{10} e^{.04x}\, dx$

20. $\int_0^1 (5 - e^x)\, dx$

Determine each area in Exercises 21–26.

21. The area under $y = x^2 + 2$ from $x = 0$ to $x = 3$.

22. The area under $y = 2 + \sqrt{x}$ from $x = 4$ to $x = 9$.

23. The area between the curves $y = 4 - x$ and $y = e^x$ from $x = 0$ to $x = 1$.

24. The area between the curve $y = x^2 + 1$ and the line $y = x$ from $x = 2$ to $x = 5$.

25. The area enclosed by the curve $y = x^2$ and the curve $y = 8 - x^2$.

26. The area enclosed by the line $y = x + 2$ and the curve $y = -x^2 + 4x + 2$.

27. (*REVENUE*) If the marginal revenue function is $R'(x) = .65 + .002x$ dollars, where x is the number of liter bottles of cola sold, determine the total revenue from the sale of 50 liter bottles of cola.

28. (*PROFIT*) Determine the profit function $P(x)$, if the marginal profit is $80 + .4x$ and $P(0) = -\$40$.

29. (*FLU OUTBREAK*) There is an outbreak of flu, and the number of people n who will have the flu after t days is

$$n = 8t + 3\sqrt{t}$$

What is the average number of people who will have the flu between the ninth and sixteenth days?

30. (*COMPOUND INTEREST*) A person deposits $5000 into a savings account that pays 7% interest per year compounded continuously for 6 years. What is the average amount of money in the account during the 6-year period?

31. Find the average value of $f(x) = x^{2/3}$ over the interval $[8, 27]$.

32. Find the average value of the function $f(x) = 1/x^2$ over the interval $[1, 5]$.

33. (*VOLUME*) Find the volume of the solid produced by revolving about the x axis the region bounded by $y = 2e^x$, $y = 0$, $x = -1$, and $x = 0$.

34. (*VOLUME*) Find the volume of the solid produced by revolving about the x axis the region bounded by $y = 1 + \sqrt[3]{x}$, $y = 0$, $x = 8$, and $x = 27$.

35. (*SURPLUS*) Suppose $S(x) = .4x + 3$ dollars and $D(x) = 15 - .2x$ dollars. Determine the equilibrium point, the consumer's surplus, and the producer's surplus.

36. (*SURPLUS*) Suppose $S(x) = .1x^2 + 2$ dollars and $D(x) = 11 - .2x^2$ dollars. Determine the equilibrium point, the consumer's surplus, and the producer's surplus.

37. Use summation notation to write the expression $x_1 f(x_1) + x_2 f(x_2) + \cdots + x_n f(x_n)$.

7

TECHNIQUES OF INTEGRATION

This chapter concentrates on what are known as *techniques of integration*. In Section 7.1, *integration by substitution* will enable you to extend the use of the familiar basic integration formulas introduced in Chapter 6. Then, *integration by parts*, a method based on the product rule for derivatives, offers an opportunity to perform integrations that are completely different from any you have seen before. Next, the use of integration tables opens up a whole new world of possibilities. Finally, numerical methods offer ways of approximating the value of definite integrals. Such methods are needed when the integral cannot be evaluated by any other means.

7.1 | *INTEGRATION BY SUBSTITUTION*

When the concept of integration was introduced in Chapter 6, several basic properties and formulas were presented. The list of basic integration formulas is summarized here.

Basic Integral Formulas

1. $\int x^n \, dx = \dfrac{x^{n+1}}{n+1} + C \qquad n \neq -1$

2. $\int k \, dx = kx + C \qquad k = \text{constant}$

3. $\int e^x \, dx = e^x + C$

4. $\int e^{kx} \, dx = \dfrac{1}{k} e^{kx} + C$

5. $\int \dfrac{1}{x} \, dx = \ln |x| + C$

These basic integration formulas were obtained by reversing known differentiation formulas. When this approach is applied to the chain rule, the process is called **integration by substitution**, which is the topic of this section.

To begin, consider $f(x) = (x^3 + 1)^5$. The derivative $f'(x)$ can be obtained by using the chain rule.

$$f'(x) = 5(x^3 + 1)^4 \cdot 3x^2$$

In this differentiation, we let $u = x^3 + 1$, and then du/dx is $3x^2$. The *integration* of $f'(x) = 5(x^3 + 1)^4 \cdot 3x^2$ can be accomplished by making the *same* substitution—let $u = x^3 + 1$. Specifically, in

$$\int 5(x^3 + 1)^4 \cdot 3x^2 \, dx$$

let

$$u = x^3 + 1$$

Then

$$\frac{du}{dx} = 3x^2$$

and

$$du = 3x^2 \, dx$$

We can now substitute within the integral. Replace $x^3 + 1$ by u and replace $3x^2 \, dx$ by du. The result is

$$\int 5\underbrace{(x^3 + 1)}_{u}^4 \cdot \underbrace{3x^2 \, dx}_{du} = \int 5u^4 \, du$$

The integration is now a familiar one, although the variable is u rather than x.

$$\int 5u^4 \, du = \frac{5u^5}{5} + C$$

$$= u^5 + C$$

To reintroduce x, use $u = x^3 + 1$ to replace u by $x^3 + 1$. The result is

$$\int 5u^4 \, du = (x^3 + 1)^5 + C$$

Thus,

$$\int 5(x^3 + 1)^4 \cdot 3x^2 \, dx = (x^3 + 1)^5 + C$$

Let's consider another example of integration by substitution.

EXAMPLE 1 Evaluate $\int 2x\sqrt{x^2 - 5} \, dx$.

SOLUTION Let $u = x^2 - 5$. Then $du/dx = 2x$ and $du = 2x \, dx$. Also, rearrange the integrand so that the substitution of u for $x^2 - 5$ and du for $2x \, dx$ can be made. The integration will then be an easy matter. Here is the entire procedure.

$$\int 2x\sqrt{x^2 - 5} \, dx = \int \sqrt{x^2 - 5} \, 2x \, dx \qquad \text{(rearranging)}$$

$$= \int \sqrt{u} \, du \qquad \text{(substituting)}$$

$$= \int u^{1/2} \, du$$

$$= \frac{2}{3} u^{3/2} + C$$

$$= \frac{2}{3}(x^2 - 5)^{3/2} + C \qquad \text{(substituting back)} \quad \blacklozenge$$

The next example demonstrates how constants can be introduced when needed in order to modify an integral.

EXAMPLE 2   Evaluate $\int (x^2 + 1)^6 \, x \, dx$.

SOLUTION   Let $u = x^2 + 1$. Then $du/dx = 2x$ and $du = 2x \, dx$. Unfortunately, the $2x \, dx$ obtained as du is not exactly the $x \, dx$ in the integral. To improve this situation, a *constant* factor can be inserted in an integral in the manner demonstrated next. Keep in mind that we want to insert a factor of 2 in order to change the $x \, dx$ to $2x \, dx$.

$$\int (x^2 + 1)^6 \, x \, dx = \int (x^2 + 1)^6 \cdot \frac{1}{2} \cdot 2 \cdot x \, dx$$

Since $\frac{1}{2} \cdot 2 = 1$, this is legitimate. In other words, we have merely multiplied the integrand by 1. Clearly, we do not want the 1/2 that now appears in the integrand. Since 1/2 is a *constant* factor, it can be placed outside the integral, just as was done in Section 6.1.

$$\int (x^2 + 1)^6 \, x \, dx = \frac{1}{2} \int (x^2 + 1)^6 \cdot 2x \, dx$$

Now the substitution can be made—u for $x^2 + 1$ and du for $2x \, dx$.

$$\int (x^2 + 1)^6 \, x \, dx = \frac{1}{2} \int u^6 \, du$$

$$= \frac{1}{2} \cdot \frac{u^7}{7} + C$$

$$= \frac{u^7}{14} + C$$

$$= \frac{(x^2 + 1)^7}{14} + C \quad \blacklozenge$$

EXAMPLE 3   A shortcut.

SOLUTION   After evaluating several integrals by substitution, you may wish to write less, since your experience enables you to visualize more. Consider the integral of Example 2.

$$\int (x^2 + 1)^6 \, x \, dx$$

Since $u = x^2 + 1$ and $du = 2x \, dx$, we need a 2 inside the integral (to create $2x$ from x) and a 1/2 in front of the integral (to balance the 2).

$$\frac{1}{2} \int (x^2 + 1)^6 \, 2x \, dx$$

Now, without actually writing another integral, *visualize* that this is in fact

$$\frac{1}{2} \int u^6 \, du$$

which will yield

$$\frac{1}{2} \cdot \frac{u^7}{7} + C$$

With all of this *in mind*, the actual steps that are written are as follows:

$$\int (x^2 + 1)^6 \, x \, dx = \frac{1}{2} \int (x^2 + 1)^6 \, 2x \, dx$$

$$= \frac{1}{2} \cdot \frac{(x^2 + 1)^7}{7} + C$$

$$= \frac{(x^2 + 1)^7}{14} + C \quad \blacklozenge$$

The shortcut suggests an easier way to evaluate such integrals. It also leads to extending the rule

$$\int x^n \, dx = \frac{x^{n+1}}{n + 1} + C \qquad n \neq -1$$

to include integrands of the form u^n, where u is a function of x.

$$\int u^n \, du = \frac{u^{n+1}}{n + 1} + C \qquad n \neq -1$$

where u is a function of x.

EXAMPLE 4   Evaluate $\int (1 - x^3)^4 \, x^2 \, dx$.

SOLUTION   Here $u = 1 - x^3$, and then $du = -3x^2 \, dx$. So a factor of -3 is needed inside the integral in order to complete the du. To balance the -3 inserted inside, we must place $-1/3$ outside the integral. The integration:

$$\int (1 - x^3)^4 \, x^2 \, dx = -\frac{1}{3} \int (1 - x^3)^4 (-3x^2) \, dx$$

$$= -\frac{1}{3} \cdot \frac{(1 - x^3)^5}{5} + C$$

$$= -\frac{(1 - x^3)^5}{15} + C \quad \blacklozenge$$

EXAMPLE 5   Evaluate $\int (x^2 + 1)^4 \, 3x \, dx$.

SOLUTION   Here $u = x^2 + 1$, and then $du = 2x \, dx$. The given integral contains $3x \, dx$, but we want $2x \, dx$. One simple way to handle this type of situation is to remove the 3. Place the constant factor 3 in front of the integral, leaving $x \, dx$ inside the integral. Then place a 2 inside the integral (to obtain the desired $2x\,dx$) and a 1/2 in front of the integral. Here is the entire procedure.

$$\int (x^2 + 1)^4\, 3x\, dx = 3\int (x^2 + 1)^4\, x\, dx$$

$$= \frac{1}{2} \cdot 3\int (x^2 + 1)^4\, 2x\, dx$$

$$= \frac{3}{2}\int (x^2 + 1)^4\, 2x\, dx$$

$$= \frac{3}{2} \cdot \frac{(x^2 + 1)^5}{5} + C$$

$$= \frac{3(x^2 + 1)^5}{10} + C \quad \blacklozenge$$

EXAMPLE 6   Evaluate the definite integral.

$$\int_0^2 \frac{x}{\sqrt{1 + 2x^2}}\, dx$$

SOLUTION

$$\int_0^2 \frac{x}{\sqrt{1 + 2x^2}}\, dx = \int_0^2 \frac{x}{(1 + 2x^2)^{1/2}}\, dx$$

$$= \int_0^2 (1 + 2x^2)^{-1/2}\, x\, dx$$

Since $u = 1 + 2x^2$, it follows that $du = 4x\, dx$. We need a 4 inside the integral and a 1/4 in front to compensate for it.

$$\int_0^2 \frac{x}{\sqrt{1 + 2x^2}}\, dx = \frac{1}{4}\int_0^2 (1 + 2x^2)^{-1/2}\, 4x\, dx$$

$$= \frac{1}{4}\left[\frac{(1 + 2x^2)^{1/2}}{1/2} \right]_0^2$$

$$= \frac{1}{4} \cdot \frac{2}{1} \left[\sqrt{1 + 2x^2} \right]_0^2$$

$$= \frac{1}{2}\left[(\sqrt{9}) - (\sqrt{1}) \right]$$

$$= 1 \quad \blacklozenge$$

Note

The definite integral of Example 6 could have been evaluated by using u in the integration and changing the limits of integration to agree with it. Since $u = 1 + 2x^2$, when (lower limit) x is 0, $u = 1 + 2(0)^2 = 1$. Similarly, when (upper limit) x is 2, $u = 1 + 2(2)^2 = 9$. Thus, we would have

$$\frac{1}{4}\int_1^9 u^{-1/2}\, du = \frac{1}{4} \cdot \frac{2}{1} \left[u^{1/2} \right]_1^9 = \frac{1}{2}(3 - 1) = 1$$

EXAMPLE 7 Evaluate $\int (x^2 + 8x + 1)^5 (x + 4)\, dx.$

SOLUTION Here $u = x^2 + 8x + 1$ and $du = (2x + 8)\, dx$. If we multiply $(x + 4)\, dx$ by 2, it will become $(2x + 8)\, dx$, which is du. So place a 2 inside the integral and 1/2 in front. The integration:

$$\int (x^2 + 8x + 1)^5 (x + 4)\, dx = \frac{1}{2}\int (x^2 + 8x + 1)^5 2(x + 4)\, dx$$

$$= \frac{1}{2} \cdot \frac{(x^2 + 8x + 1)^6}{6} + C$$

$$= \frac{(x^2 + 8x + 1)^6}{12} + C \quad \blacklozenge$$

Other integration formulas can be extended to include integration by substitution. For example, the formula

$$\int e^x \, dx = e^x + C$$

can be extended to

$$\int e^u \, du = e^u + C$$

EXAMPLE 8 Evaluate $\int xe^{x^2}\, dx.$

SOLUTION First, rearrange the integrand to put the x with the dx.

$$\int xe^{x^2}\, dx = \int e^{x^2} x \, dx$$

For this integral, $u = x^2$, which means that $du = 2x\, dx$. Consequently, a 2 must be placed inside the integral to complete the du and a 1/2 must be placed in front of the integral to compensate for it.

$$\int xe^{x^2}\, dx = \frac{1}{2}\int e^{x^2} 2x \, dx$$

The integrand is now of the form $e^u \, du$, so the integration can be completed.

$$\int xe^{x^2}\, dx = \frac{1}{2}e^{x^2} + C \quad \blacklozenge$$

Another formula that can be extended is

$$\int \frac{1}{x}\, dx = \ln |x| + C$$

The more general formula is

$$\int \frac{1}{u}\, du = \ln |u| + C$$

In view of the usual concern with obtaining du before completing any integration, this integration formula is usually considered in the form shown below. Example 9 will demonstrate why this is done.

$$\int \frac{du}{u} = \ln |u| + C$$

EXAMPLE 9 Evaluate $\displaystyle\int \frac{2x}{x^2 - 1}\, dx.$

SOLUTION *When the integrand is a fraction, the dx can be written in the numerator. Sometimes you will see integrals with the dx in the numerator rather than at the end of the integrand.* Specifically,

$$\int \frac{2x}{x^2 - 1}\, dx = \int \frac{2x\, dx}{x^2 - 1}$$

Having dx in the numerator means that the entire du will be together in the numerator. In this example, $u = x^2 - 1$, and thus $du = 2x\, dx$. This integral is clearly in the form

$$\int \frac{du}{u}$$

Thus,

$$\int \frac{2x\, dx}{x^2 - 1} = \ln |x^2 - 1| + C \quad \blacklozenge$$

EXAMPLE 10 Evaluate $\displaystyle\int \frac{e^x}{e^x - 1}\, dx.$

SOLUTION To begin,

$$\int \frac{e^x}{e^x - 1}\, dx = \int \frac{e^x\, dx}{e^x - 1}$$

Now you can see that the numerator is in fact the differential of the denominator, since $u = e^x - 1$ and $du = e^x\, dx$. Thus, the integral is of the form

$$\int \frac{du}{u}$$

Finally,

$$\int \frac{e^x}{e^x - 1} \, dx = \int \frac{e^x \, dx}{e^x - 1}$$

$$= \ln |e^x - 1| + C \quad \blacklozenge$$

7.1   Exercises

Evaluate each indefinite integral in Exercises 1–22.

1. $\int (x^2 + 3)^5 \, 2x \, dx$

2. $\int (5x + 1)^3 \, 5 \, dx$

3. $\int 2x\sqrt{x^2 - 6} \, dx$

4. $\int 6x\sqrt{1 + 3x^2} \, dx$

5. $\int (3x - 2)^6 \, dx$

6. $\int (x^2 - 10)^3 \, x \, dx$

7. $\int x\sqrt{x^2 - 3} \, dx$

8. $\int x^2\sqrt{1 + x^3} \, dx$

9. $\int (5x^3 + 1)^4 \, x^2 \, dx$

10. $\int (1 - 2x^3)^7 \, x^2 \, dx$

11. $\int \frac{1}{(7x + 2)^3} \, dx$

12. $\int \frac{x}{(1 - x^2)^4} \, dx$

13. $\int \frac{dx}{\sqrt{3x + 2}}$

14. $\int \frac{2 \, dx}{\sqrt{1 - 6x}}$

15. $\int \frac{1}{\sqrt[3]{1 + 4x}} \, dx$

16. $\int \frac{2}{(x - 4)^3} \, dx$

17. $\int 3(x + 2)^7 \, dx$

18. $\int 2(3x - 1)^6 \, dx$

19. $\int (x^2 - 4x + 1)^5 (x - 2) \, dx$

20. $\int (4x^2 + 4x - 6)^4 (2x + 1) \, dx$

21. $\int (x^{1/2} + 1)^4 \frac{1}{x^{1/2}} \, dx$

22. $\int \frac{(1 + \sqrt{x})^3}{\sqrt{x}} \, dx$

Evaluate each definite integral in Exercises 23–34.

23. $\int_0^1 (x + 1)^5 \, dx$

24. $\int_2^3 (x - 2)^4 \, dx$

25. $\int_1^2 (4x - 3)^3 \, dx$

26. $\int_0^1 (5x + 1)^2 \, dx$

27. $\int_2^3 \frac{dx}{(2x - 3)^2}$

28. $\int_1^3 \frac{dx}{(3x + 1)^3}$

29. $\int_0^4 x\sqrt{2x^2 + 4} \, dx$

30. $\int_1^3 x\sqrt{5x^2 + 4} \, dx$

31. $\int_0^4 \frac{x}{\sqrt{1 + 3x^2}} \, dx$

32. $\int_0^4 \frac{x}{\sqrt{x^2 + 9}} \, dx$

33. $\int_1^4 \frac{(1 + \sqrt{x})^2}{\sqrt{x}} \, dx$

34. $\int_1^9 \frac{(\sqrt{x} - 1)^3}{\sqrt{x}} \, dx$

Evaluate each indefinite integral in Exercises 35–46.

35. $\int 3x^2 e^{x^3} \, dx$

36. $\int x^2 e^{x^3} \, dx$

37. $\int e^{x+1} \, dx$

38. $\int e^{2t+1} \, dt$

39. $\int e^{-x} \, dx$

40. $\int e^{7t} \, dt$

41. $\int t e^{3t^2} \, dt$

42. $\int \frac{1}{x^2} e^{1/x} \, dx$

43. $\int \frac{2x}{e^{x^2}} \, dx$

44. $\int \frac{e^{\sqrt{x}}}{\sqrt{x}} \, dx$

45. $\int (e^x + e^{-x}) \, dx$

46. $\int 2x e^{1-x^2} \, dx$

Evaluate each indefinite integral in Exercises 47–56.

47. $\int \frac{1}{x + 1} \, dx$

48. $\int \frac{2}{2x + 3} \, dx$

49. $\int \frac{dx}{5x + 2}$

50. $\int \frac{dx}{3x + 4}$

51. $\int \frac{3x^2}{1 + x^3} \, dx$

52. $\int \frac{2x + 5}{x^2 + 5x - 9} \, dx$

53. $\int \frac{x^{1/2}}{1 + x^{3/2}} \, dx$

54. $\int \frac{x^2}{1 - 5x^3} \, dx$

55. $\int \frac{x^2 + 4x}{x^3 + 6x^2 - 15} \, dx$

56. $\int \frac{x^2 - 1}{x^3 - 3x + 1} \, dx$

Evaluate each indefinite integral in Exercises 57–76.

57. $\int \dfrac{x}{1 + x^2}\, dx$

58. $\int (e^x + 1)^3\, e^x\, dx$

59. $\int (1 + e^{3t})\, dt$

60. $\int \dfrac{1}{(6x - 2)^4}\, dx$

61. $\int \left(\dfrac{1}{x} + 1\right) dx$

62. $\int xe^{1 - x^2}\, dx$

63. $\int \dfrac{e^{2x}}{1 - e^{2x}}\, dx$

64. $\int \dfrac{e^x}{1 - e^x}\, dx$

65. $\int (\ln x)^4 \cdot \dfrac{1}{x}\, dx$

66. $\int \dfrac{\ln x}{x}\, dx$

67. $\int \dfrac{1}{x(\ln x)^2}\, dx$

68. $\int \dfrac{1}{x \ln x}\, dx$

69. $\int \left(1 - \dfrac{1}{e^x}\right) dx$

70. $\int \dfrac{x^2}{e^{x^3}}\, dx$

71. $\int \dfrac{\ln x^2}{x}\, dx$

72. $\int \dfrac{\sqrt{\ln x}}{x}\, dx$

73. $\int \dfrac{(\ln x)^2}{x}\, dx$

74. $\int \dfrac{e^x - e^{-x}}{e^x + e^{-x}}\, dx$

75. $\int \dfrac{1}{1 - 2x}\, dx$

76. $\int \dfrac{1}{x} \ln (2x)\, dx$

Evaluate each definite integral in Exercises 77–90.

77. $\displaystyle\int_0^1 \dfrac{2x\, dx}{1 + x^2}$

78. $\displaystyle\int_0^2 \dfrac{x\, dx}{x^2 + 1}$

79. $\displaystyle\int_0^1 \dfrac{e^x}{1 + e^x}\, dx$

80. $\displaystyle\int_0^1 e^{1 - x}\, dx$

81. $\displaystyle\int_0^1 xe^{-x^2}\, dx$

82. $\displaystyle\int_0^1 x^2 e^{x^3}\, dx$

83. $\displaystyle\int_0^{e - 1} \dfrac{1}{x + 1}\, dx$

84. $\displaystyle\int_2^{e + 1} \dfrac{1}{x - 1}\, dx$

85. $\displaystyle\int_1^e \dfrac{\ln x}{x}\, dx$

86. $\displaystyle\int_1^e \dfrac{\ln ex}{x}\, dx$

87. $\displaystyle\int_0^{\sqrt{2}} \dfrac{x\, dx}{3 - x^2}$

88. $\displaystyle\int_0^{\sqrt{5}} \dfrac{x\, dx}{6 - x^2}$

89. $\displaystyle\int_0^1 \dfrac{x^2}{e^{x^3}}\, dx$

90. $\displaystyle\int_0^1 \dfrac{dx}{e^x}$

Evaluate each indefinite integral in Exercises 91–100.

91. $\int e^2\, dx$

92. $\int 3e\, dx$

93. $\int \dfrac{\ln \sqrt{x}}{x}\, dx$

94. $\int \dfrac{1}{x} \ln \dfrac{1}{\sqrt{x}}\, dx$

95. $\int \dfrac{x^2 + 2x + 4}{x + 1}\, dx$ *Hint:* Try long division.

96. $\int \dfrac{x}{x + 1}\, dx$ *Hint:* Try long division.

97. $\int \dfrac{3x}{x - 1}\, dx$

98. $\int \dfrac{1 + \ln x}{x \ln x}\, dx$

99. $\int \dfrac{1 + \ln x}{(x \ln x)^2}\, dx$

100. $\int \dfrac{x + 1}{x + 2}\, dx$

101. Determine the equation of the curve that passes through the point (0, 5) and has slope $4xe^{x^2}$.

102. Determine the equation of the curve that passes through the point (0, 0) and has slope $x/(x^2 + 1)$.

103. Find the average value of $f(x) = \sqrt{2x + 1}$ over the interval [4, 12].

104. Find the average value of $f(x) = xe^{x^2}$ over the interval [0, 1].

105. Determine the area under the curve $y = (7x + 1)^{1/3}$ from $x = 0$ to $x = 1$.

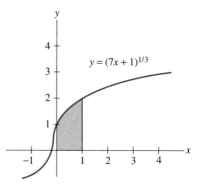

106. Determine the area under the curve $y = \sqrt{4x + 1}$ from $x = 2$ to $x = 6$.

107. (*VOLUME*) Find the volume of the solid produced by revolving about the x axis the region bounded by $y = 1/(2x + 3)$, the x axis, $x = 2$, and $x = 4$.

108. (*VOLUME*) Find the volume of the solid produced by revolving about the x axis the region bounded by $y = (2x - 1)^{1/3}$, the x axis, $x = 1$, and $x = 14$.

109. *(TREE GROWTH)* Suppose a transplanted small tree will grow to maturity according to

$$\frac{dh}{dt} = 3 + \frac{1}{2\sqrt{t+1}}$$

where h is the height of the tree in feet and t is the time in years since it was transplanted. How much does the tree grow in the first 3 years?

110. *(FLU OUTBREAK)* There is an outbreak of flu. The number of people n who will have the flu after t days is

$$n = 8t + 3\sqrt{t+2}$$

What is the average number of people who will have the flu during days 7 through 14 of the outbreak?

W 111. Compare the two integrals given next.

$$\int (x^2 + 1)^{18}\, 2\, dx \qquad \int (2x + 1)^{18}\, dx$$

(a) Which integral can be evaluated by methods studied thus far? Explain why and how it can be evaluated.

(b) Which integral cannot be evaluated by methods studied thus far? Explain why it cannot be evaluated.

W 112. As you know,

$$\int \frac{du}{u} = \ln |u| + C$$

(a) Is it necessary to retain the absolute value around u in such cases as $\ln |x^2 + 3|$ and $\ln |e^x + 1|$? Explain.

(b) Is it necessary to retain the absolute value around u in such cases as $\ln |x + 5|$ and $\ln |x^3|$? Explain.

7.2 | INTEGRATION BY PARTS

Considering the product rule from an integration point of view leads to another method of integration, called **integration by parts**. The derivative formula for a product of two functions of x, called u and v, was stated in Section 3.5 as

$$\frac{d}{dx}(u \cdot v) = u \cdot v' + v \cdot u'$$

and as

$$\frac{d}{dx}(u \cdot v) = u \cdot \frac{dv}{dx} + v \cdot \frac{du}{dx}$$

To prepare for integration, consider differential form, obtained by multiplying both sides of the latter equation by dx.

$$d(u \cdot v) = u \cdot dv + v \cdot du$$

Subtracting $v\, du$ from both sides yields

$$u\, dv = d(uv) - v\, du$$

Integration of both sides yields

$$\int u\, dv = \int d(uv) - \int v\, du$$

or

$$\int u \; dv = uv - \int v \; du$$

Integration performed by using this formula is called *integration by parts*. Such integration is set up by fitting your integral to the integral on the left side of the formula,

$$\int u \; dv$$

This means that your integral contains two parts—a function u and the differential of some other function v. An example should provide some feeling for the process.

EXAMPLE 1   Evaluate $\int xe^x \; dx$.

SOLUTION   This integral cannot be evaluated by any of the formulas we have studied so far. *If* the factor x were not present, we could readily integrate the e^x that remained. As you will see, during the process of integration by parts, a new integral will be created—one without the factor x. To begin, using the integration by parts formula,

$$\int u \; dv = uv - \int v \; du$$

consider that the integral we are asked to evaluate is always the $\int u \; dv$. Thus,

$$\int xe^x \; dx \qquad \text{is} \qquad \int u \; dv$$

One part of the integrand must be chosen as u, and what remains must then be dv. A guideline for making these choices is given after this introductory example. In this instance, we will select

$$u = x$$
$$dv = e^x \; dx$$

If $u = x$, then by differentiation, $du = dx$. Also, if $dv = e^x \; dx$, then by antidifferentiation, $v = e^x$. In summary,

$$u = x \qquad \rightarrow du = dx$$
$$dv = e^x \; dx \rightarrow \quad v = e^x$$

These four pieces—values for u, du, dv, and v—can now be used in the formula for integration by parts.

$$\int xe^x \; dx = uv - \int v \; du$$
$$= x \cdot e^x - \int e^x \; dx$$
$$= xe^x - e^x + C$$

Thus,

$$\int xe^x \, dx = xe^x - e^x + C \qquad \text{(answer)}$$

The choices of $u = x$ and $dv = e^x \, dx$ were correct. Had we chosen instead $u = e^x$ and $dv = x \, dx$, then we would have obtained from them $du = e^x \, dx$ and $v = x^2/2$. If those four pieces were used in the integration by parts formula, the result would have been

$$\int xe^x \, dx = x \cdot \frac{x^2}{2} - \int \frac{x^2}{2} e^x \, dx$$

Clearly, this result is worse than the integral we began with. Our experience suggests a guideline (given next) that is often useful. ◆

Guideline

When using integration by parts, it is often wise to let u equal the part of the integrand that you would like to eliminate—unless what remains for dv cannot be integrated.

In view of this guideline, you may want to reread the beginning of Example 1.

EXAMPLE 2 Examine each integral for integration by parts and select u.

(a) $\displaystyle\int x(x + 3)^4 \, dx$ (b) $\displaystyle\int x^2 e^x \, dx$ (c) $\displaystyle\int x^2 \ln x \, dx$

SOLUTION (a) $\displaystyle\int x(x + 3)^4 \, dx \ldots$ Let $u = x$ and $dv = (x + 3)^4 \, dx.$

(b) $\displaystyle\int x^2 e^x \, dx \ldots$ Let $u = x^2$ and $dv = e^x \, dx.$

(c) $\displaystyle\int x^2 \ln x \, dx \ldots$ Let $u = \ln x$ and $dv = x^2 \, dx.$

Note that you cannot let $u = x^2$ here, because that leaves $\ln x \, dx$ as dv, and you will not be able to integrate $\ln x$ to get v. ◆

EXAMPLE 3 Determine $\displaystyle\int x^2 e^x \, dx.$

SOLUTION From Example 2, part (b), we know to let $u = x^2$. Then

$$u = x^2 \quad \rightarrow \quad du = 2x \, dx$$
$$dv = e^x \, dx \quad \rightarrow \quad v = e^x$$

Thus,

$$\int u \, dv = uv - \int v \, du$$

is

$$\int x^2 e^x \, dx = x^2 e^x - \int 2xe^x \, dx$$
$$= x^2 e^x - 2\int xe^x \, dx$$

The integral obtained here by parts cannot be integrated *directly*, although it is an improvement. The integrand of the original was $x^2 e^x$, and this one is xe^x. Also, recall that in Example 1 we saw how this integral,

$$\int xe^x \, dx$$

can itself be evaluated by parts. Thus, to proceed from

$$\int x^2 e^x \, dx = x^2 e^x - 2\int xe^x \, dx$$

we will use a second application of integration by parts, this time to evaluate $\int xe^x \, dx$. As was done in Example 1, let $u = x$ and $dv = e^x \, dx$. Then $du = dx$ and $v = e^x$. As a result, we have

$$\int x^2 e^x \, dx = x^2 e^x - 2\left(xe^x - \int e^x \, dx \right)$$
$$= x^2 e^x - 2xe^x + 2\int e^x \, dx$$
$$= x^2 e^x - 2xe^x + 2e^x + C$$

Thus, after two applications of integration by parts,

$$\int x^2 e^x \, dx = x^2 e^x - 2xe^x + 2e^x + C \qquad \text{(answer)}$$

Incidentally, the integral

$$\int x^3 e^x \, dx$$

requires three applications of integration by parts. ◆

EXAMPLE 4   Evaluate the definite integral.

$$\int_0^1 xe^x \, dx$$

SOLUTION   The purpose of this example is to show evaluation of a *definite* integral using integration by parts. In Example 1, we obtained along the way

$$\int xe^x \, dx = xe^x - \int e^x \, dx$$

Evaluating the definite integral, we have

$$\int_0^1 xe^x \, dx = [xe^x]_0^1 - \int_0^1 e^x \, dx$$
$$= (1 \cdot e^1 - 0 \cdot e^0) - [e^x]_0^1$$
$$= e^1 - 0 - (e^1 - e^0)$$
$$= e - 0 - e + 1$$
$$= 1 \quad ◆$$

The following two important results from Section 5.2 will be needed for evaluating definite integrals.

$$\ln e = 1$$
$$\ln 1 = 0$$

Integration by parts is also used in some instances when the integrand contains a trigonometric function.

Present Value of an Income Stream

A business may generate a continuous stream of income over a period of t years. The rate of income will probably vary and thus be a function of t, say $f(t)$. The **present value of an income stream** is the amount P that must be invested now at the current interest rate r in order to produce in t years the same amount that would be produced by the income stream.

In Section 5.5, present value P was computed as $A \cdot e^{-rt}$, where A is the value t years from now. In other words, e^{-rt} is multiplied by the amount we expect to have in t years.

In considering a variable income stream, the constant A would be replaced by $f(t)$ and the present value would be the sum of (a theoretically infinite number of) terms of the form $f(t)e^{-rt}$. The present value of an income stream over T years is then given by the following definite integral.

Present Value of an Income Stream (over T years)

$$P = \int_0^T f(t)e^{-rt} \, dt$$

P = the present value

$f(t)$ = the rate at which income is generated

t = time in years

r = prevailing annual interest rate

EXAMPLE 5 ◆ *PRESENT VALUE OF AN INCOME STREAM*

A small corporation's 5-year projection shows its income t years from now as

$$f(t) = 40,000 + 10,000t \quad \text{dollars}$$

Assuming an annual interest rate of 8%, find the present value of the income stream over the 5-year period.

SOLUTION   The present value of the income stream is given by

$$P = \int_0^T f(t)e^{-rt}\, dt$$

where

$$f(t) = 40,000 + 10,000t$$
$$r = .08$$
$$T = 5$$

Thus, we have

$$P = \int_0^5 (40,000 + 10,000t)e^{-.08t}\, dt$$

$$= \int_0^5 40,000e^{-.08t}\, dt + \int_0^5 10,000te^{-.08t}\, dt$$

The first integral fits a basic formula, but the second integral requires integration by parts, with $u = 10,000t$ and $dv = e^{-.08t}\, dt$ (and then $du = 10,000\, dt$ and $v = -12.5e^{-.08t}$).

$$P = \left[40,000 \cdot \frac{1}{-.08} e^{-.08t} \right]_0^5 + \left[-125,000te^{-.08t} \right]_0^5 - \int_0^5 (-125,000e^{-.08t})\, dt$$

$$= \left[-500,000e^{-.08t} \right]_0^5 - \left[125,000te^{-.08t} \right]_0^5 + \int_0^5 125,000e^{-.08t}\, dt$$

$$= (164,840) - (418,938) + \left[-1,562,500e^{-.08t} \right]_0^5$$

$$= 164,840 - 418,938 + 515,156 \qquad \text{(using } e^{-4} \text{ to 4 decimal places)}$$

$$= 261,058$$

The present value of the income stream over the 5-year period is $261,058.   ◆

7.2   Exercises

In Exercises 1–20, evaluate each indefinite integral by using integration by parts.

1. $\displaystyle\int xe^{2x}\, dx$

2. $\displaystyle\int xe^{-x}\, dx$

3. $\displaystyle\int x \ln x\, dx$

4. $\displaystyle\int x \ln (x + 1)\, dx$

5. $\displaystyle\int x^2 \ln x\, dx$

6. $\displaystyle\int x^2 \ln 3x\, dx$

7. $\displaystyle\int (x + 1)e^x\, dx$

8. $\displaystyle\int x(e^x + 1)\, dx$

9. $\displaystyle\int \frac{\ln x}{x^2}\, dx$

10. $\displaystyle\int \ln x\, dx$

11. $\displaystyle\int x(x + 3)^4\, dx$

12. $\displaystyle\int x(x - 4)^5\, dx$

13. $\displaystyle\int \frac{x}{(x + 7)^3}\, dx$

14. $\displaystyle\int \frac{x}{(x + 2)^5}\, dx$

15. $\displaystyle\int \frac{\ln x}{\sqrt{x}}\, dx$

16. $\displaystyle\int x^3 \ln x\, dx$

17. $\displaystyle\int x^3 e^x\, dx$

18. $\displaystyle\int \sqrt{x} \ln x\, dx$

19. $\displaystyle\int x(x + 2)^{-3}\, dx$

20. $\displaystyle\int x\sqrt{1 + x}\, dx$

In Exercises 21–30, evaluate each definite integral by parts.

21. $\int_1^2 xe^x \, dx$

22. $\int_0^1 x^2 e^x \, dx$

23. $\int_1^e \ln x \, dx$

24. $\int_0^1 (x + 1)e^x \, dx$

25. $\int_0^1 x(1 + e^x) \, dx$

26. $\int_0^4 \ln (x + 1) \, dx$

27. $\int_{-1}^0 \dfrac{x}{e^x} \, dx$

28. $\int_1^e x \ln x \, dx$

29. $\int_2^3 30x(x + 1)^4 \, dx$

30. $\int_3^5 90x(x - 3)^8 \, dx$

Evaluate each indefinite integral in Exercises 31–40. Note that *only some* of the integrals require the use of integration by parts.

31. $\int xe^{x^2} \, dx$

32. $\int x(e^{2x} + 3) \, dx$

33. $\int \dfrac{\ln x}{x} \, dx$

34. $\int x \ln x \, dx$

35. $\int \ln 2x \, dx$

36. $\int \dfrac{1}{x \ln x} \, dx$

37. $\int x\sqrt{1 + x} \, dx$

38. $\int \dfrac{x}{x^2 + 3} \, dx$

39. $\int x\sqrt{1 + x^2} \, dx$

40. $\int x^{-2} \ln x \, dx$

In Exercises 41–46, evaluate each integral by parts. These integrals provide practice for work with problems involving the present value of an income stream.

41. $\int xe^{-.04x} \, dx$

42. $\int xe^{-.02x} \, dx$

43. $\int 12xe^{-.08x} \, dx$

44. $\int 4xe^{-.1x} \, dx$

45. $\int xe^{-.05x} \, dx$

46. $\int xe^{-.08x} \, dx$

47. (*INCOME STREAM*) A small corporation's 4-year projection shows its annual income t years from now as

$$f(t) = 20,000 + 3000t \quad \text{dollars}$$

Assuming an annual interest rate of 5%, find the present value of the income stream over the 4-year period.

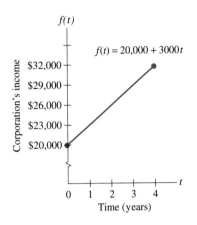

$f(t)$

$f(t) = 20,000 + 3000t$

48. (*INCOME STREAM*) Suppose your business will produce income at the rate of $f(t) = 20t$ thousand dollars per year for the next 5 years. Using an annual interest rate of 8%, find the present value of the income stream. Note that t is time in years.

49. (*INCOME STREAM*) Assume that the natural gas from a well will produce a continuous stream of income of

$$f(t) = 7000 - 300t$$

dollars per year for t years. If the prevailing annual interest rate is 8%, find the present value of the income stream over the first 6 years of operation.

50. (*INCOME STREAM*) The owners of a factory estimate that new machinery they have installed will contribute to their revenue at the rate of

$$f(t) = 8000 - 400t$$

dollars per year for t years. Determine the present value of the income stream over the first 4 years, assuming an annual interest rate of 10%.

51. Find the average value of $f(x) = xe^x$ over the interval $[0, 1]$.

52. Find the average value of $f(x) = \ln 2x$ over the interval $[e, 10]$.

53. Determine the equation of the curve that passes through the point $(1, -1)$ and has slope $\ln x$.

54. Determine the equation of the curve that passes through the point $(1, -1)$ and has slope $4x \ln x$.

55. Determine the area under the curve $y = xe^{-x}$ from $x = 1$ to $x = 3$.

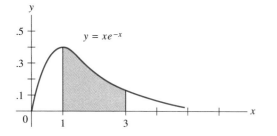

56. Determine the area under the curve $y = (x - 1)e^x$ from $x = 1$ to $x = 2$.

57. *(VOLUME)* Find the volume of the solid produced by revolving about the x axis the region bounded by $y = xe^x$, the x axis, $x = 0$, and $x = 1$.

58. *(VOLUME)* Find the volume of the solid produced by revolving about the x axis the region bounded by $y = \sqrt{x} \ln x$, the x axis, $x = e$, and $x = 5$.

W 59. Suppose we plan to use integration by parts to evaluate

$$\int \ln x \, dx$$

Is $u = 1$ and $dv = \ln x \, dx$ a good choice? Explain.

7.3 | *INTEGRATION BY TABLES*

Although you have learned ways to evaluate a variety of integrals, there are still many integrals you would be unable to evaluate. Here are a few examples of such integrals.

$$\int \sqrt{x^2 + 7} \, dx \qquad \int \frac{1}{4x^2 - 9} \, dx \qquad \int \frac{1}{1 + e^{-3x}} \, dx$$

Each of these three integrals can be evaluated by using methods that are beyond the scope of this text. But even with advanced techniques, some integration problems can be very difficult and time consuming. Consequently, lists (or tables) of integration formulas have been devised to simplify matters.

To begin the study of **integration by tables**, glance at the integrals given in appendix Table 5. This is a short list of integral formulas, intended only to help demonstrate the use of integral tables. Ordinarily, tables of integrals are several pages long. In fact, there are entire books that contain only integration formulas. Given next are a few examples of references with extensive lists of integration formulas.

1. *CRC's Standard Mathematical Tables.* The Chemical Rubber Company (CRC). 59 pages of integration formulas.

2. *Mathematical Handbook*, by Murray R. Spiegel. Schaum's Outline Series, McGraw-Hill Book Co. 44 pages of integration formulas.

3. *Table of Indefinite Integrals*, by G. Petit Bois. Dover Publications. 150 pages of integration formulas.

The three examples that follow are intended to present the process of evaluating an integral by means of a table.

EXAMPLE 1 Evaluate $\int \sqrt{x^2 + 7} \, dx$.

SOLUTION This integral fits the form of formula 8 in Table 5, namely,

$$\int \sqrt{u^2 + a^2} \, du$$

Here $u = x$ and $a = \sqrt{7}$ (from $a^2 = 7$). Since $u = x$, we have $du = dx$. The formula

$$\int \sqrt{u^2 + a^2} \, du = \frac{u}{2}\sqrt{u^2 + a^2} + \frac{a^2}{2} \ln \left| u + \sqrt{u^2 + a^2} \right| + C$$

becomes

$$\int \sqrt{x^2 + 7} \, dx = \frac{x}{2}\sqrt{x^2 + 7} + \frac{7}{2} \ln \left| x + \sqrt{x^2 + 7} \right| + C \quad \blacklozenge$$

EXAMPLE 2 Evaluate $\displaystyle\int \frac{dx}{4x^2 - 9}.$

SOLUTION This integral fits the form of formula 16 in Table 5, namely,

$$\int \frac{du}{u^2 - a^2}$$

Here $u = 2x$ (from $u^2 = 4x^2$) and $a = 3$ (from $a^2 = 9$). Note, however, that if $u = 2x$, then $du = 2 \, dx$. So we must modify the integral by inserting a factor of 2 in the integrand and 1/2 outside the integral.

$$\int \frac{dx}{4x^2 - 9} = \frac{1}{2}\int \frac{2 \, dx}{4x^2 - 9} = \frac{1}{2}\int \frac{2 \, dx}{(2x)^2 - (3)^2}$$

The formula

$$\int \frac{du}{u^2 - a^2} = \frac{1}{2a} \ln \left| \frac{u - a}{u + a} \right| + C$$

becomes

$$\int \frac{dx}{4x^2 - 9} = \frac{1}{2}\int \frac{2 \, dx}{4x^2 - 9} = \frac{1}{2}\left(\frac{1}{2(3)} \ln \left| \frac{2x - 3}{2x + 3} \right| \right) + C$$

$$= \frac{1}{12} \ln \left| \frac{2x - 3}{2x + 3} \right| + C \quad \blacklozenge$$

EXAMPLE 3 Evaluate $\displaystyle\int \frac{dx}{1 + e^{-3x}}.$

SOLUTION This integral fits the form of formula 18 in Table 5, namely,

$$\int \frac{du}{1 + e^u}$$

Here $u = -3x$. This means that $du = -3\,dx$ and our integral must be modified as follows:

$$\int \frac{dx}{1 + e^{-3x}} = -\frac{1}{3}\int \frac{-3\,dx}{1 + e^{-3x}}$$

Thus, the formula

$$\int \frac{du}{1 + e^u} = u - \ln(1 + e^u) + C$$

can be applied as

$$\int \frac{dx}{1 + e^{-3x}} = -\frac{1}{3}\int \frac{-3\,dx}{1 + e^{-3x}}$$

$$= -\frac{1}{3}[(-3x) - \ln(1 + e^{-3x})] + C$$

$$= x + \frac{1}{3}\ln(1 + e^{-3x}) + C \quad \blacklozenge$$

A special type of integration formula, called a **reduction formula**, is used to simplify an integral containing a power of some expression. The simplified integral that results will contain a *reduced* (lower) power of some expression. Such reduction formulas are used to produce integrals that are easier to integrate. Two reduction formulas appear in our integral table. The next example shows how such formulas can be used.

EXAMPLE 4   Evaluate $\int (\ln x)^2\,dx$.

SOLUTION   Using the first reduction formula (formula 19 in Table 5) with $u = x$ and $n = 2$ yields

$$\int (\ln x)^2\,dx = x(\ln x)^2 - 2\int \ln x\,dx$$

The integral

$$\int \ln x\,dx$$

can be evaluated either by parts (as suggested in Section 7.2, Exercise 10) or by an additional application of formula 19, this time with $n = 1$. Either way, the result is

$$\int \ln x\,dx = x\ln x - x$$

After substituting $x\ln x - x + C$ for $\int \ln x\,dx$ in

$$\int (\ln x)^2\,dx = x(\ln x)^2 - 2\int \ln x\,dx$$

we obtain

$$\int (\ln x)^2 \, dx = x(\ln x)^2 - 2(x \ln x - x) + C$$

This simplifies to

$$\int (\ln x)^2 \, dx = x(\ln x)^2 - 2x \ln x + 2x + C \quad \blacklozenge$$

7.3 Exercises

Use the table of integrals (Table 5) to evaluate each indefinite integral in Exercises 1–12.

1. $\displaystyle\int \sqrt{x^2 + 16} \, dx$

2. $\displaystyle\int \sqrt{x^2 + 9} \, dx$

3. $\displaystyle\int \sqrt{9x^2 + 5} \, dx$

4. $\displaystyle\int \sqrt{16x^2 + 3} \, dx$

5. $\displaystyle\int \frac{dx}{x^2 - 49}$

6. $\displaystyle\int \frac{dx}{x^2 - 4}$

7. $\displaystyle\int \frac{dx}{25x^2 - 81}$

8. $\displaystyle\int \frac{dx}{9x^2 - 64}$

9. $\displaystyle\int \frac{dx}{1 + e^{7x}}$

10. $\displaystyle\int \frac{dx}{1 + e^{4x}}$

11. $\displaystyle\int \frac{dx}{1 + e^{-4x}}$

12. $\displaystyle\int \frac{dx}{1 + e^{-2x}}$

Use Table 5 to evaluate each indefinite integral in Exercises 13–36.

13. $\displaystyle\int \frac{dx}{\sqrt{x^2 + 25}}$

14. $\displaystyle\int \frac{dx}{\sqrt{x^2 + 81}}$

15. $\displaystyle\int \frac{dx}{\sqrt{9x^2 + 16}}$

16. $\displaystyle\int \frac{dx}{\sqrt{25x^2 + 1}}$

17. $\displaystyle\int \frac{dx}{x^2 \sqrt{x^2 + 1}}$

18. $\displaystyle\int \frac{dx}{25x^2 \sqrt{25x^2 + 9}}$

19. $\displaystyle\int \frac{dx}{4x^2 \sqrt{4x^2 + 3}}$

20. $\displaystyle\int \frac{dx}{9x^2 \sqrt{9x^2 - 5}}$

21. $\displaystyle\int \frac{dx}{x^2 \sqrt{4x^2 + 3}}$

22. $\displaystyle\int \frac{dx}{x^2 \sqrt{9x^2 - 5}}$

23. $\displaystyle\int \frac{dx}{x^2 - 9}$

24. $\displaystyle\int \frac{dx}{x^2 - 16}$

25. $\displaystyle\int \frac{dx}{9x^2 - 1}$

26. $\displaystyle\int \frac{dx}{16x^2 - 25}$

27. $\displaystyle\int x^2 \ln x \, dx$

28. $\displaystyle\int x^3 \ln x \, dx$

29. $\displaystyle\int x^5 \ln x \, dx$

30. $\displaystyle\int x^4 \ln x \, dx$

31. $\displaystyle\int \frac{dx}{x \sqrt{x^2 + 1}}$

32. $\displaystyle\int \frac{dx}{x \sqrt{x^2 + 9}}$

33. $\displaystyle\int \frac{dx}{2x \sqrt{4x^2 + 25}}$

34. $\displaystyle\int \frac{dx}{4x \sqrt{16x^2 + 1}}$

35. $\displaystyle\int \frac{dx}{x \sqrt{9x^2 + 4}}$

36. $\displaystyle\int \frac{dx}{x \sqrt{4x^2 + 9}}$

Use a reduction formula from the table of integrals (Table 5) to evaluate each indefinite integral in Exercises 37–44.

37. $\displaystyle\int (\ln x)^3 \, dx$

38. $\displaystyle\int (\ln x)^4 \, dx$

39. $\displaystyle\int \ln x \, dx$

40. $\displaystyle\int (\ln x)^5 \, dx$

41. $\displaystyle\int xe^x \, dx$

42. $\displaystyle\int x^3 e^x \, dx$

43. $\displaystyle\int x^2 e^x \, dx$

44. $\displaystyle\int x^4 e^x \, dx$

45. Use properties of logarithms to show that the right-hand side of formula 18 (Table 5) can be written as

$$\ln \frac{e^u}{1 + e^u} + C$$

That is, show that

$$\ln \frac{e^u}{1 + e^u} = u - \ln(1 + e^u)$$

46. Find the average value of $f(x) = \sqrt{25x^2 + 9}$ over the interval [2, 6].

47. Find the average value of $f(x) = x^3 \ln x$ over the interval [1, 3].

48. Determine the area under the curve

$$y = \frac{1}{\sqrt{x^2 + 9}}$$

from $x = 0$ to $x = 4$.

49. Determine the area under the curve

$$y = \frac{3}{\sqrt{x^2 - 9}}$$

from $x = 5$ to $x = 7$.

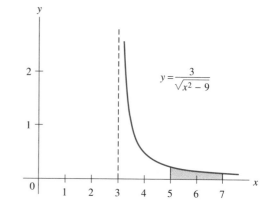

$$y = \frac{3}{\sqrt{x^2 - 9}}$$

W 50. In Example 4, a reduction formula is used to evaluate

$$\int (\ln x)^2 \, dx$$

This integral can also be evaluated by using two applications of parts. If you were to begin by parts, what would be your choice for u and dv? Explain why your choice is good by indicating what is accomplished by it.

7.4 | *NUMERICAL METHODS OF APPROXIMATION*

Sometimes we encounter definite integrals that cannot be evaluated by any of the methods of integration. In fact, even a larger table of integrals may not include them. In such instances, approximations to the value of

$$\int_a^b f(x) \, dx$$

can be obtained by using **numerical methods** that approximate the area under the curve. You may recall the development of the definite integral in Section 6.3, where *rectangles* were used to approximate the area under the curve. In this section we will seek close approximations by using *trapezoids* (by means of the "trapezoidal rule") and *parabolas* ("Simpson's rule").

A graphical example will demonstrate why trapezoids are preferable to rectangles. First, consider a rectangular approximation to the area under a curve. We will use four rectangles. See Figure 1.

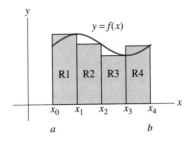

Figure 1 Rectangular approximation to area

Next, let us consider an approximation obtained by using four trapezoids rather than four rectangles. Figure 2 shows that for this function, the trapezoidal approximation is much closer to the actual area under the curve than is the rectangular approximation.

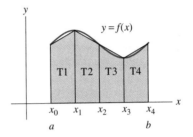

Figure 2 Trapezoidal approximation to area

The sum of the areas of all four trapezoids shown can be determined by using the formula for the area of a trapezoid, namely,

$$A = \frac{1}{2}(b_1 + b_2)h \qquad \begin{cases} b_1 = \text{one base} \\ b_2 = \text{the other base} \\ h = \text{height} \end{cases}$$

Figure 3 shows, for example, that in the first trapezoid (T1) above,

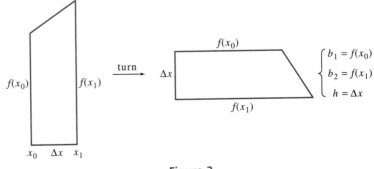

Figure 3

b_1 is $f(x_0)$, b_2 is $f(x_1)$, and h is Δx. (Note that Δx will be used, as in Section 6.3, to represent $x_1 - x_0$, $x_2 - x_1$, etc.) Thus, the areas of the four trapezoids are

$$T_1: \quad A = \frac{1}{2}[f(x_0) + f(x_1)]\,\Delta x$$

$$T_2: \quad A = \frac{1}{2}[f(x_1) + f(x_2)]\,\Delta x$$

$$T_3: \quad A = \frac{1}{2}[f(x_2) + f(x_3)]\,\Delta x$$

$$T_4: \quad A = \frac{1}{2}[f(x_3) + f(x_4)]\,\Delta x$$

The total area of all four trapezoids is the sum of these four areas.

$$A = \frac{\Delta x}{2}[f(x_0) + 2f(x_1) + 2f(x_2) + 2f(x_3) + f(x_4)]$$

In general, if n trapezoids are used, then the approximate value of the integral is given by the **trapezoidal rule**, stated next.

Trapezoidal Rule

If f is continuous on the interval $[a, b]$, then

$$\int_a^b f(x)\,dx \approx \frac{\Delta x}{2}[f(x_0) + 2f(x_1) + 2f(x_2) + \cdots + 2f(x_{n-1}) + f(x_n)]$$

where $\Delta x = \frac{b-a}{n}$, $a = x_0$, $b = x_n$, n = number of trapezoids, and $x_1 = x_0 + \Delta x$, $x_2 = x_1 + \Delta x$, $x_3 = x_2 + \Delta x$, ...

The larger the value of n, the more trapezoids will be used and the smaller will be Δx. The result of using larger n is a better approximation to the definite integral.

EXAMPLE 1   Use the trapezoidal rule with $n = 4$ to approximate the definite integral.

$$\int_0^2 \frac{1}{16 + x^2}\,dx$$

SOLUTION   For this integral,

$$f(x) = \frac{1}{16 + x^2} \qquad a = 0 \qquad b = 2$$

Thus,

$$\Delta x = \frac{b-a}{n} = \frac{2-0}{4} = \frac{1}{2} \quad \text{or} \quad .5$$

Since $\Delta x = .5$, each successive x_i is .5 larger than the preceding one. To begin, since $a = 0$, then $x_0 = 0$, and then $x_1 = .5$, $x_2 = 1$, $x_3 = 1.5$, and $x_4 = 2$. Substituting into the formula for the trapezoidal rule yields

$$\int_0^2 \frac{1}{16 + x^2} \, dx \approx \frac{.5}{2} \left[\frac{1}{16 + 0^2} + 2 \cdot \frac{1}{16 + .5^2} + \right.$$

$$\left. 2 \cdot \frac{1}{16 + 1^2} + 2 \cdot \frac{1}{16 + 1.5^2} + \frac{1}{16 + 2^2} \right]$$

$$\approx .25[.0625 + .1230 + .1176 + .1096 + .0500]$$

$$= .25[.4627]$$

$$\approx .1157$$

This approximation is quite good. The actual value of the integral correct to four decimal places is .1159, as determined by using a larger value of n or methods beyond the scope of this book. This form of integral is also available in tables of integrals. ◆

Simpson's rule approximates the area under a curve by means of parabolas and provides (in most cases) a better fit and closer approximation than that obtained by using the trapezoidal rule. The formula for Simpson's rule is given next. Some justification of the rule will follow.

Simpson's Rule

If f is continuous on the interval $[a, b]$, then

$$\int_a^b f(x) \, dx \approx \frac{\Delta x}{3} [f(x_0) + 4f(x_1) + 2f(x_2) + 4f(x_3) + \cdots +$$

$$2f(x_{n-2}) + 4f(x_{n-1}) + f(x_n)]$$

where $\Delta x = \frac{b-a}{n}$, $a = x_0$, $b = x_n$, and n must be an *even* integer. Also, $x_1 = x_0 + \Delta x$, $x_2 = x_1 + \Delta x$, $x_3 = x_2 + \Delta x$, ...

Consider x_0, $x_1 = x_0 + \Delta x$, and $x_2 = x_1 + \Delta x$. The three corresponding points on the graph of $y = f(x)$ are $(x_0, f(x_0))$, $(x_1, f(x_1))$, and $(x_2, f(x_2))$. The area under the parabola through those points can be shown to be

$$A = \frac{\Delta x}{3} [f(x_0) + 4f(x_1) + f(x_2)]$$

A second (nonoverlapping) parabola can be passed through the three points $(x_2, f(x_2))$, $(x_3, f(x_3))$, and $(x_4, f(x_4))$. Its area is

$$A = \frac{\Delta x}{3} [f(x_2) + 4f(x_3) + f(x_4)]$$

The total area of all the approximating parabolas created between $x = a$ and $x = b$ (that is, between x_0 and x_n) is the sum of several expressions of the form shown above. The grand total is shown above in the box as the formula for Simpson's rule. Note that n must be an even number, since if *one* parabola is used, the last x used is x_2; for *two* parabolas the last x is x_4; for *three* parabolas it is x_6; and so on. Figure 4 shows the exact area under a curve between x_0 and x_2 and the approximation using a parabola.

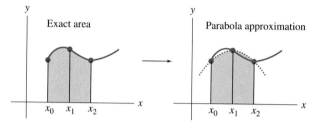

Figure 4 Exact area and approximation using a parabola

EXAMPLE 2   Use Simpson's rule with $n = 4$ to approximate the definite integral. Round all calculations to four decimal places.

$$\int_1^5 \sqrt{x^2 - 1}\, dx$$

SOLUTION   For this integral, $f(x) = \sqrt{x^2 - 1}$, $a = 1 = x_0$, and $b = 5 = x_4$. Also,

$$\Delta x = \frac{b - a}{n} = \frac{5 - 1}{4} = 1$$

and so $x_0 = 1$, $x_1 = 2$, $x_2 = 3$, $x_3 = 4$, and $x_4 = 5$. By Simpson's rule,

$$\int_1^5 \sqrt{x^2 - 1}\, dx \approx \frac{1}{3}\left[\sqrt{1^2 - 1} + 4 \cdot \sqrt{2^2 - 1} + \right.$$
$$\left. 2 \cdot \sqrt{3^2 - 1} + 4 \cdot \sqrt{4^2 - 1} + \sqrt{5^2 - 1}\,\right]$$
$$= \frac{1}{3}\left[\sqrt{0} + 4\sqrt{3} + 2\sqrt{8} + 4\sqrt{15} + \sqrt{24}\,\right]$$
$$\approx \frac{1}{3}[0 + 4(1.7321) + 2(2.8284) + 4(3.8730) + 4.8990]$$
$$= \frac{1}{3}(0 + 6.9284 + 5.6568 + 15.4920 + 4.8990)$$
$$= \frac{1}{3}(32.9762)$$
$$\approx 10.9921 \quad \blacklozenge$$

Error Estimates

When using an approximation method such as the trapezoidal rule or Simpson's rule, it is important to know how accurate an approximation the method will produce. The formulas shown next can be used to estimate the **error** E, which is the difference between the actual value and the approximate value. Notice that the formulas give the largest possible error, since they state that E is less than or equal to some number.

Error Estimates

$$\int_a^b f(x)\, dx, \quad n \text{ subintervals}$$

1. *Trapezoidal rule*

$$E \le \frac{(b-a)^3}{12n^2} \cdot M$$

where M is the maximum value of $|f''(x)|$ on the interval $[a, b]$.

2. *Simpson's rule*

$$E \le \frac{(b-a)^5}{180n^4} \cdot M$$

where M is the maximum value of $|f^{(4)}(x)|$ on the interval $[a, b]$.

EXAMPLE 3 Determine the maximum error in approximating the definite integral

$$\int_1^5 \ln x \, dx$$

(a) Using the trapezoidal rule with $n = 8$.

(b) Using Simpson's rule with $n = 8$.

SOLUTION **(a)** We will begin by determining M. Since $f(x) = \ln x$, it follows that $f'(x) = 1/x$ and $f''(x) = -1/x^2$. The function f'' has no critical numbers, and the maximum value of $|f''(x)|$ on $[1, 5]$ occurs at the interval endpoint $x = 1$. It follows that $M = |f''(1)| = 1$ and that

$$E \le \frac{(5-1)^3}{12(8)^2}\,(1)$$

or

$$E \le .0833$$

The maximum error is .0833.

(b) To obtain M, we determine $f^{(4)}(x) = -6/x^4$. The maximum value of $\left|f^{(4)}(x)\right|$ on $[1, 5]$ occurs at $x = 1$. It follows that $M = \left|f^{(4)}(1)\right| = 6$. We have then

$$E \leq \frac{(5-1)^5}{180(8)^4} \cdot 6$$

or

$$E \leq .0083$$

The maximum error is .0083. ◆

EXAMPLE 4   When using the trapezoidal rule to approximate

$$\int_1^5 \ln x$$

how large must n be so that $E \leq .01$?

SOLUTION   To guarantee that $E \leq .01$, we must choose n so that

$$\frac{(b-a)^3}{12n^2} \cdot M \leq .01$$

From Example 3, we know that M is 1. Also, $b - a = 4$. So we need

$$\frac{4^3}{12n^2} \cdot 1 \leq .01$$

or

$$16 \leq .03n^2$$

or

$$n^2 \geq 533.33$$

or

$$n \geq 23.09$$

We must use $n = 24$, since n must be an integer and 23 does not satisfy the inequality $n \geq 23.09$. ◆

In most cases, for the same n value, Simpson's rule provides a better approximation than does the trapezoidal rule. However, the error estimate for Simpson's rule requires the fourth derivative, and for some functions it can be difficult and time consuming to determine the fourth derivative. In such cases, the trapezoidal rule is the better choice. Simply use a larger n to get the accuracy you seek.

7.4   Exercises

In Exercises 1–12, consider each interval $[a, b]$ to be associated with the definite integral

$$\int_a^b f(x)\, dx$$

and determine the value of Δx to be used with a numerical approximation method such as the trapezoidal rule. Also determine $x_0, x_1, x_2, x_3, \ldots, x_n$.

1. $[0, 4]$, $n = 4$

2. $[0, 8]$, $n = 8$

3. $[1, 7]$, $n = 6$

4. $[2, 8]$, $n = 6$

5. $[0, 2]$, $n = 4$

6. $[1, 3]$, $n = 4$

7. $[2, 5]$, $n = 6$

8. $[2, 3]$, $n = 4$

9. $[2, 6]$, $n = 8$

10. $[0, 4]$, $n = 8$

11. $[3, 5]$, $n = 8$

12. $[1, 3]$, $n = 8$

In Exercises 13–24, use the trapezoidal rule to approximate each definite integral. Round calculations to four decimal places. *Use a calculator.*

13. $\int_0^2 x^2\, dx$, $n = 4$

14. $\int_0^3 x^3\, dx$, $n = 6$

15. $\int_1^3 \frac{1}{x^3}\, dx$, $n = 4$

16. $\int_1^3 \frac{1}{x^2}\, dx$, $n = 4$

17. $\int_0^3 \sqrt{x^2 + 1}\, dx$, $n = 6$

18. $\int_1^4 \sqrt{x^2 - 1}\, dx$, $n = 6$

19. $\int_{-1}^5 \frac{1}{x + 2}\, dx$, $n = 6$

20. $\int_0^2 \frac{1}{1 + x^2}\, dx$, $n = 4$

21. $\int_0^8 x\sqrt{x + 1}\, dx$, $n = 4$

22. $\int_0^4 x^2\sqrt{x + 4}\, dx$, $n = 4$

23. $\int_1^5 \ln x\, dx$, $n = 4$

24. $\int_2^5 x \ln x\, dx$, $n = 6$

In Exercises 25–36, use Simpson's rule to approximate each definite integral. Round calculations to four decimal places. *Use a calculator.*

25. $\int_1^5 (x^2 + 3)\, dx$, $n = 4$

26. $\int_0^2 (x^3 + 2)\, dx$, $n = 4$

27. $\int_0^4 \sqrt{x^2 + 1}\, dx$, $n = 4$

28. $\int_1^4 \sqrt{x^2 - 1}\, dx$, $n = 6$

29. $\int_{-1}^2 \sqrt{x^3 + 1}\, dx$, $n = 6$

30. $\int_0^2 \sqrt{x^4 + 1}\, dx$, $n = 4$

31. $\int_0^2 \frac{1}{x + 1}\, dx$, $n = 8$ **32.** $\int_0^4 \frac{1}{x^2 + 9}\, dx$, $n = 8$

33. $\int_0^2 e^x\, dx$, $n = 4$ **34.** $\int_0^1 e^{-x}\, dx$, $n = 4$

35. $\int_0^3 \frac{1}{1 + e^x}\, dx$, $n = 6$ **36.** $\int_0^2 e^{x^2}\, dx$, $n = 4$

In Exercises 37–46, use both (a) the trapezoidal rule and (b) Simpson's rule to approximate each definite integral. Round calculations to four decimal places. *Use a calculator.* Then (c) obtain the *exact* value of the integral by ordinary integration. Compare the results.

37. $\int_0^4 x^2\, dx$, $n = 4$ **38.** $\int_0^3 x^2\, dx$, $n = 6$

39. $\int_1^4 x^3\, dx$, $n = 6$ **40.** $\int_0^2 x^3\, dx$, $n = 4$

41. $\int_1^4 \sqrt{x}\, dx$, $n = 6$ **42.** $\int_1^4 \frac{1}{x^2}\, dx$, $n = 6$

43. $\int_0^1 xe^{x^2}\, dx$, $n = 4$ **44.** $\int_0^1 xe^{x^2}\, dx$, $n = 8$

45. $\int_0^2 x\sqrt{4 - x^2}\ dx, \quad n = 8$

46. $\int_0^4 x\sqrt{x^2 + 9}\ dx, \quad n = 8$

47. Determine the shaded area in the figure by using each of the following methods:
(a) The trapezoidal rule with $n = 4$.
(b) Simpson's rule with $n = 4$.
(c) The area of a circular region having radius r is πr^2. The drawing shows a quarter circle. Use $\pi \approx 3.1416$.

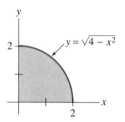

$y = \sqrt{4 - x^2}$

48. Determine the shaded semielliptical area in the figure by using each of the following:
(a) The trapezoidal rule with $n = 6$. Use the interval $[0, 3]$ and double the result.
(b) Simpson's rule with $n = 6$. Again use $[0, 3]$ and double the result.

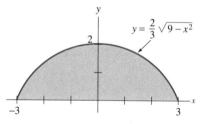

$y = \frac{2}{3}\sqrt{9 - x^2}$

W 49. Explain why in Exercise 37 Simpson's rule yielded the *exact* value of the integral. *Hint:* Think about the *graph* of the function.

In Exercises 50–55, determine the maximum error in approximating the definite integral (a) using the trapezoidal rule and (b) using Simpson's rule.

50. $\int_0^2 x^3\ dx, \quad n = 4$ **51.** $\int_0^2 x^4\ dx, \quad n = 4$

52. $\int_1^3 \ln x\ dx, \quad n = 8$ **53.** $\int_1^2 \frac{1}{x}\ dx, \quad n = 8$

54. $\int_1^3 \sqrt{x}\ dx, \quad n = 4$ **55.** $\int_0^1 e^{-x}\ dx, \quad n = 4$

In Exercises 56–59, determine how large n must be so that a trapezoidal rule approximation of the integral will have an error that does not exceed the value given as E.

56. $\int_1^2 \frac{1}{x}\ dx, \quad E = .001$ **57.** $\int_1^3 \frac{1}{x^2}\ dx, \quad E = .001$

58. $\int_0^2 x^3\ dx, \quad E = .0001$ **59.** $\int_0^2 x^4\ dx, \quad E = .0001$

W 60. What is the reason for using numerical integration methods such as the trapezoidal rule or Simpson's rule?

The Exercise Library at the back of the book contains graphing calculator and computer exercises keyed to this section.

7.5   IMPROPER INTEGRALS

Recall that the area under the graph of $y = f(x)$ between $x = a$ and $x = b$ is given by

$$\int_a^b f(x)\ dx$$

provided f is continuous and nonnegative between a and b. In probability and statistics (and in some other application areas), the interval between a and b

can extend infinitely far in either direction. The three possible types of intervals are shown next along with the corresponding integral. The integrals are called **improper integrals** because each has at least one infinite limit of integration.

interval	integral
$[a, \infty)$	$\int_a^\infty f(x)\, dx$
$(-\infty, b]$	$\int_{-\infty}^b f(x)\, dx$
$(-\infty, \infty)$	$\int_{-\infty}^\infty f(x)\, dx$

In order to motivate a definition for the improper integral

$$\int_a^\infty f(x)\, dx$$

let us consider the following definite integral:

$$\int_a^b f(x)\, dx$$

Figure 5 shows the area under the curve for larger and larger values of b.

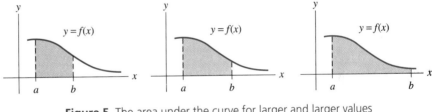

Figure 5 The area under the curve for larger and larger values of b

It seems reasonable to consider the improper integral in question (with upper limit ∞) to be the limit of the integral from a to b as b approaches infinity. The formal definition is given next.

Improper Integral from a to ∞

$$\int_a^\infty f(x)\, dx = \lim_{b \to \infty} \int_a^b f(x)\, dx$$

provided that f is continuous on the interval $[a, \infty)$.

As you can see, evaluating improper integrals will involve determining limits at infinity. Because of this, some special limits at infinity are listed here. Most of these limits also appeared in Chapter 2 or in Chapter 5. You may want to review Section 2.4 on limits at infinity, since variations of these limits may appear. In Exercises 1–22 at the end of this section, you will have a chance to practice determining limits before you need to use them in the improper integrals presented later in the exercises.

Some Limits at Infinity

$$\lim_{x \to \infty} \frac{1}{x^n} = 0 \qquad n > 0$$

$$\lim_{x \to -\infty} \frac{1}{x^n} = 0 \qquad n > 0$$

$$\lim_{x \to \infty} x^n = \infty \qquad n > 0$$

$$\lim_{x \to -\infty} x^n = \begin{cases} \infty & n = 2, 4, 6, \ldots \\ -\infty & n = 1, 3, 5, \ldots \end{cases}$$

$$\lim_{x \to \infty} \sqrt{x} = \infty \qquad \lim_{x \to \infty} \frac{1}{\sqrt{x}} = 0$$

$$\lim_{x \to \infty} e^x = \infty \qquad \lim_{x \to -\infty} e^x = 0$$

$$\lim_{x \to \infty} \frac{1}{e^x} = 0 \qquad \lim_{x \to \infty} e^{-x} = 0$$

$$\lim_{x \to \infty} \ln x = \infty$$

EXAMPLE 1 Evaluate the improper integral.

$$\int_1^\infty \frac{1}{x^2} \, dx$$

SOLUTION To begin, use the definition of the improper integral. Then perform the integration. Finally, evaluate the limit. The steps:

$$\int_1^\infty \frac{1}{x^2} \, dx = \lim_{b \to \infty} \int_1^b \frac{1}{x^2} \, dx$$

$$= \lim_{b \to \infty} \int_1^b x^{-2} \, dx$$

$$= \lim_{b \to \infty} \left[\frac{x^{-1}}{-1} \right]_1^b$$

$$= \lim_{b \to \infty} \left[-\frac{1}{x} \right]_1^b$$

$$= \lim_{b \to \infty} \left[\left(-\frac{1}{b} \right) - \left(-\frac{1}{1} \right) \right]$$

$$= \lim_{b \to \infty} \left(-\frac{1}{b} + 1 \right)$$

Now that the integration has been completed, the limit can be determined. And since $\lim_{b \to \infty} (1/b) = 0$, the final result is simply 1.

$$\int_1^\infty \frac{1}{x^2} \, dx = 1$$

Note that this integral represents the area under the curve $y = 1/x^2$ for $x \geq 1$. Figure 6 shows this area.

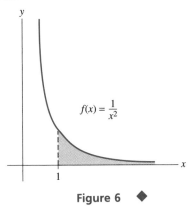

$f(x) = \dfrac{1}{x^2}$

Figure 6 ◆

EXAMPLE 2   Evaluate the improper integral.

$$\int_1^\infty \frac{1}{x} \, dx$$

SOLUTION   This *appears* to be an integral very similar to the integral of the previous example. Even the graph of the function is nearly the same. See Figure 7. However, evaluating the integral will show that the two integrals are quite different indeed.

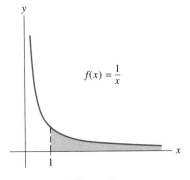

$f(x) = \dfrac{1}{x}$

Figure 7

$$\int_1^\infty \frac{1}{x}\,dx = \lim_{b \to \infty} \int_1^b \frac{1}{x}\,dx$$

$$= \lim_{b \to \infty} \left[\ln |x|\right]_1^b$$

$$= \lim_{b \to \infty} \left(\ln b - \ln 1\right)$$

$$= \lim_{b \to \infty} \left(\ln b\right) \qquad\qquad \text{since ln 1 = 0}$$

$$= \infty \quad \blacklozenge$$

Note

The integral just evaluated is said to be **divergent**, since it has no finite numerical value. By contrast, the integral of Example 1,

$$\int_1^\infty \frac{1}{x^2}\,dx$$

is said to be **convergent**, since it has a finite value. There is indeed a finite area under the curve.

EXAMPLE 3 ◆ *POLLUTION OF A RIVER*

A factory has been dumping large quantities of waste into a nearby river. But now, to meet the new government environmental standard, the polluting has been decreased to $e^{-.4t}$ tons per year, where t is the number of years from now. What is the total amount of waste dumped into the river if the polluting continues indefinitely according to the formula?

SOLUTION   The total amount of waste dumped into the river will be

$$\int_0^\infty e^{-.4t}\,dt \quad \text{tons}$$

This integral is evaluated next.

$$\int_0^\infty e^{-.4t}\,dt = \lim_{b \to \infty} \int_0^b e^{-.4t}\,dt$$

$$= \frac{1}{-.4} \lim_{b \to \infty} \int_0^b e^{-.4t}(-.4)\,dt$$

$$= -2.5 \lim_{b \to \infty} \int_0^b e^{-.4t}(-.4)\,dt$$

$$= -2.5 \lim_{b \to \infty} \left[e^{-.4t}\right]_0^b$$

$$= -2.5 \lim_{b \to \infty} \left(e^{-.4b} - e^0\right)$$

$$= -2.5 \lim_{b \to \infty} \left(\frac{1}{e^{.4b}} - 1\right)$$

$$= -2.5(0 - 1)$$

$$= 2.5$$

The total amount of waste dumped into the river will be 2.5 tons, assuming the polluting continues indefinitely into the future. See Figure 8.

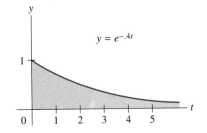

Figure 8 The graph of the function shows the (decreasing) rate of pollution. The area under the graph gives the total pollution that will occur. ◆

EXAMPLE 4 ◆ **CAPITAL VALUE OF A RENTAL PROPERTY**

The **capital value** of a rental property, assuming that it will last indefinitely, is

$$\int_0^\infty Re^{-kt}\,dt$$

where R is the annual rent and k is the current annual rate of interest. (The amount received in rent will then earn interest at this rate.) Determine the capital value of a rental property if the annual rent is \$10,000 and the annual rate of interest is 8%.

SOLUTION The capital value is

$$\int_0^\infty 10{,}000e^{-.08t}\,dt = \lim_{b\to\infty}\int_0^b 10{,}000e^{-.08t}\,dt$$

$$= \frac{10{,}000}{-.08}\lim_{b\to\infty}\int_0^b e^{-.08t}(-.08)\,dt$$

$$= -125{,}000\lim_{b\to\infty}\left(e^{-.08b} - e^0\right)$$

$$= -125{,}000(0 - 1)$$

$$= 125{,}000$$

The capital value is \$125,000. ◆

Improper integrals of the type

$$\int_{-\infty}^b f(x)\,dx$$

are defined and evaluated in a manner comparable to that for the improper integrals you have seen already.

$$\int_{-\infty}^b f(x)\,dx = \lim_{a\to-\infty}\int_a^b f(x)\,dx$$

provided that f is continuous on the interval $(-\infty, b]$.

EXAMPLE 5   Evaluate the improper integral.

$$\int_{-\infty}^{0} e^x \, dx$$

SOLUTION

$$\int_{-\infty}^{0} e^x \, dx = \lim_{a \to -\infty} \int_{a}^{0} e^x \, dx$$

$$= \lim_{a \to -\infty} \left[e^x \right]_{a}^{0}$$

$$= \lim_{a \to -\infty} \left(e^0 - e^a \right)$$

$$= \lim_{a \to -\infty} \left(1 - e^a \right)$$

$$= 1 - 0$$

$$= 1$$

Keep in mind that $e^a \to 0$ as $a \to -\infty$.   ◆

Improper integrals of the type

$$\int_{-\infty}^{\infty} f(x) \, dx$$

are defined and evaluated in a way that involves both types of improper integrals you have seen already.

$$\int_{-\infty}^{\infty} f(x) \, dx = \int_{-\infty}^{c} f(x) \, dx + \int_{c}^{\infty} f(x) \, dx$$

provided that f is continuous on the interval $(-\infty, \infty)$. The number c used in the two integrals can be any real number, although 0 or 1 is often a good choice. If either (or both) of the two integrals diverges, then the original integral also diverges. Otherwise, the original integral converges.

EXAMPLE 6   Evaluate the improper integral.

$$\int_{-\infty}^{\infty} xe^{-x^2} \, dx$$

SOLUTION   We will use 0 for c.

$$\int_{-\infty}^{\infty} xe^{-x^2} \, dx = \int_{-\infty}^{0} xe^{-x^2} \, dx + \int_{0}^{\infty} xe^{-x^2} \, dx$$

$$= \lim_{a \to -\infty} \int_{a}^{0} xe^{-x^2} \, dx + \lim_{b \to \infty} \int_{0}^{b} xe^{-x^2} \, dx$$

$$= -\frac{1}{2} \lim_{a \to -\infty} \int_a^0 e^{-x^2}(-2x)\, dx - \frac{1}{2} \lim_{b \to \infty} \int_0^b e^{-x^2}(-2x)\, dx$$

$$= -\frac{1}{2} \lim_{a \to -\infty} \left[e^{-x^2} \right]_a^0 - \frac{1}{2} \lim_{b \to \infty} \left[e^{-x^2} \right]_0^b$$

$$= -\frac{1}{2} \lim_{a \to -\infty} (1 - e^{-a^2}) - \frac{1}{2} \lim_{b \to \infty} (e^{-b^2} - 1)$$

$$= -\frac{1}{2}(1 - 0) - \frac{1}{2}(0 - 1)$$

$$= 0 \quad \blacklozenge$$

7.5   Exercises

Evaluate each limit in Exercises 1–22.

1. $\lim\limits_{b \to \infty} \dfrac{1}{b^2}$

2. $\lim\limits_{b \to \infty} \dfrac{1}{b}$

3. $\lim\limits_{a \to -\infty} \dfrac{2}{a}$

4. $\lim\limits_{a \to -\infty} \dfrac{1}{a^2}$

5. $\lim\limits_{b \to \infty} \sqrt{b+1}$

6. $\lim\limits_{b \to \infty} \sqrt{b}$

7. $\lim\limits_{b \to \infty} \dfrac{1}{\sqrt{2b}}$

8. $\lim\limits_{b \to \infty} \dfrac{1}{\sqrt{b+1}}$

9. $\lim\limits_{b \to \infty} (b^{1/2} + 1)$

10. $\lim\limits_{b \to \infty} \dfrac{4}{b^3}$

11. $\lim\limits_{b \to \infty} \ln b$

12. $\lim\limits_{b \to \infty} 3 \ln b$

13. $\lim\limits_{b \to \infty} e^b$

14. $\lim\limits_{b \to \infty} e^{-b}$

15. $\lim\limits_{a \to -\infty} (e^a + 3)$

16. $\lim\limits_{a \to -\infty} e^{-a}$

17. $\lim\limits_{b \to \infty} b^3$

18. $\lim\limits_{b \to \infty} b^2$

19. $\lim\limits_{a \to -\infty} \dfrac{1}{e^a}$

20. $\lim\limits_{b \to \infty} \dfrac{1}{e^b}$

21. $\lim\limits_{b \to \infty} \left(2 - \dfrac{1}{\sqrt{b}} \right)$

22. $\lim\limits_{a \to -\infty} \left(3 + \dfrac{1}{a} \right)$

In Exercises 23–44, evaluate each improper integral, if it is convergent. If the integral is divergent, so indicate.

23. $\displaystyle\int_1^\infty \dfrac{1}{x^3}\, dx$

24. $\displaystyle\int_2^\infty \dfrac{1}{x^3}\, dx$

25. $\displaystyle\int_1^\infty x^{-4}\, dx$

26. $\displaystyle\int_1^\infty x^{-2}\, dx$

27. $\displaystyle\int_1^\infty \dfrac{1}{2x}\, dx$

28. $\displaystyle\int_1^\infty \sqrt{x}\, dx$

29. $\displaystyle\int_1^\infty \dfrac{1}{\sqrt{x}}\, dx$

30. $\displaystyle\int_1^\infty \dfrac{1}{x^{3/2}}\, dx$

31. $\displaystyle\int_0^\infty e^x\, dx$

32. $\displaystyle\int_1^\infty e^{2x}\, dx$

33. $\displaystyle\int_1^\infty e^{-x}\, dx$

34. $\displaystyle\int_0^\infty e^{-2x}\, dx$

35. $\displaystyle\int_0^\infty \dfrac{dx}{x+1}$

36. $\displaystyle\int_{-2}^\infty \dfrac{dx}{x+3}$

37. $\displaystyle\int_0^\infty \sqrt{2x+9}\, dx$

38. $\displaystyle\int_0^\infty \sqrt[3]{x}\, dx$

39. $\displaystyle\int_0^\infty \dfrac{x}{1+x^2}\, dx$

40. $\displaystyle\int_0^\infty \dfrac{x^2}{x^3+4}\, dx$

41. $\displaystyle\int_1^\infty \dfrac{\ln x}{x}\, dx$

42. $\displaystyle\int_e^\infty \dfrac{dx}{x \ln x}$

43. $\displaystyle\int_0^\infty e^{-x/2}\, dx$

44. $\displaystyle\int_0^\infty e^{-.01x}\, dx$

In Exercises 45–50, determine the requested area.

45. The area under the curve $y = e^{-x}$ for $x \geq 0$

46. The area under the curve $y = e^{-.1x}$ for $x \geq 0$

47. The area under the curve $y = x^{-3}$ for $x \geq 2$

48. The area under the curve $y = \dfrac{1}{x^2}$ for $x \geq 10$

49. The area under the curve $y = \dfrac{1}{(x-1)^2}$ for $x \geq 2$

50. The area under the curve $y = \dfrac{x}{(1+x^2)^2}$ for $x \geq 0$

51. *(WATER POLLUTION)* Let t be the number of years from now. What is the total amount of waste dumped into a lake by a factory that dumps waste at the rate of $5e^{-.1t}$ pounds per year indefinitely?

52. *(AIR POLLUTION)* If radioactive material enters the earth's atmosphere at the rate of $100e^{-.02t}$ pounds per year and this continues indefinitely, what will be the total amount of radioactive material released to the atmosphere?

53. *(REAL ESTATE)* A real estate investment pays the investor indefinitely at the rate of $\$4000e^{-.08t}$ per year t years from now. Find the total amount received by the investor. The curve in the figure below shows the rate at which the real estate investment pays the investor. The area under the curve is the total amount received by the investor.

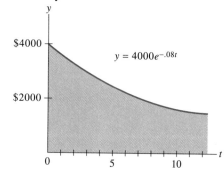

54. *(CAPITAL VALUE)* Determine the capital value of a rental property if the annual rent is $\$10{,}920$ and the annual interest rate is 7%.

55. *(CAPITAL VALUE)* Determine the capital value of a rental property if the annual rent is $\$6390$ and the annual interest rate is 9%.

56. In reference to capital value (Example 4), show that

$$\int_0^\infty R e^{-kt}\,dt = \frac{R}{k}$$

by evaluating the improper integral. Then use the R and k values of Example 4 to obtain the result of

that example from R/k rather than from the actual integration shown in that example.

In Exercises 57–68, evaluate each improper integral if it is convergent. If the integral is divergent, so indicate.

57. $\displaystyle\int_{-\infty}^{-1} \frac{1}{x^3}\,dx$

58. $\displaystyle\int_{-\infty}^{-2} 2x^{-3}\,dx$

59. $\displaystyle\int_{-\infty}^{0} e^{3x}\,dx$

60. $\displaystyle\int_{-\infty}^{0} e^{4x}\,dx$

61. $\displaystyle\int_{-\infty}^{0} e^{-2x}\,dx$

62. $\displaystyle\int_{-\infty}^{0} e^{-x}\,dx$

63. $\displaystyle\int_{-\infty}^{0} \frac{x}{1+x^2}\,dx$

64. $\displaystyle\int_{-\infty}^{0} \frac{2x}{x^2+10}\,dx$

65. $\displaystyle\int_{-\infty}^{0} \frac{dx}{(x-2)^3}$

66. $\displaystyle\int_{-\infty}^{0} \frac{dx}{(x-1)^3}$

67. $\displaystyle\int_{-\infty}^{-2} \frac{dx}{(4-x)^2}$

68. $\displaystyle\int_{-\infty}^{-1} \frac{dx}{(1-x)^2}$

In Exercises 69–74, determine the requested area.

69. The area under the curve $y = e^x$ for $x \leq -1$

70. The area under the curve $y = e^{2x}$ for $x \leq 0$

71. The area under the curve $y = \dfrac{1}{(x-1)^2}$ for $x \leq -2$

72. The area under the curve $y = \dfrac{1}{(x+1)^2}$ for $x \leq -2$

73. The area under the curve $y = \dfrac{1}{x^3}$ for $x \leq -4$

74. The area under the curve $y = \dfrac{1}{(x+2)^3}$ for $x \leq -3$

Evaluate each improper integral in Exercises 75–86. If the integral is divergent, so indicate.

75. $\displaystyle\int_{-\infty}^{\infty} x\,dx$

76. $\displaystyle\int_{-\infty}^{\infty} x^3\,dx$

77. $\displaystyle\int_{-\infty}^{\infty} e^{-x}\,dx$

78. $\displaystyle\int_{-\infty}^{\infty} e^x\,dx$

79. $\displaystyle\int_{-\infty}^{\infty} x^2 e^{-x^3}\,dx$

80. $\displaystyle\int_{-\infty}^{\infty} x^3 e^{-x^4}\,dx$

81. $\displaystyle\int_{-\infty}^{\infty} \frac{x}{(1+x^2)^2}\,dx$

82. $\displaystyle\int_{-\infty}^{\infty} \frac{x}{(x^2+4)^2}\,dx$

83. $\displaystyle\int_{-\infty}^{\infty} xe^{-2x^2}\, dx$

84. $\displaystyle\int_{-\infty}^{\infty} e^{-3x}\, dx$

85. $\displaystyle\int_{-\infty}^{\infty} \frac{x}{\sqrt{x^2+1}}\, dx$

86. $\displaystyle\int_{-\infty}^{\infty} \frac{x}{\sqrt[3]{x^2+1}}\, dx$

87. In Example 1, the improper integral converges, thus giving the area under the curve $y = 1/x^2$, beginning at $x = 1$ and continuing infinitely far to the right. How many square units is that area?

W 88. In Example 2, the improper integral diverges. Give an area interpretation for this situation.

The Exercise Library at the back of the book contains graphing calculator and computer exercises keyed to this section.

7.6 DIFFERENTIAL EQUATIONS

Interpreting the derivative as a rate of change has led to a variety of applications, including

1. Marginal cost, marginal revenue, and marginal profit

2. Velocity and acceleration

3. Rate of growth of bacteria, a city's population, or money invested with continuous compounding of interest

4. Increasing and decreasing functions

5. Tests for concavity and relative extrema

Since there are many applications involving rates of change, it would not be surprising to encounter equations that include and describe such rates of change. Such equations would contain derivatives, since rates of change are described by derivatives.

An equation that contains one or more derivatives of some function is called a **differential equation**. The following are two examples of differential equations.

$$\frac{dy}{dx} + 2xy = x^3$$

$$y'' - 3y' - 4y = 0$$

A **solution** of a differential equation is any function $y = f(x)$ that satisfies the differential equation.

EXAMPLE 1 Show that $y = x^2$ is a solution of the differential equation

$$xy' = 2y$$

SOLUTION $y = x^2$ is a solution if it satisfies the differential equation. In other words, if x^2 is substituted for y and $2x$ for y' in the given equation $xy' = 2y$, an identity (always true equation) will result. To begin,

$$xy' = 2y$$

After substitution of x^2 for y and $2x$ for y', we have

$$x \cdot 2x = 2 \cdot x^2$$

Upon simplification, we obtain an obvious identity.

$$2x^2 = 2x^2$$

Thus, $y = x^2$ is indeed a solution. ◆

Example 1 can be continued by noting that $y = x^2$ is not the only solution of $xy' = 2y$. Specifically, $y = 3x^2$ and $y = 10x^2$ are also solutions. (You may want to verify this.) In fact, $y = cx^2$ is a solution for any real number c. We call $y = cx^2$ the **general solution** and a solution such as $y = x^2$ or $y = 3x^2$ a **particular solution**. Figure 9 shows graphs of some particular solutions of $xy' = 2y$.

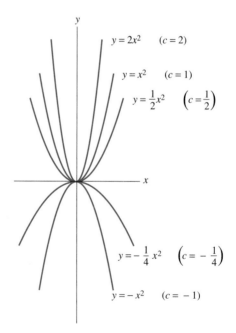

Figure 9 Some particular solutions $y = cx^2$

It is interesting to note that the solutions of differential equations are themselves equations. Solving a differential equation amounts to finding all functions that satisfy the differential equation.

Sometimes a differential equation is presented in a form that enables you to solve it simply by antidifferentiating, as shown in the next example.

EXAMPLE 2   Solve the differential equation $y' = 4x + 1$.

SOLUTION   Since the derivative of the desired function is $4x + 1$, the function we seek is the antiderivative of $4x + 1$. That is, since

$$\frac{dy}{dx} = 4x + 1$$

it follows that

$$y = \int (4x + 1)\, dx$$

or

$$y = 2x^2 + x + C$$

The general solution of the differential equation is $y = 2x^2 + x + C$.   ◆

In practice, when you solve a differential equation, the solution you obtain will be a general solution. *Then* there may be additional conditions that the function and its derivative(s) must satisfy. When such **initial conditions** are given, a particular solution can be obtained.

If Example 2 had included initial conditions, then a specific value could have been determined for C and a particular solution would have been obtained. Suppose, for example, that we had known that $y = 7$ when $x = 0$. Then in

$$y = 2x^2 + x + C$$

we would have

$$7 = 2(0)^2 + 0 + C$$

or

$$C = 7$$

The particular solution would then be

$$y = 2x^2 + x + 7$$

An initial condition such as $y = 7$ when $x = 0$ is usually written in the functionlike notation $y(0) = 7$.

EXAMPLE 3   Find the particular solution of the differential equation $y' = 12x^2 - 8$, with initial condition $y(1) = 6$.

SOLUTION   The nature of this differential equation enables us to solve it directly by anti-differentiation. From

$$y' = 12x^2 - 8$$

we obtain

$$y = 4x^3 - 8x + C$$

Since $y(1) = 6$, it follows that

$$6 = 4(1)^3 - 8(1) + C$$
$$6 = 4 - 8 + C$$

Clearly $C = 10$, and we have the particular solution

$$y = 4x^3 - 8x + 10 \quad \blacklozenge$$

In order to solve differential equations, it is often useful to separate the derivative dy/dx into the two differentials dy and dx. (Hence the name "differential" equations.) In Example 2 we had

$$\frac{dy}{dx} = 4x + 1$$

which may be written as

$$dy = (4x + 1)\, dx$$

Integration can then be performed on each side:

$$\int dy = \int (4x + 1)\, dx$$

The result is

$$y = 2x^2 + x + C$$

Note

Technically, a constant is produced on each side (by each integration)— say C_1 on the left and C_2 on the right. We combine them into one constant C on the right.

EXAMPLE 4 Solve the differential equation.

$$\frac{dy}{dx} = \frac{x^2}{y^3}$$

SOLUTION In this case it is *necessary* to use differential form, since we cannot anti-differentiate x^2/y^3 with respect to x. The y^3 poses a problem. However, using differential form, we can write

$$dy = \frac{x^2}{y^3}\, dx$$

Then we can remove the y^3 from the $x^2\ dx$ by multiplying both sides by y^3. The result is

$$y^3\ dy\ =\ x^2\ dx$$

The variables are now separated—y^3 is with dy and x^2 is with dx. Integration is now possible.

$$\int y^3\ dy\ =\ \int x^2\ dx$$

or

$$\frac{y^4}{4}\ =\ \frac{x^3}{3}\ +\ C\qquad\text{(solution)}$$

The method used to separate the variables and solve the equation is called **separation of variables**. ◆

The purpose of this section has been to introduce the idea of differential equations. There are entire courses and sequences of courses devoted to the methods of solving such equations. While most applications involve differential equations that are beyond the scope of this presentation, the next example offers an application opportunity.

EXAMPLE 5 ◆ *BACTERIA GROWTH*

Suppose the rate of growth of bacteria in a culture is 12% per hour. Let A represent the amount of bacteria present at any time t (in hours). Then we can say that the rate of change of A with respect to t is .12 times A, or

$$\frac{dA}{dt}\ =\ .12A$$

To solve this equation for A, we begin by writing it in differential form.

$$dA\ =\ .12A\ dt$$

Separating the variables leads to

$$\frac{dA}{A}\ =\ .12\ dt$$

The integration is

$$\int \frac{dA}{A}\ =\ \int .12\ dt$$

or

$$\ln |A|\ =\ .12t\ +\ C_1$$

Since $A > 0$, we have

$$\ln A\ =\ .12t\ +\ C_1$$

Changing the equation from logarithmic to exponential form will yield A:

$$A = e^{.12t+C_1}$$

The right-hand side can be simplified. To begin,

$$A = e^{.12t}e^{C_1} \qquad \text{(using a law of exponents)}$$

Now, since both e and C_1 are constants, it follows that e^{C_1} is a constant. Call it C. Then we have

$$A = Ce^{.12t}$$

Notice that we have derived the exponential growth formula $A = Ce^{kt}$ (from Chapter 5) for the specific case that k is .12.   ◆

7.6   Exercises

In Exercises 1–10, show that the specified function is a solution of the given differential equation.

1. $xy' = 2y; \ y = 5x^2$

2. $y' + y = (x + 1)^2; \ y = 1 + x^2$

3. $y' - 4y = 0; \ y = e^{4x}$

4. $y' + 2y = 0; \ y = e^{-2x}$

5. $xy' - 1.5y = 3; \ y = 4x^{3/2} - 2$

6. $3xy' - 4y = 0; \ y = 3x^{4/3}$

7. $y'' - 3y' + y = x^2; \ y = x^2 + 6x + 16$

8. $xy'' - y' = 5; \ y = 3x^2 - 5x$

9. $(y')^2 - 4y - 8x = 0; \ y = 1 - 2x$

10. $x(y')^2 - 4y + 4 = 0; \ y = 4x + 1$

In Exercises 11–20, determine the general solution of each differential equation.

11. $y' = 6x + 19$ **12.** $y' = 12x^3 - 18x + 1$

13. $y' = e^x + 1$ **14.** $y' = \dfrac{1}{\sqrt{x}}$

15. $f'(x) = 2 + \sqrt{x}$ **16.** $f'(x) = e^{.01x}$

17. $\dfrac{dy}{dt} = e^{-2t}$ **18.** $\dfrac{dy}{dt} = 4t^{1/3}$

19. $f'(t) = 1 + \dfrac{3}{t}$ **20.** $f'(t) = t + 6\sqrt{t}$

In Exercises 21–30, determine the particular solution of each differential equation.

21. $y' = 6x^2 - 2x; \ y(0) = 5$

22. $y' = 1 - 4x; \ y(0) = 1$

23. $f'(x) = x^7 + 3; \ f(0) = 14$

24. $g'(x) = \sqrt[3]{x}; \ g(0) = 3$

25. $\dfrac{dy}{dx} = 10x - e^x; \ y(0) = 0$

26. $\dfrac{dy}{dx} = 2e^{.01x}; \ y(0) = 250$

27. $\dfrac{dy}{dx} = 3\sqrt{x}; \ y(4) = 11$

28. $\dfrac{dy}{dx} = 1 + x^{-.5}; \ y(1) = 7$

29. $\dfrac{dy}{dt} = \dfrac{8}{t}; \ y(1) = 3$

30. $\dfrac{dy}{dt} = \dfrac{1}{2}\left(1 + \dfrac{1}{t}\right); \ y(1) = 5/2$

In Exercises 31–42, use *separation of variables* to solve each differential equation.

31. $\dfrac{dy}{dx} = 2x$

32. $\dfrac{dy}{dx} = 1 + \sqrt{x}$

33. $\dfrac{dy}{dx} = xy$

34. $\dfrac{dy}{dx} = \dfrac{x}{y}$

35. $\dfrac{dy}{dt} = te^{-y}$

36. $\dfrac{dy}{dt} = ye^t$

37. $y' = 6xy^2 + 5y^2$

38. $y' = x^2y - y$

39. $x\,dy + y\,dx = 0$

40. $2t\,dy - y^3\,dt = 0$

41. $\dfrac{dy}{e^x} = \dfrac{dx}{e^y}$

42. $dy = \dfrac{dx}{\sqrt{y}}$

Differential equations may describe marginal cost, marginal revenue, and marginal profit. In Exercises 43–48, solve the differential equation to find the requested cost, revenue, or profit.

43. (*Cost*) $C'(x) = 50 - .06x$, $C(0) = 150$. Find $C(x)$.

44. (*Cost*) $C'(x) = \dfrac{20}{\sqrt{x}}$, $C(0) = 35$. Find $C(x)$.

45. (*Revenue*) $R'(x) = 5 + .0002x$, $R(0) = 0$. Find $R(x)$.

46. (*Revenue*) $R'(x) = 70 - .4x$, $R(0) = 0$. Find $R(x)$.

47. (*Profit*) $P'(x) = 80 - .3\sqrt{x}$, $P(0) = -50$. Find $P(x)$.

48. (*Profit*) $P'(x) = 200 + .6x - .08x^2$, $P(0) = 0$. Find $P(x)$.

In Exercises 49–54, write the differential equation that describes the situation. *Do not solve* the equation. Use t for time, unless the situation indicates otherwise.

49. The rate of change of y with respect to x is 3 less than x.

50. The rate of change of y with respect to x is a constant times the product of x and y.

51. (*BACTERIAL INFECTION*) A bacterial infection is growing at the rate of 5% per hour. Let B be the number of bacteria in the infection.

52. (*FUNGUS GROWTH*) The rate of growth of a fungus is 3% per hour. Let F be the size of the fungus.

53. (*RADIOACTIVITY*) A mass of radioactive substance *loses* its radioactivity at the rate of 2% per year. Let M be the amount that is radioactive.

54. The value of x is *decreasing* at the rate of twice the square root of x per second.

55. (*BACTERIA GROWTH*) Suppose the rate of growth of bacteria in a culture is 9% per hour. Let A represent the amount of bacteria present at any time t (in hours). Following Example 5,
(a) Establish a differential equation that describes the number of bacteria present in the culture at any time t.
(b) Solve the differential equation for A.
(c) Determine the value of C assuming there are 3000 bacteria at the beginning.

56. (*INTEREST*) The amount of money M in an account is increasing at the rate of 9% per year.
(a) Write a differential equation that describes the amount at any time t.
(b) Solve the differential equation for A. Assume that the account began with $2000 in it.

57. (*GNP GROWTH*) Suppose the gross national product (GNP) of a small country is $100,000,000, and it increases at the rate of 3% per year.
(a) Establish a differential equation that describes the amount A of the GNP at any time t.
(b) Solve the differential equation for A.

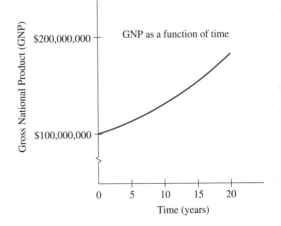

A graph of the solution of the differential equation of Exercise 57.

58. (*INVESTMENT LOSS*) An investment is losing value at the rate of 4% per year. Let A be the amount of the investment, and let t be the time in years. The investment was worth $5000 originally.
  (a) Write a differential equation that describes the value of the investment at any time t.
  (b) Solve the differential equation for A.

59. (*DRUG CONCENTRATION*) A patient is given 10 milligrams of a drug intravenously, after which the amount present in the bloodstream decreases at the rate of 8% per hour. Let A represent the amount of the drug present at any time t (in hours).
  (a) Establish a differential equation that describes the number of milligrams of the drug in the bloodstream at any time t.
  (b) Solve the differential equation for A. Be sure to use the given initial condition to determine the constant.

60. (*STIMULUS AND RESPONSE*) The Weber-Fechner law (in psychology) offers a relationship between stimulus (S) and response (R).

$$\frac{dR}{dS} = k \cdot \frac{1}{S} \qquad k = \text{constant}$$

  (a) Determine R by solving the differential equation.
  (b) Solve for the added constant C by using the initial condition that response R is 0 for some small stimulus s.
  (c) Use your knowledge of algebra to rewrite the solution [from part (b)] as

$$R = k \ln \frac{S}{s}$$

W 61. Could $y = 3$ be a solution to a differential equation? Explain.

W 62. How many solutions are there for the differential equation $y' = 6x$? Explain.

7.7 | PROBABILITY AND CALCULUS

Perhaps it is human nature to wonder about the chances that a particular event will occur. How likely is it to rain today? Since our first child was a boy, how likely is it that our second child will be a girl? What are my chances of winning the lottery this week?

The formal study of probability is considered to have begun about 1654, when Blaise Pascal and Pierre de Fermat studied games of chance and the gambling associated with them.

An English merchant named John Graunt studied data available from over 50 years of burial records and then drew conclusions about life expectancy, publishing his results in 1662. His work provided a foundation for the development of life insurance companies, which followed soon afterward.

Our everyday life provides casual experiences with probability. The chance of rain may be 70% or 50% or 100% or 0%. When such percents are converted to fractions or decimal numbers, we have

percent	other form
70%	$\frac{7}{10}$ or .7
50%	$\frac{1}{2}$ or .5
100%	1
0%	0

To say that it will rain (that rain is a sure thing) is to say that the chance of rain is 100%. By contrast, if an event cannot occur, then its chance of occurring is 0%.

This example suggests that the largest value for a probability is 100%, or 1, and the smallest value is 0%, or 0.

> The probability that an event occurs will always be a number between 0 and 1 inclusive. The probability of an impossible event is 0. The probability of a certain event is 1.

Sometimes probabilities can be represented by areas under a curve. In such instances, calculus (integration, in particular) can be used to find the desired probabilities.

Before we introduce formal definitions and the associated calculus, let us consider a simple example that will provide some useful orientation.

Consider an overnight delivery company that guarantees delivery between 9 a.m. and 4 p.m. the next day. In theory, the chance of delivery between 9 a.m. and 4 p.m. is 100%. Furthermore, the company claims that the chance of a morning delivery is 60% (or .6). This means the chance of an afternoon delivery is 40% (or .4). We will assume that a morning delivery (9 a.m.–12 noon) can come at any time in that interval with equal likelihood. Similarly, an afternoon delivery (12 noon–4 p.m.) can come at any time in that interval with equal likelihood. A graph, using a 24-hour clock, that describes this situation is shown in Figure 10.

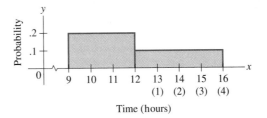

Figure 10 Overnight delivery

Notice that the area under the graph from $x = 9$ to $x = 12$ is indeed .6—calculated as the area of the rectangle having length 3 (from $12 - 9$) and width .2. Also, note that the area under the graph from $x = 12$ to $x = 16$ is the .4 we expect. It is calculated as $(4)(.1) = .4$. (The y values .2 and .1 were selected to make this work out.)

We shall use a special notation for probability. $P(9 \leq x \leq 12)$ will mean "the probability that x is between 9 and 12." Thus,

$$P(9 \leq x \leq 12) = .6$$

and

$$P(12 \leq x \leq 16) = .4$$

You can also see from the graph such probabilities as

$$P(9 \leq x \leq 10) = .2$$
$$P(9 \leq x \leq 11) = .4$$
$$P(13 \leq x \leq 16) = .3$$
$$P(9 \leq x \leq 16) = 1$$
$$P(10 \leq x \leq 13) = .5$$

Calculus provides the means to determine probabilities when the function is neither constant nor composed of parts that are constants. (The function we considered consisted of two parts, each of which was constant over an interval.) If the graph of f were such that $P(c \leq x \leq d)$ is the area under f from $x = c$ to $x = d$, then, using calculus, we would want to have

$$P(c \leq x \leq d) = \int_c^d f(x) \, dx$$

A function that can be used in this manner to determine probabilities is called a **probability density function**. The definition of a probability density function is given next. A graph of such a function is shown in Figure 11.

Probability Density Function

Let f be defined over the interval $[a, b]$. Then f is a *probability density function* if

1. $f(x) \geq 0$ for $a \leq x \leq b$

2. $P(a \leq x \leq b) = 1$

The probability that x is between c and d is given by

$$P(c \leq x \leq d) = \int_c^d f(x) \, dx$$

Note

Since $\int_c^d f(x) \, dx$ gives the area under the curve between $x = c$ and $x = d$ (and thus the probability that x is between c and d), condition 2 of the definition can be stated as

$$\int_a^b f(x) \, dx = 1$$

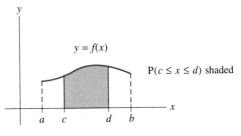

Figure 11 A probability density function

EXAMPLE 1 Consider $f(x) = \dfrac{1}{21}x^2$ for x in the interval $[1, 4]$.

(a) Show that f is a probability density function.

(b) Determine $P(2 \le x \le 3)$, the probability that x is between 2 and 3.

SOLUTION (a) We must show that

  1. $f(x) \ge 0$ for all x in the interval $[1, 4]$

  2. $\displaystyle\int_1^4 f(x)\,dx = 1$

First, $x^2 \ge 0$ for all x in $[1, 4]$, and thus $\frac{1}{21}x^2 \ge 0$ for all x in $[1, 4]$.
Second,

$$\int_1^4 \frac{1}{21}x^2\,dx = \left[\frac{1}{21}\cdot\frac{x^3}{3}\right]_1^4$$
$$= \left[\frac{1}{63}x^3\right]_1^4$$
$$= \frac{1}{63}(64 - 1)$$
$$= 1$$

Thus, f is a probability density function.

(b) The probability that x is between 2 and 3 is

$$P(2 \le x \le 3) = \int_2^3 \frac{1}{21}x^2\,dx$$
$$= \left[\frac{1}{63}x^3\right]_2^3$$
$$= \frac{1}{63}(27 - 8)$$
$$= \frac{19}{63}$$

Thus, $P(2 \le x \le 3) = 19/63 \approx .3016$ ◆

EXAMPLE 2   ◆   *LIGHT BULB LIFE*

Suppose the total hours x that a light bulb will burn is given by the probability density function

$$f(x) = .002e^{-.002x} \qquad x \geq 0$$

(a) Verify that f is indeed a probability density function.

(b) Find the probability that a light bulb will burn 300 hours or less.

(c) Find the probability that a light bulb will burn between 100 and 400 hours.

SOLUTION   **(a)** Since $e^u > 0$ for any real number u, it follows that

$$.002e^{-.002x} \geq 0 \qquad \text{for } x \geq 0$$

which satisfies condition 1 for a probability density function. Second, since the inequality $x \geq 0$ describes the interval $[0, \infty)$, we must show that

$$\int_0^\infty .002e^{-.002x} \, dx = 1$$

Here is the step-by-step process of evaluation of this improper integral.

$$\int_0^\infty .002e^{-.002x} \, dx = .002 \int_0^\infty e^{-.002x} \, dx$$

$$= .002 \lim_{b \to \infty} \int_0^b e^{-.002x} \, dx$$

$$= (.002) \cdot \frac{1}{-.002} \lim_{b \to \infty} \left[e^{-.002x}\right]_0^b$$

$$= -\lim_{b \to \infty} \left(e^{-.002b} - e^{-.002(0)}\right)$$

$$= -\lim_{b \to \infty} \left(e^{-.002b} - 1\right)$$

$$= -\lim_{b \to \infty} e^{-.002b} + 1$$

$$= 0 + 1$$

$$= 1$$

(b) The probability that a light bulb will burn 300 hours or less is

$$\int_0^{300} .002e^{-.002x} \, dx = .002 \int_0^{300} e^{-.002x} \, dx$$

$$= -\left[e^{-.002x}\right]_0^{300}$$

$$= -\left(e^{-.6} - e^0\right)$$

$$= -e^{-.6} + 1$$

$$\approx -.5488 + 1$$

$$= .4512$$

(c) The probability that a light bulb will burn between 100 and 400 hours is

$$\int_{100}^{400} .002 e^{-.002x} \, dx = .002 \int_{100}^{400} e^{-.002x} \, dx$$
$$= -\left[e^{-.002x} \right]_{100}^{400}$$
$$= -\left[e^{-.8} - e^{-.2} \right]$$
$$\approx -.4493 + .8187$$
$$= .3694$$

See Figure 12.

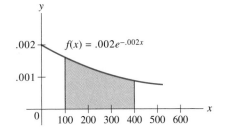

Figure 12 The area of the shaded region is the probability that a light bulb will burn between 100 and 400 hours ◆

EXAMPLE 3 The function defined by $f(x) = 4x^2$ on $[2, 5]$ is not a probability density function. Multiply f by a constant to create a probability density function.

SOLUTION Although $f(x) = 4x^2 \geq 0$ for all x in the interval $[2, 5]$, f is not a probability density function because

$$\int_{2}^{5} 4x^2 \, dx = \left[\frac{4x^3}{3} \right]_{2}^{5} = 156 \neq 1$$

The integral must be equal to 1 for f to be a probability density function. If the original function is divided by 156, it will then be a probability density function. Thus,

$$g(x) = \frac{4x^2}{156} = \frac{x^2}{39}$$

is a probability density function. (Note that $g(x) \geq 0$ for all x in the interval.) ◆

7.7 Exercises

In Exercises 1–4, obtain the requested probabilities from the given graph. You may want to refer back to the overnight delivery example.

1. (a) $P(5 \le x \le 6)$
 (b) $P(6 \le x \le 8)$
 (c) $P(6 \le x \le 9)$
 (d) $P(5 \le x \le 9)$

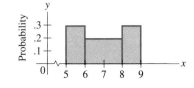

2. (a) $P(2 \le x \le 4)$
 (b) $P(4 \le x \le 7)$
 (c) $P(5 \le x \le 7)$
 (d) $P(2 \le x \le 7)$

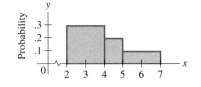

3. (a) $P(7 \le x \le 8)$
 (b) $P(5 \le x \le 7)$
 (c) $P(4 \le x \le 8)$
 (d) $P(4 \le x \le 7)$

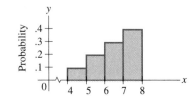

4. (a) $P(6 \le x \le 7)$
 (b) $P(7 \le x \le 9)$
 (c) $P(6 \le x \le 8)$
 (d) $P(6 \le x \le 9)$

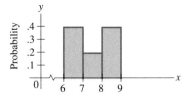

In Exercises 5–12, show that each function f is a probability density function on the given interval.

5. $f(x) = \dfrac{1}{30}x$ on $[2, 8]$ **6.** $f(x) = \dfrac{2x}{15}$ on $[1, 4]$

7. $f(x) = \dfrac{1}{12}(3x^2 + 2x)$ on $[0, 2]$

8. $f(x) = \dfrac{1}{42}(2x + 1)$ on $[0, 6]$

9. $f(x) = \dfrac{3}{38}\sqrt{x}$ on $[4, 9]$ **10.** $f(x) = \dfrac{x^3}{4}$ on $[0, 2]$

11. $f(x) = \dfrac{10}{9x^2}$ on $[1, 10]$ **12.** $f(x) = \dfrac{1}{6}$ on $[4, 10]$

In Exercises 13–18, find the required probabilities associated with the given probability density function.

13. $f(x) = \dfrac{3}{64}x^2$ on $[0, 4]$
 (a) $P(1 \le x \le 4)$ **(b)** $P(0 \le x \le 2)$

14. $f(x) = \dfrac{1}{16}x$ on $[2, 6]$
 (a) $P(2 \le x \le 5)$ **(b)** $P(3 \le x \le 6)$

15. $f(x) = \dfrac{1}{10}$ on $[0, 10]$
 (a) $P(1 \le x \le 6)$ **(b)** $P(7 \le x \le 10)$

16. $f(x) = \dfrac{1}{8}(4 - x)$ on $[0, 4]$
 (a) $P(0 \le x \le 2)$ **(b)** $P(1 \le x \le 4)$

17. $f(x) = \dfrac{1}{x}$ on $[1, e]$
 (a) $P(1 \le x \le 2)$ **(b)** $P(2 \le x \le e)$

18. $f(x) = \dfrac{3}{76}(1 + \sqrt{x})$ on $[1, 9]$

   (a) $P(1 \le x \le 4)$ **(b)** $P(4 \le x \le 9)$

In Exercises 19–24, modify the given function f to create a function g that will be a probability density function on the given interval (see Example 3, if needed).

19. $f(x) = 2x$ on $[1, 8]$ **20.** $f(x) = \sqrt{x}$ on $[1, 4]$

21. $f(x) = x^3$ on $[0, 1]$ **22.** $f(x) = \dfrac{1}{x^2}$ on $[1, 2]$

23. $f(x) = x - 5$ on $[6, 10]$ **24.** $f(x) = x^2$ on $[3, 7]$

25. (*NEWSPAPER SALES*) The probability density function for the number of newspapers x a convenience store will sell in a day is

$$f(x) = .0008x \qquad 0 \le x \le 50$$

What is the probability the store will sell at least 40 newspapers tomorrow?

26. (*WAITING IN LINE*) The wait in line (x minutes) on the first day of registration is described by the probability density function

$$f(x) = .0002x \qquad 0 \le x \le 100$$

   (a) Find the probability that the wait will be 10 minutes or less.
   (b) Find the probability that the wait will be between 30 minutes and an hour.
   (c) Find the probability that the wait will be at least an hour.

27. (*SMOKE DETECTOR*) A smoke detector flashes every 30 seconds to show that it is operational. The probability density function is

$$f(x) = \dfrac{1}{30} \qquad 0 \le x \le 30$$

where x is the number of seconds before the next flash. Find the probability that you will see a flash within 6 seconds of looking at the smoke detector.

28. (*WAITING FOR A BUS*) The wait (x minutes) for a bus is described by the probability density function

$$f(x) = \dfrac{1}{25} \qquad 0 \le x \le 25$$

   (a) Find the probability that the wait will be no more than 10 minutes.

   (b) Find the probability that the wait will be at least 15 minutes.
   (c) Verify that f is a probability density function.

29. (*LENGTH OF CALL*) A local phone company determines that the probability density function for the length of time (x minutes) of a phone call is

$$f(x) = .25e^{-.25x} \qquad x \ge 0$$

   (a) Find the probability that a phone call will last no more than 4 minutes.
   (b) Find the probability that a phone call will last between 4 and 10 minutes.

30. (*LIGHT BULB LIFE*) The probability density function for the total time (x hours) that a light bulb will burn is

$$f(x) = .001e^{-.001x} \qquad x \ge 0$$

   (a) Verify that f is indeed a probability density function.
   (b) Find the probability that a light bulb will burn between 200 and 500 hours.

31. (*LEARNING EXPERIMENT*) A psychologist asks the subjects of her experiment to learn a task that she has devised. The probability density function is determined to be

$$f(x) = \dfrac{30}{13x^2} \qquad 2 \le x \le 15$$

where x is the number of minutes needed to learn the task. (The task cannot be learned in less than 2 minutes.)

   (a) Find the probability that a subject chosen at random would learn the task in 5 minutes or less.
   (b) Find the probability that a subject chosen at random would take at least 10 minutes to learn the task.

32. (*LEARNING EXPERIMENT*) A psychologist devises a task similar to the one used in Exercise 31. He finds that the probability density function has the form

$$f(x) = \dfrac{k}{x^2} \qquad 4 \le x \le 20$$

where x is the number of minutes needed to learn the task. The fastest time to learn the task is 4 minutes, and the slowest is 20 minutes.

   (a) Determine k.
   (b) What is the probability that a participant learns the task in 10 minutes or less?

W 33. Let f be defined on the interval $[0, 10]$ and let $f(x) \geq 0$ for all x in the interval. Also, suppose that

$$\int_{1}^{4} f(x) \, dx = 2.5$$

Can function f be a probability density function? Explain.

W 34. Suppose you are given a function defined on a closed interval. Explain using words (no mathematical symbols) how you would determine whether the function is a probability density function.

Chapter List *Important terms and ideas*

integration by substitution
integration by parts
income stream (money flow)
present value of an income stream
integration by tables
reduction formula
numerical integration

trapezoidal rule
Simpson's rule
error estimates
improper integrals
divergent integral
convergent integral
capital value

differential equation
particular solution
general solution
initial condition
separation of variables
probability density function

Review Exercises for Chapter 7

Evaluate each indefinite integral in Exercises 1–28. Use substitution, if possible. Otherwise, use parts or the tables.

1. $\int x\sqrt{1 + x^2} \, dx$

2. $\int 2x\sqrt{1 - x^2} \, dx$

3. $\int \dfrac{x \, dx}{\sqrt{1 + 2x}}$

4. $\int \dfrac{x}{(x + 1)^2} \, dx$

5. $\int (x - 2)^3 \, dx$

6. $\int (x - 5)^4 \, dx$

7. $\int x(x - 2)^3 \, dx$

8. $\int x(x + 1)^4 \, dx$

9. $\int xe^{x^2} \, dx$

10. $\int \dfrac{x}{e^{x^2}} \, dx$

11. $\int xe^{3x} \, dx$

12. $\int \dfrac{x}{e^{3x}} \, dx$

13. $\int \dfrac{x}{\sqrt{x^2 + 1}} \, dx$

14. $\int \dfrac{1}{\sqrt{x^2 - 1}} \, dx$

15. $\int \dfrac{1}{\sqrt{x^2 + 1}} \, dx$

16. $\int \dfrac{1}{x\sqrt{x^2 + 1}} \, dx$

17. $\int \ln 2x \, dx$

18. $\int x \ln 3x \, dx$

19. $\int \dfrac{\ln 2x}{x} \, dx$

20. $\int \dfrac{x + 1}{x} \, dx$

21. $\int \dfrac{e^{2x}}{1 + e^{2x}} \, dx$

22. $\int \dfrac{e^{3x}}{1 - e^{3x}} \, dx$

23. $\int \dfrac{1}{1 + e^{2x}} \, dx$

24. $\int \dfrac{e^{2x} + 1}{e^x} \, dx$

25. $\int 100xe^{-.05x} \, dx$

26. $\int (x - e^{-.02x}) \, dx$

27. $\int \dfrac{2x + 3}{x - 1} \, dx$

28. $\int \dfrac{x + 1}{e^x} \, dx$

In Exercises 29–32, use (a) the trapezoidal rule and (b) Simpson's rule to approximate each definite integral. Round calculations to four decimal places. *Use a calculator.*

29. $\int_{0}^{2} \sqrt{9 - x^2} \, dx, \, n = 4$

30. $\int_{0}^{2} e^{2x} \, dx, \, n = 4$

31. $\int_1^4 \dfrac{1}{x^2 + 4}\, dx,\ n = 6$

32. $\int_2^5 x \ln x,\ n = 6$

35. $\int_{-\infty}^0 e^{-.5x}\, dx$

36. $\int_{-\infty}^0 e^{.5x}\, dx$

In Exercises 33–40, evaluate each improper integral if it is convergent. If the integral is divergent, so indicate.

37. $\int_2^\infty x^{-2/3}\, dx$

38. $\int_0^\infty \dfrac{dx}{x + 2}$

39. $\int_{-\infty}^\infty \dfrac{x}{x^2 + 2}\, dx$

40. $\int_{-\infty}^\infty x^2\, dx$

33. $\int_1^\infty \dfrac{1}{x^4}\, dx$

34. $\int_1^\infty \dfrac{1}{\sqrt[3]{x}}\, dx$

MULTIVARIABLE CALCULUS

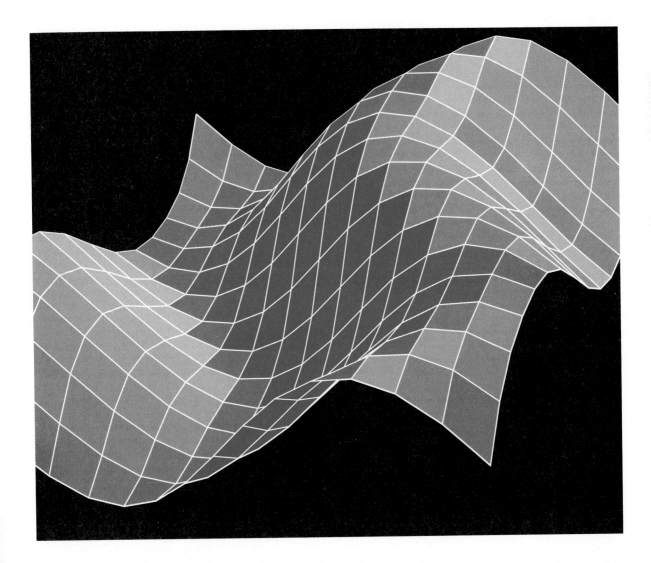

$\bigcup$p to this point we have worked exclusively with functions of one variable. Yet there are many applications of functions of two or more variables. For example, costs, revenues, and profits often depend on more than one variable.

After we introduce functions, graphs, and their applications, we will consider the calculus of functions of two or more variables. The presentation will include differentiation, maximum/minimum concepts, differentials, integrals, and other ideas and applications.

8.1   *FUNCTIONS OF SEVERAL VARIABLES*

The notation $f(x)$ used for functions of one variable can be extended to $f(x, y)$ for **functions of two variables** and $f(x, y, z)$ for **functions of three variables**. Functions of four or more variables will not be used in this text.

Let us consider an example of a function $z = f(x, y)$ of two variables.

$$z = x^2 + 5xy \qquad \text{or} \qquad f(x, y) = x^2 + 5xy$$

For every x and y supplied, there is a corresponding $f(x, y)$ value (or z value). If $x = 2$ and $y = 4$, then

$$z = (2)^2 + 5(2)(4) = 44$$

or

$$f(2, 4) = 44$$

EXAMPLE 1   Let $f(x, y) = 10xy + 3$. Find $f(4, 9)$.

SOLUTION $f(4, 9)$ is the number obtained when 4 is used for x and 9 is used for y. If

$$f(x, y) = 10xy + 3$$

then

$$f(4, 9) = 10(4)(9) + 3$$
$$= 363 \quad \blacklozenge$$

EXAMPLE 2   Let $z = 5e^{xy}$. Find z when $x = 4$ and $y = 0$.

SOLUTION   Using 4 for x and 0 for y yields

$$z = 5e^{xy}$$
$$= 5e^{(4)(0)}$$
$$= 5e^0$$
$$= 5 \cdot 1$$
$$= 5 \quad \blacklozenge$$

EXAMPLE 3 ◆ **VOLUME OF A CONE**

The volume of a cone is $\frac{1}{3}\pi r^2 h$, where r is the radius of the base of the cone and h is its height. Clearly, volume V is a function of two variables, r and h. The function can be given either as $V = \frac{1}{3}\pi r^2 h$ or as $V(r, h) = \frac{1}{3}\pi r^2 h$. Notice that π is a constant; $\pi \approx 3.14$. (See Figure 1.)

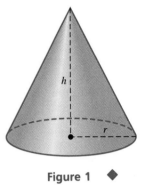

Figure 1 ◆

EXAMPLE 4 ◆ **COST OF MAKING CHAIRS**

A furniture manufacturer makes chairs from oak and from maple. If the oak chairs cost \$75 each to make and the maple chairs cost \$50 each, determine the cost function for x oak chairs and y maple chairs.

SOLUTION Since it costs \$75 to make each oak chair, it will cost $75x$ dollars to make x oak chairs. Similarly, it costs $50y$ dollars to make y maple chairs. Thus, it costs $75x + 50y$ dollars to make all the chairs. Using C as the cost function, we have

$$C(x, y) = 75x + 50y \quad ◆$$

EXAMPLE 5 ◆ **CEPHALIC INDEX**

Anthropologists use head shape as one means of classifying humans. The *cephalic index C* is given by

$$C(w, l) = 100\frac{w}{l}$$

where w is the width and l is the length of the person's head. (Both measurements are made across the top of the head.)

(a) Find the cephalic index of a person whose head has width 6 inches and length 8 inches.

(b) Find $C(16, 20)$, where the measurements are in centimeters.

SOLUTION **(a)** We seek $C(6, 8)$.

$$C(6, 8) = 100 \cdot \frac{6}{8}$$
$$= 75$$

The cephalic index is 75.

(b) We calculate $C(16, 20)$.

$$C(16, 20) = 100 \cdot \frac{16}{20}$$

$$= 80$$

The cephalic index is 80. ◆

The Cobb-Douglas Production Function

In economics, the **Cobb-Douglas production function** specifies how the number of units of a product manufactured depends upon the amounts spent on labor and capital. "Capital" means buildings, machinery, and other things needed in the production process. If x represents the number of units of labor and y represents the number of units of capital, then the number of units produced, $f(x, y)$, is given by

$$f(x, y) = Cx^a y^{1-a}$$

The number C is a positive constant, and a is a positive number less than 1. The exponents are a and $1 - a$, which means the sum of the exponents is 1. Here are two examples of Cobb-Douglas production functions.

$$f(x, y) = 20x^{.4}y^{.6}$$
$$f(x, y) = x^{2/3}y^{1/3}$$

EXAMPLE 6 ◆ **COBB-DOUGLAS PRODUCTION FUNCTION**

Given the Cobb-Douglas production function $f(x, y) = 20x^{.4}y^{.6}$, determine the number of units produced using 1200 units of labor and 2000 units of capital.

SOLUTION The number of units produced is $f(1200, 2000)$.

$$f(1200, 2000) = 20(1200)^{.4}(2000)^{.6}$$
$$\approx 20(17.05)(95.64)$$
$$\approx 32,613 \text{ units}$$

The calculations $(1200)^{.4}$ and $(2000)^{.6}$ were done using the $\boxed{y^x}$ key of a calculator. ◆

Graphs

A function $z = f(x, y)$ can be graphed in a **three-dimensional rectangular coordinate system**. When an x value and a y value are supplied, a corresponding z value is determined. The points are of the form (x, y, z). Because for every x-y pair there is a corresponding z value, the collection of all the points (x, y, z) of $z = f(x, y)$ forms a surface. An example of such a surface is shown in Figure 2, graphed in a three-dimensional rectangular coordinate system.

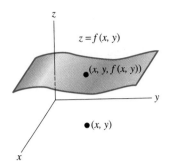

Figure 2 The graph of a surface, $z = f(x, y)$

Next, we will plot two points in order to provide some feeling for three-dimensional plotting and graphing. First, consider the point (2, 3, 4). To plot the point, begin at the origin (where the x, y, and z axes meet). Go 2 units down the x axis, then 3 units parallel to the y axis, and finally go up 4 units parallel to the z axis. See Figure 3.

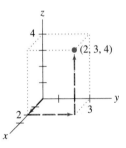

Figure 3 Plotting the point (2, 3, 4)

Only the positive coordinate axes are shown in Figure 3. The positive z axis goes upward and is perpendicular to the plane formed by the x and y axes. (The negative z axis is an extension of the z axis below the xy plane. The negative y axis is a leftward extension of the y axis. The negative x axis is an extension of the x axis back through the plane created by the y and z axes.) As a second example, consider the point (3, 4, −2), which is plotted in Figure 4.

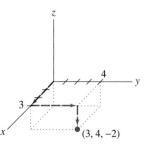

Figure 4 Plotting the point (3, 4, −2)

Although we will not pursue three-dimensional graphing, the graph of the function $z = x^2 + y^2$ is shown in Figure 5 as an example. The graph is a surface called a paraboloid. Computers can be used to obtain graphs of surfaces—functions of the form $z = f(x, y)$. In fact, many personal computers (PCs) have such "computer graphics" capability.

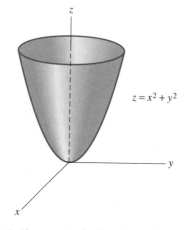

$z = x^2 + y^2$

Figure 5 The graph of a function of two variables

Domain

The **domain** of a function $z = f(x, y)$ is the set of all ordered pairs of real numbers (x, y) for which $f(x, y)$ is defined.

EXAMPLE 7   Determine the domain of each function.

(a) $z = \dfrac{3x}{2y - 16}$ **(b)** $f(x, y) = \sqrt{y - 5x}$ **(c)** $z = x^2 - y^2 + 10$

SOLUTION   **(a)** The function

$$z = \frac{3x}{2y - 16}$$

is defined for all (x, y) except those that result in division by zero. Thus, $2y - 16$ cannot be zero, which means $y \neq 8$. The domain of this function consists of all ordered pairs of real numbers (x, y) except those for which $y = 8$.

(b) The function

$$f(x, y) = \sqrt{y - 5x}$$

is defined for all (x, y) except those that give the square root of a negative number. Thus, $y - 5x$ must be nonnegative:

$$y - 5x \geq 0$$

This is better expressed by solving for y in terms of x:

$$y \geq 5x$$

The domain of this function consists of all ordered pairs of real numbers (x, y) in which $y \geq 5x$.

(c) The function

$$z = x^2 - y^2 + 10$$

is defined for all (x, y). There are no choices of x or y that will lead to division by zero or to the square root (or other even root) of a negative number. Thus, the domain of this function consists of all ordered pairs of real numbers (x, y). ◆

Level Curves

You have probably seen *topographic maps* of mountainous regions. In such maps, the rugged three-dimensional surface is represented by a series of two-dimensional curves. Each of the curves represents a horizontal slice of the mountain's surface at a particular level of elevation. The topographic map gives a "bird's-eye view" of the mountain. In Figure 6, c is the elevation in feet.

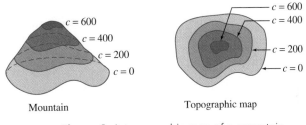

Figure 6 A topographic map of a mountain

The topographic map is intended as an introductory example. In general, the graph of a three-dimensional surface $z = f(x, y)$ can be described by a collection of two-dimensional curves called **level curves**. The level curves are obtained by letting $f(x, y)$ be equal to "height" c and then selecting values for c.

EXAMPLE 8 Sketch some level curves of the function $f(x, y) = 9 - x^2 - y^2$. Specifically, let $c = 0, 2, 4, 6$, and 8.

SOLUTION Although a three-dimensional graph of the function is neither requested nor needed, it is shown in Figure 7 to increase your understanding of the level curve concept.

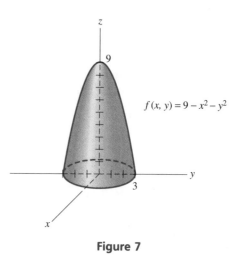

$f(x, y) = 9 - x^2 - y^2$

Figure 7

To obtain level curves for $f(x, y) = 9 - x^2 - y^2$, we will use the equation $9 - x^2 - y^2 = c$ with various c values. Specifically, we will use $c = 0, 2, 4, 6,$ and 8 (the arbitrary choices in the statement of the example).

c	intermediate result	level curve
0	$9 - x^2 - y^2 = 0$	$x^2 + y^2 = 9$
2	$9 - x^2 - y^2 = 2$	$x^2 + y^2 = 7$
4	$9 - x^2 - y^2 = 4$	$x^2 + y^2 = 5$
6	$9 - x^2 - y^2 = 6$	$x^2 + y^2 = 3$
8	$9 - x^2 - y^2 = 8$	$x^2 + y^2 = 1$

Note
The graph of $x^2 + y^2 = r^2$ is a *circle* centered at the origin and having radius r.

The graph of $x^2 + y^2 = 9$ is a circle centered at the origin and having radius 3. The graph of $x^2 + y^2 = 7$ is a circle with center at $(0, 0)$ and radius $\sqrt{7}$. Figure 8 shows the level curves together with the three-dimensional graph of $f(x, y) = 9 - x^2 - y^2$.

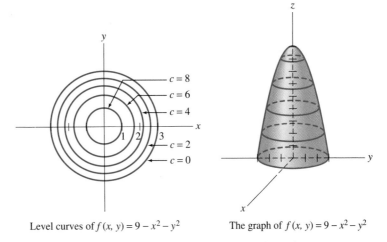

Level curves of $f(x, y) = 9 - x^2 - y^2$ The graph of $f(x, y) = 9 - x^2 - y^2$

Figure 8 The level curves and graph of the function $f(x, y) = 9 - x^2 - y^2$ ◆

EXAMPLE 9   Sketch some level curves of the function $f(x, y) = x^2 - y$. Use $c = 0, 1, 2,$ and 3.

SOLUTION

c	intermediate result	level curve
0	$x^2 - y = 0$	$y = x^2$
1	$x^2 - y = 1$	$y = x^2 - 1$
2	$x^2 - y = 2$	$y = x^2 - 2$
3	$x^2 - y = 3$	$y = x^2 - 3$

Each of these level curves is a parabola, as shown in Figure 9.

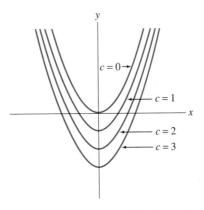

Figure 9 Level curves of $f(x, y) = x^2 - y$ ◆

If f is a production function, then the points (x, y) of each level curve are points at which the production level is constant. Such curves are called **isoquants** ("equal quantities"). Let us consider the Cobb-Douglas production function $f(x, y) = 20x^{2/3}y^{1/3}$ and the associated level curve (isoquant) $20x^{2/3}y^{1/3} = 300$. This level curve contains all points (x, y) for which the production level is 300. The points of the isoquant are all the combinations of labor x and capital y that will yield 300 as a production level. See Figure 10.

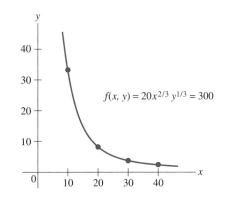

Figure 10 An isoquant (level curve) associated with a Cobb-Douglas production function

8.1 Exercises

For each function given in Exercises 1–10, compute $f(4, 2)$, $f(-1, 6)$, $f(0, 2)$, $f(1, 0)$, and $f(0, 1)$.

1. $f(x, y) = x^2 + y^2$

2. $f(x, y) = x^2 - y^2 + 1$

3. $f(x, y) = \dfrac{x}{x + y}$

4. $f(x, y) = \sqrt{x^2 + y^2}$

5. $f(x, y) = 1 - x - y$

6. $f(x, y) = -x^2 + y^2$

7. $f(x, y) = 3x^2 - 7y$

8. $f(x, y) = 3y - 7x$

9. $f(x, y) = \sqrt{5 + x^2 + y^2}$

10. $f(x, y) = (x + y)^2$

In Exercises 11–14, compute $g(1, 4)$ and $g(e, 0)$ for each function.

11. $g(x, y) = y \ln x$

12. $g(x, y) = xe^y$

13. $g(x, y) = y + x \ln x$

14. $g(x, y) = xy \ln x^2$

Compute $f(0, 1, 2)$ for each function in Exercises 15–18.

15. $f(x, y, z) = x^2 + y^2 - z^2 + 7$

16. $f(x, y, z) = \sqrt{xyz + 1}$

17. $f(x, y, z) = \dfrac{x + y}{2y}$

18. $f(x, y, z) = e^x + 5 \ln y - 3z$

19. (*COMPOUND INTEREST*) If an amount p dollars (principal) is invested at annual interest rate $r\%$ compounded continuously for t years, the total amount accumulated is given by the function

$$f(p, r, t) = pe^{.01rt}$$

The factor .01 is placed in the exponent in order to convert the percent ($r\%$) to a decimal form.
(a) Determine $f(1000, 8, 2)$.

(b) How much money will accumulate if the account begins with $3000, the interest rate is 10%, and the money is left in the account for 7 years?

20. *(COMPOUND INTEREST)* Using $f(p, r, t) = pe^{.01rt}$, find
   (a) $f(100, 6, 5)$
   (b) $f(2000, 9, 3)$

21. *(PRODUCTION FUNCTION)* Given the Cobb-Douglas production function $f(x, y) = 20x^{.4}y^{.6}$, determine the number of units produced when using 2400 units of labor and 4000 units of capital.

22. *(PRODUCTION FUNCTION)* Given the Cobb-Douglas production function $f(x, y) = 4x^{1/3}y^{2/3}$, determine the number of units produced when 60 units of labor and 150 units of capital are used.

23. *(IQ)* A person's IQ is defined as

$$f(a, m) = 100 \cdot \frac{m}{a}$$

where a is the person's actual age in years and m is the person's mental age (based on test results) in years.
   (a) What is the IQ of a 10-year-old child whose test score suggests a mental age of 13?
   (b) What is the IQ of a 10-year-old child whose test score yields a mental age of 8?
   W (c) According to the function defined here, what is the meaning of an IQ of 100?

24. *(POINTS IN FOOTBALL)* The number of points scored by a kicker in football depends on the number of field goals (3 points each) and points after touchdown (1 point each) made by the kicker. Determine the total point function $P(x, y)$ for a player who makes x field goals and y point-after-touchdown kicks.

25. *(POINTS IN BASKETBALL)* The number of points scored by a basketball player depends on the number of free throws (1 point each), regular field goals (2 points each), and "long" field goals (3 points each) made by the player. Determine the total point function $P(x, y, z)$ for a player who makes x free throws, y regular field goals, and z long field goals.

26. *(CEPHALIC INDEX)* Refer to Example 5. What is the cephalic index of a person whose head width is the same as its length?

27. *(POISEUILLE'S LAW)* When Poiseuille's law is applied to the flow of blood in capillaries, the resistance to blood flow is given by

$$R(l, r) = k \cdot \frac{l}{r^4}$$

where k is a constant, l is the length of capillary, and r is the radius of the capillary (see figure).
   (a) Compute $R(64, 2)$ and $R(64, 1)$.
   W (b) What happens to the resistance within the capillary when the radius of the capillary is reduced to half its original size?

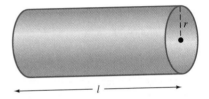

28. *(FLAG COST)* A souvenir maker produces two sizes of flags. The large flags cost $4 each to produce, and the small flags cost $3 each to produce. Determine the cost function $C(x, y)$ for making x large flags and y small flags.

29. *(FLAG PROFIT)* Assume that the souvenir maker of Exercise 28 sells large flags for $6.50 each and small ones for $5 each.
   (a) Determine the revenue function.
   (b) Determine the profit function.

30. *(VOLUME)* The volume of a cylinder is $\pi r^2 h$, where r is the radius and h is the height (see figure). We can write $V(r, h) = \pi r^2 h$.
   (a) Compute $V(10, 3)$.
   W (b) Explain in words the meaning of $V(10, 3)$.
   (c) If the radius and height are known to be the same, that is, $r = h$, express the volume of the cylinder as a function of one variable. In other words, determine a function $V(r)$ or $V(h)$.

31. *(VOLUME OF A BOX)* The volume of a rectangular box is computed as length times width times height.
   (a) If a rectangular box has length x, width y, and height z, obtain the function $V(x, y, z)$ that gives its volume (see figure).

(b) If the height z of the box [mentioned in part (a)] is twice the width y, obtain the function $V(x, y)$ that expresses the volume of the box in terms of variables x and y only.

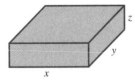

32. (*COST OF A BOX*) If the box in Exercise 31 has a top, the surface area of the box (that is, the *area* of all four sides plus the top and bottom) is

$$S(x, y, z) = 2xy + 2xz + 2yz$$

If the cost to make such a box is 5¢ per square unit for the sides and 7¢ per square unit for the top and bottom, determine the function $C(x, y, z)$, the cost in cents of making such a box.

In Exercises 33–50, determine the domain of each function.

33. $f(x, y) = x + y$

34. $f(x, y) = 5x - 2y + 3$

35. $f(x, y) = x^2 + y^2 - 9$

36. $f(x, y) = 20 - x^2 + y^2$

37. $z = \sqrt{y - x}$

38. $z = \sqrt{3x - y}$

39. $z = \dfrac{x}{1 - y}$

40. $z = \dfrac{2y + 9}{3x}$

41. $f(x, y) = \dfrac{9}{7xy}$

42. $f(x, y) = \dfrac{1}{y - 2x}$

43. $g(x, y) = \dfrac{x^2 + y^2}{x - y}$

44. $g(x, y) = x\sqrt{y - 4}$

45. $f(x, y) = \dfrac{\sqrt{5x}}{2y + 3}$

46. $f(x, y) = \dfrac{\sqrt{y}}{1 - 9x}$

47. $z = 5xe^y$

48. $z = e^{x-y}$

49. $z = y \ln (x - 1)$

50. $z = 3x \ln (y + 2)$

51. Using $c = 1, 3, 5,$ and 7, obtain level curves for the function $f(x, y) = 9 - x^2 - y^2$ of Example 8.

W 52. Describe the level "curve" obtained for the function $f(x, y) = 9 - x^2 - y^2$ if $c = 9$ is used.

In Exercises 53–66, sketch the level curves of the given function for the specific c values listed.

53. $f(x, y) = 16 - x^2 - y^2$; $c = 0, 7, 12, 15$

54. $f(x, y) = 10 - x^2 - y^2$; $c = 1, 6, 9$

55. $f(x, y) = x^2 - y + 1$; $c = 0, 1, 2$

56. $f(x, y) = x^2 - y - 1$; $c = 0, 1, 2, 3$

57. $f(x, y) = x - y + 2$; $c = 0, 2, 4, 6$

58. $f(x, y) = x + y$; $c = -2, 0, 2, 4$

59. $f(x, y) = 1 - x - y$; $c = 1, 3, 5, 7$

60. $f(x, y) = 3 - x - y$; $c = 0, 1, 2, 3$

61. $f(x, y) = \dfrac{y}{x}$; $c = 1, 2, 3, 4$

62. $f(x, y) = -\dfrac{y}{x}$; $c = 1, 2, 3, 4$

63. $f(x, y) = e^x - y$; $c = -3, -2, -1, 0$

64. $f(x, y) = e^x + y$; $c = 0, 1, 2, 3$

65. $f(x, y) = \ln x - y$; $c = -2, -1, 0, 1$

66. $f(x, y) = \ln x + y$; $c = -1, 0, 1, 2$

The Exercise Library at the back of the book contains graphing calculator and computer exercises keyed to this section.

8.2 | *PARTIAL DERIVATIVES*

We have covered a variety of interpretations and applications of the derivative of functions of one variable. One interpretation is that the derivative of f is the rate of change of $f(x)$ compared with the change in the variable x. Recall the definition of the derivative of $f(x)$ with respect to x,

$$f'(x) = \lim_{\Delta x \to 0} \frac{f(x + \Delta x) - f(x)}{\Delta x}$$

To extend the rate of change interpretation to a function $f(x, y)$ of two variables, we consider how the function values change as x changes (that is, with y held constant) and how the function values change as y changes (that is, with x held constant). There are two separate derivatives, called **partial derivatives**. First, consider informal definitions.

Partial Derivatives—Informal Definitions

1. The **partial derivative of $f(x, y)$ with respect to x** is obtained by differentiating $f(x, y)$ with respect to x, treating y as a constant. The notation for this partial derivative is f_x or $\partial f / \partial x$.

2. The **partial derivative of $f(x, y)$ with respect to y** is obtained by differentiating $f(x, y)$ with respect to y, treating x as a constant. The notation for this partial derivative is f_y or $\partial f / \partial y$.

Next we give the formal definitions.

Partial Derivatives—Formal Definitions

1. The **partial derivative of $f(x, y)$ with respect to x** is

$$\frac{\partial f}{\partial x} = \lim_{\Delta x \to 0} \frac{f(x + \Delta x, y) - f(x, y)}{\Delta x}$$

In $\partial f / \partial x$, y is held constant. This means there is no change in y and no Δy appears. The change (Δx) is in x only.

2. The **partial derivative of $f(x, y)$ with respect to y** is

$$\frac{\partial f}{\partial y} = \lim_{\Delta y \to 0} \frac{f(x, y + \Delta y) - f(x, y)}{\Delta y}$$

In $\partial f / \partial y$, x is held constant. This means there is no change in x and no Δx appears. The change (Δy) is in y only.

EXAMPLE 1   Use the formal definition to determine $\partial f/\partial x$ for $f(x, y) = x^2 y$.

SOLUTION

$$\frac{\partial f}{\partial x} = \lim_{\Delta x \to 0} \frac{f(x + \Delta x, y) - f(x, y)}{\Delta x}$$

$$= \lim_{\Delta x \to 0} \frac{(x + \Delta x)^2 y - x^2 y}{\Delta x}$$

$$= \lim_{\Delta x \to 0} \frac{[x^2 + 2x(\Delta x) + (\Delta x)^2] y - x^2 y}{\Delta x}$$

$$= \lim_{\Delta x \to 0} \frac{x^2 y + 2x(\Delta x)y + (\Delta x)^2 y - x^2 y}{\Delta x}$$

$$= \lim_{\Delta x \to 0} \frac{2x(\Delta x)y + (\Delta x)^2 y}{\Delta x}$$

$$= \lim_{\Delta x \to 0} \frac{\Delta x[2xy + (\Delta x)y]}{\Delta x}$$

$$= \lim_{\Delta x \to 0} [2xy + (\Delta x)y]$$

$$= 2xy \quad \blacklozenge$$

Note

We do not actually need to use the formal definition to determine partial derivatives. Since we already know the rules for ordinary differentiation, all we need to do is keep in mind which variable is being treated as a constant.

The derivative determined in Example 1 can be obtained using the informal definition, as demonstrated in the next example.

EXAMPLE 2   If $f(x, y) = x^2 y$, determine $\partial f/\partial x$.

SOLUTION   Since we seek the partial derivative of $f(x, y)$ with respect to x, consider y to be a constant. Then

$$\frac{\partial}{\partial x}(x^2 y) = 2x \cdot y = 2xy \qquad \text{(Answer)}$$

Since y is treated as a constant, the derivative of $x^2 y$ is y times the derivative of x^2, or y times $2x$. As an alternative, you may prefer to factor out the "constant" y, placing it in front of the $\partial/\partial x$ before continuing the differentiation.

$$\frac{\partial}{\partial x}(x^2 y) = y \cdot \frac{\partial}{\partial x} x^2 = y \cdot 2x = 2xy \quad \blacklozenge$$

EXAMPLE 3 If $f(x, y) = x^2y$, determine $\partial f/\partial y$.

SOLUTION Here x is treated as a constant. This means x^2 is a constant. So we are differentiating a constant times y.

$$\frac{\partial f}{\partial y} = \frac{\partial}{\partial y}(x^2y) = x^2 \cdot 1 = x^2 \quad \blacklozenge$$

Here is another example, this time using the notation f_x and f_y for the partial derivatives.

EXAMPLE 4 Let $f(x, y) = x^3 + x^2y^4$.   Find **(a)** f_x **(b)** f_y

SOLUTION **(a)** $f_x = 3x^2 + 2x \cdot y^4 = 3x^2 + 2xy^4$

Keep in mind that y is treated as a constant when determining f_x.

(b) $f_y = 0 + x^2 \cdot 4y^3 = 4x^2y^3$

Keep in mind that x is treated as a constant when determining f_y. $\blacklozenge$

EXAMPLE 5 Determine $\partial f/\partial x$ and $\partial f/\partial y$ for each function.

     (a) $f(x, y) = xe^y$ **(b)** $f(x, y) = \ln x$

SOLUTION **(a)** $\dfrac{\partial}{\partial x}(xe^y) = e^y$

            $\dfrac{\partial}{\partial y}(xe^y) = xe^y$

      (b) $\dfrac{\partial}{\partial x}(\ln x) = \dfrac{1}{x}$

            $\dfrac{\partial}{\partial y}(\ln x) = 0$ $\blacklozenge$

Geometric Interpretation of the Partial Derivative

The geometric interpretation of partial derivatives resembles the original geometric interpretation of the derivative of a function that was presented in Section 3.1. For $z = f(x, y)$, $\partial f/\partial y$ is the derivative of $f(x, y)$ with respect to y, treating x as a constant. Thus, $\partial f/\partial y$ is the slope of a line tangent to the graph of the surface $z = f(x, y)$. Since x is held constant, the tangent line lies in a plane parallel to the yz plane. See Figure 11. Similarly, $\partial f/\partial x$ is the slope of a line tangent to the graph of $z = f(x, y)$. In this case y is held constant, so the tangent line lies in a plane parallel to the xz plane. See Figure 12.

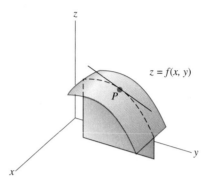

Figure 11 The slope at point P is $\partial f/\partial y$ at P

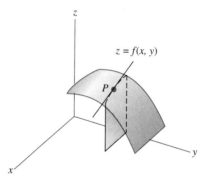

Figure 12 The slope at point P is $\partial f/\partial x$ at P

Evaluating a Partial Derivative

Sometimes it is necessary to evaluate a partial derivative of $f(x, y)$ at a point (a, b), just as we have previously evaluated ordinary derivatives of $f(x)$ at a number a. The partial derivatives of $f(x, y)$ evaluated at (a, b) are written as

$$\frac{\partial f}{\partial x}(a, b) \quad \text{or} \quad f_x(a, b)$$

$$\frac{\partial f}{\partial y}(a, b) \quad \text{or} \quad f_y(a, b)$$

EXAMPLE 6   For $f(x, y) = x^2 + 3xy + y + 7$, determine $f_x(5, 8)$ and $f_y(5, 8)$.

SOLUTION
$$f_x = 2x + 3y, \text{ which means } f_x(5, 8) = 2(5) + 3(8) = 34$$
$$f_y = 3x + 1, \text{ which means } f_y(5, 8) = 3(5) + 1 = 16 \quad \blacklozenge$$

With the use of partial differentiation, information analogous to marginal cost and marginal revenue can be obtained from cost functions $C(x, y)$ and revenue functions $R(x, y)$.

EXAMPLE 7 ◆ REVENUE

Suppose that revenue from the sale of x model A10 stereo speakers and y model A20 stereo speakers is given by

$$R(x, y) = 100x + 150y - .03x^2 - .02y^2 \quad \text{dollars}$$

Determine the rate at which revenue will change with respect to the change in the number of model A10 speakers sold, when 50 model A10 speakers and 40 model A20 speakers have been sold.

SOLUTION The rate of change we seek is

$$\frac{\partial R}{\partial x}(50, 40)$$

Since

$$R(x, y) = 100x + 150y - .03x^2 - .02y^2$$

it follows that

$$\frac{\partial R}{\partial x} = 100 - .06x$$

and thus

$$\frac{\partial R}{\partial x}(50, 40) = 100 - .06(50) = 97 \qquad \text{(answer)}$$

The answer (97) means that the additional revenue that will be obtained from the sale of the next A10 speaker is approximately $97, assuming that 50 A10 speakers and 40 A20 speakers have been sold. ◆

A function $f(x, y, z)$ of three variables has partial derivatives f_x, f_y, and f_z. The partial derivative f_x is determined by treating both y and z as constants. In general, to determine the partial derivative with respect to any one of the variables, consider the other two variables as constants.

EXAMPLE 8 Let $f(x, y, z) = x^2y^3z + x \ln z + y$. Determine

(a) f_x (b) f_y (c) f_z

SOLUTION (a) $f_x = 2x \cdot y^3z + 1 \cdot \ln z + 0 = 2xy^3z + \ln z$

(b) $f_y = x^2z \cdot 3y^2 + 0 + 1 = 3x^2y^2z + 1$

(c) $f_z = x^2y^3 \cdot 1 + x \cdot \dfrac{1}{z} + 0 = x^2y^3 + \dfrac{x}{z}$ ◆

Higher-Order Partial Derivatives

Just as there are higher-order ordinary derivatives, there are higher-order partial derivatives. We will now consider second partial derivatives of a function $f(x, y)$ of two variables.

To begin, there are two first partial derivatives, $\partial f/\partial x$ and $\partial f/\partial y$. (Using alternative notation, they are f_x and f_y.) Each of these two derivatives can be differentiated with respect to either x or y to obtain a second partial derivative. This means there are four second partial derivatives, as shown below.

Second Partial Derivatives

$$\frac{\partial}{\partial x}\left(\frac{\partial f}{\partial x}\right) = \frac{\partial^2 f}{\partial x^2} = f_{xx}$$

$$\frac{\partial}{\partial x}\left(\frac{\partial f}{\partial y}\right) = \frac{\partial^2 f}{\partial x \partial y} = f_{yx}$$

$$\frac{\partial}{\partial y}\left(\frac{\partial f}{\partial x}\right) = \frac{\partial^2 f}{\partial y \partial x} = f_{xy}$$

$$\frac{\partial}{\partial y}\left(\frac{\partial f}{\partial y}\right) = \frac{\partial^2 f}{\partial y^2} = f_{yy}$$

Notice a possible source of confusion with the so-called "mixed partials" f_{yx} and f_{xy}.

f_{yx} means that f is differentiated *first* with respect to y and *then* with respect to x. This notation follows from $(f_y)_x = f_{yx}$.

and

$\dfrac{\partial^2 f}{\partial x \partial y}$ means that f is differentiated *first* with respect to y and *then* with respect to x.

This means that

$$f_{yx} = \frac{\partial^2 f}{\partial x \partial y} \quad \text{and} \quad f_{xy} = \frac{\partial^2 f}{\partial y \partial x}$$

EXAMPLE 9   Let $f(x, y) = x^3 + 4xy^2 + 15$. Find all four second partials.

SOLUTION

$$\frac{\partial f}{\partial x} = 3x^2 + 4y^2 \quad \text{and} \quad \frac{\partial f}{\partial y} = 8xy$$

Now,

$$f_{xx} = \frac{\partial^2 f}{\partial x^2} = \frac{\partial}{\partial x}\left(\frac{\partial f}{\partial x}\right) = \frac{\partial}{\partial x}(3x^2 + 4y^2) = 6x$$

$$f_{yy} = \frac{\partial^2 f}{\partial y^2} = \frac{\partial}{\partial y}\left(\frac{\partial f}{\partial y}\right) = \frac{\partial}{\partial y}(8xy) = 8x$$

$$f_{yx} = \frac{\partial^2 f}{\partial x \partial y} = \frac{\partial}{\partial x}\left(\frac{\partial f}{\partial y}\right) = \frac{\partial}{\partial x}(8xy) = 8y$$

$$f_{xy} = \frac{\partial^2 f}{\partial y \partial x} = \frac{\partial}{\partial y}\left(\frac{\partial f}{\partial x}\right) = \frac{\partial}{\partial y}(3x^2 + 4y^2) = 8y \quad \blacklozenge$$

Notice that $f_{yx} = f_{xy}$ in Example 9. As a matter of fact, for all functions $z = f(x, y)$ used in this book, it will be true that $f_{yx} = f_{xy}$.

This concludes the introduction to partial derivatives. These concepts will be used in a variety of applications throughout the remainder of the chapter.

8.2 Exercises

Determine both $\partial f/\partial x$ and $\partial f/\partial y$ for each function in Exercises 1–20. Use the informal definition (as done in Examples 2–5).

1. $f(x,y) = 2x + 5y$ **2.** $f(x,y) = 2xy$

3. $f(x,y) = x^3 - 4y^2$ **4.** $f(x,y) = x - x^2y^2 + y^3$

5. $f(x,y) = \dfrac{x}{y}$ **6.** $f(x,y) = \dfrac{y^3}{x^2}$

7. $f(x,y) = ye^x + 1$ **8.** $f(x,y) = x^2e^y$

9. $f(x,y) = y \ln x$ **10.** $f(x,y) = e^x \ln y$

11. $f(x,y) = x^2 \ln y^3$ **12.** $f(x,y) = \ln xy$

13. $f(x,y) = e^{3xy}$ **14.** $f(x,y) = 1 - e^{-xy}$

15. $f(x,y) = x\sqrt{y}$ **16.** $f(x,y) = \sqrt{xy}$

17. $f(x,y) = \sqrt{x^2 + y^2}$ **18.** $f(x,y) = x^2\sqrt{1 + y^2}$

19. $f(x,y) = \dfrac{x}{x+y}$ **20.** $f(x,y) = \dfrac{xy}{x+y}$

Determine f_x and f_y for each function in Exercises 21–30. Use the informal definition (as done in Examples 2–5).

21. $f(x,y) = x^3y^5$ **22.** $f(x,y) = 5x^2 - 2y^3$

23. $f(x,y) = 1 - xy^4$ **24.** $f(x,y) = xy^3 + x^3y$

25. $f(x,y) = \ln(x^2 + y^2)$ **26.** $f(x,y) = e^{x^2+y^2}$

27. $f(x,y) = xe^{-y}$ **28.** $f(x,y) = y\sqrt{x}$

29. $f(x,y) = \sqrt{xy^3}$ **30.** $f(x,y) = \dfrac{x+y}{x-y}$

In Exercises 31–38, use the *formal definition* of the partial derivative to determine $\partial f/\partial x$ for each function. Refer to Example 1.

31. $f(x,y) = xy$ **32.** $f(x,y) = xy^2$

33. $f(x,y) = x^2$ **34.** $f(x,y) = y^2$

35. $f(x,y) = 3y$ **36.** $f(x,y) = 5x$

37. $f(x,y) = x^2y^2$ **38.** $f(x,y) = x^2y^2 + 1$

39–46. Use the functions of Exercises 31–38, but determine $\partial f/\partial y$ instead. Again, use the *formal definition* of partial derivative.

47. Let $f(x, y) = x^2y^2 - 3x + 2y$. Determine

(a) $f_x(5, 4)$ (b) $f_y(-1, 8)$

(c) $\dfrac{\partial f}{\partial x}(1, -5)$ (d) $\dfrac{\partial f}{\partial y}(0, 9)$

48. Let $f(x, y) = xe^y + x^3y$. Determine

(a) $f_x(4, 3)$ (b) $f_y(3, 0)$

(c) $\dfrac{\partial f}{\partial x}(-1, 6)$ (d) $\dfrac{\partial f}{\partial y}(2, 1)$

49. Let $g(x, y) = x \ln y + xy$. Determine

(a) $g_y(8, 2)$ (b) $g_x(3, 1)$

(c) $\dfrac{\partial g}{\partial x}(4, e)$ (d) $\dfrac{\partial g}{\partial y}(2e, e)$

50. Let $g(x, y) = y^2 \ln x$. Determine

(a) $g_y(e, 3)$ (b) $g_x(2, 10)$

(c) $\dfrac{\partial g}{\partial x}(1, e)$ (d) $\dfrac{\partial g}{\partial y}(1, e)$

For each function in Exercises 51–60, find f_x, f_y, and f_z.

51. $f(x,y,z) = x^2 + y^2 + z^2$ **52.** $f(x,y,z) = x^2y^2z^2$

53. $f(x,y,z) = xyz - x + y$ **54.** $f(x,y,z) = \dfrac{\ln x}{y} + z$

55. $f(x,y,z) = xye^z$ **56.** $f(x,y,z) = xz\sqrt{y}$

57. $f(x,y,z) = \sqrt{2x + 2y + 2z}$

58. $f(x,y,z) = 3x^{-1}y^2z$

59. $f(x,y,z) = \dfrac{x}{y+z}$ **60.** $f(x,y,z) = \dfrac{xy}{z^2}$

For each function in Exercises 61–70, find all four second partials.

61. $f(x,y) = x^2 + xy + y^2$ **62.** $f(x,y) = x^3 + x^2y^2 - 10$

63. $f(x,y) = y \ln x$ **64.** $f(x,y) = x^2 + \ln y$

65. $g(x,y) = 3xe^y$ **66.** $g(x,y) = e^{xy}$

67. $h(x,y) = x^2 \ln y$ **68.** $h(x,y) = \dfrac{x}{y}$

69. $f(x,y) = \ln(x + y^2)$ **70.** $f(x,y) = \sqrt{x - y}$

71. *(REVENUE)* Refer to Example 7.
(a) Determine the approximate additional revenue obtained from the sale of the next A20 speaker, assuming that 50 A10 speakers and 40 A20 speakers have been sold.
(b) Determine the approximate additional revenue obtained from the sale of the next A10 speaker, assuming that 60 A10 speakers and 38 A20 speakers have been sold.

72. *(COST)* Suppose that the cost of producing x amplifiers and y receivers is given by

$$C(x, y) = .1x^2 + .2y^2 + 90x + 140y$$

Assume that 30 amplifiers and 70 receivers have been produced.
(a) What is the approximate cost of producing the 31st amplifier, assuming that the production of receivers remains at 70?
(b) What is the approximate cost of producing the 71st receiver, assuming that the production of amplifiers remains at 30?

73. *(PRODUCTION FUNCTION)* Consider the Cobb-Douglas production function $f(x, y) = 2x^{.4}y^{.6}$, where x is the number of units of labor and y is the number of units of capital. (See Example 6 in Section 8.1, if needed.)
(a) Determine $\partial f/\partial x$, the *marginal productivity of labor*.
(b) Determine the *marginal productivity of capital*.

74. *(PRODUCTION FUNCTION)* Redo Exercise 73 using $f(x, y) = 20x^{2/3}y^{1/3}$.

75. *(PRODUCTION FUNCTION)* Consider $f(x, y) = 20x^{1/4}y^{3/4}$, where x is the number of units of labor, y is the number of units of capital, and $f(x, y)$ is the number of units produced. Assume that the current production

level uses 256 units of labor and 81 units of capital. If the number of units of capital is held fixed and labor is increased by one unit, determine the approximate increase in the number of units produced—the *marginal productivity of labor*.

76. *(PRODUCTION FUNCTION)* Refer to Exercise 75. If the number of units of labor is held fixed and the number of units of capital is increased by 1, determine the approximate increase in the number of units produced—the *marginal productivity of capital*.

77. *(TEMPERATURE VARIATION)* A thin metal plate is being heated in such a way that the temperature T at any point (x, y) on the surface of the plate is given by

$$T(x, y) = 350 - x^2 - y^2$$

T is measured in degrees, and x and y are measured in centimeters (see figure).

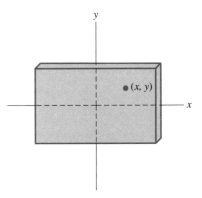

(a) What is the temperature at the origin?
W (b) Explain, based on the equation for temperature, why the temperature will be greatest at the origin.
(c) Determine the rate of change of temperature with respect to distance at the point $(10, 6)$, assuming x is allowed to vary and y is held constant.
(d) Determine the rate of change of temperature with respect to distance at the point $(10, 6)$, assuming y is allowed to vary and x is held constant.
W (e) Explain why the rates of change in parts (c) and (d) are negative.

78. *(COST)* Suppose the cost of making a product is given by $C(x, y) = 15 + 3x + xy + 8y$, where x is the number of units of labor and y is the number of units of capital.

(a) Determine the rate of change of cost with respect to labor when $x = 5$. Assume the number of units of capital is kept constant at $y = 3$.

(b) Determine the rate of change of cost with respect to capital when $y = 3$. Assume the number of units of labor is kept constant at $x = 5$.

79. *(Volume)* Determine the instantaneous rate of change of the volume of a cylinder with respect to its height, if the radius remains constant $(V = \pi r^2 h)$.

80. *(Volume)* Determine the instantaneous rate of change of the volume of a cone with respect to its radius, if the height remains constant $(V = \frac{1}{3}\pi r^2 h)$.

81. *(Gas Law)* The ideal gas law states that

$$P = k \cdot \frac{T}{V}$$

where P is the pressure exerted by the gas, T is the temperature of the gas, and V is the volume of the gas. The number k is a constant.

(a) Determine the rate of change of pressure with respect to temperature, assuming the volume is kept constant.

(b) Determine $\partial P/\partial V$.

W (c) Tell in words what $\partial P/\partial V$ represents.

82. *(Ohm's Law)* Ohm's law gives a formula for the relationship of the electric current I (in amperes), resistance R (in ohms), and voltage V (in volts) in an electric circuit.

$$I = \frac{V}{R}$$

(a) Find the rate of change of current with respect to voltage when resistance is held constant.

(b) Find the rate of change of current with respect to resistance when voltage is kept constant.

(c) Find the rate of change of voltage with respect to resistance when current is kept constant.

83. *(Physiology)* Physiological concern about body heat loss led to the development of a practical formula for measuring the surface area S of a person's body based on the individual's weight w and height h.

$$S(w, h) = .0072w^{.425}h^{.725} \text{ square meters}$$

where w is in kilograms and h is in centimeters.

(a) Determine $S(80, 178)$. *Use a calculator.*

(b) Find $S_w(w, h)$ and $S_h(w, h)$.

(c) Find $S_w(80, 178)$ and $S_h(80, 178)$. *Use a calculator.*

W (d) Explain the meaning of S_w and S_h.

8.3 | MAXIMUM AND MINIMUM

Since the graph of $z = f(x, y)$ is a surface in three dimensions, a *relative maximum point* stands out as a peak at the top of a hill. A *relative minimum point* stands out as being the bottom of a valley. See Figure 13.

In Chapter 4 we began the search for relative extrema of functions of one variable by finding critical numbers. Now, with functions of two variables, we will seek *critical points* (a, b), since we want the x and y values needed to maximize or minimize a function $z = f(x, y)$. Note how the definition given next is an extension of the definition of critical number given in Section 4.1.

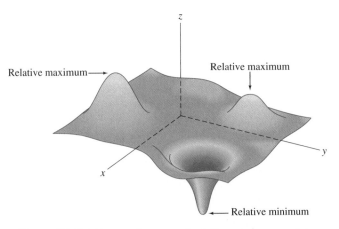

Figure 13 Relative maximum and relative minimum points of a surface $z = f(x, y)$

Critical Point

Let $z = f(x, y)$ be defined at point (a, b). Then (a, b) is a **critical point** of $f(x, y)$ if $f_x(a, b) = 0$ *and* $f_y(a, b) = 0$.

(Also, (a, b) is a critical point if either $f_x(a, b)$ or $f_y(a, b)$ fails to exist. However, we will not study such situations.)

With functions of one variable, the derivative is the slope of the tangent line. With functions of two variables, there are two tangent lines to consider. The partial derivatives f_x and f_y are the slopes of these two tangent lines. For (a, b) to be a critical point, we insist that $f_x(x, y)$ and $f_y(x, y)$ both be zero at (a, b). There must be two horizontal tangent lines at (a, b), as shown in Figure 14.

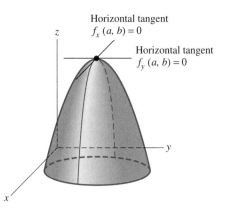

Figure 14 A critical point

EXAMPLE 1   Find the critical points of $f(x, y) = 8x + 2y - x^2 + 2y^2$.

SOLUTION   First, find the partial derivatives.

$$f_x(x, y) = 8 - 2x$$
$$f_y(x, y) = 2 + 4y$$

Next, solve the system of equations created by setting each partial derivative equal to zero. In other words, solve the system

$$\begin{cases} 8 - 2x = 0 \\ 2 + 4y = 0 \end{cases}$$

Since each equation contains only one variable, the solution is readily obtained. From $8 - 2x = 0$, we get $x = 4$. From $2 + 4y = 0$, we get $y = -1/2$. The solution, then, is

$$x = 4, y = -\frac{1}{2}$$

Thus, the critical point is $(4, -1/2)$.   ◆

EXAMPLE 2   Find the critical points of $f(x, y) = x^2 - xy + y^2 + 4x - 23y$.

SOLUTION   First, find the partial derivatives.

$$f_x(x, y) = 2x - y + 4 \qquad \text{and} \qquad f_y(x, y) = 2y - x - 23$$

Next, solve the system

$$\begin{cases} 2x - y + 4 = 0 \\ 2y - x - 23 = 0 \end{cases}$$

We can write the first equation as $y = 2x + 4$ and then substitute $2x + 4$ for y in the second equation.

$$2(2x + 4) - x - 23 = 0$$
$$4x + 8 - x - 23 = 0$$
$$3x - 15 = 0$$
$$x = 5$$

Since $y = 2x + 4$, we know that

$$y = 2(5) + 4$$
$$y = 14$$

Thus, $(5, 14)$ is a critical point, the only critical point of this function.   ◆

Just as with functions of one variable, critical points do not always lead to relative extrema. Figure 15 shows the graph of a function f such that $f_x(a, b) = 0$ and $f_y(a, b) = 0$. Yet the point $(a, b, f(a, b))$ is not a relative maximum or minimum. The point is called a **saddle point**. The saddle-shaped surface is produced if there is a minimum when approaching from one direction and a maximum when approaching from another direction.

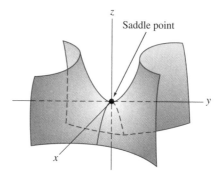

Figure 15 A saddle point

There is a second derivative test that can be used to determine whether critical point (a, b) is associated with a relative maximum, a relative minimum, or neither.

Second Partials Test

For $z = f(x, y)$, if $f_x(a, b) = 0$ and $f_y(a, b) = 0$, then consider

$$D = (f_{xx})(f_{yy}) - (f_{xy})^2 \qquad \text{evaluated at } (a, b)$$

1. If $D > 0$ and $f_{xx} > 0$, then f has a *relative minimum* at (a, b).

2. If $D > 0$ and $f_{xx} < 0$, then f has a *relative maximum* at (a, b).

3. If $D < 0$, then f has a *saddle point* at (a, b).

4. If $D = 0$, then no conclusion can be drawn.

Note

As an aid in remembering the second partials test, consider that when $D > 0$, the test for relative maximum or relative minimum comes down to whether $f_{xx} > 0$ (minimum) or $f_{xx} < 0$ (maximum), which resembles the second derivative test for functions of one variable.

EXAMPLE 3   Determine the relative extrema of $f(x, y) = 3x^3 + y^2 - 9x - 6y + 1.$

SOLUTION First, determine the critical points—points (x, y) for which both $f_x = 0$ and $f_y = 0$. In this example, $f_x = 9x^2 - 9$ and $f_y = 2y - 6$. We then need to solve the system

$$\begin{cases} 9x^2 - 9 = 0 \\ 2y - 6 = 0 \end{cases}$$

Solving the system yields $x = \pm 1$ and $y = 3$. Thus, the critical points are $(1, 3)$ and $(-1, 3)$.

To apply the second partials test, we need to determine D for the points $(1, 3)$ and $(-1, 3)$. So we will evaluate

$$D = (f_{xx})(f_{yy}) - (f_{xy})^2$$

Note that for the function of this example, $f_{xx} = 18x$, $f_{yy} = 2$, and $f_{xy} = 0$.

For (1, 3), we have

$$\begin{aligned} D &= f_{xx}(1, 3) \cdot f_{yy}(1, 3) - [f_{xy}(1, 3)]^2 \\ &= 18(1) \cdot 2 - 0 \\ &= 36 \end{aligned}$$

Thus, $D > 0$ and $f_{xx}(1, 3) = 18 > 0$. Therefore, $f(1, 3)$ is a relative minimum (by case 1 of the four cases given in the second partials test). The actual relative minimum functional value is -14, determined by using 1 for x and 3 for y in the original function.

$$f(1, 3) = 3(1)^3 + (3)^2 - 9(1) - 6(3) + 1 = -14$$

For (-1, 3), we have

$$\begin{aligned} D &= f_{xx}(-1, 3) \cdot f_{yy}(-1, 3) - [f_{xy}(-1, 3)]^2 \\ &= 18(-1) \cdot 2 - 0 \\ &= -36 \end{aligned}$$

Thus, $D < 0$. By case 3 of the second partials test, this means that $f(-1, 3)$ is neither a relative maximum nor a relative minimum. Rather, f has a saddle point at $(-1, 3, -2)$, with -2 being the value of $f(-1, 3)$. ◆

The examples that follow show applications of the ideas that have been presented.

EXAMPLE 4 ◆ **MAXIMUM PROFIT**

A company makes two kinds of umbrellas. One kind sells for $8, and the other sells for $12. The revenue (in hundreds of dollars) from the sale of x hundred $8 umbrellas and y hundred $12 umbrellas is $R = 8x + 12y$. Suppose the cost of making the umbrellas is given by $C = x^2 - xy + y^2 - 4x + 6y + 2$ hundred dollars. Determine how many of each kind of umbrella should be made in order to maximize profit.

SOLUTION   Since profit = revenue − cost, we have

$$P(x, y) = (8x + 12y) − (x^2 − xy + y^2 − 4x + 6y + 2)$$

or

$$P(x, y) = −x^2 + xy − y^2 + 12x + 6y − 2$$

To determine the critical points of function P, obtain the partial derivatives.

$$P_x = −2x + y + 12$$
$$P_y = x − 2y + 6$$

Solving the equations $P_x = 0$ and $P_y = 0$,

$$\begin{cases} −2x + y + 12 = 0 \\ x − 2y + 6 = 0 \end{cases}$$

yields $x = 10$ and $y = 8$. Thus, $(10, 8)$ is the only critical point. We now test it to be sure it produces a maximum value for P. To evaluate D, first determine the values of P_{xx}, P_{yy}, and P_{xy} at $(10, 8)$.

$$P_{xx}(10, 8) = −2, \qquad P_{yy}(10, 8) = −2 \qquad P_{xy}(10, 8) = 1$$

Now,

$$D = P_{xx}(10, 8) \cdot P_{yy}(10, 8) − [P_{xy}(10, 8)]^2$$
$$= (−2)(−2) − (1)^2$$
$$= 3$$

Since $D > 0$ and $P_{xx}(10, 8) < 0$, P does indeed have a *maximum* value when $x = 10$ and $y = 8$. Keeping in mind that the numbers 10 and 8 represent *hundreds* of umbrellas, we conclude that the company should make 1000 of the $8 umbrellas and 800 of the $12 umbrellas in order to maximize its profit.   ◆

Note

In the applied problems of this section, the relative maximum or minimum will always be the absolute maximum or minimum as well. You will not have to check further. It is beyond the scope of this book to present such methods of verification.

EXAMPLE 5   ◆   *MINIMUM COST*

A manufacturer of aquariums wants to make a large rectangular box-shaped aquarium that will hold 64 cubic feet of water. If the material for the base costs

$20 per square foot and the material for the sides costs $10 per square foot, find the dimensions for which the cost of the materials will be the least.

SOLUTION   Let the length, width, and height of the aquarium be x feet, y feet, and z feet, respectively. (See Figure 16.)

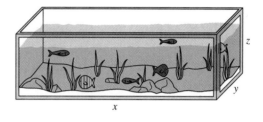

Figure 16

The area of the base is $x \cdot y$ square feet, and the cost per square foot is $20. This means the base costs $20xy$ dollars to make. Similarly, there are two sides of area xz square feet and two sides of area yz square feet. So the total area of all four sides is $2xz + 2yz$. Since their cost per square foot is $10, the cost of all four sides is $10(2xz + 2yz)$ dollars. Putting all this together, we see that the total cost of materials is

$$C(x, y, z) = 20xy + 10(2xz + 2yz)$$

or

$$C(x, y, z) = 20xy + 20xz + 20yz$$

This function of three variables can be changed to a function of two variables by using the fact that the volume $(x \cdot y \cdot z)$ must be 64 cubic feet.

$$xyz = 64$$

can be written as

$$z = \frac{64}{xy}$$

and then z can be replaced by $64/xy$ in $C(x, y, z) = 20xy + 20xz + 20yz$. The resulting function of two variables x and y is

$$C(x, y) = 20xy + 20x \cdot \frac{64}{xy} + 20y \cdot \frac{64}{xy}$$

which can be simplified to

$$C(x, y) = 20xy + \frac{1280}{y} + \frac{1280}{x}$$

We can proceed to minimize this function by getting $\partial C/\partial x = 0$ and $\partial C/\partial y = 0$.

$$\begin{cases} \dfrac{\partial C}{\partial x} = 20y - \dfrac{1280}{x^2} = 0 \\[4mm] \dfrac{\partial C}{\partial y} = 20x - \dfrac{1280}{y^2} = 0 \end{cases}$$

The first equation yields $y = 64/x^2$. Substituting $64/x^2$ for y in the second equation produces

$$20x - \dfrac{1280}{\left(\dfrac{64}{x^2}\right)^2} = 0$$

which can be simplified to

$$20x - \dfrac{20x^4}{64} = 0$$

or

$$x - \dfrac{x^4}{64} = 0$$

After factoring out an x, we have

$$(x)\left(1 - \dfrac{x^3}{64}\right) = 0$$

The product is zero if $x = 0$ (which we reject as an impossible dimension for an aquarium) or if the other factor, $1 - x^3/64$, is zero. Continuing,

$$1 - \dfrac{x^3}{64} = 0$$

leads to

$$\dfrac{x^3}{64} = 1$$
$$x^3 = 64$$
$$x = 4$$

To obtain y, return to either of the original equations ($\partial C/\partial x = 0$ or $\partial C/\partial y = 0$) involving x and y and use 4 for x. We shall use the equation

$$20y - \dfrac{1280}{x^2} = 0$$

Substituting 4 for x, we obtain

$$20y - \frac{1280}{16} = 0$$

Then

$$20y - 80 = 0$$
$$20y = 80$$
$$y = 4$$

Finally, since volume xyz is 64, we have $xyz = 64$, or $(4)(4)z = 64$. This yields: $z = 4$.

Thus, the cost of the aquarium will be minimized if we make the length, width, and height all 4 feet. In Exercise 34, you will have the opportunity to verify that these dimensions do indeed minimize the cost (rather than maximize it). ◆

8.3 Exercises

In Exercises 1–16, find the critical points of each function.

1. $f(x, y) = x^2 + y^2 - 6x + 2y$

2. $f(x, y) = x^2 - 2y^2 - 5x + 4y - 9$

3. $f(x, y) = 3x^3 + y^2 - 36x - 10y + 7$

4. $f(x, y) = x^3 + 3y^2 - 48x + 3y - 2$

5. $g(x, y) = 2x^3 + 3x^2 + y^2 - 8y + 5$

6. $g(x, y) = 4x^3 - 6x^2 + 3y^2 - 12y - 1$

7. $f(x, y) = x^3 + y^3 - 3x - 27y + 4$

8. $f(x, y) = x^2 - 4y^2 + 2$

9. $f(x, y) = x^2 + 9y^2 - 17$

10. $g(x, y) = -x^2 - y^2 + 4x + 4y - 8$

11. $h(x, y) = x^2 + 6xy + 2y^2 - 6x + 10y + 1$

12. $f(x, y) = x^2 + xy + y^2 + 3x - 3y + 16$

13. $f(x, y) = x^3 + y^3 - 3x^2 - 3y^2 - 9x + 3$

14. $f(x, y) = x^2 + xy + 3x + 2y$

15. $f(x, y) = \frac{1}{3}x^3 - xy + \frac{1}{2}y^2 - 3$

16. $f(x, y) = x^2 + 4xy + y^2 + 2$

In Exercises 17–29, find the relative maximum and minimum values of each function, if there are any.

17. $f(x, y) = x^2 + y^2 + 8x - 2y + 5$

18. $f(x, y) = x^2 + y^2 + 6$

19. $f(x, y) = 1 + 4x - 6y - x^2 - y^2$

20. $g(x, y) = x^2 + xy + y^2 + 8x - 8y$

21. $f(x, y) = 4x^2 + 2y^2 - 2xy - 10y - 2x + 1$

22. $f(x, y) = x^2 + 6xy - y^2 + 9$

23. $f(x, y) = -x^2 + xy - y^2 - 2x - 2y + 3$

24. $g(x, y) = 4x^3 - 6x^2 + 3y^2 - 12y - 1$

25. $f(x, y) = x^2 - 4xy + y^3 + 4y - 7$

26. $f(x, y) = x^2 + xy + y^2 + 3x - 3y$

27. $f(x, y) = x^3 + y^3 - 3x^2 - 3y^2 - 9x + 4$

28. $g(x, y) = 4x^3 + y^3 - 12x - 3y + 5$

29. $f(x, y) = x^2 + 3xy - y^2 + 4y - 6x + 1$

W 30. (*REVENUE*) Attempt to maximize the revenue function

$$R(x, y) = 2xy - \frac{1}{2}x^2 - y^2 - 3x - 4y + 20$$

What is your conclusion?

31. (*PROFIT*) A local company advertises on the radio and in the newspaper. Let x and y represent the amounts (in thousands of dollars) spent on radio and newspaper advertising, respectively. The company's profit based on this advertising has been determined to be (in thousands of dollars)

$$P(x, y) = -2x^2 - xy - y^2 + 8x + 9y + 10$$

Determine the amount of money this company should spend on each type of advertising in order to maximize its profit.

32. (*REVENUE*) A manufacturer makes x thousand radios and y thousand tape recorders. The resulting revenue (in thousands of dollars) is

$$R(x, y) = xy - 2x^2 - y^2 + 30x - 4y + 20$$

How many of each product should be made in order to have maximum revenue? Also, how much is that maximum revenue?

33. (*PROFIT*) A company makes two kinds of personal computers. One sells for $1200 and the other sells for $1500. If the company sells x of the $1200 computers and y of the $1500 computers, the revenue (in hundreds of dollars) is $R = 12x + 15y$. Suppose that the cost of making the computers is known to be $C = x^2 + y^2 - xy$ hundred dollars. Determine how many of each kind of computer should be made in order to maximize the *profit* made by the company.

34. (*COST*) Use $C(x, y)$ and the second partials test to show that the values $x = 4$ and $y = 4$ do indeed minimize (rather than maximize) the cost function of Example 5.

35. Find three positive numbers that satisfy both of these conditions:
   (i) Their sum is 27.
   (ii) The sum of the squares of the numbers is as small as possible.
   Hint: Call the numbers x, y, and z. Obtain an equation from condition (i) and a function from condition (ii). Then use the equation to substitute for z in the

function. Finally, proceed to minimize the resulting function of x and y.

36. (*MINIMUM MATERIAL*) A rectangular cardboard box (with a top) is being made to contain a volume of 27 cubic feet. Find the dimensions that will minimize the amount of material used to make the box.

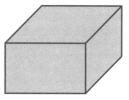

37. (*USPS PACKAGE SIZE*) The United States Postal Service (USPS) insists that the length plus girth of a package to be mailed cannot exceed 84 inches. The *girth* is the distance around the middle. The package shown here is taped along the girth, which is $2x + 2z$ inches. Determine the dimensions (x, y, and z) of the largest volume package that can be mailed (see figure). The length of the package is y.

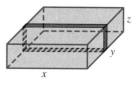

38. (*STORAGE FACILITY*) A farmers' cooperative plans to construct a rectangular storage facility for grain. The building will be made to contain a volume of 16,000 cubic feet. The cost for the roof and the floor is $10 per square foot. The cost for the sides is $5 per square foot.
   (a) What should be the dimensions of the storage facility in order to minimize the cost of materials used?
   (b) What is the cost of materials for a storage facility built to the specifications determined in part (a)?

W 39. (*MINIMUM COST*) While it is often very important for a manufacturer to minimize the cost of making a product, sometimes other considerations can override this concern. Use the aquarium example (Example 5) and give reasons why you may not want to produce the shape indicated by the lowest-cost approach.

8.4 | *LAGRANGE MULTIPLIERS*

Some applied problems require that we maximize or minimize a function subject to some additional condition or constraint. In Example 3 of Section 4.5, a family wanted to determine the dimensions that would maximize the area of their patio. The *constraint* was that they had only 120 feet of fence to enclose the patio. We solved that problem and other applied maximum/minimum problems by inserting the constraint information into the function to be maximized or minimized—either by immediate substitution or by manipulating a constraint equation and then substituting. Unfortunately, such equation manipulation cannot always be done easily, and sometimes it can produce an expression that is difficult to work with.

The French-Italian mathematician Joseph Louis Lagrange (1736–1813) developed a method that uses partial derivatives to solve such constrained maximum/minimum problems. In his method, called **Lagrange multipliers**, the constraint is stated as a function and is multiplied by a new variable called a Lagrange multiplier.

Figure 17 Joseph Louis Lagrange (1736–1813)

We may wish to maximize a function $z = f(x, y)$. The graph of such a function is a surface in three dimensions. The constraint, being an equation involving x and y (only), is a curve in the xy plane. Finding the maximum of the function subject to the constraint means finding the highest point on the surface that lies directly above the graph of the constraint equation. See Figure 18.

We now state the method of Lagrange multipliers for functions of two variables. Examples are given immediately after the statement. Note the use of the Greek letter *lambda* (λ). The number λ is the Lagrange multiplier.

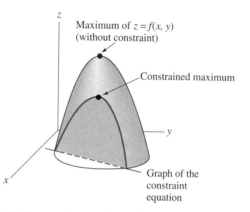

Figure 18 The maximum value of $z = f(x, y)$, subject to a constraint

The Method of Lagrange Multipliers

The relative extrema of $z = f(x, y)$ subject to the constraint $g(x, y) = 0$ are found from the new function

$$F(x, y, \lambda) = f(x, y) + \lambda \cdot g(x, y)$$

by solving the system

$$\begin{cases} F_x(x, y, \lambda) = 0 \\ F_y(x, y, \lambda) = 0 \\ F_\lambda(x, y, \lambda) = 0 \end{cases}$$

for λ and then for points (x, y). The relative extrema are included among the solution points (x, y) obtained.

The method of Lagrange multipliers (as stated above) does not specify how you know that the point obtained is a maximum, a minimum, or neither. Unfortunately, advanced techniques are needed to make the distinction. For problems given in this book, however, it is safe to assume that the point obtained is the extremum requested. If more than one point is obtained, functional values can be calculated to compare and determine which is the desired maximum or minimum.

Proving that the method of Lagrange multipliers does indeed work requires advanced calculus. However, a geometric illustration using level curves offers justification for the method. Consider various level curves $f(x, y) = c$ of the function $z = f(x, y)$. If the level curves are graphed in the xy plane along with the graph of the constraint equation $g(x, y) = 0$, then the desired constrained extremum will occur at the point where the constraint curve is tangent to a level curve. (In the case of a maximum, it will be the highest level curve that intersects the constraint curve.) The method of Lagrange multipliers selects this point from among all points of intersection of the constraint curve and the level curves. Figure 19 shows the level curves of a function similar to the one graphed in Figure 18. The line is the constraint curve. The coordinates of the point of tangency will yield the constrained maximum. Looking at both Figure 19 and Figure 18 should help you to see this idea.

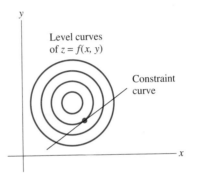

Figure 19 The coordinates of the point of tangency (big dot) will yield the constrained maximum

EXAMPLE 1   Determine the minimum value of $f(x, y) = 2x^2 + y^2 + 7$ subject to the constraint $x + y = 18$.

SOLUTION   Here $f(x, y) = 2x^2 + y^2 + 7$. The constraint $x + y = 18$ must be written in the form $g(x, y) = 0$. Thus we have

$$f(x, y) = 2x^2 + y^2 + 7$$
$$g(x, y) = x + y - 18$$

Now we can easily construct the function $F(x, y, \lambda) = f(x, y) + \lambda \cdot g(x, y)$. It is

$$F(x, y, \lambda) = (2x^2 + y^2 + 7) + \lambda \cdot (x + y - 18)$$

or

$$F(x, y, \lambda) = 2x^2 + y^2 + 7 + \lambda x + \lambda y - 18\lambda$$

It follows that the three partial derivatives are

$$F_x(x, y, \lambda) = 4x + \lambda$$
$$F_y(x, y, \lambda) = 2y + \lambda$$
$$F_\lambda(x, y, \lambda) = x + y - 18$$

The system we must solve is

$$\begin{cases} 4x + \lambda = 0 \\ 2y + \lambda = 0 \\ x + y - 18 = 0 \end{cases}$$

From the first two equations, we obtain $\lambda = -4x$ and $\lambda = -2y$. Consequently, $-4x = -2y$, or $y = 2x$. And since we now know that $y = 2x$, we can substitute $2x$ for y in the third equation, $x + y - 18 = 0$. As a result, we obtain the equation $x + 2x - 18 = 0$, or $3x = 18$. It follows readily that

$$x = 6$$
$$y = 12$$

In this example only one point (x, y) is produced, namely, $(6, 12)$. Consequently, the minimum value of $f(x, y) = 2x^2 + y^2 + 7$ comes from $(6, 12)$. Specifically,

$$f(6, 12) = 2 \cdot 6^2 + 12^2 + 7 = 223$$

We conclude that the minimum value of $f(x, y) = 2x^2 + y^2 + 7$, subject to the given constraint $x + y = 18$, is 223. ◆

EXAMPLE 2 ◆ COBB-DOUGLAS PRODUCTION FUNCTION

Suppose that for a particular product the number of units manufactured is given by the Cobb-Douglas production function

$$f(x, y) = 600x^{2/3}y^{1/3}$$

where x is the number of units of labor and y is the number of units of capital. The cost of labor is \$400 per unit, and the cost of capital is \$200 per unit. If the company wants to make 54,000 units of their product at the lowest possible cost, determine

(a) The number of units of labor and the number of units of capital they should use to minimize their cost, *and*

(b) The actual minimum cost of producing the 54,000 units.

SOLUTION Since x units of labor cost \$400 each and y units of capital cost \$200 each, the combined cost of labor and capital is

$$C(x, y) = 400x + 200y$$

And *this* is the function we want to minimize. The *constraint* is that production $600x^{2/3}y^{1/3}$ be 54,000 units. Thus, we seek to minimize $C(x, y)$ subject to $600x^{2/3}y^{1/3} = 54,000$. Here is a brief summary.

Minimize: $C(x, y) = 400x + 200y$

Subject to: $g(x, y) = 600x^{2/3}y^{1/3} - 54,000 = 0$

It follows that

$$F(x, y, \lambda) = 400x + 200y + \lambda \cdot (600x^{2/3}y^{1/3} - 54,000)$$

or

$$F(x, y, \lambda) = 400x + 200y + 600\lambda x^{2/3}y^{1/3} - 54{,}000\lambda$$

Obtaining partial derivatives, we have

$$\begin{cases} F_x = 400 + 400\lambda x^{-1/3}y^{1/3} = 0 \\ F_y = 200 + 200\lambda x^{2/3}y^{-2/3} = 0 \\ F_\lambda = 600x^{2/3}y^{1/3} - 54{,}000 = 0 \end{cases}$$

Solving the $F_x = 0$ equation for λ yields

$$\lambda = \frac{-400x^{1/3}}{400y^{1/3}}$$

Similarly, solving $F_y = 0$ for λ yields

$$\lambda = \frac{-200y^{2/3}}{200x^{2/3}}$$

The two expressions that are equal to λ can be set equal to each other. The result is

$$\frac{-400x^{1/3}}{400y^{1/3}} = \frac{-200y^{2/3}}{200x^{2/3}}$$

or

$$\frac{x^{1/3}}{y^{1/3}} = \frac{y^{2/3}}{x^{2/3}}$$

Multiplying both sides by common denominator $x^{2/3}y^{1/3}$ produces

$$x = y$$

We can now substitute x for y in the third equation, $600x^{2/3}y^{1/3} - 54{,}000 = 0$. The result is the following equation in one unknown, x:

$$600x^{2/3}x^{1/3} - 54{,}000 = 0$$

or

$$600x - 54{,}000 = 0$$

This equation leads to $x = 90$. Then, since $x = y$, it follows that that $y = 90$ also. We can now give answers to parts (a) and (b).

(a) The producer should use 90 units of labor and 90 units of capital in order to minimize the cost of making 54,000 units.

(b) The actual cost of making 54,000 units is $C(x, y)$ when $x = 90$ and $y = 90$. Since $C(x, y) = 400x + 200y$, we have

$$C(90, 90) = 400 \cdot 90 + 200 \cdot 90$$
$$= 36{,}000 + 18{,}000$$
$$= 54{,}000$$

The minimum cost of making 54,000 units is \$54,000. ◆

8.4 Exercises

Use Lagrange multipliers to find the requested maximum or minimum in each exercise of this section.

1. Maximize: $f(x, y) = xy$
Subject to: $x + y = 10$.

2. Maximize: $f(x, y) = 4xy$
Subject to: $x + 2y = 16$.

3. Minimize: $f(x, y) = x^2 + y^2 + 3$
Subject to: $2x + y = 5$.

4. Minimize: $f(x, y) = x^2 + y^2 - 7$
Subject to: $x + 2y = 10$.

5. Maximize: $f(x, y) = 4 - x^2 - y^2$
Subject to: $2x + y = 10$.

6. Minimize: $f(x, y) = x + y^2 + 1$
Subject to: $x + 4y = 6$.

7. Minimize: $f(x, y) = x^2 + 2y^2 - xy$
Subject to: $2x - y = 4$.

8. Minimize: $f(x, y) = 2x^2 + y^2 - xy$
Subject to: $x + y = 8$.

9. Maximize: $f(x, y) = 2xy - 4x$
Subject to: $x + y = 12$.

10. Minimize: $f(x, y) = 8y - 2xy$
Subject to: $x + 2y = 12$.

11. Minimize: $f(x, y) = x^2 + xy + y^2$
Subject to: $x + y = 20$.

12. Find two numbers (x and y) whose sum is 90 and whose product is maximum.

13. Find two numbers whose sum is 1000 and whose product is maximum.

14. (*GARDEN AREA*) If the area of a rectangular garden must be 1000 square feet, what should be its length and width in order to minimize the amount of fencing needed to enclose it? (See figure.)

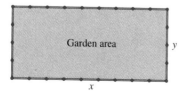

Garden area

y

x

15. (*FARMING*) A farmer has 200 meters of fencing. Determine the largest rectangular area he can enclose as a pig pen.

16. (*FARMING*) Assume that the pig pen of Exercise 15 is constructed using a wall of the barn as one side, so that only three sides need to be fenced in. Determine the largest area possible.

17. (*PRODUCTION FUNCTION*) Redo Example 2 (the Cobb-Douglas production function) using the function $f(x, y) = 300x^{2/3}y^{1/3}$. Assume that the cost of labor is $200 per unit and the cost of capital is $100 per unit. Also assume that the manufacturer wants to make 21,000 units.

18. (*PRODUCTION FUNCTION*) Redo Example 2 (the Cobb-Douglas production function) using the function $f(x, y) = 90x^{1/3}y^{2/3}$. Assume that the cost of labor is $50 per unit and the cost of capital is $100 per unit. Also assume that the manufacturer wants to make 6300 units.

19. (*PRODUCTION FUNCTION*) Determine the minimum cost of producing 24,000 units of a product, assuming the cost is $20x + 80y$ and the production function is $f(x, y) = 200x^{1/2}y^{1/2}$, where x is the number of units of labor and y is the number of units of capital.

20. (*PRODUCTION FUNCTION*) Determine the minimum cost of producing 1000 units of a product, assuming the cost is $20x + 60y$ and the production function is $f(x, y) = 40x^{1/4}y^{3/4}$, where x is the number of units of labor and y is the number of units of capital.

21. (*PRODUCTION FUNCTION*) Consider the production function $f(x, y) = 600x^{2/3}y^{1/3}$ and the cost constraint $400x + 200y = 30,000$ dollars. How many units of labor (x) and how many units of capital (y) should be used in order to maximize production?

22. (*PROFIT*) Suppose a producer's profit on the sale of x radios and y tape recorders is

$$P(x, y) = 60x + 80y - x^2 - y^2$$

Assuming that the producer must make a total of 40 units, how many of each should be made in order to maximize profit, and what is that maximum profit?

23. *(Container size)* A potato chip company needs cylindrical tin containers that hold 200 cubic inches. What should be the radius and the height of the container, if the company wishes to use the least amount of material to make it? The total surface area of the can is $2\pi xy + 2\pi x^2$, where x is the radius and y is the height.

POTATO CHIPS

24. *(Carpet area)* Professor Konnick has a house with an entrance foyer in the shape of an ellipse. The equation $2x^2 + y^2 = 64$ defines the elliptical border of the floor. (x and y are in feet.) Find the dimensions of the largest (area) rectangular rug that can be placed in the foyer (see figure). Note that the area of the rug is not xy.

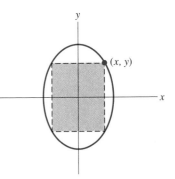

8.5 | THE METHOD OF LEAST SQUARES

Up until now we have solved a variety of problems by performing calculus operations on functions that were provided. However, there can be situations in which we are not actually given a function. Rather, we may only have *points* that describe a relationship between two variables. The problem, then, is to construct a function that represents the points, at least approximately.

The points themselves usually come from data that has been collected. For example, a study of people may show the following heights (in inches) and corresponding average weights (in pounds).

height (x)	weight (y)
62	110
64	130
68	140
70	160
72	170

The points:

(x, y)
(62, 110)
(64, 130)
(68, 140)
(70, 160)
(72, 170)

When the points are plotted, the resulting graph is called a **scatter diagram**. Figure 20 shows a scatter diagram of our data points.

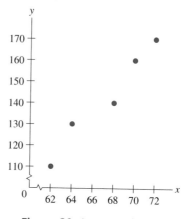

Figure 20 A scatter diagram

Although the points do not lie on a straight line, a straight line can be drawn that will be near these points. In other words, a line fits the data reasonably well. See Figure 21.

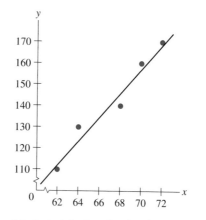

Figure 21 A straight line fits the data reasonably well

Other lines can be drawn that also fit this data fairly well. The straight line that is considered to be the "best" is the one that minimizes the sum of the squares of the vertical distances from each point to that line. The line is called the **least squares line** or the **regression line**.

Now let us proceed to determine the least squares line $y = mx + b$ for any set of points $(x_1, y_1), (x_2, y_2), \ldots, (x_n, y_n)$. The distance from each point to the line is measured vertically. This means that the distance is the difference in the y coordinates. Thus the distances are $y_1 - (mx_1 + b), y_2 - (mx_2 + b), \ldots,$ $y_n - (mx_n + b)$. The sum of the squares of these distances is

$$[y_1 - (mx_1 + b)]^2 + [y_2 - (mx_2 + b)]^2 + \cdots + [y_n - (mx_n + b)]^2$$

which we can simplify to

$$(y_1 - mx_1 - b)^2 + (y_2 - mx_2 - b)^2 + \cdots + (y_n - mx_n - b)^2$$

Since we seek the values of m and b that minimize this sum, we will call the sum $f(m, b)$, obtain $\partial f/\partial m = 0$ and $\partial f/\partial b = 0$, and then solve for m and b. This is the method used in Section 8.3 to minimize functions.

$$f(m,b) = (y_1 - mx_1 - b)^2 + (y_2 - mx_2 - b)^2 + \cdots + (y_n - mx_n - b)^2$$

$$\frac{\partial f}{\partial m} = -2x_1(y_1 - mx_1 - b) - 2x_2(y_2 - mx_2 - b) - \cdots - 2x_n(y_n - mx_n - b) = 0$$

$$\frac{\partial f}{\partial b} = -2(y_1 - mx_1 - b) - 2(y_2 - mx_2 - b) - \cdots - 2(y_n - mx_n - b) = 0$$

With some effort, we can obtain m and b. The result is

$$m = \frac{n(x_1y_1 + x_2y_2 + \cdots + x_ny_n) - (x_1 + x_2 + \cdots + x_n)(y_1 + y_2 + \cdots + y_n)}{n(x_1^2 + x_2^2 + \cdots + x_n^2) - (x_1 + x_2 + \cdots + x_n)^2}$$

$$b = \frac{(y_1 + y_2 + \cdots + y_n) - m(x_1 + x_2 + \cdots + x_n)}{n}$$

These formulas for m and b can be condensed by using summation notation. In summary,

---

Least Squares Line

The **least squares line** for the n points $(x_1, y_1), (x_2, y_2), \ldots, (x_n, y_n)$ is

$$y = mx + b$$

where

$$m = \frac{n \cdot \Sigma x_i y_i - (\Sigma x_i)(\Sigma y_i)}{n \cdot \Sigma x_i^2 - (\Sigma x_i)^2}$$

$$b = \frac{\Sigma y_i - m \cdot \Sigma x_i}{n}$$

> *Note*
>
> Once the equation of the least squares line is determined, predictions can be made by considering other x values and the y values that will correspond based on the equation. Keep in mind that such predictions are only approximations or good guesses, since the data points used to obtain the equation of the line are only near the line rather than on the line.

EXAMPLE 1   Determine the equation of the least squares line for the data given below *and* estimate the y value when x is 6.

x	2	3	4	7	8
y	5	7	9	8	12

SOLUTION   To obtain the least squares line $y = mx + b$, we use the boxed formulas to determine m and b. In this example, n is 5, because there are 5 data points. We proceed then to determine the values of the various sums that must be substituted into the formulas for m and b.

$$\Sigma x_i = 2 + 3 + 4 + 7 + 8 = 24$$
$$\Sigma y_i = 5 + 7 + 9 + 8 + 12 = 41$$
$$\Sigma x_i y_i = 2 \cdot 5 + 3 \cdot 7 + 4 \cdot 9 + 7 \cdot 8 + 8 \cdot 12 = 219$$
$$\Sigma x_i^2 = 2^2 + 3^2 + 4^2 + 7^2 + 8^2 = 142$$
$$(\Sigma x_i)^2 = (24)^2 = 576$$

Substituting these numbers into the formulas yields

$$m = \frac{5 \cdot 219 - 24 \cdot 41}{5 \cdot 142 - 576} \approx .83$$

and

$$b = \frac{41 - .83(24)}{5} \approx 4.22$$

Thus, the least squares line for this data is

$$y = .83x + 4.22 \qquad \text{(answer)}$$

Also, when $x = 6$,

$$y = .83(6) + 4.22 = 9.2$$

Thus, y is approximately 9.2 when x is 6. ◆

Note

It is generally not advisable to use a least squares line for *extrapolation*, that is, for estimating y values that correspond to x values outside the range of the x values in the given data. The flaw in the reasoning is not always obvious, but in some cases it is strikingly clear. In our section-opening example that used heights and weights, would it be reasonable to obtain the least squares line and then use it to estimate the weight of a person who is 9 feet tall? For another example, consider data on the depreciation of automobiles. Would it be reasonable to look at the values of cars that are 1 or 2 or 3 years old and then use the least squares line to estimate the value of an 8-year-old car? The least squares line will not predict the fact that the depreciation eventually slows down or stops. Extrapolation may even produce a negative value for an automobile!

8.5   Exercises

In Exercises 1–12, determine the equation of the least squares line for each collection of data points.

1.

x	3	5	6
y	8	11	12

2.

x	1	4	7
y	2	9	13

3.

x	8	9	10	12
y	0	3	5	6

4.

x	6	8	10	12
y	3	5	6	10

5.

x	0	1	2	3	4
y	3	4	6	6	8

6.

x	0	3	4	5	6
y	15	9	8	6	2

7. (1, 4), (2, 5), (3, 8)

8 (1, 2), (2, 5), (3, 3)

9. (0, 4), (1, 5), (2, 6), (4, 10)

10. (1, 0), (3, 4), (4, 4), (5, 7)

11. (1, 5), (2, 6), (3, 3), (4, 1), (5, 0)

12. (1, 5), (2, 5), (3, 4), (4, 2), (5, 1)

13. Consider the data given.

x	2	3	4	6
y	7	6	6	2

(a) Determine the equation of the least squares line.
(b) Estimate y for $x = 5$.
(c) Estimate x for $y = 4$.

14. Consider the data given.

x	6	8	9	10
y	1	5	6	8

(a) Determine the equation of the least squares line.
(b) Estimate y for $x = 7$.
(c) Estimate x for $y = 3$.

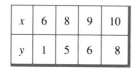 **15. (ACT AND GPA)** A college advisor is comparing the pre-college ACT scores of five students with the grade point average (GPA) they obtained in college. The data:

ACT (x)	GPA (y)
15	2.0
18	2.4
20	2.6
21	2.8
24	3.2

(a) Find the equation of the least squares line.
(b) Based on this study, what GPA might you expect from a student with an ACT score of 23?

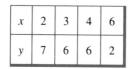 **16. (ADVERTISING AND REVENUE)** The table given here shows the amount x (in thousands of dollars) that a company has spent monthly on advertising and the corresponding amount y (in thousands of dollars) that the company grossed in revenue that month. Determine the least squares line.

advertising	revenue
3	80
4	100
5	110
6	140
8	170

17. (CHEMISTRY EXPERIMENT) A student in chemistry lab conducted an experiment in which he was able to dissolve different amounts (y) of a compound in water by varying the temperature (x) of the water. (The amount of water used was kept constant.) His data:

temperature (°C)	10	20	30	40	50	60
grams dissolved	61	65	72	77	85	90

(a) Determine the equation of the least squares line.
(b) Use the equation obtained to estimate how many grams of the compound will dissolve at 35°C.
(c) Estimate what temperature is needed to dissolve 80 grams of the compound.

18. (HEIGHT AND WEIGHT) Find the equation of the least squares line for the height and weight data given at the beginning of this section (and repeated here for easy reference).

height (x)	weight (y)
62	110
64	130
68	140
70	160
72	170

19. Show, in steps, how the least squares formula for b (in $y = mx + b$) can be written as $b = \bar{y} - m\bar{x}$.

W 20. The note at the end of the section warns of the danger of extrapolation. Two cases are cited—unrealistic heights and negative values for automobiles. Think of another example in which extrapolation may lead to absurd results, and write a brief explanation of it.

The Exercise Library at the back of the book contains graphing calculator and computer exercises keyed to this section.

8.6 | TOTAL DIFFERENTIALS

Section 3.10 presented the concept of the differential of a function of one variable. For $y = f(x)$, the differential dy was defined as

$$dy = f'(x)\, dx$$

with $dx = \Delta x$ and $dy \approx \Delta y$. The differential dy was used to approximate Δy for small values of dx. We estimated such things as the change in revenue associated with small changes in advertising expenditures and the change in price that would cause a small change in demand.

The differential concept can be extended to functions of two variables by the following definition.

Total Differential

Let $z = f(x, y)$. The **total differential** dz is

$$dz = f_x(x, y) \cdot dx + f_y(x, y) \cdot dy$$

where $dx = \Delta x$, $dy = \Delta y$, and $dz \approx \Delta z$ and

$$\Delta z = f(x + \Delta x, y + \Delta y) - f(x, y)$$

EXAMPLE 1 Let $z = f(x, y) = x^2 y + 3x - 7y$. Determine dz.

SOLUTION
$$\begin{aligned} dz &= f_x(x, y) \cdot dx + f_y(x, y) \cdot dy \\ &= (2xy + 3)\, dx + (x^2 - 7)\, dy \quad \blacklozenge \end{aligned}$$

Since $dz \approx \Delta z$, the total differential dz gives an approximation to the change in z (that is, Δz) corresponding to small changes in x and y. Here are two examples.

EXAMPLE 2 Let $z = f(x, y) = xy^2 + 7x - 1$.

(a) Use differentials to find dz (the approximate change in z) when x changes from 4 to 4.01 and y changes from 5 to 5.03.

(b) Determine the exact change in z (that is, Δz) when x changes from 4 to 4.01 and y changes from 5 to 5.03.

SOLUTION   **(a)** First, obtain the differential. Since

$$dz = f_x(x, y)\, dx + f_y(x, y)\, dy$$

we have

$$dz = (y^2 + 7)\, dx + 2xy\, dy$$

The x value begins at 4 and changes to 4.01. This means that $x = 4$ and $dx = .01$. Similarly, since y begins at 5 and changes to 5.03, we have $y = 5$ and $dy = .03$. Using these values of x, dx, y, and dy, we find

$$dz = (5^2 + 7)(.01) + 2(4)(5)(.03)$$
$$= .32 + 1.20$$
$$= 1.52$$

The change in z is approximately 1.52.

(b) The exact change Δz is $f(4.01, 5.03) - f(4, 5)$.

$$f(4.01, 5.03) - f(4, 5)$$
$$= [(4.01)(5.03)^2 + 7(4.01) - 1] - [(4)(5)^2 + 7(4) - 1]$$
$$= 128.5266 - 127$$
$$= 1.5266$$

Notice that the total differential provides a quick and good approximation—1.52 versus 1.5266.   ◆

EXAMPLE 3   ◆   *REVENUE INCREASE ASSOCIATED WITH ADVERTISING*

A company advertises on radio and in the newspaper. The function

$$R(x, y) = -.25x^2 + 22x - .5y^2 + 18y + 300$$

gives the revenue in thousands of dollars that is associated with an expenditure of x hundred dollars on radio advertising and y hundred dollars on newspaper advertising. If the amount being spent on advertising now stands at $2000 for radio and $1200 for newspapers, use the total differential to determine the approximate increase in revenue associated with a $150 increase in radio advertising and a $100 increase in newspaper advertising.

SOLUTION   The approximate increase in revenue will be the value of the total differential dR for $x = 20$, $y = 12$, $dx = 1.5$, and $dy = 1$.

$$dR = R_x(x, y) \cdot dx + R_y(x, y) \cdot dy$$
$$= (-.5x + 22)\, dx + (-1.0y + 18)\, dy$$
$$= [-.5(20) + 22](1.5) + [-1.0(12) + 18](1)$$
$$= 24$$

The approximate increase in revenue associated with a $150 increase in radio advertising and a $100 increase in newspaper advertising is $24,000.

Exercise 25 asks you to show that the exact increase in revenue is $22,937.50.   ◆

8.6   Exercises

In Exercises 1–10, determine dz.

1. $z = f(x, y) = x^2 + y^2 + 10$

2. $z = f(x, y) = 3x^2 - 8y^2 + 5$

3. $z = f(x, y) = xy^2 + 3x^2 - 2y^2$

4. $z = f(x, y) = 5x^2 + 3x^2y - 8y$

5. $z = f(x, y) = xe^y + ye^x + 1$

6. $z = f(x, y) = 4e^{xy} - 12$

7. $z = f(x, y) = \dfrac{y}{x} - 8x + 3y$

8. $z = f(x, y) = \dfrac{x}{y} + 4xy + 5$

9. $z = f(x, y) = x \ln y - x^2$

10. $z = f(x, y) = y \ln x + y^3$

In Exercises 11–16, determine dz for the given values of x, dx, y, and dy.

11. $z = f(x, y) = 4x^2y^3$; $x = 5$, $dx = .01$, $y = 9$, $dy = .02$

12. $z = f(x, y) = 3xy + y$; $x = 12$, $dx = .03$, $y = 17$, $dy = .01$

13. $z = f(x, y) = 6e^{1/3}y^{2/3}$; $x = 1$, $dx = .02$, $y = 8$, $dy = .03$

14. $z = f(x, y) = 4x^{1/2}y^{1/3}$; $x = 9$, $dx = .001$, $y = 8$, $dy = .015$

15. $z = f(x, y) = x\sqrt{y}$; $x = 12$, $dx = .015$, $y = 64$, $dy = .01$

16. $z = f(x, y) = y + \sqrt{x}$; $x = 100$, $dx = .8$, $y = 5$, $dy = .01$

In Exercises 17–22, use the total differential to approximate the change in f corresponding to the given changes in x and y.

17. $f(x, y) = 5x^3y^2 + y$; x changes from 4 to 4.01, y changes from 6 to 6.02.

18. $f(x, y) = x + x^2y^2$; x changes from 3 to 3.02, y changes from 10 to 10.01.

19. $f(x, y) = x \ln y$; x changes from 10 to 10.1, y changes from 1 to 1.2.

20. $f(x, y) = 3y \ln (x - 1)$; x changes from 2 to 2.01, y changes from 20 to 20.5.

21. $f(x, y) = 12x^{2/3}y^{1/3}$; x changes from 8 to 9, y changes from 27 to 29.

22. $f(x, y) = 10x^{1/2}y^{1/2}$; x changes from 16 to 17, y changes from 25 to 28.

23. (*REVENUE AND ADVERTISING*) Redo Example 3 assuming that the amount spent on radio advertising will be increased from $1800 to $2000 and that the amount spent on newspaper advertising will be increased from $1400 to $1500.

24. (*REVENUE AND ADVERTISING*) Find the *exact* increase in revenue for the situation described in Exercise 23.

25. (*REVENUE AND ADVERTISING*) Show that the *exact* increase in revenue for the situation described in Example 3 is $22,937.50.

26. (*VOLUME CHANGE*) A manufacturer of paper drinking cups decides to make its standard cup slightly smaller than before. The cups are conical and hold volume $V = \frac{1}{3}\pi r^2h$, where r is the radius of the top and h is the height (see figure). The radius will be changed from 1.5 inches to 1.4 inches, and the height will be changed from 4 inches to 3.9 inches. Use differentials to approximate the reduction in volume that results from these changes.

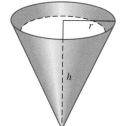

27. (*VOLUME REDUCTION*) A juice drink producer decides to reduce the size of the cylindrical can used to package the product for sale to the public. The volume of a can is $V = \pi r^2 h$, where r is the radius of the base and h is the height. The radius will be reduced from 2 inches to 1.9 inches, and the height will be reduced from 9 inches to 8.8 inches. Use differentials to approximate the reduction in volume as a result of these changes.

28. (*INCREASING PRODUCTION*) The cost of producing x units of one product (product M) and y units of another product (product N) is given by

$$C(x, y) = 500 + 30x + 60y - .1x^2 - .3y^2 \text{ dollars}$$

Use differentials to determine the approximate cost of raising the production levels one unit each—from 80 to 81 units of product M and from 93 to 94 units of product N.

29. (*INCREASING PRODUCTION*) The cost of producing x units of one product (product A) and y units of another product (product B) is given by

$$C(x, y) = 900 + 80x + 45y - .02xy \text{ dollars}$$

Use differentials to determine the approximate cost of raising the production levels one unit each—from 120 to 121 units of product A and from 200 to 201 units of product B.

30. (*PRODUCTION FUNCTION*) Consider the Cobb-Douglas production function

$$f(x, y) = 40x^{1/3}y^{2/3}$$

where x is the number of units of labor, y is the number of units of capital, and $f(x, y)$ is the number of units produced. Use differentials to estimate the change in production if the number of units of labor is increased from 27 to 28 and the number of units of capital is increased from 64 to 66.

31. (*PRODUCTION FUNCTION*) Consider the Cobb-Douglas production function

$$f(x, y) = 24x^{1/2}y^{1/2}$$

where x is the number of units of labor, y is the number of units of capital, and $f(x, y)$ is the number of units produced. Use differentials to estimate the change in production if the number of units of labor is increased from 25 to 26 and capital is increased from 36 to 39 units.

32. (*AREA CHANGE*) The area of a rectangle having length x and width y is given by

$$A(x, y) = xy$$

(a) Use differentials to approximate the increase in area dA if the length is increased from 50 meters to 50.03 meters and the width is increased from 40 to 40.15 meters.

(b) Determine the exact increase in area suggested by the increases given in part (a).

(c) Show by using calculations that the difference between the exact increase in area [part (b)] and the approximate increase in area [part (a)] is $(dx)(dy)$. The drawing shown below illustrates this idea.

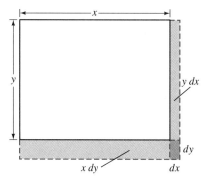

8.7 | *DOUBLE INTEGRALS*

After studying partial differentiation earlier in the chapter, it is natural to wonder if there isn't a corresponding process for antidifferentiation (integration). Indeed there is a process called **partial integration**, in which the integration is performed with respect to one variable while treating the other variable(s) as constant. Here next are two examples of such integration, using definite integrals.

EXAMPLE 1   Evaluate $\int_1^2 6x^2y \, dx$.

SOLUTION   As indicated by the dx, the integration is with respect to x while treating y as a constant.

$$\int_1^2 6x^2y \, dx = \left[2x^3y\right]_{x=1}^{x=2}$$
$$= (16y) - (2y)$$
$$= 14y \quad \blacklozenge$$

EXAMPLE 2   Evaluate $\int_0^4 6x^2y \, dy$.

SOLUTION   As indicated by the dy, the integration is with respect to y while treating x as a constant.

$$\int_0^4 6x^2y \, dy = \left[3x^2y^2\right]_{y=0}^{y=4}$$
$$= (48x^2) - (0)$$
$$= 48x^2 \quad \blacklozenge$$

In Example 1, integrating with respect to x produced a function of y.

$$\int_a^b f(x, y) \, dx \qquad \text{is a function of } y$$

Since the integral yields a function of y, it could be integrated again—this time with respect to y. Similarly, the integral of Example 2 yields a function of x, which could be integrated again—this time with respect to x. With all of this in mind, we proceed to perform two such integrations.

EXAMPLE 3   Evaluate.   **(a)** $\int_0^4 \left[\int_1^2 6x^2y \, dx\right] dy$ **(b)** $\int_1^2 \left[\int_0^4 6x^2y \, dy\right] dx$

SOLUTION   **(a)** $$\int_0^4 \left[\int_1^2 6x^2 y \, dx \right] dy = \int_0^4 [14y] \, dy \qquad \text{Done in Example 1}$$

$$= \left[7y^2 \right]_0^4$$

$$= 112$$

(b) $$\int_1^2 \left[\int_0^4 6x^2 y \, dy \right] dx = \int_1^2 [48x^2] \, dx \qquad \text{Done in Example 2}$$

$$= \left[16x^3 \right]_1^2$$

$$= 112 \quad \blacklozenge$$

The two integrals of Example 3 yield the same result. That is, we have seen that

$$\int_1^2 \left[\int_0^4 6x^2 y \, dy \right] dx = \int_0^4 \left[\int_1^2 6x^2 y \, dx \right] dy$$

The equality just obtained is an example of an important result, which is stated next.

Order of Integration

Double integrations such as those of Example 3 can be performed with respect to either variable first. The order of integration will not matter. That is,

$$\int_a^b \left[\int_c^d f(x, y) \, dy \right] dx = \int_c^d \left[\int_a^b f(x, y) \, dx \right] dy$$

The brackets can be removed, in which case we have

$$\int_a^b \int_c^d f(x, y) \, dy \, dx = \int_c^d \int_a^b f(x, y) \, dx \, dy$$

It is understood that the inner integration is done first.

We shall see next that the limits of integration lead to inequalities that define a rectangular region. We have

$$a \le x \le b$$

and

$$c \le y \le d$$

The region that includes all x between a and b and all y between c and d is rectangular, as illustrated in Figure 22. We will call such a region R.

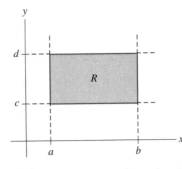

Figure 22 A rectangular region *R*

A definition of the double integral can now be given.

Double Integral

The **double integral** of $f(x, y)$ over the rectangular region *R* defined by $a \le x \le b$ and $c \le y \le d$ is given by

$$\iint_R f(x, y)\, dy\, dx \qquad or \qquad \iint_R f(x, y)\, dx\, dy$$

and is the same as

$$\int_a^b \int_c^d f(x, y)\, dy\, dx \qquad or \qquad \int_c^d \int_a^b f(x, y)\, dx\, dy$$

This definition of the double integral is less formal than one you might find in a more advanced mathematics text. It is a practical definition that provides us the means to evaluate a variety of double integrals.

EXAMPLE 4   Evaluate the double integral $\displaystyle\int_3^4 \int_0^6 (3x + 2y)\, dx\, dy.$

SOLUTION   The inner integration (with respect to *x*) is done first.

$$\int_3^4 \int_0^6 (3x + 2y)\, dx\, dy = \int_3^4 \left[\frac{3x^2}{2} + 2xy \right]_{x=0}^{x=6} dy$$

$$= \int_3^4 [(54 + 12y) - (0 + 0)]\, dy$$

$$= \int_3^4 (54 + 12y)\, dy$$

$$= \left[54y + 6y^2 \right]_3^4$$

$$= (216 + 96) - (162 + 54)$$

$$= 96 \quad \blacklozenge$$

EXAMPLE 5   Evaluate $\displaystyle\int_0^7 \int_1^2 4xy \, dy \, dx.$

SOLUTION   The inner integration (with respect to y) is done first.

$$\int_0^7 \int_1^2 4xy \, dy \, dx = \int_0^7 \left[2xy^2 \right]_{y=1}^{y=2} dx$$

$$= \int_0^7 (8x - 2x) \, dx$$

$$= \int_0^7 6x \, dx$$

$$= \left[3x^2 \right]_0^7$$

$$= 147 \quad \blacklozenge$$

Volume

There is a geometric interpretation of the double integral. Recall that with functions of one variable, the definite integral represents the area under the curve when the function is nonnegative. Specifically,

$$\int_a^b f(x) \, dx$$

is the area under the graph of $y = f(x)$ between $x = a$ and $x = b$, when $f(x) \geq 0$. Furthermore, the integral notation represents the limit of the sum of the areas of approximating rectangles that have length $f(x_i)$ and width Δx.

$$\int_a^b f(x) \, dx = \lim_{\Delta x \to 0} \sum_{i=1}^n f(x_i) \, \Delta x$$

> **Note**
>
> If you have forgotten this idea or you would like to see the drawing and development of the area under a curve as the definite integral, refer to Section 6.3.

With functions of two variables, the graph is a surface. Consider $z = f(x, y)$ over the region R defined by $a \leq x \leq b$ and $c \leq y \leq d$, and assume that $f(x, y) \geq 0$ for all (x, y) in region R. See Figure 23. The rectangular region can be subdivided into small rectangles having length Δx and width Δy. The area of each rectangle is then $\Delta x \cdot \Delta y$. See Figure 24. Select some point (x_i, y_j) in the rectangular region. Then form a box by using as the base the Δx by Δy subregion containing (x_i, y_j) and using $f(x_i, y_j)$ as the height. The box will reach upward as far as the graph of the surface. Its volume will be $f(x_i, y_j) \, \Delta x \, \Delta y$—height times length times width. See Figure 25.

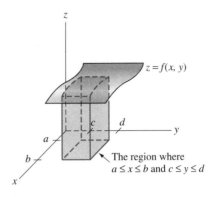

Figure 23 Considering the volume under a surface

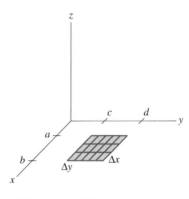

Figure 24 The subdivided rectangular region

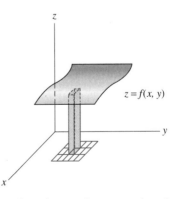

Figure 25 A box that reaches upward to the surface

Thus, $f(x_i, y_j)\, \Delta x\, \Delta y$ is the approximate volume between the surface $z = f(x, y)$ and one small region Δx by Δy in the xy plane. But there are many such small regions—based on, say, n Δx's and m Δy's. If you sum up all such box volumes involving all the subregions Δx by Δy, the volume is

$$V_{\text{boxes}} = \sum_{j=1}^{m} \sum_{i=1}^{n} f(x_i, y_j)\, \Delta x\, \Delta y$$

This is *approximately* the volume under the surface. The exact volume under the surface is the limit of this sum as $\Delta x \to 0$ and $\Delta y \to 0$. Using integral notation as done in Sections 6.3–6.5, we have the following result.

Volume Under a Surface

Let $z = f(x, y)$ be nonnegative for all (x, y) in the region R defined by $a \le x \le b$ and $c \le y \le d$. Then the **volume** between the surface $z = f(x, y)$ and the region R is given by

$$V = \int_{c}^{d} \int_{a}^{b} f(x, y)\, dx\, dy$$

EXAMPLE 6   Determine the volume between the surface $f(x, y) = x^3 + xy$ and the region R defined by $0 \le x \le 2$ and $1 \le y \le 4$.

SOLUTION   Since x goes from 0 to 2 and y goes from 1 to 4, the volume is

$$V = \int_{1}^{4} \int_{0}^{2} (x^3 + xy)\, dx\, dy$$

$$= \int_{1}^{4} \left[\frac{x^4}{4} + \frac{x^2 y}{2} \right]_{x=0}^{x=2} dy$$

$$= \int_{1}^{4} [(4 + 2y) - (0 + 0)]\, dy$$

$$= \int_{1}^{4} (4 + 2y)\, dy$$

$$= \left[4y + y^2 \right]_{1}^{4}$$

$$= (16 + 16) - (4 + 1)$$

$$= 27 \text{ cubic units} \quad \blacklozenge$$

Average Value

In Section 6.5 the average value of a function f of one variable x over an interval $a \le x \le b$ was found to be

$$\frac{1}{b - a} \int_{a}^{b} f(x)\, dx$$

There is a comparable formula for the average value of a function $f(x, y)$ of two variables over the region R.

> **Average Value of a Function**
>
> The **average value of** $f(x, y)$ over the region R defined by $a \le x \le b$ and $c \le y \le d$ is
>
> $$\frac{1}{\text{area of } R} \cdot \int_c^d \int_a^b f(x, y)\, dx\, dy$$
>
> Since the length of the region is $b - a$ and the width is $d - c$, the *area of the region R* is $(b - a)(d - c)$.

EXAMPLE 7 Determine the average value of the function $f(x, y) = 2x + 3y$ over the region defined by $1 \le x \le 4$ and $0 \le y \le 5$.

SOLUTION The area of the region is $(4 - 1)(5 - 0) = 15$. The average value is then

$$\frac{1}{15} \int_0^5 \int_1^4 (2x + 3y)\, dx\, dy = \frac{1}{15} \int_0^5 \left[x^2 + 3xy \right]_{x=1}^{x=4} dy$$

$$= \frac{1}{15} \int_0^5 [(16 + 12y) - (1 + 3y)]\, dy$$

$$= \frac{1}{15} \int_0^5 (15 + 9y)\, dy$$

$$= \frac{1}{15} \left[15y + \frac{9y^2}{2} \right]_0^5$$

$$= \frac{1}{15} \left[\left(75 + \frac{225}{2} \right) - (0 + 0) \right]$$

$$= \frac{1}{15} \cdot \frac{375}{2}$$

$$= \frac{25}{2} \quad \blacklozenge$$

Nonrectangular Regions

Thus far, double integrals have been defined over rectangular regions where $a \le x \le b$ and $c \le y \le d$. In such instances all four limits of integration are constants. On the other hand, if either or both of the inner limits of integration were made to be variables or functions of a variable, then the region R over which we are integrating would no longer be a rectangle. The next example illustrates how the evaluation of such integrals resembles that for others already presented. The example also demonstrates how the nonrectangular region can be drawn.

EXAMPLE 8   Evaluate $\displaystyle\int_1^2\int_0^x (2y + 3)\ dy\ dx.$

SOLUTION
$$\int_1^2\int_0^x (2y + 3)\ dy\ dx = \int_1^2 \left[y^2 + 3y\right]_{y=0}^{y=x} dx$$

$$= \int_1^2 [(x^2 + 3x) - (0 + 0)]\ dx$$

$$= \int_1^2 (x^2 + 3x)\ dx$$

$$= \left[\frac{x^3}{3} + \frac{3x^2}{2}\right]_1^2$$

$$= \frac{41}{6}$$

Figure 26 illustrates the region over which the integration was done. It is the region bounded by $y = 0$, $y = x$, $x = 1$, and $x = 2$.

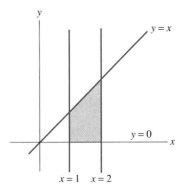

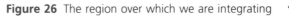

Figure 26 The region over which we are integrating   ◆

EXAMPLE 9   Evaluate $\displaystyle\int_0^1\int_{\sqrt{y}}^{y^2} 2x\ dx\ dy.$

SOLUTION
$$\int_0^1\int_{\sqrt{y}}^{y^2} 2x\ dx\ dy = \int_0^1 \left[x^2\right]_{x=\sqrt{y}}^{x=y^2} dy$$

$$= \int_0^1 (y^4 - y)\ dy$$

$$= \left[\frac{y^5}{5} - \frac{y^2}{2}\right]_0^1$$

$$= -\frac{3}{10} \quad ◆$$

8.7   Exercises

In Exercises 1–10, use partial integration to evaluate each integral.

1. $\int_1^3 12x^2y\,dx$

2. $\int_0^2 12x^2y\,dy$

3. $\int_0^3 8xy\,dy$

4. $\int_1^2 8xy^2\,dx$

5. $\int_4^7 (1 + 3x)\,dy$

6. $\int_4^7 (1 + 3x)\,dx$

7. $\int_2^5 (2x + 2y)\,dx$

8. $\int_1^6 (2x + 2y)\,dy$

9. $\int_2^4 (x^2 + y)\,dy$

10. $\int_0^3 (x^2 + y)\,dx$

Evaluate each double integral in Exercises 11–26.

11. $\int_1^3 \int_0^2 (2x + 6y)\,dx\,dy$

12. $\int_1^4 \int_0^2 3x^2\,dx\,dy$

13. $\int_0^2 \int_1^4 8xy\,dy\,dx$

14. $\int_2^3 \int_1^2 (1 + 2xy)\,dy\,dx$

15. $\int_1^4 \int_0^1 6xy^2\,dx\,dy$

16. $\int_1^3 \int_0^2 6x^2y\,dx\,dy$

17. $\int_0^3 \int_2^4 (3x^2y^2 - 7)\,dy\,dx$

18. $\int_{-1}^6 \int_0^4 dy\,dx$

19. $\int_0^1 \int_0^2 (x^2 + y^2)\,dy\,dx$

20. $\int_0^4 \int_0^1 (x^3 + y^3)\,dx\,dy$

21. $\int_3^8 \int_0^6 \sqrt{1 + x}\,dy\,dx$

22. $\int_0^6 \int_3^8 \sqrt{1 + x}\,dx\,dy$

23. $\int_0^2 \int_0^1 e^{x+y}\,dx\,dy$

24. $\int_0^1 \int_2^7 e^y\,dx\,dy$

25. $\int_0^8 \int_1^e \frac{y}{x}\,dx\,dy$

26. $\int_1^e \int_0^8 \frac{y}{x}\,dy\,dx$

27. As you know, when all four limits of integration are constants (that is, when integrating over a rectangular region), the order of integration can be reversed. Reverse the order of integration in Exercise 11 and show that the result is the same as the original.

28. Do the same as in Exercise 27, except use the integrals of Exercises 13, 15, and 23.

29. Determine the limits of integration and then evaluate the double integral below over the rectangular region

R whose vertices are (2, 1), (4, 1), (2, 3), and (4, 3). A drawing of the region may be helpful.

$$\iint_R (x + 2y)\,dx\,dy$$

30. Use the same integral as in Exercise 29, but let the vertices of the rectangular region be (0, 2), (8, 2), (0, 5), and (8, 5).

In Exercises 31–40, determine the volume between the given surface $z = f(x, y)$ and the rectangular region R defined in each case.

31. $f(x, y) = x + y$; R: $0 \le x \le 2$, $1 \le y \le 3$

32. $f(x, y) = 6x^2y$; R: $1 \le x \le 4$, $0 \le y \le 2$

33. $f(x, y) = 1$; R: $-1 \le x \le 1$, $-1 \le y \le 1$

34. $f(x, y) = 3$; R: $-2 \le x \le 3$, $-1 \le y \le 0$

35. $f(x, y) = x^2 + y^2$; R: $0 \le x \le 2$, $1 \le y \le 2$

36. $f(x, y) = x^2 + y^2 + 1$; R: $0 \le x \le 3$, $0 \le y \le 3$

37. $f(x, y) = 1 + x^2$; R: $1 \le x \le 2$, $0 \le y \le 4$

38. $f(x, y) = y^2 + 2$; R: $0 \le x \le 5$, $0 \le y \le 4$

39. $f(x, y) = \dfrac{1}{y}$; R: $1 \le x \le 10$, $1 \le y \le e$

40. $f(x, y) = e^x$; R: $0 \le x \le 1$, $1 \le y \le 5$

In Exercises 41–48, determine the average value of the function $f(x, y)$ over the rectangular region defined.

41. $f(x, y) = 4x + 3y$; R: $2 \le x \le 5$, $0 \le y \le 4$

42. $f(x, y) = x + y + 1$; R: $1 \le x \le 3$, $2 \le y \le 5$

43. $f(x, y) = 3x^2y$; R: $2 \le x \le 5$, $1 \le y \le 4$

44. $f(x, y) = 12xy^2$; R: $0 \le x \le 4$, $1 \le y \le 4$

45. $f(x, y) = 1 - 2x + y^2$; R: $0 \le x \le 1$, $4 \le y \le 10$

46. $f(x, y) = 10$; R: $1 \le x \le 5$, $2 \le y \le 5$

47. $f(x, y) = x^2 + y^2$; R: $0 \le x \le 2$, $1 \le y \le 3$

48. $f(x, y) = \sqrt{1 + x}$; R: $1 \le x \le 6$, $0 \le y \le 4$

Evaluate each double integral in Exercises 49–62.

49. $\int_0^1 \int_0^y e^x \, dx \, dy$

50. $\int_1^2 \int_0^{y-1} y^2 \, dx \, dy$

51. $\int_2^3 \int_1^{x^2} 4x \, dy \, dx$

52. $\int_1^2 \int_1^{y^2} y \, dx \, dy$

53. $\int_0^2 \int_x^{x^2+1} xy \, dy \, dx$

54. $\int_2^4 \int_{x/2}^{\sqrt{x}} xy \, dy \, dx$

55. $\int_0^2 \int_0^{2x} e^{x^2} \, dy \, dx$

56. $\int_0^2 \int_{x-1}^0 (1 - x + y) \, dy \, dx$

57. $\int_1^2 \int_{-y}^{2y} 3x^2 y \, dx \, dy$

58. $\int_0^{1/4} \int_{y^2}^{y/2} (x + y) \, dx \, dy$

59. $\int_0^1 \int_y^{\sqrt{y}} 2xy \, dx \, dy$

60. $\int_0^1 \int_{x^2}^x xy \, dy \, dx$

61. $\int_0^1 \int_x^{x^3} 2y \, dy \, dx$

62. $\int_2^3 \int_x^{x^2+1} (3x + 2y) \, dy \, dx$

In Exercises 63–68, draw the nonrectangular region over which integration was performed.

63. The region in Exercise 51.

64. The region in Exercise 53.

65. The region in Exercise 54.

66. The region in Exercise 55.

67. The region in Exercise 56.

68. The region in Exercise 60.

Chapter List *Important terms and ideas*

function of two variables
function of three variables
Cobb-Douglas production function
three-dimensional rectangular
   coordinate system
domain of $z = f(x, y)$
level curves
isoquants
partial derivative

second partial derivatives
critical point
relative maximum
relative minimum
saddle point
second partials test
Lagrange multipliers
constraint
method of least squares

least squares line
regression line
total differential
partial integration
double integral
volume under a surface
average value of a function

Review Exercises for Chapter 8

1. Let $f(x, y) = \dfrac{x^2 + y^2}{2x}$. Compute $f(4, 8)$ and $f(-6, 0)$.

2. Let $f(x, y) = \dfrac{2y + 9}{1 + 5x}$. Compute $f(5, 2)$ and $f(0, 0)$.

3. Determine the domain of $f(x, y) = \dfrac{y - 3}{x - 1}$.

4. Determine the domain of $f(x, y) = \sqrt{2x + y}$.

In Exercises 5–8, sketch the level curves of the given function for the specific c values listed.

5. $f(x, y) = x^2 - y + 2$; $c = -2, 0, 1$

6. $f(x, y) = x^2 + y + 4$; $c = 2, 3, 4$

7. $f(x, y) = 2x - y$; $c = 1, 3, 5$

8. $f(x, y) = 2x + y$; $c = -3, -2, -1, 0$

Determine f_x and f_y for each function in Exercises 9–12.

9. $f(x,y) = x^4 + x^2y^2 - y^4$ **10.** $f(x, y) = x(1 + y)^6$

11. $f(x, y) = e^{3x} + x \ln y$ **12.** $f(x, y) = \dfrac{2y}{3x}$

13. For $f(x, y, z) = xe^y + 3z^5$, find f_x, f_y, and f_z.

14. For $f(x, y, z) = 5xy^2z^3 - 1$, find f_x, f_y, and f_z.

15. Find all four second partials of the function defined by $f(x, y) = 5x^2 - xy + y^3$.

16. Find all four second partials of $f(x, y) = ye^{2x}$.

17. Find the critical points of the function defined by $f(x, y) = x^2 + 2xy - y^2 + 8x - 16y$.

18. Find the critical points of $f(x, y) = 4x^2 - y^2 + 17$.

19. Find the relative maximum or minimum value (if any) of $f(x, y) = 2x^2 + 12xy - 2y^2 - 11$.

20. Find the relative maximum or minimum value (if any) of $f(x, y) = 2x^2 + y^2 - xy - 5y - x + 6$.

21. (*BASEBALL, ERA*) In baseball, a pitcher's earned run average (ERA) is given by

$$E(r, i) = \frac{9r}{i}$$

where r is the number of earned runs scored during the number of innings i that he or she pitched.
(a) Matt pitched 4 innings and gave up 1 earned run. What is his ERA?
(b) Tanya pitched 12 innings and gave up no earned runs. What is her ERA?

22. (*SURFACE OF BOX*) The surface area of an open box (a box with no top) having length x, width y, and height z is

$$S(x, y, z) = xy + 2xz + 2yz$$

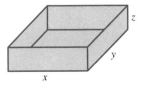

(a) Compute the value of $S(3, 5, 2)$.

W (b) Explain in words exactly what you determined in part (a). In your explanation, *do not use* the letters x, y, z, or S or the notation $S(3, 5, 2)$.

23. (*REVENUE*) A microwave oven producer makes x hundred of its regular model and y hundred of its deluxe model. The resulting revenue (in hundreds of dollars) is

$$R(x, y) = xy - x^2 - 2y^2 - x + 25y$$

How many of each model should be made in order to maximize revenue?

24. Use Lagrange multipliers to maximize the function $f(x, y) = 4xy - 8x$ subject to $x + y = 30$.

25. Use Lagrange multipliers to minimize the function $f(x, y) = x^2 + 2y^2 - 2x + 5$ subject to $x + 2y = 10$.

26. (*CHOLESTEROL LEVEL*) Blood cholesterol level is known to increase as you get older. Find the equation of the least squares line for the data given here from a random sample of 7 men.

age (x)	23	28	31	46	47	53	61
cholesterol (y)	164	201	180	235	220	305	260

27. (*ASSESSMENT TESTING*) The college gives all students an assessment test before they begin studying calculus. At the end of the calculus course, the instructor notes each student's final exam score (percent). Given the following data for 8 students, find the equation of the least squares line.

assessment score (x)	20	21	23	25	26	27	28	30
calculus exam (y)	72	90	76	87	80	98	85	92

28. (*PRODUCTION FUNCTION*) Given the Cobb-Douglas production function

$$f(x, y) = 50x^{1/4}y^{3/4}$$

use differentials to estimate the change in production if the number of units of labor (x) is decreased from

26 to 25 and the number of units of capital (y) is decreased from 40 to 38.

29. (*VOLUME INCREASE*) The voume of a box with a square base is given by

$$V(x, y) = x^2 y$$

where the length and width of the box are each x and the height is y. Use differentials to approximate the increase in volume obtained by increasing both length and width from 120 centimeters to 124 centimeters and increasing the height from 70 centimeters to 73 centimeters.

Evaluate each double integral in Exercises 30–39.

30. $\displaystyle\int_{4}^{10}\int_{2}^{3} 12xy\,dx\,dy$

31. $\displaystyle\int_{2}^{8}\int_{1}^{3} (4x - 3y)\,dx\,dy$

32. $\displaystyle\int_{-4}^{9}\int_{1}^{e} \frac{1}{y}\,dy\,dx$

33. $\displaystyle\int_{0}^{5}\int_{0}^{1/2} 4e^{2x}\,dx\,dy$

34. $\displaystyle\int_{2}^{5}\int_{1}^{6} \sqrt{y + 3}\,dy\,dx$

35. $\displaystyle\int_{-2}^{1}\int_{1}^{4} 8\,dy\,dx$

36. $\displaystyle\int_{-3}^{1}\int_{2}^{2y} (x - y)\,dx\,dy$

37. $\displaystyle\int_{0}^{6}\int_{0}^{\sqrt{y}} (2x + 4x^3)\,dx\,dy$

38. $\displaystyle\int_{-1}^{0}\int_{0}^{x+1} e^y\,dy\,dx$

39. $\displaystyle\int_{0}^{1}\int_{x}^{x^2} 4y\,dy\,dx$

9

TRIGONOMETRIC FUNCTIONS

Trigonometry was invented over 2000 years ago to solve the basic application problems of that era. As society's needs have changed, mathematics has evolved as well. Today we find that applications of trigonometry can include calculus ideas as well as some of the ancient fundamental concepts. In this chapter, you will have the opportunity to see a wide range of ideas and applications of trigonometry.

9.1 | *RIGHT TRIANGLES*

Trigonometry was developed in the second century B.C. by Greek mathematicians. Their purpose was to solve problems of astronomy and surveying in which they needed to determine distances and angles that could not be measured directly. The word *trigonometry* comes from Greek words that mean "triangle measurement." Much of their theory and applications were based on the use of right triangles. A **right triangle** is a triangle that contains a 90° (right) angle.

Right triangle trigonometry is still used today for various applications. Greek letters are still used frequently to name angles within triangles. The most popular letters are θ (theta) and ϕ (phi). In the right triangle shown in Figure 1, we will focus attention on angle θ.

Figure 1 A right triangle

The side opposite the right (90°) angle is called the **hypotenuse**. The side opposite the angle of interest (θ) is called the **opposite side**. The remaining side, adjacent to angle θ, is called the **adjacent side**. (See Figure 2.) Three trigonometric functions—sine, cosine, and tangent—can be defined in terms of the lengths of the sides of a right triangle.

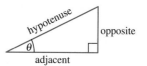

Figure 2 The sides of a right triangle

> ## Three Trigonometric Functions
>
> $$\text{sine of } \theta = \mathbf{sin}\ \boldsymbol{\theta} = \frac{\text{opposite}}{\text{hypotenuse}}$$
>
> $$\text{cosine of } \theta = \mathbf{cos}\ \boldsymbol{\theta} = \frac{\text{adjacent}}{\text{hypotenuse}}$$
>
> $$\text{tangent of } \theta = \mathbf{tan}\ \boldsymbol{\theta} = \frac{\text{opposite}}{\text{adjacent}}$$

Values of sin θ, cos θ, and tan θ are available from calculators or from tables. In most cases, calculators and tables provide approximations rather than exact values.

Note that sin θ, cos θ, and tan θ are indeed *functions of* θ. The value of tan θ, for example, depends on the value of θ. Consider a right triangle that contains an angle called θ. As the angle θ gets larger, so does the side opposite angle θ, and thus tan θ gets larger. See Figure 3.

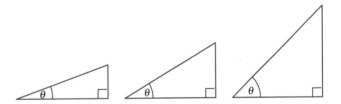

Figure 3 As θ gets larger, so too does the side opposite angle θ (the length of the adjacent side stays constant)

EXAMPLE 1 Use a calculator to obtain the value of each expression.

(a) sin 23°   **(b)** cos 52°   **(c)** tan 61°

SOLUTION Be sure that your calculator is set for degrees. When so set, the screen will usually display "DEG" or "D."

(a) *The steps for sin 23°:*   *The display:*

    1. Enter 23   23.

    2. Press ⬚sin⬚   0.390731128

We will round such results to four decimal places. Thus,

$$\sin 23° \approx .3907$$

(b) *The steps for cos 52°:*   *The display:*

    1. Enter 52   52.

    2. Press [**cos**]   0.615661475

    We will round the result to .6157. Thus,

$$\cos 52° \approx .6157$$

(c) *The steps for tan 61°:*   *The display:*

    1. Enter 61   61.

    2. Press [**tan**]   1.804047755

    Thus,

$$\tan 61° \approx 1.8040$$

The procedure used here will work for most basic calculators. It is possible that your calculator may require a different procedure. ◆

    The section ends with an application that shows how right triangle trigonometry can be used to solve problems of indirect measurement.

EXAMPLE 2 ◆ **HEIGHT OF A TREE**

To determine the height of a tree, a park ranger walks out 30 feet from the base of the tree. The ranger then determines the angle of elevation to the top of the tree to be 40°. (See Figure 4.) Use this information to determine the height of the tree.

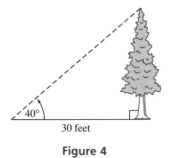

40°

30 feet

Figure 4

SOLUTION If we let x be the height of the tree, then we have the following right triangle (Figure 5).

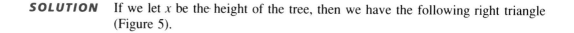

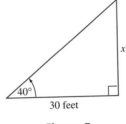

Figure 5

The sides involved are the side opposite the 40° angle (the side labeled *x*) and the adjacent side (30 feet). The trigonometric function that involves the opposite and adjacent sides is the tangent. This means we have

$$\tan 40° = \frac{\text{opposite}}{\text{adjacent}}$$

or

$$\tan 40° = \frac{x}{30}$$

By calculator (or Table 3), tan 40° ≈ .8391. Continuing,

$$.8391 \approx \frac{x}{30}$$

Multiplying both sides by 30 yields

$$x \approx 30(.8391)$$

or

$$x \approx 25.173$$

The tree is approximately 25 feet high. ◆

9.1 Exercises

In Exercises 1–12, use a calculator to obtain the approximate value of each expression. Round all results to four decimal places.

1. sin 24°

2. sin 78°

3. cos 37°

4. cos 12°

5. tan 62°

6. tan 14°

7. sin 17°

8. sin 74°

9. cos 33°

10. cos 71°

11. tan 41°

12. tan 31°

13. (*Kite Flying*) A child is flying a kite. 400 feet of string have been let out. The angle formed between the string and the ground is 53°. How high is the kite (to the nearest foot)? (See figure on next page.)

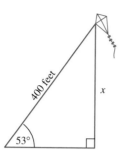

16. (*RIVER WIDTH*) Use the right triangle in the figure below to determine the distance across the river (to the nearest foot).

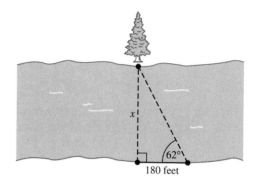

14. (*KITE FLYING*) A man is flying a kite. If the string makes a 47° angle with the ground and the kite is flying at a height of 150 feet, how much string has been let out (to the nearest foot)?

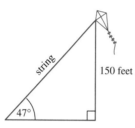

15. (*STREETLIGHT HEIGHT*) How tall is the streetlight (to the nearest foot) if it casts a 14-foot shadow when the sun shines from an angle of 36° with respect to the ground?

17. (*CEILOMETER USE*) Airports measure the height of a cloud cover by using *ceilometers*. In this way, they can determine if it is safe for planes to land and take off. As shown in the figure below, a projector directs a beam of light vertically up into the clouds. The detector senses the spot where the light beam hits the clouds, after which the angle θ between the ground and the light spot is determined. Finally, right triangle trigonometry is used to determine the height of the cloud cover. If the distance between the projector and the detector is 250 feet and the angle θ is determined to be 41°, how high (to the nearest foot) is the cloud cover?

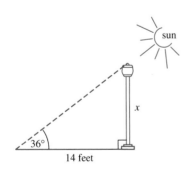

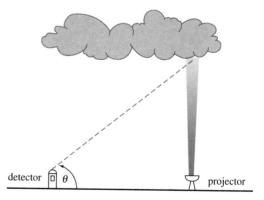

9.2	*RADIANS AND THE TRIGONOMETRY FOR CALCULUS*

When calculus involves trigonometry, angles measured in degrees are not used. Instead, a newer unit called *radians* is used. The radian concept was introduced in the 1870s.

Consider an angle θ with each side being a radius of a circle, as shown in Figure 6. Angle θ has a measure of **one radian** when the arc between the two radii has the same length as the radius. In Figure 7, angle θ has a measure of one radian. We will soon see that one radian is approximately equal to 57°.

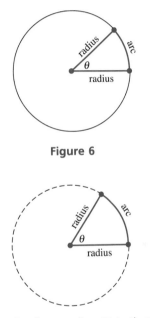

Figure 6

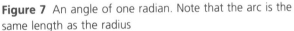

Figure 7 An angle of one radian. Note that the arc is the same length as the radius

The circumference of a circle is $2\pi r$. Dividing $2\pi r$ by arc length r shows us that there are 2π arcs of length r on a circle. Since each arc of length r is associated with an angle of 1 radian, it follows that a full revolution of 360° is the same as 2π radians. In other words,

$$360° = 2\pi \text{ radians}$$

Dividing both sides of this equation by 2 yields an important result.

$$180° = \pi \text{ radians}$$

Dividing both sides of this equation by π shows that 1 radian is $180/\pi$ degrees.

$$1 \text{ radian} = \frac{180°}{\pi}$$

Using 3.14 as an approximation for π leads to the result

$$1 \text{ radian} \approx \frac{180°}{3.14} \approx 57.3°$$

Returning to the equation $180° = \pi$, dividing both sides by 180 shows that 1 degree is equal to $\pi/180$ radian.

$$1° = \frac{\pi}{180} \text{ radian}$$

Using 3.14 as an approximation for π leads to the result

$$1° \approx .017 \text{ radian}$$

The last two boxed formulas can be used to convert angles from radians to degrees and vice versa.

EXAMPLE 1 Change $\pi/4$ and $2\pi/3$ from radians to degrees.

SOLUTION Multiply each expression by $180°/\pi$.

$$\frac{\pi}{4} = \frac{\pi}{4} \cdot \frac{180°}{\pi} = 45°$$

$$\frac{2\pi}{3} = \frac{2\pi}{3} \cdot \frac{180°}{\pi} = 120° \quad \blacklozenge$$

EXAMPLE 2 Change $60°$ and $315°$ to radians.

SOLUTION Multiply each expression by $\pi/180°$.

$$60° = 60° \cdot \frac{\pi}{180°} = \frac{\pi}{3}$$

$$315° = 315° \cdot \frac{\pi}{180°} = \frac{7\pi}{4} \quad \blacklozenge$$

Alternative definitions of sine and cosine that do not depend on right triangles can be established by using the unit circle. The **unit circle** has its center at the origin and a radius of 1. Its equation is $x^2 + y^2 = 1$. See Figure 8. If we begin at the point $(1, 0)$ and then move around the circle through an angle of θ radians, the x coordinate obtained is cos θ and the y coordinate is sin θ. (See Figure 9.) These are the unit circle definitions of sine and cosine.

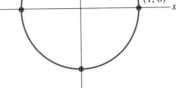

x	y
1	0
0	1
−1	0
0	−1

Figure 8 The unit circle

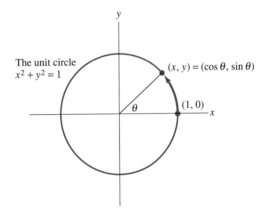

Figure 9 The unit circle definitions of sine and cosine

Trigonometric Functions

The unit circle definitions of sine, cosine, and tangent are as follows. (See Figure 9.)

$$\cos \theta = x$$
$$\sin \theta = y$$
$$\tan \theta = \frac{y}{x} = \frac{\sin \theta}{\cos \theta} \qquad x \neq 0$$

where (x, y) is a point on the unit circle.

To see that these two new definitions are indeed consistent with the right triangle definitions, consider Figure 10 and the work that follows.

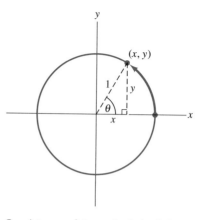

Figure 10 Consistency of the unit circle definitions with the right triangle definitions

Using the right triangle definitions, we have

$$\cos \theta = \frac{\text{adjacent}}{\text{hypotenuse}} = \frac{x}{1} = x$$

$$\sin \theta = \frac{\text{opposite}}{\text{hypotenuse}} = \frac{y}{1} = y$$

$$\tan \theta = \frac{\text{opposite}}{\text{adjacent}} = \frac{y}{x} = \frac{\sin \theta}{\cos \theta}$$

The letter θ used here represents an angle in radians. Letters other than θ can be used. For example, we could have $\sin t$, $\cos u$, $\sin x$, $\cos x$, and others.

Key values of $\cos \theta$ and $\sin \theta$ can be obtained from the unit circle by using the unit circle definitions of sine and cosine. Figure 11 shows the values of $\sin 0$, $\cos 0$, $\sin \pi/2$, $\cos \pi/2$, and others obtained from the unit circle.

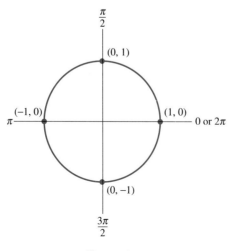

	0	$\dfrac{\pi}{2}$	π	$\dfrac{3\pi}{2}$	2π
$\cos t$	1	0	-1	0	1
$\sin t$	0	1	0	-1	0

Figure 11

It is also important to know the values of $\cos \theta$ and $\sin \theta$ for $\theta = \pi/6$, $\pi/4$, and $\pi/3$. The following table provides them. Background on the origin of the values is provided in Exercise 45.

	$\dfrac{\pi}{6}$	$\dfrac{\pi}{4}$	$\dfrac{\pi}{3}$
$\cos \theta$	$\dfrac{\sqrt{3}}{2}$	$\dfrac{\sqrt{2}}{2}$	$\dfrac{1}{2}$
$\sin \theta$	$\dfrac{1}{2}$	$\dfrac{\sqrt{2}}{2}$	$\dfrac{\sqrt{3}}{2}$

(Quadrant I)

In principle, corresponding values in other quadrants can be determined by symmetry. This approach is demonstrated in Figure 12 for $\cos 2\pi/3$ and $\sin 2\pi/3$.

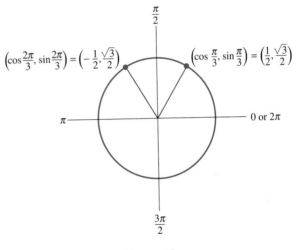

Figure 12

A table is provided here for reference. It includes the exact values of $\sin \theta$ and $\cos \theta$ for special θ values in quadrants II, III, and IV.

	$\dfrac{2\pi}{3}$	$\dfrac{3\pi}{4}$	$\dfrac{5\pi}{6}$	$\dfrac{7\pi}{6}$	$\dfrac{5\pi}{4}$	$\dfrac{4\pi}{3}$	$\dfrac{5\pi}{3}$	$\dfrac{7\pi}{4}$	$\dfrac{11\pi}{6}$
$\cos \theta$	$-\dfrac{1}{2}$	$-\dfrac{\sqrt{2}}{2}$	$-\dfrac{\sqrt{3}}{2}$	$-\dfrac{\sqrt{3}}{2}$	$-\dfrac{\sqrt{2}}{2}$	$-\dfrac{1}{2}$	$\dfrac{1}{2}$	$\dfrac{\sqrt{2}}{2}$	$\dfrac{\sqrt{3}}{2}$
$\sin \theta$	$\dfrac{\sqrt{3}}{2}$	$\dfrac{\sqrt{2}}{2}$	$\dfrac{1}{2}$	$-\dfrac{1}{2}$	$-\dfrac{\sqrt{2}}{2}$	$-\dfrac{\sqrt{3}}{2}$	$-\dfrac{\sqrt{3}}{2}$	$-\dfrac{\sqrt{2}}{2}$	$-\dfrac{1}{2}$

II III IV

Values of tan θ can be computed as sin θ divided by cos θ, as demonstrated in the next example.

EXAMPLE 3   Determine the exact value of each expression.

(a) $\tan \dfrac{7\pi}{6}$ (b) $\tan \dfrac{\pi}{4}$ (c) $\tan \dfrac{\pi}{2}$

SOLUTION   We will use the fact that tangent is equal to sine divided by cosine.

(a) $\tan \dfrac{7\pi}{6} = \dfrac{\sin \dfrac{7\pi}{6}}{\cos \dfrac{7\pi}{6}} = \dfrac{-\dfrac{1}{2}}{-\dfrac{\sqrt{3}}{2}} = \dfrac{1}{\sqrt{3}}$ or $\dfrac{\sqrt{3}}{3}$

(b) $\tan \dfrac{\pi}{4} = \dfrac{\sin \dfrac{\pi}{4}}{\cos \dfrac{\pi}{4}} = \dfrac{\dfrac{\sqrt{2}}{2}}{\dfrac{\sqrt{2}}{2}} = 1$

(c) $\tan \dfrac{\pi}{2} = \dfrac{\sin \dfrac{\pi}{2}}{\cos \dfrac{\pi}{2}} = \dfrac{1}{0}$ (undefined)

Observe that $\tan \dfrac{\pi}{2}$ is undefined.   ◆

In addition to the sine, cosine, and tangent, there are three other trigonometric functions. These additional functions can be defined in terms of the sine, cosine, and tangent.

secant	$\sec \theta = \dfrac{1}{\cos \theta}$	$\cos \theta \neq 0$
cosecant	$\csc \theta = \dfrac{1}{\sin \theta}$	$\sin \theta \neq 0$
cotangent	$\cot \theta = \dfrac{\cos \theta}{\sin \theta}$	$\sin \theta \neq 0$

Notice, too, that cot $\theta = 1/\tan \theta$ when tan θ is defined and not zero.

EXAMPLE 4   Determine the exact value of each expression.

(a) $\sec \dfrac{\pi}{3}$ (b) $\csc \dfrac{3\pi}{2}$ (c) $\cot \dfrac{5\pi}{6}$

SOLUTION   We will use the definitions of secant, cosecant, and tangent.

(a) $\sec \dfrac{\pi}{3} = \dfrac{1}{\cos \dfrac{\pi}{3}} = \dfrac{1}{1/2} = 2$

(b) $\csc \dfrac{3\pi}{2} = \dfrac{1}{\sin \dfrac{3\pi}{2}} = \dfrac{1}{-1} = -1$

(c) $\cot \dfrac{5\pi}{6} = \dfrac{\cos \dfrac{5\pi}{6}}{\sin \dfrac{5\pi}{6}} = \dfrac{-\sqrt{3}/2}{1/2} = -\sqrt{3}$ ◆

Calculators can be used to obtain approximate values of trigonometric functions. Such an approach is particularly useful when the angle involved is not one of the special, familiar ones for which the exact value of the sine or cosine is known.

EXAMPLE 5   Use a calculator to obtain the approximate value of each expression.

(a) $\sin 1.25$ **(b)** $\cos 2.09$ **(c)** $\tan 2$ **(d)** $\sec 2.6$

(e) $\csc 1.8$ **(f)** $\cot 3.07$

SOLUTION   Notice that degrees are not indicated, which means the numbers 1.25, 2.09, 2, 2.6, 1.8, and 3.07 are radians rather than degrees. *Be sure your calculator is set for radians (R or RAD), not degrees.* (The keys to change the setting from degrees to radians or radians to degrees vary from calculator to calculator.)

(a) *The steps for sin 1.25:*   *The display:*

  1. Enter 1.25   1.25

  2. Press ⟨ sin ⟩   0.948984619

  Rounded to four decimal places, the result is

$$\sin 1.25 \approx .9490$$

(b) *The steps for cos 2.09:*   *The display:*

  1. Enter 2.09   2.09

  2. Press ⟨ cos ⟩   −0.496188912

  Rounded to four decimal places, the result is

$$\cos 2.09 \approx -.4962$$

(c) *The steps for tan 2:*   *The display:*

    1. Enter 2   2.

    2. Press [tan]   −2.185039863

Rounded to four decimal places, the result is

$$\tan 2 \approx -2.1850$$

(d) *The steps for sec 2.6:*   *The display:*

    1. Enter 2.6   2.6

    2. Press [cos]   −0.856888753

    3. Press [1/x]   −1.167012633

Notice that the secant is evaluated as the reciprocal of the cosine, which is exactly how the secant is defined. Step 2 produces cos 2.6. Step 3 produces 1/cos 2.6, which is sec 2.6. Rounded to four decimal places, the result is

$$\sec 2.6 \approx -1.1670$$

(e) *The steps for csc 1.8:*   *The display:*

    1. Enter 1.8   1.8

    2. Press [sin]   0.97384763

    3. Press [1/x]   1.026854683

The cosecant is evaluated (by its definition) as the reciprocal of the sine. Step 2 produces sin 1.8. Step 3 produces 1/sin 1.8, which is csc 1.8. Rounded to four decimal places, the result is

$$\csc 1.8 \approx 1.0269$$

(f) *The steps for cot 3.07:*   *The display:*

    1. Enter 3.07   3.07

    2. Press [tan]   −0.071715221

    3. Press [1/x]   −13.94404123

The cotangent is evaluated as the reciprocal of the tangent. Step 2 produces tan 3.07. Step 3 produces 1/tan 3.07, which is cot 3.07. Rounded to four decimal places, the result is

$$\cot 3.07 \approx -13.9440 \quad \blacklozenge$$

Radians and Real Numbers

When we begin at the point (1, 0) and move around the unit circle through an *angle* of θ radians, we are also moving on the circle itself through a *distance* of θ units. The length of arc is the same as the number of radians, because the circumference ($C = 2\pi r$) of the unit circle is $C = 2\pi(1) = 2\pi$, which is the same as the number of radians in one complete revolution. Thus, not only is θ a measure of an angle in radians, it is also a distance—a real number. This means that when you see an expression such as sin 1.06, it may be the sine of 1.06 radians *or* the sine of the real number 1.06. In calculus, the 1.06 is often a real number rather than an angle in radians.

Sine and Cosine Graphs

A graph of the sine function $y = \sin x$ can be obtained by supplying various values of x and computing the corresponding y values. You can also anticipate the general appearance of the graph by studying the unit circle. Keep in mind that the sine is the second coordinate of each point on the unit circle. In the first quadrant (between 0 and $\pi/2$), the sine gradually increases from 0 to 1. In the second quadrant (between $\pi/2$ and π), sine values decrease from 1 to 0. In the third quadrant, sine values decrease from 0 to -1. In the fourth quadrant, sine values increase from -1 to 0. (See Figure 13.)

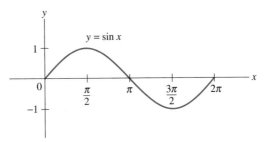

Figure 13 One cycle of the sine curve

We could continue around the unit circle past 2π or go around the circle in the clockwise (negative) direction. Either way, additional cycles of the basic sine curve would be obtained. See Figure 14.

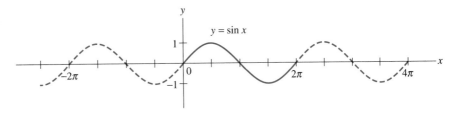

Figure 14 Additional cycles of the basic sine curve

If a similar approach is taken for $y = \cos x$, we will obtain the graph shown in Figure 15.

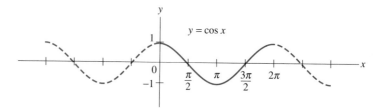

Figure 15 The cosine curve

From both the unit circle and the graphs of $y = \sin x$ and $y = \cos x$, we can see that the values of sine and cosine vary between -1 and 1.

$$-1 \le \sin x \le 1$$
$$-1 \le \cos x \le 1$$

The graphs also show that $y = \sin x$ and $y = \cos x$ are continuous functions.

The graphs show clearly the periodic nature of the sine and cosine functions. This characteristic leads to a variety of applications to situations in which function values fluctuate or cycle.

EXAMPLE 6 ◆ **BLOOD PRESSURE**

A person's blood pressure is constantly fluctuating and cycling. Someone's blood pressure at any time t (in seconds) may be given by the function

$$P(t) = 110 + 25 \cos 5.8t$$

(a) Find the *systolic pressure*, which is the maximum blood pressure.

(b) Find the *diastolic pressure*, which is the minimum blood pressure.

(c) What is this person's blood pressure after 2 seconds?

SOLUTION **(a)** The blood pressure will be maximum when $\cos 5.8t$ is 1, since 1 is the largest value that the cosine can be. At that time,

$$P(t) = 110 + 25(1) = 135$$

This person's systolic pressure is 135.

(b) The blood pressure will be minimum when $\cos 5.8t$ is -1, since -1 is the smallest value that the cosine can be. At that time,

$$P(t) = 110 + 25(-1) = 85$$

This person's diastolic pressure is 85.

(c) After 2 seconds, the blood pressure is $P(2)$.

$$P(2) = 110 + 25 \cos 5.8(2)$$
$$= 110 + 25 \cos 11.6$$
$$\approx 110 + 25(.5683) \qquad \text{(by calculator)}$$
$$\approx 110 + 14 \qquad \text{(nearest whole number)}$$
$$= 124$$

We see that after 2 seconds the blood pressure is 124. Figure 16 shows a graph of this blood pressure function.

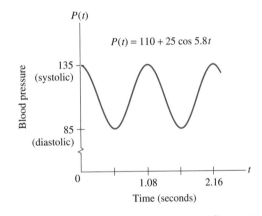

Figure 16 A graph showing the continual fluctuation in blood pressure ◆

9.2 Exercises

In Exercises 1–10, use the method of Example 1 to change each angle measure from radians to degrees.

1. $\pi/6$

2. $\pi/3$

3. $\pi/2$

4. $\pi/10$

5. $5\pi/6$

6. $3\pi/4$

7. $5\pi/4$

8. $11\pi/6$

9. $4\pi/3$

10. $5\pi/3$

In Exercises 11–20, use the method of Example 2 to change each angle measure from degrees to radians.

11. 90°

12. 30°

13. 45°

14. 270°

15. 120°

16. 150°

17. 210°

18. 240°

19. 135°

20. 330°

In Exercises 21–32, use the method of Examples 3 and 4 to determine the exact value requested.

21. $\tan \dfrac{\pi}{3}$　　　　　　　**22.** $\tan \dfrac{2\pi}{3}$

23. $\tan \dfrac{7\pi}{4}$　　　　　　　**24.** $\tan \pi$

25. $\tan 0$　　　　　　　**26.** $\tan \dfrac{5\pi}{4}$

27. $\cot \dfrac{\pi}{2}$　　　　　　　**28.** $\cot \dfrac{\pi}{6}$

29. $\sec \pi$　　　　　　　**30.** $\sec \dfrac{2\pi}{3}$

31. $\csc \dfrac{\pi}{6}$　　　　　　　**32.** $\csc \dfrac{\pi}{4}$

In Exercises 33–44, use a calculator to obtain the approximate value of each expression. Refer to Example 5.

33. $\sin 1.1$　　　　　　　**34.** $\sin 3$

35. $\cos .8$　　　　　　　**36.** $\cos 2.4$

37. $\tan 1.73$　　　　　　　**38.** $\tan 4$

39. $\sec 2$　　　　　　　**40.** $\sec 1.2$

41. $\csc .5$　　　　　　　**42.** $\csc 1.03$

43. $\cot 1.9$　　　　　　　**44.** $\cot 2.3$

45. The values of $\sin t$ and $\cos t$ for $t = \pi/6$, $\pi/4$, and $\pi/3$ can be obtained by using right triangles. Keep in mind that $\pi/6 = 30°$, $\pi/4 = 45°$, and $\pi/3 = 60°$.

(a) In a right triangle with a 45° angle, the angles are 45°, 45°, and 90° (since the sum of the angles in a triangle is 180°). The sides opposite the 45° angles are the same size. We'll let them be 1 unit each. The Pythagorean theorem yields a hypotenuse of length $\sqrt{2}$.

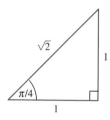

Use the right triangle to show that $\sin \pi/4 = \sqrt{2}/2$ and $\cos \pi/4 = \sqrt{2}/2$.

(b) In a right triangle with a 30° angle, the angles are 30°, 60°, and 90°. In such a triangle, the side opposite the 30° angle is half the length of the hypotenuse. If we let the hypotenuse be 2 units, then the side opposite the 30° angle will be 1 unit. The Pythagorean theorem yields $\sqrt{3}$ as the third side.

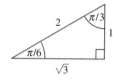

Use the right triangle to show that $\sin \pi/6 = 1/2$, $\cos \pi/6 = \sqrt{3}/2$, $\sin \pi/3 = \sqrt{3}/2$, and $\cos \pi/3 = 1/2$.

46. (*BLOOD PRESSURE*) Refer to Example 6. What was the person's blood pressure at the beginning of the test?

47. (*BLOOD PRESSURE*) Hamid's blood pressure is given by the function

$$P(t) = 105 + 25 \cos 6t$$

where t is the time in seconds.
(a) What is Hamid's systolic (maximum) pressure?
(b) What is his diastolic (minimum) pressure?
(c) What is his blood pressure at the beginning of the test?
(d) What is his blood pressure after 1 second?

48. (*BLOOD PRESSURE*) A person's blood pressure will vary according to

$$P = a + b \cos ct$$

where P is the blood pressure and t is the time in seconds. The number of heartbeats H per minute (pulse rate) is then

$$H = \frac{30c}{\pi}$$

Suppose that Elwin's blood pressure is given by $P = 90 + 20 \cos 6.5t$. What is his pulse rate?

49. (*SALES CYCLE*) A lawn mower distributor finds that last year's sales were cyclical and were given by

$$S(t) = 3000 + 2000 \cos \frac{\pi t}{6}$$

where S is the number of lawn mowers sold and t is the time in months (1 is July, 2 is August, . . . , 12 is June).

(a) How many lawn mowers were sold in August?

(b) How many lawn mowers did the distributor sell in December?

(c) What is the greatest number of lawn mowers sold in any month?

(d) What is the fewest number of lawn mowers sold in any month?

50. (*SALES CYCLE*) Suppose that the number of items sold in a particular month is given by

$$S(t) = 1200 + 500 \sin \frac{\pi t}{6}$$

where $t = 1$ means January, $t = 2$ means February, and so on.

(a) How many units were sold in January?

(b) How many units were sold in June?

(c) What is the maximum number of units sold in any month?

(d) What is the minimum number of units sold in any month?

W 51. Consider the graphs of $y = \sin x$ and $y = \cos x$ in order to evaluate the limits.

(a) $\lim\limits_{x \to \infty} \sin x$

(b) $\lim\limits_{x \to \infty} \cos x$

In each case, you seek the number toward which sin x (and then cos x) tends as x gets larger and larger. *Explain the results.*

W 52. Using the unit circle, explain why the value of the sine (the y coordinate) cannot be greater than 1.

W 53. Using the unit circle, explain why the value of the cosine' (the x coordinate) cannot be less than -1.

The Exercise Library at the back of the book contains graphing calculator and computer exercises keyed to this section.

9.3 | *DIFFERENTIATION OF TRIGONOMETRIC FUNCTIONS*

In this section we obtain the differentiation formulas for all six trigonometric functions. The definition of the derivative is used to obtain the derivative of the sine function. The other five derivatives follow readily after that.

We will need to make use of some trigonometric **identities**, that is, equations that are true for all values that the variables can take on. Here are two identities that we will use.

$$\cos^2 t + \sin^2 t = 1$$

$$\sin (u + v) = \sin u \cos v + \cos u \sin v$$

Note that $\cos^2 t$ means $(\cos t)^2$ and $\sin^2 t$ means $(\sin t)^2$. (Exercise 60 provides the opportunity to prove the first of these identities.)

To determine the derivative of $y = \sin x$, we will make use of the following limits.

$$\lim\limits_{t \to 0} \frac{\sin t}{t} = 1 \qquad (t \text{ in radians})$$

$$\lim\limits_{t \to 0} \frac{\cos t - 1}{t} = 0 \qquad (t \text{ in radians})$$

Exercises 61 and 62 provide the opportunity to verify these results by using a calculator to complete tables in which t approaches 0.

To obtain the derivative of $f(x) = \sin x$, we begin with the definition of the derivative, namely,

$$f'(x) = \lim_{\Delta x \to 0} \frac{f(x + \Delta x) - f(x)}{\Delta x}$$

Using $f(x) = \sin x$, we have

$$f'(x) = \lim_{\Delta x \to 0} \frac{\sin (x + \Delta x) - \sin x}{\Delta x}$$

The expression $\sin (x + \Delta x)$ can be expanded by using the identity given for $\sin (u + v)$. Specifically,

$$\sin (x + \Delta x) = \sin x \cos \Delta x + \cos x \sin \Delta x$$

Substituting this expanded form for $\sin (x + \Delta x)$ yields

$$f'(x) = \lim_{\Delta x \to 0} \frac{\sin x \cos \Delta x + \cos x \sin \Delta x - \sin x}{\Delta x}$$

Interchanging the last two terms in the numerator produces

$$f'(x) = \lim_{\Delta x \to 0} \frac{\sin x \cos \Delta x - \sin x + \cos x \sin \Delta x}{\Delta x}$$

Now, $\sin x$ can be factored out of the first two terms of the numerator.

$$f'(x) = \lim_{\Delta x \to 0} \frac{\sin x (\cos \Delta x - 1) + \cos x \sin \Delta x}{\Delta x}$$

The fraction can be split into two equivalent fractions.

$$f'(x) = \lim_{\Delta x \to 0} \left[\frac{\sin x (\cos \Delta x - 1)}{\Delta x} + \frac{\cos x \sin \Delta x}{\Delta x} \right]$$

The limit of the sum can be written as the sum of the limits.

$$f'(x) = \lim_{\Delta x \to 0} \frac{\sin x (\cos \Delta x - 1)}{\Delta x} + \lim_{\Delta x \to 0} \frac{\cos x \sin \Delta x}{\Delta x}$$

Within each limit, the factors can be rearranged. Then the two limits that were listed earlier can be used to obtain the final result.

$$f'(x) = \lim_{\Delta x \to 0} (\sin x) \left(\frac{\cos \Delta x - 1}{\Delta x} \right) + \lim_{\Delta x \to 0} (\cos x) \left(\frac{\sin \Delta x}{\Delta x} \right)$$

$$= (\sin x)(0) + (\cos x)(1)$$

$$= \cos x$$

Thus, we have shown that

$$\frac{d}{dx}(\sin x) = \cos x$$

The chain rule can be used to obtain the following formula for differentiating $y = \sin u$, where u is a function of x.

$$\frac{d}{dx}(\sin u) = (\cos u)\frac{du}{dx}$$

EXAMPLE 1   Differentiate $y = \sin 8x$.

SOLUTION   The function has the form $y = \sin u$, where $u = 8x$.

$$\frac{dy}{dx} = \frac{d}{dx}(\sin 8x)$$

$$= (\cos 8x)\frac{d}{dx}(8x)$$

$$= (\cos 8x) \cdot 8$$

$$= 8 \cos 8x \quad \blacklozenge$$

EXAMPLE 2   Determine $\dfrac{d}{dx}(\sin^3 x)$.

SOLUTION   Keeping in mind that $\sin^3 x$ means $(\sin x)^3$, we have

$$\frac{d}{dx}(\sin^3 x) = \frac{d}{dx}(\sin x)^3$$

$$= 3(\sin x)^2 \cdot \frac{d}{dx}(\sin x)$$

$$= 3(\sin x)^2 \cdot \cos x$$

$$= 3 \sin^2 x \cos x \quad \blacklozenge$$

EXAMPLE 3   Find the derivative of $y = x \sin x$.

SOLUTION   Note that $x \sin x$ is the *product* of x and $\sin x$. Using the product rule,

$$\frac{dy}{dx} = \frac{d}{dx}(x \cdot \sin x)$$

$$= (x)(\cos x) + (\sin x)(1)$$

$$= x \cos x + \sin x \quad \blacklozenge$$

The derivative of $y = \cos x$ can be obtained by using the derivative of the sine function together with the following two trigonometric identities.

$$\sin\left(\frac{\pi}{2} - x\right) = \cos x$$

$$\cos\left(\frac{\pi}{2} - x\right) = \sin x$$

The first identity is used to begin the procedure. The second identity is used later on to simplify the result.

$$\frac{d}{dx}(\cos x) = \frac{d}{dx}\sin\left(\frac{\pi}{2} - x\right)$$

$$= \cos\left(\frac{\pi}{2} - x\right) \cdot (-1)$$

$$= (\sin x)(-1)$$

$$= -\sin x$$

Thus,

$$\frac{d}{dx}(\cos x) = -\sin x$$

By the chain rule, if u is a function of x, we have

$$\frac{d}{dx}(\cos u) = (-\sin u) \cdot \frac{du}{dx}$$

EXAMPLE 4   Find the derivative of $y = \cos 4x$.

SOLUTION   The function has the form $y = \cos u$, where $u = 4x$.

$$\frac{dy}{dx} = \frac{d}{dx}(\cos 4x)$$

$$= (-\sin 4x) \cdot \frac{d}{dx}(4x)$$

$$= (-\sin 4x)(4)$$

$$= -4\sin 4x \quad \blacklozenge$$

EXAMPLE 5   Differentiate $y = \dfrac{3x}{\cos x}$.

SOLUTION   The expression is the *quotient* of $3x$ and $\cos x$. Using the quotient rule,

$$\frac{dy}{dx} = \frac{d}{dx}\left(\frac{3x}{\cos x}\right)$$

$$= \frac{(\cos x)(3) - (3x)(-\sin x)}{(\cos x)^2}$$

$$= \frac{3\cos x + 3x\sin x}{\cos^2 x} \quad \blacklozenge$$

The next two examples offer applications involving the sine and cosine functions and their derivatives.

EXAMPLE 6   ◆   *SALES CYCLE*

Lawn mower sales are cyclical, being low in the winter, increasing in early spring, and continuing to increase through spring. Some decline will occur in summer, and fall sales will show a dramatic decline. The cyclical sales pattern suggests that the distributor's sales may be approximated by a function involving the sine or cosine, such as

$$L(t) = 1350 - 1200 \cos \frac{\pi t}{6}$$

where L is the number of lawnmowers sold and t is the time in months (1 is January, 2 is February, and so on).

We can see that March sales are 1350 mowers, since for $t = 3$ (for March) we have

$$L(3) = 1350 - 1200 \cos \frac{\pi(3)}{6}$$

$$= 1350 - 1200 \cos \frac{\pi}{2}$$

$$= 1350 - 1200\,(0)$$

$$= 1350$$

The derivative, dL/dt, gives the rate at which monthly sales are changing.

$$\frac{dL}{dt} = \frac{d}{dt}\left(1350 - 1200 \cos \frac{\pi}{6}t\right)$$

$$= 0 - 1200 \cdot \left(-\sin \frac{\pi}{6}t\right) \cdot \frac{\pi}{6}$$

which simplifies to

$$\frac{dL}{dt} = 200\pi \sin\frac{\pi}{6}t$$

In March, monthly sales are changing (increasing) at the rate of

$$\frac{dL}{dt} = 200\pi \sin\frac{\pi(3)}{6} \approx 628 \text{ mowers/month}$$

In October, monthly sales are changing at the rate of

$$\frac{dL}{dt} = 200\pi \sin\frac{\pi(10)}{6} \approx -544 \text{ mowers/month}$$

The minus sign indicates that the sales rate is *decreasing*. ◆

EXAMPLE 7 ◆ *PREDATOR-PREY RELATIONSHIP*

The population of some wild animal species may fluctuate because of a predator-prey relationship. One species may be prey for another (predator) species. (For example, in northern Canada, the snowshoe rabbit is prey for the lynx.) The prey population declines as the predator population increases. But such a decline in the prey population (which is food for the predator) eventually results in a decline in the predator population. Then, a declining predator population enables the prey population to increase. Both populations fluctuate in a cyclical manner.

Suppose that in some region, the predator population and the prey population are given by the following functions, where t is time in months:

$$predator: \quad D(t) = 1400 + 600 \sin\frac{\pi t}{12}$$

$$prey: \quad Y(t) = 2500 + 1500 \cos\frac{\pi t}{12}$$

$D(t)$ gives the predator population at any time t. $Y(t)$ gives the prey population at any time t. For example, after 6 months ($t = 6$), we have

$$D(6) = 1400 + 600 \sin\frac{6\pi}{12} = 2000$$

$$Y(6) = 2500 + 1500 \cos\frac{6\pi}{12} = 2500$$

$D'(t)$ and $Y'(t)$ can be obtained and then used to determine the rate at which either population is increasing or decreasing at any particular time. See Exercises 58 and 59.

A graph of the predator and prey population functions and their relationship is shown in Figure 17.

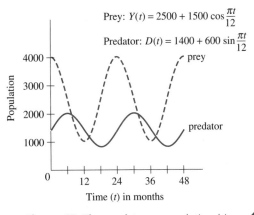

Prey: $Y(t) = 2500 + 1500 \cos \dfrac{\pi t}{12}$

Predator: $D(t) = 1400 + 600 \sin \dfrac{\pi t}{12}$

Figure 17 The predator-prey relationship   ◆

We proceed next to find the derivative formulas for the other four trigono-metric functions. The derivative of $y = \tan x$ is readily obtained by considering $\tan x$ as $\sin x \,/\, \cos x$ and differentiating the quotient.

$$
\begin{aligned}
D_x(\tan x) &= D_x\!\left(\frac{\sin x}{\cos x}\right) \\[2mm]
&= \frac{(\cos x)(\cos x) - (\sin x)(-\sin x)}{(\cos x)^2} \\[2mm]
&= \frac{\cos^2 x + \sin^2 x}{\cos^2 x}
\end{aligned}
$$

Using the identity $\cos^2 x + \sin^2 x = 1$ (from the beginning of this section) and then noting that $1/\cos x$ is the same as $\sec x$ will yield the final result. In steps,

$$
\begin{aligned}
D_x(\tan x) &= \frac{1}{\cos^2 x} \\[2mm]
&= \sec^2 x
\end{aligned}
$$

Thus,

$$
\frac{d}{dx}(\tan x) = \sec^2 x
$$

and

$$
\frac{d}{dx}(\tan u) = \sec^2 u \cdot \frac{du}{dx}
$$

The remaining differentiation formulas are given next. Proofs of the rules are left as exercises, with hints. (See Exercises 65–67.)

$$\frac{d}{dx}(\cot x) = -\csc^2 x$$

$$\frac{d}{dx}(\sec x) = \sec x \tan x$$

$$\frac{d}{dx}(\csc x) = -\csc x \cot x$$

and

$$\frac{d}{dx}(\cot u) = -\csc^2 u \, \frac{du}{dx}$$

$$\frac{d}{dx}(\sec u) = \sec u \tan u \, \frac{du}{dx}$$

$$\frac{d}{dx}(\csc u) = -\csc u \cot u \, \frac{du}{dx}$$

EXAMPLE 8   Determine the derivative of each function.

(a) $y = \tan 3x^2$ **(b)** $y = \cot^2 x$ **(c)** $f(x) = \sec 2x$

(d) $f(x) = \csc x^2$

SOLUTION   Each differentiation makes use of one of our new formulas.

(a) $\dfrac{dy}{dx} = \dfrac{d}{dx}(\tan 3x^2) = (\sec^2 3x^2)(6x) = 6x \sec^2 3x^2$

(b) $\dfrac{dy}{dx} = \dfrac{d}{dx}(\cot^2 x) = \dfrac{d}{dx}(\cot x)^2 = 2(\cot x)^1(-\csc^2 x) = -2 \csc^2 x \cot x$

(c) $f'(x) = \dfrac{d}{dx}(\sec 2x) = (\sec 2x \tan 2x)(2) = 2 \sec 2x \tan 2x$

(d) $f'(x) = \dfrac{d}{dx}(\csc x^2) = (-\csc x^2 \cot x^2)(2x) = -2x \csc x^2 \cot x^2$ ◆

9.3 Exercises

In Exercises 1–16, find the derivative of each function.

1. $y = \sin 2x$

2. $y = \sin 6x$

3. $y = 7 \sin x$

4. $y = -3 \sin x$

5. $y = \sin x^2$

6. $y = \sin x^3$

7. $y = \sin^2 x$

8. $y = \sin^4 x$

9. $f(x) = 5 \sin 3x$

10. $f(x) = 4 \sin 2x$

11. $f(x) = x \sin 4x$

12. $f(x) = x \sin 5x$

13. $g(t) = 4t \sin t$

14. $g(t) = 7t \sin t$

15. $y = \dfrac{\sin x}{x}$

16. $y = \dfrac{x}{\sin x}$

In Exercises 17–36, find the derivative of each function. Note that π is a constant.

17. $y = \cos 2x$

18. $y = \cos 6x$

19. $f(x) = \cos \pi x$

20. $f(x) = \cos 2\pi x$

21. $y = \sin \dfrac{\pi t}{3}$

22. $y = \sin \dfrac{\pi t}{6}$

23. $y = \cos \dfrac{\pi t}{4}$

24. $y = \cos \dfrac{\pi t}{2}$

25. $y = \dfrac{\sin x}{3}$

26. $y = \dfrac{\cos x}{2}$

27. $y = 1 + \cos^2 2x$

28. $y = \cos^3 4x$

29. $g(x) = x \cos x$

30. $g(x) = x \cos x^2$

31. $y = \sqrt{\cos 4x}$

32. $y = \sqrt{\sin 2x}$

33. $f(x) = e^x \sin x$

34. $f(x) = e^x \cos x$

35. $y = \dfrac{\cos x}{x}$

36. $y = \dfrac{\cos 2x}{x}$

In Exercises 37–54, find the derivative of each function.

37. $y = \tan 4x$

38. $y = 4 - \tan 2x$

39. $y = 8 \csc 2t$

40. $y = \csc^2 3t$

41. $y = 1 + \sec \pi t$

42. $y = 8 \sec \pi t$

43. $y = 2 \csc x^3$

44. $y = -\csc (x + 1)$

45. $f(x) = \sqrt{\cot x}$

46. $y = \sqrt{\tan x}$

47. $y = -5 \cot 2\pi x$

48. $y = -2 \tan 4\pi x$

49. $y = \tan^2 3x$

50. $y = \cot^2 2x$

51. $y = e^x \sec x$

52. $y = e^x \csc x$

53. $f(x) = 2x \tan x$

54. $f(x) = 4x \cot x$

55. (SALES CYCLE) A snow shovel distributor finds that sales are cyclical and can be approximated by the function

$$S(t) = 10{,}000 + 9000 \cos \frac{\pi t}{6}$$

where S is the number of snow shovels sold and t is the time in months (1 is January, 2 is February, and so on).
(a) How many snow shovels do they sell in June?
(b) How many snow shovels do they sell in the month of December?
(c) Determine dS/dt.
(d) Find the rate at which monthly sales are changing in September. Are sales increasing or decreasing?
(e) Find the rate at which monthly sales are changing in March. Are sales increasing or decreasing?

56. (SALES CYCLE) A boat rental company has cyclical sales with revenue defined by

$$R(t) = 2000 - 1700 \cos \frac{\pi t}{26}$$

where R is the weekly revenue in dollars. Time t is in weeks, with 1 being the first week of the year, 2 being the second week of the year, etc.
(a) Compare the company's revenue the last week of the year ($t = 52$, late December) with that in the middle of the year ($t = 26$, late June).

(b) Obtain $R'(t)$.

(c) Compare the rate of change of revenue when $t = 13$ (early spring) and $t = 39$ (fall), and explain the difference.

57. (*TIDAL FLUCTUATION*) Tides have a cyclical nature. If the height of water in a harbor varies from a low of 2 feet to a high of 10 feet, the height at any time of the day is given by

$$H(t) = 6 + 4 \sin \frac{\pi t}{6}$$

H is the height of the tide in feet and t is the time, using a 24-hour clock (0 is midnight, 1 is 1 a.m., 13 is 1 p.m.).

(a) Show that the high tides (10 feet) will occur at 3 a.m. and 3 p.m.

(b) Show that the low tides (2 feet) will occur at 9 a.m. and 9 p.m.

(c) Obtain dH/dt.

(d) Show that dH/dt is zero at the high tide and low tide times.

W (e) Explain from a calculus perspective why dH/dt is zero at high and low tide.

58. (*PREDATOR-PREY*) Refer to the predator-prey example (Example 7).

(a) Find the rate at which the predator population is changing after 24 months. Is the population increasing or decreasing?

(b) Determine the rate at which the prey population is changing after 6 months. Is the population increasing or decreasing?

59. (*PREDATOR-PREY*) Consider a predator-prey relationship in which the populations are given as follows, with t being the time in months.

$$\text{predator:} \quad D(t) = 1000 + 500 \sin \frac{\pi t}{12}$$

$$\text{prey:} \quad Y(t) = 4000 + 2000 \cos \frac{\pi t}{12}$$

(a) Determine the predator population after 12 months.

(b) Determine the prey population after 12 months.

(c) Is the predator population increasing or decreasing after 12 months, and at what rate?

W (d) After 12 months, the prey population has reached its lowest level. Determine $Y'(t)$ when $t = 12$ and explain how the calculus result is consistent with this observation.

60. The identity $\cos^2 t + \sin^2 t = 1$ can be obtained from the unit circle $x^2 + y^2 = 1$ by using the unit circle definition of $\sin t$ and $\cos t$ (use t rather than θ) to make appropriate substitutions. Try it.

61. Use a calculator to complete the following table, which suggests that

$$\lim_{t \to 0} \frac{\sin t}{t} = 1$$

By sure your calculator is set for *radians*. Keep as many digits as your calculator produces.

t	$\sin t$	$\dfrac{\sin t}{t}$
.1		
.01		
.001		
.0001		
−.1		
−.01		
−.001		
−.0001		

62. Use a calculator to complete the following table, which suggests that

$$\lim_{t \to 0} \frac{\cos t - 1}{t} = 0$$

Be sure your calculator is set for *radians*. Keep as many digits as your calculator produces.

t	$\cos t$	$\cos t - 1$	$\dfrac{\cos t - 1}{t}$
.1			
.01			
.001			
.0001			
.00001			
$-.1$			
$-.01$			
$-.001$			
$-.0001$			
$-.00001$			

63. Use a calculator to complete the table in order to show that the identity is true for the six numbers given in the table. Set your calculator for *radians*. Use four decimal places.

$$\sin\left(\frac{\pi}{2} - x\right) = \cos x$$

x	$\dfrac{\pi}{2} - x$	$\sin\left(\dfrac{\pi}{2} - x\right)$	$\cos x$
.3			
.7			
1.2			
2.3			
3.5			
4.1			

64. Use a calculator to complete the table in order to show that the identity is true for the six numbers given in the table. Set your calculator for *radians*. Use four decimal places.

$$\cos\left(\frac{\pi}{2} - x\right) = \sin x$$

x	$\dfrac{\pi}{2} - x$	$\cos\left(\dfrac{\pi}{2} - x\right)$	$\sin x$
.4			
.9			
1.4			
2.7			
3.2			
4.0			

65. Prove the differentiation formula

$$\frac{d}{dx}\cot x = -\csc^2 x$$

Hint: Follow the pattern for the proof that $D_x(\tan x) = \sec^2 x$.

66. Prove the differentiation formula

$$\frac{d}{dx}\sec x = \sec x \tan x$$

Hint: Consider sec x as $1/\cos x$. When needed, note that

$$\frac{\sin x}{\cos^2 x} = \frac{\sin x}{\cos x \cos x} = \frac{1}{\cos x}\cdot\frac{\sin x}{\cos x}$$

67. Follow the idea of Exercise 66 in order to prove the differentiation formula

$$\frac{d}{dx}\csc x = -\csc x \cot x$$

68. Is the graph of $y = \cos 4x$ concave up or concave down at $x = \pi/3$?

69. Is the graph of $y = \sin 2x$ concave up or concave down at $x = \pi/3$?

70. Determine the slope of the line tangent to the graph of $y = 1 - 5 \sin 3x$ at the point $(\pi/2, 6)$.

71. Determine the slope of the line tangent to the graph of $y = 3 + 8 \cos 2x$ at the point $(\pi/4, 3)$.

In Exercises 72–77, obtain dy/dx by using implicit differentiation.

72. $x = 1 + \cos y$

73. $2x = \sin y$

74. $x = 3 \sin y^2$

75. $x = e^y \cos x$

76. $\sin x = x \sin y$

77. $x = \tan xy$

The Exercise Library at the back of the book contains graphing calculator and computer exercises keyed to this section.

9.4 | INTEGRATION OF TRIGONOMETRIC FUNCTIONS

Formulas for antidifferentiation of the sine and cosine can be obtained from the known differentiation formulas. From

$$\frac{d}{dx}(\sin x) = \cos x$$

and

$$\frac{d}{dx}(\cos x) = -\sin x$$

it follows that

$$\int \cos x \, dx = \sin x + C$$

$$\int \sin x \, dx = -\cos x + C$$

As verification, note that differentiating $\sin x + C$ produces the integrand $\cos x$ and differentiating $-\cos x + C$ produces the integrand $\sin x$.

In consideration of the chain rule, with u being a function of x, we have

$$\int \cos u \, du = \sin u + C$$

$$\int \sin u \, du = -\cos u + C$$

EXAMPLE 1   Evaluate $\int \cos 4x \, dx$.

SOLUTION   Here $u = 4x$, from which it follows that $du = 4 \, dx$. We introduce the needed 4 into the integrand (and place 1/4 in front of the integral).

$$\int \cos 4x \, dx = \frac{1}{4} \int \cos 4x \cdot 4 \, dx$$

$$= \frac{1}{4} \sin 4x + C \quad \blacklozenge$$

EXAMPLE 2   Evaluate $\int x \sin x^2 \, dx$.

SOLUTION   Here $u = x^2$ and $du = 2x \, dx$. We can arrange the integrand so that x goes with the dx. Then we can introduce the needed 2 into the integrand (and place 1/2 in front of the integral).

$$\int x \sin x^2 \, dx = \int \sin x^2 \cdot x \, dx$$

$$= \frac{1}{2} \int \sin x^2 \cdot 2x \, dx$$

$$= -\frac{1}{2} \cos x^2 + C \quad \blacklozenge$$

EXAMPLE 3   Evaluate $\int \sin^3 x \cos x \, dx$.

SOLUTION   Since $\sin^3 x$ means $(\sin x)^3$ and $\cos x$ is the derivative of $\sin x$, we have the form

$$\int u^n \, du$$

with $u = \sin x$ and $du = \cos x \, dx$. Thus,

$$\int \sin^3 x \cos x \, dx = \int (\sin x)^3 \cos x \, dx$$

$$= \frac{\sin^4 x}{4} + C \quad \blacklozenge$$

EXAMPLE 4   Evaluate $\int \dfrac{\cos x \, dx}{2 + \sin x}$.

SOLUTION   To begin, observe that $\cos x$ is the derivative of $2 + \sin x$. This means that the integral is of the form

$$\int \frac{du}{u} = \ln |u| + C$$

where

$$u = 2 + \sin x$$
$$du = \cos x \, dx$$

Thus,

$$\int \frac{\cos x \, dx}{2 + \sin x} = \ln |2 + \sin x| + C \quad \blacklozenge$$

EXAMPLE 5 Evaluate the definite integral $\displaystyle\int_0^{\pi/6} \cos 3x \, dx$.

SOLUTION Here $u = 3x$ and $du = 3 \, dx$.

$$\int_0^{\pi/6} \cos 3x \, dx = \frac{1}{3} \int_0^{\pi/6} \cos 3x \cdot 3 \, dx$$

$$= \frac{1}{3} \left[\sin 3x \right]_0^{\pi/6}$$

$$= \frac{1}{3} \left(\sin \frac{\pi}{2} - \sin 0 \right)$$

$$= \frac{1}{3}(1 - 0)$$

$$= \frac{1}{3} \quad \blacklozenge$$

EXAMPLE 6 Find the area under the curve $y = \sin x$ from $x = 0$ to $x = \pi/2$.

SOLUTION The desired area is shown shaded in Figure 18.

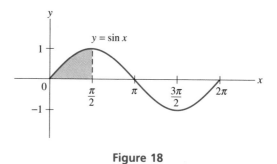

Figure 18

The area is given by the definite integral of the function $y = \sin x$ from $x = 0$ to $x = \pi/2$, namely

$$\int_0^{\pi/2} \sin x \, dx = \left[-\cos x \right]_0^{\pi/2}$$

$$= \left(-\cos \frac{\pi}{2} \right) - (-\cos 0)$$

$$= (-0) - (-1)$$

$$= 1$$

The area is 1 square unit.   ◆

EXAMPLE 7   ◆   **PREDATOR-PREY RELATIONSHIP**

In Example 7 of Section 9.3, the prey population at any time t was given by

$$Y(t) = 2500 + 1500 \cos \frac{\pi t}{12}$$

Determine the average prey population during the first 18 months.

SOLUTION   The average prey population during the first 18 months is the average value of the function from $t = 0$ to $t = 18$, namely.

$$\frac{1}{18 - 0} \int_0^{18} Y(t) \, dt = \frac{1}{18} \int_0^{18} \left(2500 + 1500 \cos \frac{\pi}{12} t \right) dt$$

$$= \frac{500}{18} \int_0^{18} \left(5 + 3 \cos \frac{\pi}{12} t \right) dt \qquad \text{(factoring out 500)}$$

$$= \frac{500}{18} \left[5t + \frac{12}{\pi} \cdot 3 \sin \frac{\pi}{12} t \right]_0^{18}$$

$$= \frac{250}{9} \left[5t + \frac{36}{\pi} \sin \frac{\pi}{12} t \right]_0^{18}$$

$$= \frac{250}{9} \left[\left(5 \cdot 18 + \frac{36}{\pi} \sin \frac{3\pi}{2} \right) - \left(5 \cdot 0 + \frac{36}{\pi} \sin 0 \right) \right]$$

$$= \frac{250}{9} \left[90 + \frac{36}{\pi} (-1) \right]$$

$$\approx 27.8(90 - 11.5)$$

$$\approx 2182$$

The average prey population during the first 18 months is approximately 2182.   ◆

Recall the four differentiation formulas,

$$D_x(\tan x) = \sec^2 x$$
$$D_x(\cot x) = -\csc^2 x$$
$$D_x(\sec x) = \sec x \tan x$$
$$D_x(\csc x) = -\csc x \cot x$$

Four corresponding integration formulas follow by considering $\sec^2 x$, $\csc^2 x$, $\sec x \tan x$, and $\csc x \cot x$ as integrands.

$$\int \sec^2 x \, dx = \tan x + C \qquad\qquad \int \sec x \tan x \, dx = \sec x + C$$

$$\int \csc^2 x \, dx = -\cot x + C \qquad\qquad \int \csc x \cot x \, dx = -\csc x + C$$

Also,

$$\int \sec^2 u \, du = \tan u + C \qquad\qquad \int \sec u \tan u \, du = \sec u + C$$

$$\int \csc^2 u \, du = -\cot u + C \qquad\qquad \int \csc u \cot u \, du = -\csc u + C$$

A formula for integrating $\tan x$ can be obtained by considering $\tan x$ as $\sin x / \cos x$ and then noting that if $u = \cos x$, then $du = -\sin x \, dx$.

$$\int \tan x \, dx = \int \frac{\sin x}{\cos x} \, dx$$

$$= -\int \frac{-\sin x \, dx}{\cos x} \qquad \leftarrow du = -\sin x \, dx$$
$$\qquad\qquad\qquad\qquad\quad \leftarrow u = \cos x$$

$$= -\int \frac{du}{u}$$

$$= -\ln |u| + C$$

$$= -\ln |\cos x| + C$$

A similar approach will yield an integration formula for $\cot x$. As a result, we have the integration formulas shown next.

$$\int \tan x \, dx = -\ln |\cos x| + C \qquad\qquad \int \cot x \, dx = \ln |\sin x| + C$$

Also,

$$\int \tan u \; du = -\ln |\cos u| + C \qquad \int \cot u \; du = \ln |\sin u| + C$$

EXAMPLE 8 Evaluate $\int \tan 2x \; dx$.

SOLUTION Here $u = 2x$ and $du = 2 \; dx$.

$$\int \tan 2x \; dx = \frac{1}{2} \int \tan 2x \cdot 2 \; dx$$

$$= \frac{1}{2} (-\ln |\cos 2x|) + C$$

$$= -\frac{1}{2} \ln |\cos 2x| + C \quad \blacklozenge$$

EXAMPLE 9 Evaluate $\int \sec^2 5x \; dx$.

SOLUTION Here $u = 5x$ and $du = 5 \; dx$.

$$\int \sec^2 5x \; dx = \frac{1}{5} \int \sec^2 5x \cdot 5 \; dx$$

$$= \frac{1}{5} \tan 5x + C \quad \blacklozenge$$

9.4   Exercises

Evaluate each indefinite integral in Exercises 1–18.

1. $\int \sin 2x \; dx$

2. $\int \sin 6x \; dx$

3. $\int \cos 5x \; dx$

4. $\int \cos 3x \; dx$

5. $\int x \cos x^2 \; dx$

6. $\int x^2 \sin x^3 \; dx$

7. $\int \sin (x + 1) \; dx$

8. $\int \cos (x - 2) \; dx$

9. $\int \sin^2 x \cos x \; dx$

10. $\int \sin x \cos^4 x \; dx$

11. $\int \cos x \sin x \; dx$

12. $\int \sin 2x \cos 2x \; dx$

13. $\int \frac{\sin x}{\cos^2 x} \; dx$

14. $\int \frac{\cos x}{\sqrt{\sin x}} \; dx$

15. $\int \sqrt{1 + \cos x} \sin x \; dx$

16. $\int \sqrt{1 + \sin x} \cos x \; dx$

17. $\int \frac{\sin x}{1 + \cos x} \; dx$

18. $\int \frac{\cos x \; dx}{1 - \sin x}$

Evaluate each definite integral in Exercises 19–24.

19. $\int_0^{\pi/4} \cos 2x \; dx$

20. $\int_0^{\pi/6} \sin 3x \; dx$

21. $\int_{\pi/4}^{\pi/2} \sin 4x \; dx$

22. $\int_{\pi/4}^{\pi/2} \cos 4x \; dx$

23. $\int_0^{\pi/3} \cos^2 x \sin x \; dx$

24. $\int_{\pi/6}^{\pi/2} \sin^3 x \cos x \; dx$

Evaluate each indefinite integral in Exercises 25–38.

25. $\int \tan 4x \, dx$

26. $\int \tan 6x \, dx$

27. $\int x \tan x^2 \, dx$

28. $\int x^2 \tan x^3 \, dx$

29. $\int \sec^2 2x \, dx$

30. $\int \csc^2 3x \, dx$

31. $\int \sec 3x \tan 3x \, dx$

32. $\int \csc 4x \cot 4x \, dx$

33. $\int \dfrac{\sec^2 x}{1 + \tan x} \, dx$

34. $\int \dfrac{\sec x \tan x}{1 + \sec x} \, dx$

35. $\int \cot 2x \, dx$

36. $\int x \cot x^2 \, dx$

37. $\int \tan^5 x \sec^2 x \, dx$

38. $\int \tan^2 x \sec^2 x \, dx$

In Exercises 39–44, determine the area under the given curve.

39. $y = \sin x$ from $x = \pi/6$ to $x = \pi$

40. $y = \cos x$ from $x = 0$ to $x = \pi/3$

41. $y = \cos 2x$ from $x = 0$ to $x = \pi/4$

42. $y = \sin 2x$ from $x = 0$ to $x = \pi/4$

43. $y = \sin 3x$ from $x = \pi/6$ to $x = \pi/3$

44. $y = \cos 4x$ from $x = 0$ to $x = \pi/8$

45. Find the average value of the function
$$f(x) = 12 \sin 2x$$
on the interval from $x = 0$ to $x = \pi$.

46. Find the average value of the function
$$f(x) = 10 \cos 3x$$
on the interval from $x = 0$ to $x = \pi/6$.

47. *(PREDATOR-PREY)* In Example 7 of Section 9.3, the predator population at t months was given by
$$D(t) = 1400 + 600 \sin \frac{\pi t}{12}$$
Determine the average predator population during the first 18 months.

48. *(PREDATOR-PREY)* Refer to Exercise 47 and determine the average predator population during the first two years.

49. *(PREDATOR-PREY)* Exercise 59, Section 9.3, gives the predator and prey populations at time t months as

$$predator: \quad D(t) = 1000 + 500 \sin \frac{\pi t}{12}$$

$$prey: \quad Y(t) = 4000 + 2000 \cos \frac{\pi t}{12}$$

Determine the average predator and prey populations during the first two years.

50. *(PREDATOR-PREY)* Refer to Example 7 of Section 9.3 and determine the average predator and prey populations during the first year.

51. *(MARGINAL REVENUE)* A producer has determined that the marginal revenue for its product is given by
$$R'(x) = 200 + 15 \cos \pi x \quad \text{dollars}$$
Assume that revenue from the sale of zero units is 0 dollars. Find the revenue function.

52. *(MARGINAL COST)* A manufacturer anticipates that the marginal cost for its product will be given by
$$C'(x) = 80 - 6 \sin \frac{\pi x}{2} \quad \text{dollars}$$
The fixed cost is $200. Find the cost function.

53. *(CONSUMPTION)* Suppose the rate at which a particular soft drink is consumed in Florida is estimated to be
$$c'(t) = 90{,}000 + 14{,}000 \sin \frac{\pi t}{6}$$
cans per month from the beginning of April through the end of March. Let $t = 0$ be the beginning of April, $t = 1$ the end of April, $t = 2$ the end of May, and so on, with $t = 12$ being the end of March.
(a) Determine the total number of cans of soft drink consumed from the beginning of April through the end of September.
(b) Determine the total number of cans of soft drink consumed from the end of September through the end of March.

Solve each differential equation in Exercises 54–59.

54. $y' = 1 + \tan x$

55. $y' = 2x - \sec^2 x$

56. $dy = e^{-y} \sin x \, dx$

57. $\cos x \, dy = \sin x \, dx$

58. $\dfrac{dy}{\sin x} = y \, dx$

59. $\sec x \, dy = \csc y \, dx$

60. (a) Evaluate $\displaystyle\int_0^{2\pi} \sin x \, dx$.

(b) Find the area between the graph of $y = \sin x$ and the x axis from $x = 0$ to $x = 2\pi$. The graph in Figure 18 (Example 6) will be helpful.

W (c) Explain why parts (a) and (b) yield different results.

W 61. In Example 4, the result of integration was

$$\ln |2 + \sin x| + C$$

Decide whether this expression can or cannot be simplified to

$$\ln (2 + \sin x) + C$$

Give a complete explanation to support your decision.

9.5 | *INTEGRATION BY PARTS REVISITED*

The method of integration by parts, presented in Section 7.2, can be applied to certain integrals containing trigonometric functions. This section provides two examples of this application.

EXAMPLE 1   Evaluate $\displaystyle\int x \cos x \, dx$.

SOLUTION   Consider the integration by parts formula,

$$\int u \, dv = uv - \int v \, du$$

Select $u = x$, since du will then equal dx and this choice will eliminate the x and assure a simpler integral. Thus,

$$u = x$$
$$dv = \cos x \, dx$$

which leads to

$$du = dx$$
$$v = \sin x$$

and

$$\int x \cos x \, dx = x \sin x - \int \sin x \, dx$$
$$= x \sin x - (-\cos x) + C$$
$$= x \sin x + \cos x + C \quad \blacklozenge$$

EXAMPLE 2   Evaluate $\displaystyle\int x^2 \cos x \, dx$.

SOLUTION Let $u = x^2$ and $dv = \cos x\, dx$. Then $du = 2x\, dx$ and $v = \sin x$. This leads to

$$\int x^2 \cos x\, dx = uv - \int v\, du$$

$$= x^2 \sin x - \int \sin x \cdot 2x\, dx$$

$$= x^2 \sin x - 2\int x \sin x\, dx$$

The integral $\int x \sin x\, dx$ can be evaluated by using parts in a manner similar to that used in Example 1. In this integral, let $u = x$ and $dv = \sin x\, dx$. It follows that $du = dx$ and $v = -\cos x$.

$$\int x \sin x\, dx = -x \cos x - \int (-\cos x)\, dx$$

$$= -x \cos x + \int \cos x\, dx$$

$$= -x \cos x + \sin x$$

Now we can return to

$$\int x^2 \cos x\, dx = x^2 \sin x - 2\int x \sin x\, dx$$

and replace the integral on the right by $-x \cos x + \sin x$. This leads to

$$\int x^2 \cos x\, dx = x^2 \sin x - 2(-x \cos x + \sin x) + C$$

Finally,

$$\int x^2 \cos x\, dx = x^2 \sin x + 2x \cos x - 2 \sin x + C \quad \blacklozenge$$

9.5 Exercises

In Exercises 1–18, evaluate each indefinite integral. Some integrals require *integration by parts*.

1. $\int x \sin x\, dx$

2. $\int x \sin 3x\, dx$

3. $\int x \sin x^2\, dx$

4. $\int x^2 \cos x^3\, dx$

5. $\int (x + \sin x)\, dx$

6. $\int x^2 \cos 2x\, dx$

7. $\int x \cos 4x\, dx$

8. $\int 2x \cos x\, dx$

9. $\int \sin x \cos x\, dx$

10. $\int (x - \cos x)\, dx$

11. $\int x^2 \sin x\, dx$

12. $\int \sqrt{\sin x} \cos x\, ax$

13. $\int (1 + \sec^2 x)\, dx$

14. $\int \tan (x - 2)\, dx$

15. $\int x \sec^2 x\, dx$

16. $\int x \csc^2 x\, dx$

17. $\int \dfrac{1 + \sec^2 x}{x + \tan x}\, dx$

18. $\int 4 \sec x \tan x\, dx$

Chapter List   *Important terms and ideas*

right triangle	cosine	cosecant
opposite side	tangent	cotangent
hypotenuse	radian measure	differentiation rules
adjacent side	unit circle	integration rules
sine	secant	

Review Exercises for Chapter 9

In Exercises 1–4, change the angle measure as requested.

1. Change $7\pi/6$ to degrees.

2. Change $7\pi/4$ to degrees.

3. Change $225°$ to radians.

4. Change $300°$ to radians.

In Exercises 5–8, determine the exact value of each expression. *Do not use a calculator.*

5. $\tan \pi/6$

6. $\cot \pi/4$

7. $\sec \pi/4$

8. $\csc \pi/3$

9. Let $f(t) = 4 \sin \dfrac{\pi t}{4}$. Determine

    (a) $f(1)$

    (b) $f(2)$

10. (**BLOOD PRESSURE**) Jane's blood pressure is given by the function $P(t) = 100 + 20 \cos 6t$. What is her maximum blood pressure?

Determine the derivative of each function in Exercises 11–24.

11. $y = \sin (2x + 1)$

12. $y = \cos x^2$

13. $y = \cos^2 3x$

14. $y = \cot 3x^2$

15. $f(x) = x \sec x$

16. $f(x) = x^2 \csc x$

17. $y = \dfrac{\tan x}{x}$

18. $y = \dfrac{x}{\sec x}$

19. $y = (1 + \tan x)^3$

20. $y = e^x \tan x$

21. $f(x) = \sin e^x$

22. $y = \sec \dfrac{\pi x}{4}$

23. $y = \ln (\sin x)$

24. $y = \cos (\ln x)$

Evaluate each integral in Exercises 25–38.

25. $\displaystyle\int \cos 4x \, dx$

26. $\displaystyle\int \sin 7x \, dx$

27. $\displaystyle\int e^x \sin e^x \, dx$

28. $\displaystyle\int \dfrac{\sin \sqrt{x}}{\sqrt{x}} \, dx$

29. $\displaystyle\int 5 \sin 3x \, dx$

30. $\displaystyle\int 2 \cos 4x \, dx$

31. $\displaystyle\int \cos (x + \pi) \, dx$

32. $\displaystyle\int \sin (x + 1) \, dx$

33. $\displaystyle\int_0^{\pi/6} \sin 2x \, dx$

34. $\displaystyle\int_{\pi/3}^{\pi/2} \cos x \, dx$

35. $\displaystyle\int_{\pi/3}^{\pi} \sin \dfrac{1}{2}x \, dx$

36. $\displaystyle\int_0^{\pi} \cos \dfrac{1}{4}x \, dx$

37. $\displaystyle\int (x + 1) \sin x \, dx$

38. $\displaystyle\int (x + 1) \cos x \, dx$

In Exercises 39–42, determine the area under the curve.

39. $y = \cos x$ from $x = 0$ to $x = \pi/3$

40. $y = \sin x$ from $x = \pi/3$ to $x = \pi/2$

41. $y = \sin 2x$ from $x = 0$ to $x = \pi/6$

42. $y = \cos 2x$ from $x = 0$ to $x = \pi/6$

APPENDIX TABLES

APPENDIX TABLE 1

Powers of e

x	e^x	e^{-x}	x	e^x	e^{-x}	x	e^x	e^{-x}
.00	1.0000	1.0000	.21	1.2337	.8106	.42	1.5220	.6570
.01	1.0101	.9900	.22	1.2461	.8025	.43	1.5373	.6505
.02	1.0202	.9802	.23	1.2586	.7945	.44	1.5527	.6440
.03	1.0305	.9704	.24	1.2712	.7866	.45	1.5683	.6376
.04	1.0408	.9608	.25	1.2840	.7788	.46	1.5841	.6313
.05	1.0513	.9512	.26	1.2969	.7711	.47	1.6000	.6250
.06	1.0618	.9418	.27	1.3100	.7634	.48	1.6161	.6188
.07	1.0725	.9324	.28	1.3231	.7558	.49	1.6323	.6126
.08	1.0833	.9231	.29	1.3364	.7483	.50	1.6487	.6065
.09	1.0942	.9139	.30	1.3499	.7408	.51	1.6653	.6005
.10	1.1052	.9048	.31	1.3634	.7334	.52	1.6820	.5945
.11	1.1163	.8958	.32	1.3771	.7261	.53	1.6989	.5886
.12	1.1275	.8869	.33	1.3910	.7189	.54	1.7160	.5827
.13	1.1388	.8781	.34	1.4049	.7118	.55	1.7333	.5769
.14	1.1503	.8694	.35	1.4191	.7047	.56	1.7507	.5712
.15	1.1618	.8607	.36	1.4333	.6977	.57	1.7683	.5655
.16	1.1735	.8521	.37	1.4477	.6907	.58	1.7860	.5599
.17	1.1853	.8437	.38	1.4623	.6839	.59	1.8040	.5543
.18	1.1972	.8353	.39	1.4770	.6771	.60	1.8221	.5488
.19	1.2092	.8270	.40	1.4918	.6703	.61	1.8404	.5434
.20	1.2214	.8187	.41	1.5068	.6637	.62	1.8589	.5379

APPENDIX TABLE 1 (*Continued*)

Powers of *e*

x	e^x	e^{-x}	x	e^x	e^{-x}	x	e^x	e^{-x}
.63	1.8776	.5326	.84	2.3164	.4317	1.05	2.8577	.3499
.64	1.8965	.5273	.85	2.3396	.4274	1.06	2.8864	.3465
.65	1.9155	.5220	.86	2.3632	.4232	1.07	2.9154	.3430
.66	1.9348	.5169	.87	2.3869	.4190	1.08	2.9447	.3396
.67	1.9542	.5117	.88	2.4109	.4148	1.09	2.9743	.3362
.68	1.9739	.5066	.89	2.4351	.4107	1.10	3.0042	.3329
.69	1.9937	.5016	.90	2.4596	.4066	1.11	3.0344	.3296
.70	2.0138	.4966	.91	2.4843	.4025	1.12	3.0649	.3263
.71	2.0340	.4916	.92	2.5093	.3985	1.13	3.0957	.3230
.72	2.0544	.4868	.93	2.5345	.3946	1.14	3.1268	.3198
.73	2.0751	.4819	.94	2.5600	.3906	1.15	3.1582	.3166
.74	2.0959	.4771	.95	2.5857	.3867	1.16	3.1899	.3135
.75	2.1170	.4724	.96	2.6117	.3829	1.17	3.2220	.3104
.76	2.1383	.4677	.97	2.6379	.3791	1.18	3.2544	.3073
.77	2.1598	.4630	.98	2.6645	.3753	1.19	3.2871	.3042
.78	2.1815	.4584	.99	2.6912	.3716	1.20	3.3201	.3012
.79	2.2034	.4538	1.00	2.7183	.3679	1.21	3.3535	.2982
.80	2.2255	.4493	1.01	2.7456	.3642	1.22	3.3872	.2952
.81	2.2479	.4449	1.02	2.7732	.3606	1.23	3.4212	.2923
.82	2.2705	.4404	1.03	2.8011	.3570	1.24	3.4556	.2894
.83	2.2933	.4360	1.04	2.8292	.3535	1.25	3.4903	.2865

APPENDIX TABLE 1 (*Continued*)

Powers of *e*

x	e^x	e^{-x}	x	e^x	e^{-x}	x	e^x	e^{-x}
1.26	3.5254	.2837	1.47	4.3492	.2299	1.68	5.3656	.1864
1.27	3.5609	.2808	1.48	4.3929	.2276	1.69	5.4195	.1845
1.28	3.5966	.2780	1.49	4.4371	.2254	1.70	5.4739	.1827
1.29	3.6328	.2753	1.50	4.4817	.2231	1.71	5.5290	.1809
1.30	3.6693	.2725	1.51	4.5267	.2209	1.72	5.5845	.1791
1.31	3.7062	.2698	1.52	4.5722	.2187	1.73	5.6407	.1773
1.32	3.7434	.2671	1.53	4.6182	.2165	1.74	5.6973	.1755
1.33	3.7810	.2645	1.54	4.6646	.2144	1.75	5.7546	.1738
1.34	3.8190	.2618	1.55	4.7115	.2122	1.76	5.8124	.1720
1.35	3.8574	.2592	1.56	4.7588	.2101	1.77	5.8709	.1703
1.36	3.8962	.2567	1.57	4.8066	.2080	1.78	5.9299	.1686
1.37	3.9354	.2541	1.58	4.8550	.2060	1.79	5.9895	.1670
1.38	3.9749	.2516	1.59	4.9037	.2039	1.80	6.0496	.1653
1.39	4.0149	.2491	1.60	4.9530	.2019	1.81	6.1104	.1637
1.40	4.0552	.2466	1.61	5.0028	.1999	1.82	6.1719	.1620
1.41	4.0960	.2441	1.62	5.0531	.1979	1.83	6.2339	.1604
1.42	4.1371	.2417	1.63	5.1039	.1959	1.84	6.2965	.1588
1.43	4.1787	.2393	1.64	5.1552	.1940	1.85	6.3598	.1572
1.44	4.2207	.2369	1.65	5.2070	.1920	1.86	6.4237	.1557
1.45	4.2631	.2346	1.66	5.2593	.1901	1.87	6.4883	.1541
1.46	4.3060	.2322	1.67	5.3122	.1882	1.88	6.5535	.1526

APPENDIX TABLE 1 (*Continued*)

Powers of *e*

x	e^x	e^{-x}	x	e^x	e^{-x}	x	e^x	e^{-x}
1.89	6.6194	.1511	3.00	20.0855	.0498	5.10	164.022	.0061
1.90	6.6859	.1496	3.10	22.1980	.0450	5.20	181.272	.0055
1.91	6.7531	.1481	3.20	24.5325	.0408	5.30	200.337	.0050
1.92	6.8210	.1466	3.30	27.1126	.0369	5.40	221.406	.0045
1.93	6.8895	.1451	3.40	29.9641	.0334	5.50	244.692	.0041
1.94	6.9588	.1437	3.50	33.1155	.0302	5.60	270.426	.0037
1.95	7.0287	.1423	3.60	36.5982	.0273	5.70	298.867	.0033
1.96	7.0993	.1409	3.70	40.4473	.0247	5.80	330.300	.0030
1.97	7.1707	.1395	3.80	44.7012	.0224	5.90	365.038	.0027
1.98	7.2427	.1381	3.90	49.4024	.0202	6.00	403.429	.0025
1.99	7.3155	.1367	4.00	54.5982	.0183	6.10	445.858	.0022
2.00	7.3891	.1353	4.10	60.3403	.0166	6.20	492.749	.0020
2.10	8.1662	.1225	4.20	66.6863	.0150	6.30	544.572	.0018
2.20	9.0250	.1108	4.30	73.6998	.0136	6.40	601.845	.0017
2.30	9.9742	.1003	4.40	81.4509	.0123	6.50	665.142	.0015
2.40	11.0232	.0907	4.50	90.0171	.0111	6.60	735.095	.0014
2.50	12.1825	.0821	4.60	99.4843	.0101	6.70	812.406	.0012
2.60	13.4637	.0743	4.70	109.9472	.0091	6.80	897.847	.0011
2.70	14.8797	.0672	4.80	121.5104	.0082	6.90	992.275	.0010
2.80	16.4446	.0608	4.90	134.2898	.0074	7.00	1096.63	.00091
2.90	18.1741	.0550	5.00	148.413	.0067	7.10	1211.97	.00083

APPENDIX TABLE 1 (*Continued*)

Powers of *e*

x	e^x	e^{-x}	x	e^x	e^{-x}	x	e^x	e^{-x}
7.20	1339.43	.00075	8.20	3640.95	.00027	9.20	9897.13	.00010
7.30	1480.30	.00068	8.30	4023.87	.00025	9.30	10938.02	.00009
7.40	1635.98	.00061	8.40	4447.07	.00022	9.40	12088.38	.00008
7.50	1808.04	.00055	8.50	4914.77	.00020	9.50	13359.73	.00007
7.60	1998.20	.00050	8.60	5431.66	.00018	9.60	14764.78	.00007
7.70	2208.35	.00045	8.70	6002.91	.00017	9.70	16317.61	.00006
7.80	2440.60	.00041	8.80	6634.24	.00015	9.80	18033.74	.00006
7.90	2697.28	.00037	8.90	7331.97	.00014	9.90	19930.37	.00005
8.00	2980.96	.00034	9.00	8103.08	.00012			
8.10	3294.47	.00030	9.10	8955.29	.00011			

APPENDIX TABLE 2

Natural Logarithms

x	ln x	x	ln x	x	ln x	x	ln x	x	ln x
.01	−4.6052	.22	−1.5141	.43	−.8440	.64	−.4463	.85	−.1625
.02	−3.9120	.23	−1.4697	.44	−.8210	.65	−.4308	.86	−.1508
.03	−3.5066	.24	−1.4271	.45	−.7985	.66	−.4155	.87	−.1393
.04	−3.2189	.25	−1.3863	.46	−.7765	.67	−.4005	.88	−.1278
.05	−2.9957	.26	−1.3471	.47	−.7550	.68	−.3857	.89	−.1165
.06	−2.8134	.27	−1.3093	.48	−.7340	.69	−.3711	.90	−.1054
.07	−2.6593	.28	−1.2730	.49	−.7133	.70	−.3567	.91	−.0943
.08	−2.5257	.29	−1.2379	.50	−.6931	.71	−.3425	.92	−.0834
.09	−2.4079	.30	−1.2040	.51	−.6733	.72	−.3285	.93	−.0726
.10	−2.3026	.31	−1.1712	.52	−.6539	.73	−.3147	.94	−.0619
.11	−2.2073	.32	−1.1394	.53	−.6349	.74	−.3011	.95	−.0513
.12	−2.1203	.33	−1.1087	.54	−.6162	.75	−.2877	.96	−.0408
.13	−2.0402	.34	−1.0788	.55	−.5978	.76	−.2744	.97	−.0305
.14	−1.9661	.35	−1.0498	.56	−.5798	.77	−.2614	.98	−.0202
.15	−1.8971	.36	−1.0217	.57	−.5621	.78	−.2485	.99	−.0101
.16	−1.8326	.37	−.9943	.58	−.5447	.79	−.2357	1.00	.0000
.17	−1.7720	.38	−.9676	.59	−.5276	.80	−.2231	1.01	.0100
.18	−1.7148	.39	−.9416	.60	−.5108	.81	−.2107	1.02	.0198
.19	−1.6607	.40	−.9163	.61	−.4943	.82	−.1985	1.03	.0296
.20	−1.6094	.41	−.8916	.62	−.4780	.83	−.1863	1.04	.0392
.21	−1.5606	.42	−.8675	.63	−.4620	.84	−.1744	1.05	.0488

APPENDIX TABLE 2 (*Continued*)

Natural Logarithms

x	ln x	x	ln x	x	ln x	x	ln x	x	ln x
1.06	.0583	1.27	.2390	1.48	.3920	1.69	.5247	1.90	.6419
1.07	.0677	1.28	.2469	1.49	.3988	1.70	.5306	1.91	.6471
1.08	.0770	1.29	.2546	1.50	.4055	1.71	.5365	1.92	.6523
1.09	.0862	1.30	.2624	1.51	.4121	1.72	.5423	1.93	.6575
1.10	.0953	1.31	.2700	1.52	.4187	1.73	.5481	1.94	.6627
1.11	.1044	1.32	.2776	1.53	.4253	1.74	.5539	1.95	.6678
1.12	.1133	1.33	.2852	1.54	.4318	1.75	.5596	1.96	.6729
1.13	.1222	1.34	.2927	1.55	.4383	1.76	.5653	1.97	.6780
1.14	.1310	1.35	.3001	1.56	.4447	1.77	.5710	1.98	.6831
1.15	.1398	1.36	.3075	1.57	.4511	1.78	.5766	1.99	.6881
1.16	.1484	1.37	.3148	1.58	.4574	1.79	.5822	2.00	.6931
1.17	.1570	1.38	.3221	1.59	.4637	1.80	.5878	2.05	.7178
1.18	.1655	1.39	.3293	1.60	.4700	1.81	.5933	2.10	.7419
1.19	.1740	1.40	.3365	1.61	.4762	1.82	.5988	2.15	.7655
1.20	.1823	1.41	.3436	1.62	.4824	1.83	.6043	2.20	.7885
1.21	.1906	1.42	.3507	1.63	.4886	1.84	.6098	2.25	.8109
1.22	.1989	1.43	.3577	1.64	.4947	1.85	.6152	2.30	.8329
1.23	.2070	1.44	.3646	1.65	.5008	1.86	.6206	2.35	.8544
1.24	.2151	1.45	.3716	1.66	.5068	1.87	.6259	2.40	.8755
1.25	.2231	1.46	.3784	1.67	.5128	1.88	.6313	2.45	.8961
1.26	.2311	1.47	.3853	1.68	.5188	1.89	.6366	2.50	.9163

APPENDIX TABLE 2 (*Continued*)

Natural Logarithms

x	ln x	x	ln x	x	ln x	x	ln x	x	ln x
2.55	.9361	3.60	1.2809	4.65	1.5369	5.70	1.7405	6.75	1.9095
2.60	.9555	3.65	1.2947	4.70	1.5476	5.75	1.7492	6.80	1.9169
2.65	.9746	3.70	1.3083	4.75	1.5581	5.80	1.7579	6.85	1.9242
2.70	.9933	3.75	1.3218	4.80	1.5686	5.85	1.7664	6.90	1.9315
2.75	1.0116	3.80	1.3350	4.85	1.5790	5.90	1.7750	6.95	1.9387
2.80	1.0296	3.85	1.3481	4.90	1.5892	5.95	1.7834	7.00	1.9459
2.85	1.0473	3.90	1.3610	4.95	1.5994	6.00	1.7918	7.05	1.9530
2.90	1.0647	3.95	1.3737	5.00	1.6094	6.05	1.8001	7.10	1.9601
2.95	1.0818	4.00	1.3863	5.05	1.6194	6.10	1.8083	7.15	1.9671
3.00	1.0986	4.05	1.3987	5.10	1.6292	6.15	1.8165	7.20	1.9741
3.05	1.1151	4.10	1.4110	5.15	1.6390	6.20	1.8245	7.25	1.9810
3.10	1.1314	4.15	1.4231	5.20	1.6487	6.25	1.8326	7.30	1.9879
3.15	1.1474	4.20	1.4351	5.25	1.6582	6.30	1.8405	7.35	1.9947
3.20	1.1632	4.25	1.4469	5.30	1.6677	6.35	1.8485	7.40	2.0015
3.25	1.1787	4.30	1.4586	5.35	1.6771	6.40	1.8563	7.45	2.0082
3.30	1.1939	4.35	1.4702	5.40	1.6864	6.45	1.8641	7.50	2.0149
3.35	1.2090	4.40	1.4816	5.45	1.6956	6.50	1.8718	7.55	2.0215
3.40	1.2238	4.45	1.4929	5.50	1.7047	6.55	1.8795	7.60	2.0281
3.45	1.2384	4.50	1.5041	5.55	1.7138	6.60	1.8871	7.65	2.0347
3.50	1.2528	4.55	1.5151	5.60	1.7228	6.65	1.8946	7.70	2.0412
3.55	1.2669	4.60	1.5261	5.65	1.7317	6.70	1.9021	7.75	2.0477

APPENDIX TABLE 2 (*Continued*)

Natural Logarithms

x	$\ln x$	x	$\ln x$	x	$\ln x$	x	$\ln x$	x	$\ln x$
7.80	2.0541	9.70	2.2721	19.0	2.9444	39.0	3.6636	60.0	4.0943
7.85	2.0605	9.80	2.2824	19.5	2.9704	40.0	3.6889	61.0	4.1109
7.90	2.0669	9.90	2.2925	20.0	2.9957	41.0	3.7136	62.0	4.1271
7.95	2.0732	10.0	2.3026	21.0	3.0445	42.0	3.7377	63.0	4.1431
8.00	2.0794	10.5	2.3514	22.0	3.0910	43.0	3.7612	64.0	4.1589
8.10	2.0919	11.0	2.3979	23.0	3.1355	44.0	3.7842	65.0	4.1744
8.20	2.1041	11.5	2.4423	24.0	3.1781	45.0	3.8067	66.0	4.1897
8.30	2.1163	12.0	2.4849	25.0	3.2189	46.0	3.8286	67.0	4.2047
8.40	2.1282	12.5	2.5257	26.0	3.2581	47.0	3.8501	68.0	4.2195
8.50	2.1401	13.0	2.5649	27.0	3.2958	48.0	3.8712	69.0	4.2341
8.60	2.1518	13.5	2.6027	28.0	3.3322	49.0	3.8918	70.0	4.2485
8.70	2.1633	14.0	2.6391	29.0	3.3673	50.0	3.9120	71.0	4.2627
8.80	2.1748	14.5	2.6741	30.0	3.4012	51.0	3.9318	72.0	4.2767
8.90	2.1861	15.0	2.7081	31.0	3.4340	52.0	3.9512	73.0	4.2905
9.00	2.1972	15.5	2.7408	32.0	3.4657	53.0	3.9703	74.0	4.3041
9.10	2.2083	16.0	2.7726	33.0	3.4965	54.0	3.9890	75.0	4.3175
9.20	2.2192	16.5	2.8034	34.0	3.5264	55.0	4.0073	76.0	4.3307
9.30	2.2300	17.0	2.8332	35.0	3.5553	56.0	4.0254	77.0	4.3438
9.40	2.2407	17.5	2.8622	36.0	3.5835	57.0	4.0431	78.0	4.3567
9.50	2.2513	18.0	2.8904	37.0	3.6109	58.0	4.0604	79.0	4.3694
9.60	2.2618	18.5	2.9178	38.0	3.6338	59.0	4.0775	80.0	4.3820

APPENDIX TABLE 2 (*Continued*)

Natural Logarithms

x	ln x	x	ln x	x	ln x	x	ln x	x	ln x
81.0	4.3944	120.	4.7875	330.	5.7991	540.	6.2916	750.	6.6201
82.0	4.4067	130.	4.8675	340.	5.8289	550.	6.3099	760.	6.6333
83.0	4.4188	140.	4.9416	350.	5.8579	560.	6.3279	770.	6.6464
84.0	4.4308	150.	5.0106	360.	5.8861	570.	6.3456	780.	6.6593
85.0	4.4427	160.	5.0752	370.	5.9135	580.	6.3630	790.	6.6720
86.0	4.4543	170.	5.1358	380.	5.9402	590.	6.3801	800.	6.6846
87.0	4.4659	180.	5.1930	390.	5.9661	600.	6.3969	810.	6.6970
88.0	4.4773	190.	5.2470	400.	5.9915	610.	6.4135	820.	6.7093
89.0	4.4886	200.	5.2983	410.	6.0162	620.	6.4297	830.	6.7214
90.0	4.4998	210.	5.3471	420.	6.0403	630.	6.4457	840.	6.7334
91.0	4.5109	220.	5.3936	430.	6.0638	640.	6.4615	850.	6.7452
92.0	4.5218	230.	5.4381	440.	6.0868	650.	6.4770	860.	6.7569
93.0	4.5326	240.	5.4806	450.	6.1092	660.	6.4922	870.	6.7685
94.0	4.5433	250.	5.5215	460.	6.1312	670.	6.5073	880.	6.7799
95.0	4.5539	260.	5.5607	470.	6.1527	680.	6.5221	890.	6.7912
96.0	4.5643	270.	5.5984	480.	6.1738	690.	6.5367	900.	6.8024
97.0	4.5747	280.	5.6348	490.	6.1944	700.	6.5511	910.	6.8134
98.0	4.5850	290.	5.6699	500.	6.2146	710.	6.5653	920.	6.8244
99.0	4.5951	300.	5.7038	510.	6.2344	720.	6.5793	930.	6.8352
100.	4.6052	310.	5.7366	520.	6.2538	730.	6.5930	940.	6.8459
110.	4.7005	320.	5.7683	530.	6.2729	740.	6.6067	950.	6.8565

APPENDIX TABLE 2 (*Continued*)

Natural Logarithms

x	$\ln x$	x	$\ln x$	x	$\ln x$	x	$\ln x$	x	$\ln x$
960.	6.8669	1200.	7.0901	1800.	7.4955	4000.	8.2940	7000.	8.8537
970.	6.8773	1300.	7.1701	1900.	7.5496	4500.	8.4118	7500.	8.9227
980.	6.8876	1400.	7.2442	2000.	7.6009	5000.	8.5172	8000.	8.9872
990.	6.8977	1500.	7.3132	2500.	7.8240	5500.	8.6125	8500.	9.0478
1000.	6.9078	1600.	7.3778	3000.	8.0064	6000.	8.6995	9000.	9.1050
1100.	7.0031	1700.	7.4384	3500.	8.1605	6500.	8.7796	9500.	9.1590

APPENDIX TABLE 3

Trigonometric Functions—Degrees

θ	$\sin \theta$	$\cos \theta$	$\tan \theta$	θ	$\sin \theta$	$\cos \theta$	$\tan \theta$	θ	$\sin \theta$	$\cos \theta$	$\tan \theta$
0°	.0000	1.0000	.0000	21°	.3584	.9336	.3839	42°	.6691	.7431	.9004
1°	.0175	.9998	.0175	22°	.3746	.9272	.4040	43°	.6820	.7314	.9325
2°	.0349	.9994	.0349	23°	.3907	.9205	.4245	44°	.6947	.7193	.9657
3°	.0523	.9986	.0524	24°	.4067	.9135	.4452	45°	.7071	.7071	1.0000
4°	.0698	.9976	.0699	25°	.4226	.9063	.4663	46°	.7193	.6947	1.0355
5°	.0872	.9962	.0875	26°	.4384	.8988	.4877	47°	.7314	.6820	1.0724
6°	.1045	.9945	.1051	27°	.4540	.8910	.5095	48°	.7431	.6691	1.1106
7°	.1219	.9925	.1228	28°	.4695	.8829	.5317	49°	.7547	.6561	1.1504
8°	.1392	.9903	.1405	29°	.4848	.8746	.5543	50°	.7660	.6428	1.1918
9°	.1564	.9877	.1584	30°	.5000	.8660	.5774	51°	.7771	.6293	1.2349
10°	.1736	.9848	.1763	31°	.5150	.8572	.6009	52°	.7880	.6157	1.2799
11°	.1908	.9816	.1944	32°	.5299	.8480	.6249	53°	.7986	.6018	1.3270
12°	.2079	.9781	.2126	33°	.5446	.8387	.6494	54°	.8090	.5878	1.3764
13°	.2250	.9744	.2309	34°	.5592	.8290	.6745	55°	.8192	.5736	1.4281
14°	.2419	.9703	.2493	35°	.5736	.8192	.7002	56°	.8290	.5592	1.4826
15°	.2588	.9659	.2679	36°	.5878	.8090	.7265	57°	.8387	.5446	1.5399
16°	.2756	.9613	.2867	37°	.6018	.7986	.7536	58°	.8480	.5299	1.6003
17°	.2924	.9563	.3057	38°	.6157	.7880	.7813	59°	.8572	.5150	1.6643
18°	.3090	.9511	.3249	39°	.6293	.7771	.8098	60°	.8660	.5000	1.7321
19°	.3256	.9455	.3443	40°	.6428	.7660	.8391	61°	.8746	.4848	1.8040
20°	.3420	.9397	.3640	41°	.6561	.7547	.8693	62°	.8829	.4695	1.8807

APPENDIX TABLE 3 (*Continued*)

Trigonometric Functions—Degrees

θ	sin θ	cos θ	tan θ	θ	sin θ	cos θ	tan θ	θ	sin θ	cos θ	tan θ
63°	.8910	.4540	1.9626	73°	.9563	.2924	3.2709	83°	.9925	.1219	8.1443
64°	.8988	.4384	2.0503	74°	.9613	.2756	3.4874	84°	.9945	.1045	9.5144
65°	.9063	.4226	2.1445	75°	.9659	.2588	3.7321	85°	.9962	.0872	11.4301
66°	.9135	.4067	2.2460	76°	.9703	.2419	4.0108	86°	.9976	.0698	14.3007
67°	.9205	.3907	2.3559	77°	.9744	.2250	4.3315	87°	.9986	.0523	19.0811
68°	.9272	.3746	2.4751	78°	.9781	.2079	4.7046	88°	.9994	.0349	28.6363
69°	.9336	.3584	2.6051	79°	.9816	.1908	5.1446	89°	.9998	.0175	57.2900
70°	.9397	.3420	2.7475	80°	.9848	.1736	5.6713	90°	1.0000	.0000	—
71°	.9455	.3256	2.9042	81°	.9877	.1564	6.3138				
72°	.9511	.3090	3.0777	82°	.9903	.1392	7.1154				

APPENDIX TABLE 4

Trigonometric Functions—Radians

t	$\sin t$	$\cos t$	$\tan t$	t	$\sin t$	$\cos t$	$\tan t$	t	$\sin t$	$\cos t$	$\tan t$
.00	.0000	1.0000	.0000	.21	.2085	.9780	.2131	.42	.4078	.9131	.4466
.01	.0100	1.0000	.0100	.22	.2182	.9759	.2236	.43	.4169	.9090	.4586
.02	.0200	.9998	.0200	.23	.2280	.9737	.2341	.44	.4259	.9048	.4708
.03	.0300	.9996	.0300	.24	.2377	.9713	.2447	.45	.4350	.9004	.4831
.04	.0400	.9992	.0400	.25	.2474	.9689	.2553	.46	.4439	.8961	.4954
.05	.0500	.9988	.0500	.26	.2571	.9664	.2660	.47	.4529	.8916	.5080
.06	.0600	.9982	.0601	.27	.2667	.9638	.2768	.48	.4618	.8870	.5206
.07	.0699	.9976	.0701	.28	.2764	.9611	.2876	.49	.4706	.8823	.5334
.08	.0799	.9968	.0802	.29	.2860	.9582	.2984	.50	.4794	.8776	.5463
.09	.0899	.9960	.0902	.30	.2955	.9553	.3093	.51	.4882	.8727	.5594
.10	.0998	.9950	.1003	.31	.3051	.9523	.3203	.52	.4969	.8678	.5726
.11	.1098	.9940	.1104	.32	.3146	.9492	.3314	.53	.5055	.8628	.5859
.12	.1197	.9928	.1206	.33	.3240	.9460	.3425	.54	.5141	.8577	.5994
.13	.1296	.9916	.1307	.34	.3335	.9428	.3537	.55	.5227	.8525	.6131
.14	.1395	.9902	.1409	.35	.3429	.9394	.3650	.56	.5312	.8473	.6269
.15	.1494	.9888	.1511	.36	.3523	.9359	.3764	.57	.5396	.8419	.6310
.16	.1593	.9872	.1614	.37	.3616	.9323	.3879	.58	.5480	.8365	.6552
.17	.1692	.9856	.1717	.38	.3709	.9287	.3994	.59	.5564	.8309	.6696
.18	.1790	.9838	.1820	.39	.3802	.9249	.4111	.60	.5646	.8253	.6841
.19	.1889	.9820	.1923	.40	.3894	.9211	.4228	.61	.5729	.8196	.6989
.20	.1987	.9801	.2027	.41	.3986	.9171	.4346	.62	.5810	.8139	.7139

APPENDIX TABLE 4 (*Continued*)

Trigonometric Functions—Radians

t	sin t	cos t	tan t	t	sin t	cos t	tan t	t	sin t	cos t	tan t
.63	.5891	.8080	.7291	.84	.7446	.6675	1.1156	1.05	.8674	.4976	1.7433
.64	.5972	.8021	.7445	.85	.7513	.6600	1.1383	1.06	.8724	.4889	1.7844
.65	.6052	.7961	.7602	.86	.7578	.6524	1.1616	1.07	.8772	.4801	1.8270
.66	.6131	.7900	.7761	.87	.7643	.6448	1.1853	1.08	.8820	.4713	1.8712
.67	.6210	.7838	.7923	.88	.7707	.6372	1.2097	1.09	.8866	.4625	1.9171
.68	.6288	.7776	.8087	.89	.7771	.6294	1.2346	1.10	.8912	.4536	1.9648
.69	.6365	.7712	.8253	.90	.7833	.6216	1.2602	1.11	.8957	.4447	2.0143
.70	.6442	.7648	.8423	.91	.7895	.6137	1.2864	1.12	.9001	.4357	2.0660
.71	.6518	.7584	.8595	.92	.7956	.6058	1.3133	1.13	.9044	.4267	2.1198
.72	.6594	.7518	.8771	.93	.8016	.5978	1.3409	1.14	.9086	.4176	2.1759
.73	.6669	.7452	.8949	.94	.8076	.5898	1.3692	1.15	.9128	.4085	2.2345
.74	.6743	.7358	.9131	.95	.8134	.5817	1.3984	1.16	.9168	.3993	2.2958
.75	.6816	.7317	.9316	.96	.8192	.5735	1.4284	1.17	.9208	.3902	2.3600
.76	.6889	.7248	.9505	.97	.8249	.5653	1.4592	1.18	.9246	.3809	2.4273
.77	.6961	.7179	.9697	.98	.8305	.5570	1.4910	1.19	.9284	.3717	2.4979
.78	.7033	.7109	.9893	.99	.8360	.5487	1.5237	1.20	.9320	.3624	2.5722
.79	.7104	.7038	1.0092	1.00	.8415	.5403	1.5574	1.21	.9356	.3530	2.6503
.80	.7174	.6967	1.0296	1.01	.8468	.5319	1.5922	1.22	.9391	.3436	2.7328
.81	.7243	.6895	1.0505	1.02	.8521	.5234	1.6281	1.23	.9425	.3342	2.8198
.82	.7311	.6822	1.0717	1.03	.8573	.5148	1.6652	1.24	.9458	.3248	2.9119
.83	.7379	.6749	1.0934	1.04	.8624	.5062	1.7036	1.25	.9490	.3153	3.0096

APPENDIX TABLE 4 (*Continued*)

Trigonometric Functions—Radians

t	sin t	cos t	tan t	t	sin t	cos t	tan t	t	sin t	cos t	tan t
1.26	.9521	.3058	3.1133	1.37	.9799	.1994	4.9131	1.48	.9959	.0907	10.9834
1.27	.9551	.2963	3.2236	1.38	.9819	.1896	5.1774	1.49	.9967	.0807	12.3499
1.28	.9580	.2867	3.3413	1.39	.9837	.1798	5.4707	1.50	.9975	.0707	14.1014
1.29	.9608	.2771	3.4672	1.40	.9854	.1700	5.7979	1.51	.9982	.0608	16.4281
1.30	.9636	.2675	3.6021	1.41	.9871	.1601	6.1654	1.52	.9987	.0508	19.6695
1.31	.9662	.2579	3.7471	1.42	.9887	.1502	6.5811	1.53	.9992	.0408	24.4984
1.32	.9687	.2482	3.9033	1.43	.9901	.1403	7.0555	1.54	.9995	.0308	32.4611
1.33	.9711	.2385	4.0723	1.44	.9915	.1304	7.6018	1.55	.9998	.0208	48.0785
1.34	.9735	.2288	4.2556	1.45	.9927	.1205	8.2381	1.56	.9999	.0108	92.6205
1.35	.9757	.2190	4.4552	1.46	.9939	.1106	8.9886	1.57	1.0000	.0008	1255.7656
1.36	.9779	.2092	4.6734	1.47	.9949	.1006	9.8874				

APPENDIX TABLE 5

Table of Integrals

Familiar Formulas

1. $\int u^n \, du = \dfrac{u^{n+1}}{n+1} + C$

2. $\int k \, du = ku + C$

3. $\int \dfrac{1}{u} \, du = \ln |u| + C$

4. $\int e^u \, du = e^u + C$

New Formulas   (Note that $a > 0$ in formulas involving a^2.)

5. $\int \dfrac{u \, du}{a + bu} = \dfrac{1}{b^2} (bu - a \ln |a + bu|) + C$

6. $\int \dfrac{du}{u(a + bu)} = \dfrac{1}{a} \ln \left| \dfrac{u}{a + bu} \right| + C$

7. $\int \dfrac{u \, du}{\sqrt{a + bu}} = \dfrac{2(bu - 2a)}{3b^2} \sqrt{a + bu} + C$

8. $\int \sqrt{u^2 + a^2} \, du = \dfrac{u}{2} \sqrt{u^2 + a^2} + \dfrac{a^2}{2} \ln \left| u + \sqrt{u^2 + a^2} \right| + C$

9. $\int \dfrac{du}{\sqrt{u^2 + a^2}} = \ln \left| u + \sqrt{u^2 + a^2} \right| + C$

10. $\int \dfrac{du}{\sqrt{u^2 - a^2}} = \ln \left| u + \sqrt{u^2 - a^2} \right| + C$

11. $\int \dfrac{du}{u\sqrt{u^2 + a^2}} = -\dfrac{1}{a} \ln \left| \dfrac{a + \sqrt{u^2 + a^2}}{u} \right| + C$

12. $\int \dfrac{du}{u\sqrt{a^2 - u^2}} = -\dfrac{1}{a} \ln \left| \dfrac{a + \sqrt{a^2 - u^2}}{u} \right| + C$

13. $\int \dfrac{du}{u^2\sqrt{u^2 + a^2}} = -\dfrac{\sqrt{u^2 + a^2}}{a^2 u} + C$

14. $\int \dfrac{du}{u^2\sqrt{a^2 - u^2}} = -\dfrac{\sqrt{a^2 - u^2}}{a^2 u} + C$

APPENDIX TABLE 5 (*Continued*)

Table of Integrals

15. $\displaystyle\int \frac{\sqrt{a^2 - u^2}}{u}\, du = \sqrt{a^2 - u^2} - a \ln \left| \frac{a + \sqrt{a^2 - u^2}}{u} \right| + C$

16. $\displaystyle\int \frac{du}{u^2 - a^2} = \frac{1}{2a} \ln \left| \frac{u - a}{u + a} \right| + C$

17. $\displaystyle\int u^n \ln u\, du = \frac{u^{n+1}}{n + 1} \ln u - \frac{u^{n+1}}{(n + 1)^2} + C \qquad n \ne -1$

18. $\displaystyle\int \frac{du}{1 + e^u} = u - \ln (1 + e^u) + C$

Two Reduction Formulas

19. $\displaystyle\int (\ln u)^n\, du = u(\ln u)^n - n \int (\ln u)^{n-1}\, du$

20. $\displaystyle\int u^n e^u\, du = u^n e^u - n \int u^{n-1} e^u\, du$

Trigonometric Integrals

21. $\displaystyle\int \sin u\, du = -\cos u + C$

22. $\displaystyle\int \cos u\, du = \sin u + C$

23. $\displaystyle\int \sec^2 u\, du = \tan u + C$

24. $\displaystyle\int \csc^2 u\, du = -\cot u + C$

25. $\displaystyle\int \sec u \tan u\, du = \sec u + C$

26. $\displaystyle\int \csc u \cot u\, du = -\csc u + C$

27. $\displaystyle\int \tan u\, du = -\ln |\cos u| + C$

28. $\displaystyle\int \cot u\, du = \ln |\sin u| + C$

EXERCISE LIBRARY

This Exercise Library contains a variety of graphing calculator exercises that can be used to supplement the study of calculus. The exercises can also be done using computer software. (A few exercises are designed for computers only and are identified accordingly.)

The exercises direct your attention to the calculus concept, without the need for algebraic manipulation. The library includes some exercises that would be very difficult to work without the assistance of a graphing calculator or computer. You will also find exercises that offer an opportunity to solve problems in new ways, using approaches that become both feasible and appealing when a graphing calculator or computer is available.

Within the main body of the book, the special symbol

is placed at the end of the exercises in each section for which graphing calculator and computer exercises have been created.

Answers to all Library exercises are given at the end of the Library.

SECTION 1.5

1. Study the graph of $f(x) = x^3 + 5x^2 + 3x - 5$ on the interval $[-10,10]$.
   (a) How many zeros does f have in the interval? (The concept of a *zero* of a function was introduced in Section 1.3, pages 18–19.)
   (b) To the nearest integer, what is the smallest zero of f in the interval?

2. Study the graph of $f(x) = x^3 + 3x^2 - x - 10$ on the interval $[-10,10]$.
   (a) How many zeros does f have in the interval?
   (b) How many positive zeros does f have in the interval?
   (c) How many negative zeros does f have in the interval?

3. Study the graph of $f(x) = 2x^3 - x^2 + 5x + 1$ on the interval $[-5, 5]$.
   (a) How many zeros of f are found between 0 and 1?
   (b) Estimate to the nearest tenth the negative zero of f.

4. For what values of x is the given function positive?

$$f(x) = 1 - x^{1.5}$$

5. To the nearest tenth, determine all x values for which $f(x) = g(x)$, where $f(x) = x^2 - 4x + 5$ and $g(x) = x + 2$.

6. To the nearest tenth, determine all x values for which $f(x) = g(x)$, where $f(x) = x - 1.7\sqrt{x}$ and $g(x) = 4.8 - .6x$.

7. Graph $y = x^3 - 4x^2 + 3x - 5$. To the nearest tenth, what is the smallest value of y (lowest that the graph reaches) for x values chosen in the interval $[0, 4]$?

8. Graph $y = x^3 + x^2 - 8x - 9$. To the nearest tenth, what is the largest value of y (highest that the graph reaches) for x values chosen in $[-5, 2]$?

SECTION 1.7

1. Assume a company's profit is given by $P(x) = x^{1.5} - 4x - 2$ hundred dollars, where x is the number of units sold. Use a graph to determine (to the nearest whole number) how many units must be sold in order to break even.

2. Graph revenue function $R(x) = .2x^2$ and cost function $C(x) = .9x + 4$.
   (a) Determine the break-even quantity (to the nearest whole number).
   (b) Determine, to the nearest tenth, the y-coordinate of the break-even point.

3. Graph cost function $C(x) = 2.1 + .3x$ and revenue function $R(x) = 1.2x^{.5}$. Is there anything unusual about the break-even point?

4. Suppose profit in hundreds of dollars is given by $P(x) = -x^3 + 6x^2 - 19$. To the nearest hundred dollars, what is the largest profit attainable?

5. Let the demand equation be $p = .2x^2 - 8.4x + 120$ and the supply equation be $p = 3.4 + .2x^2$. Determine the equilibrium point (both coordinates to the nearest whole number).

SECTION 2.1

1. Study the graph of the function in order to determine the given limit.

$$f(x) = \frac{x^2 - 1}{x - 1} \qquad \lim_{x \to 1} f(x)$$

2. Study the graph of the function in order to determine the given limit.

$$f(x) = \frac{3x^3 - .48x}{x - .4} \qquad \lim_{x \to .4} f(x)$$

3. Study the graph of the function in order to determine the given limit.

$$f(x) = \frac{x^{2.5} - .04x^{.5}}{x^{1.5} - .2x^{.5}} \qquad \lim_{x \to .2} f(x)$$

4. Study the graph of the function in order to determine the given limit.

$$f(x) = \frac{\dfrac{1}{x} - \dfrac{1}{2}}{x - 2} \qquad \lim_{x \to 2} f(x)$$

SECTION 2.3

In Exercises 1 and 2, graph the function to determine if the given limit exists. If the limit exists, find it. If the limit does not exist, so indicate.

1. $f(x) = 1 + \sqrt{x^2 - 4x + 3}$

    (a) $\lim\limits_{x \to 2} f(x)$ **(b)** $\lim\limits_{x \to 3^-} f(x)$ **(c)** $\lim\limits_{x \to 3^+} f(x)$ **(d)** $\lim\limits_{x \to 3} f(x)$ **(e)** $\lim\limits_{x \to 4} f(x)$

2. $f(x) = 5 - \sqrt{8 - x^{1.5}}$

    (a) $\lim\limits_{x \to 4^-} f(x)$ **(b)** $\lim\limits_{x \to 4^+} f(x)$ **(c)** $\lim\limits_{x \to 4} f(x)$

SECTION 2.4

In Exercises 1 and 2, graph the function to determine the given limit at infinity.

1. $f(x) = \dfrac{10 + x + x^3}{x^3}$

    (a) $\lim\limits_{x \to \infty} f(x)$ **(b)** $\lim\limits_{x \to -\infty} f(x)$

2. $f(x) = \dfrac{3x^2 + x}{x^2 - 4}$

    (a) $\lim\limits_{x \to \infty} f(x)$ **(b)** $\lim\limits_{x \to -\infty} f(x)$

SECTION 2.5

In Exercises 1–3, study the graph of the function in order to determine whether the required limit is finite, ∞, or $-\infty$. (If finite, determine the number.)

1. $f(x) = \dfrac{x^3 - 3x + 4}{1 - x}$

    (a) $\lim\limits_{x \to 1^+} f(x)$ **(b)** $\lim\limits_{x \to 1^-} f(x)$

2. $f(x) = \dfrac{x^2}{x^2 - 9}$

    (a) $\lim\limits_{x \to 3^+} f(x)$ **(b)** $\lim\limits_{x \to 3^-} f(x)$ **(c)** $\lim\limits_{x \to -3^+} f(x)$ **(d)** $\lim\limits_{x \to -3^-} f(x)$

3. $f(x) = \dfrac{x^2 - x}{x^2 + 3x - 4}$

    (a) $\lim\limits_{x \to 1^+} f(x)$ **(b)** $\lim\limits_{x \to 1^-} f(x)$ **(c)** $\lim\limits_{x \to -4^+} f(x)$ **(d)** $\lim\limits_{x \to -4^-} f(x)$

In Exercises 4–6, look at the graph of the function in order to determine the vertical asymptote(s) of the function.

4. $f(x) = \dfrac{10x}{x^2 - 25}$

5. $f(x) = \dfrac{x + 5}{x + 6}$

6. $f(x) = \dfrac{x^3}{x^4 - 64x}$

SECTION 3.1

1. From the graph of the function

$$f(x) = \frac{18}{x^2 - 2x + 4}$$

determine whether the slope of the tangent line would be positive, negative, or zero at the points given below.
(**a**) $(-2, 1.5)$ (**b**) $(1, 6)$ (**c**) $(0, 4.5)$ (**d**) $(2, 4.5)$

2. Graph $y = x^2 + 5x + 1$ and $y = -5.25$. Does the line appear to be tangent to the curve?

3. Graph $y = 1 - 3x^2 - x^3$ and $y = 2x - 1$. Does the line appear to be tangent to the curve?

4. Graph $y = 3 - x^4$ and $y = -4x + 6$. Does the line appear to be tangent to the curve?

SECTION 3.3

1. A ball is thrown upward from the ground and travels so that its distance s (in feet) from the ground after t seconds is

$$s(t) = -16t^2 + 75t \qquad t \geq 0$$

Graph the distance function.
(**a**) Estimate, to the nearest tenth of a second, when the velocity of the ball will be zero. (*Hint:* The derivative is both the slope of the tangent line and the velocity.)
(**b**) Estimate to the nearest foot the maximum height that the ball will reach.

SECTION 3.4

1. Graph the cost function

$$C(x) = 43 + 5.2x - .1x^2 \qquad x \geq 0$$

Determine to the nearest unit, the number of units (x) for which marginal cost is zero.

2. Graph the profit function

$$P(x) = 8.5x - .2x^2 \qquad x \geq 0$$

Determine to the nearest unit, the number of units (x) for which marginal profit is zero.

SECTION 3.5

1. Graph the function defined by

$$f(x) = \frac{x + 5}{x + 2}$$

Determine whether the slope of the tangent line is positive, negative, or zero at the given value of x.
(a) $x = 4$ **(b)** $x = -1$ **(c)** $x = -3$

SECTION 3.6

1. Graph function f and determine whether $f'(20)$ is positive or negative.

$$f(x) = (x^2 + 1)^{1/2} (x - 1)^{3/4}$$

2. Graph the function defined by

$$f(x) = \sqrt{\frac{x + 6}{x + 2}}$$

Is $f'(0) > f'(1)$?

3. Graph $\quad y = x^2 - x\sqrt{2x + 1}$.
     (a) Find a positive value of x for which $y' > 0$.
     (b) Find a value of x for which $y' = 0$.
     (c) Find a positive value of x for which $y' < 0$.

SECTION 3.7

1. Use the graph of $f'(x) = .2x\sqrt{x^2 - 9}$ to determine whether $f''(x)$ is positive, negative, zero, or undefined at the given value of x.
     (a) $x = 4.2$ **(b)** $x = 1.7$ **(c)** $x = -5.1$

2. Graph the marginal profit function

$$MP(x) = 8.03\sqrt{x} - 1.1x$$

and use the graph to determine if the rate of change of marginal profit is positive or negative for the given number of units sold (x).

(a) x between 5 and 9 units

(b) x between 24 and 30 units

SECTION 3.8

1. Graph the circle given by the equation

$$x^2 + y^2 = 25$$

Begin by manipulating the equation into $y = \pm\sqrt{25 - x^2}$ and then graphing the two functions $y = \sqrt{25 - x^2}$ and $y = -\sqrt{25 - x^2}$. Then, use implicit differentiation to obtain the slope of the tangent line to the circle. Next, obtain the equation of the tangent line at the point $(-3, 4)$. Finally, graph the tangent line together with the circle. (The equation of the tangent line is given in the answer section.)

SECTION 4.1

1. Graph the given function in order to determine where it is increasing and where it is decreasing.

$$f(x) = \frac{20}{x^2 + 4}$$

2. Graph the given function in order to determine where it is increasing and where it is decreasing.

$$f(x) = \frac{x^2}{16 - x^2}$$

3. Graph $f(x) = -x^3 + 3x^2 + 1$ to determine where the function is increasing and where it is decreasing.

4. For $f(x) = x^3 - 3x^2 + 20$, obtain $f'(x)$. Then graph $f'(x)$ to determine where f is increasing and where it is decreasing.

5. For $f(x) = 15 - 6x^{2/3}$, obtain $f'(x)$. Then graph $f'(x)$ to determine where f is increasing and where it is decreasing.

SECTION 4.2

1. Graph $f(x) = x^4 + 4x^3 + 6x^2 + 4x - 2$ and estimate to the nearest integer the coordinates of all relative maximum and relative minimum points.

2. Graph $f(x) = -x^3 - 3x^2 + 6$ and estimate to the nearest integer the coordinates of all relative maximum and relative minimum points.

3. Graph $f(x) = x^3 - 6x^2 + 6x + 7$ and estimate to the nearest tenth the coordinates of all relative maximum and relative minimum points.

4. Graph $f(x) = .2x^3 + x^2 - 3x + 5$ and estimate to the nearest tenth the coordinates of all relative maximum and relative minimum points.

5. From the graph of $f'(x) = 3x^2 - 6x - 9$, estimate to the nearest integer the x coordinate of each relative extreme point of function f. Indicate whether the x is associated with a relative maximum or relative minimum.

6. What can you conclude about the relative extreme points of f when you examine the graph of $f'(x) = 3x^2 + 12x + 12$?

SECTION 4.3

1. Graph $f(x) = x^3 + 9x^2 + 27x + 29$ in order to determine where it is concave up and where it is concave down.

2. Graph the function in order to determine where it is concave up and where it is concave down.

$$f(x) = \frac{.4x^2}{x^2 - 25}$$

3. Graph $f(x) = x^3 - 5x^2 + 7x - 4$ and estimate the x coordinate of the point of inflection. Your estimate should be within .2 of the exact value.

4. Graph $f(x) = 6 - 1.5x^2 - .2x^3$ and estimate the coordinates of the point of inflection (each to the nearest integer).

5. Examine the graph of each function in order to determine how many points of inflection it has (none, one, two).
   (a) $f(x) = x^3 - 3x^2 + 3x + 1$
   (b) $f(x) = x^4 - 8x^3 + 24x^2 - 32x + 17$
   (c) $f(x) = x^4 - x^3 - 4x^2 + 4x + 3$
   (d) $f(x) = .3x^4 + x^3 - 4x^2 + x + 1$

SECTION 4.4

1. Study the graph of $f(x) = 6 - .2x^2 + 3x^{.5}$ to estimate (to the nearest tenth) the absolute maximum value of the function on the interval $[1, 7]$.

2. Study the graph of $f(x) = -x^3 + 1.7x^2 + 3.2$ to estimate (to the nearest tenth) the absolute maximum value of the function on the specified intervals.
   (a) $[-1, 3]$ (b) $[0, 4]$

3. Study the graph of $f(x) = x^3 - 3.4x^2 + 4$ to estimate (to the nearest tenth) the absolute minimum value of the function on the specified intervals.
   (a) $[0, 4]$ (b) $[-1.5, 4.5]$

4. Study the graph of $f(x) = 2x - \sqrt{1 + x^3}$.
   (a) Estimate (to the nearest tenth) the absolute minimum value of the function on the interval $[-1, 6]$.
   (b) Estimate (to the nearest tenth) the absolute maximum value of the function on the interval $[-1, 6]$.

SECTION 5.1

1. In Section 5.1, Exercise 52, the function

$$y = \frac{20}{1 + 9e^{-.2x}}$$

gave the number of fruit flies y present after x days.
   (a) Use the graph of the function to determine the approximate number of fruit flies present after 9 days.
   (b) Use the graph to determine how many days it will take before there will be 18 fruit flies.
   (c) Is the x coordinate of the point of inflection closer to 6, 12, 18, 21, or 24?
   (d) Write the equation of the horizontal asymptote.

2. Use the graphs of $y = xe^x$ and $y = 12 - x$ to find to the nearest tenth all x such that $xe^x = 12 - x$.

SECTION 5.3

1. Graph $f(x) = x - e^{x-4}$.
   (a) Find to the nearest tenth the coordinates of the relative maximum point.
   (b) Where is the graph concave up?

2. Graph $f(x) = 7 - .6e^x - .6e^{-x}$ and determine the interval on which the function is increasing.

3. Graph $f(x) = e^{-x}/x^2$.
   (a) Is $f'(1)$ positive, negative, zero, or undefined?
   (b) Is $f''(-4)$ positive or negative?
   (c) Determine any horizontal asymptote(s).

SECTION 5.4

1. Graph $f(x) = 3.2 + 1.8x - 2 \ln x$.
   (a) Find to the nearest tenth the coordinates of the relative minimum point.
   (b) Where is the graph concave up?

2. Graph $f(x) = x + e^x - 20 \ln x$. Determine if the graph is increasing or decreasing at the given points.

   (a) $(3, f(3))$ (b) $(1, f(1))$ (c) $(2, f(2))$

3. Graph the function

$$f(x) = \frac{x^2 - 10 \ln x}{2x}$$

   (a) Is $f'(2)$ positive, negative, zero, or undefined?
   (b) Is $f''(1)$ positive or negative?

SECTION 7.4

1. Use a *computer* and the trapezoidal rule to approximate the definite integral of Example 1, namely,

$$\int_0^2 \frac{1}{16 + x^2} \, dx$$

   Begin with $n = 4$ and continue by using larger and larger n, until your approximation is correct to 7 decimal places.

2. Redo Exercise 1, using instead Simpson's rule.

3. Use the approach of Exercise 1 to approximate the definite integral correct to 7 decimal places.

$$\int_3^7 \frac{1}{x^2 - 1} \, dx$$

4. Redo Exercise 3, using instead Simpson's rule.

5. Use the approach of Exercise 1 to approximate the definite integral correct to 7 decimal places.

$$\int_5^7 \frac{1}{\sqrt{x^2 - 9}} \, dx$$

6. Redo Exercise 5, using instead Simpson's rule.

SECTION 7.5

1. Recall the integral from Example 6.

$$\int_{-\infty}^{\infty} xe^{-x^2} \, dx$$

   Graph the function $y = xe^{-x^2}$ and study it to suggest an intuitive explanation of why the integral of Example 6 is equal to 0.

SECTION 8.1

1. Consider the function

$$f(x, y) = x^2 - y$$

Graph the level curve for $c = -2$. Also graph the lines $y = x + 2$, $y = 6$, and $y = 1 - 2x$. Which line is tangent to the level curve?

2. Consider the function

$$f(x, y) = 2\sqrt{x} - y$$

Graph the level curve for $c = -3$. Also graph the lines $y = 2x + 3$, $y = x + 4$, and $y = 6 - x$. Which line is tangent to the level curve?

3. Consider the Cobb-Douglas production function

$$f(x, y) = 4x^{2/3}y^{1/3}$$

Three associated isoquants are

(i) $4x^{2/3}y^{1/3} = 12$

(ii) $4x^{2/3}y^{1/3} = 16$

(iii) $4x^{2/3}y^{1/3} = 20$

Each of these three equations can be manipulated into a form more suitable for graphing, namely,

(i) $y_1 = \dfrac{27}{x^2}$

(ii) $y_2 = \dfrac{64}{x^2}$

(iii) $y_3 = \dfrac{125}{x^2}$

Graph the three functions for $x > 0$ and answer the questions that follow.
(a) Which function (y_1, y_2, or y_3) will have the smallest y value for any particular x value?
(b) Of the three functions (y_1, y_2, y_3), which one is associated with the largest production level?

SECTION 8.5

1. In Example 1, the least squares line for the points (2, 5), (3, 7), (4, 9), (7, 8), and (8, 12) was determined to be $y = .83x + 4.22$. Graph the least squares line *and* plot the five given points.
(a) From the graph, which point is closest (vertically) to the line?
(b) From the graph, which point is furthest (vertically) from the line?

2. Redo Exercise 1 using instead the points (2, 7), (3, 6), (4, 6), and (6, 2) and the least squares line $y = -1.23x + 9.86$ (from Section 8.5, Exercise 13).

SECTION 9.2

1. Graph $y = \sin x$ and $y = \cos\left(x - \dfrac{\pi}{2}\right)$. From the graphs, determine an equation

that gives the relationship between $\sin x$ and $\cos\left(x - \dfrac{\pi}{2}\right)$.

2. Repeat Exercise 1 using instead $y = \cos x$ and $y = \sin\left(x + \dfrac{\pi}{2}\right)$.

3. Repeat Exercise 1 using instead $y = \sin x$ and $y = \cos\left(x + \dfrac{\pi}{2}\right)$.

4. Repeat Exercise 1 using instead $y = \cos x$ and $y = \sin\left(x - \dfrac{\pi}{2}\right)$.

5. Graph $y = \cos x$ and $y = \sec x$. (Recall that $\sec x = 1/\cos x$.)
   (a) Describe what happens to $\sec x$ when $\cos x$ approaches 0. Consider that $\cos x$ approaches 0 through positive values and through negative values.
   (b) Write the equation of the two vertical asymptotes of $y = \sec x$ that are closest to the y axis.

6. Graph $y = \sin x$ and $y = \csc x$. (Recall that $\csc x = 1/\sin x$.)
   (a) Describe what happens to $\csc x$ when $\sin x$ approaches 0. Consider that $\sin x$ approaches 0 through positive values and through negative values.
   (b) Write the equations of the three vertical asymptotes of $y = \csc x$ that are closest to the y axis.

7. Use graphs to determine whether the statements below are true or false.
   (a) $\sin (x + \pi) = \sin x$ **(b)** $\sin x = -\sin (x + \pi)$
   (c) $\cos (x - \pi) = -\cos x$ **(d)** $\cos x = \cos (x + \pi)$
   (e) $\sin\left(x + \dfrac{\pi}{2}\right) = \cos (x - \pi)$ **(f)** $\cos\left(x + \dfrac{\pi}{2}\right) = \sin\left(x - \dfrac{\pi}{2}\right)$
   (g) $\cos\left(x - \dfrac{\pi}{2}\right) = -\sin (x + \pi)$ **(h)** $\tan x = -\cot x$

SECTION 9.3

1. Study the graph of $y = \sin x$ to determine whether it is concave up or concave down at each of the following values of x.
   (a) $x = 1$ **(b)** $x = 3.5$ **(c)** $x = 5.3$ **(d)** $x = 8$
   (e) $x = -2$ **(f)** $x = -5$

2. Repeat Exercise 1 using instead $y = \cos x$.

3. Use graphs to determine whether the specified value of $f'(x)$ or $f''(x)$ is positive or negative.

    (a) $f'(2)$ when $f(x) = \cos x$ **(b)** $f''(2)$ when $f(x) = \cos x$
    (c) $f'(7)$ when $f(x) = \sin x$ **(d)** $f''(7)$ when $f(x) = \sin x$
    (e) $f'(9)$ when $f(x) = \sin x$ **(f)** $f''(9)$ when $f(x) = \sin x$

4. Graph $f(x) = 3 \cos (x + \pi)$ and determine the point of inflection having the smallest positive x coordinate.

5. Use graphs to compare the slope, concavity, and relative extrema of the functions $y = \sin x$ and $y = 5 \sin x$.

ANSWERS TO EXERCISE LIBRARY

SECTION 1.5

1. (a) 3 **(b)** -4 **2. (a)** 1 **(b)** 1   **(c)** none   **3. (a)** none **(b)** $-.2$
4. x in $[0,1)$ **5.** .7 and 4.3   **6.** 5.5   **7.** -7.1 **8.** 3.0

SECTION 1.7

1. 17   **2. (a)** 7   **(b)** 10.5   **3.** There is no break-even point. Cost is always greater than revenue, which means the company will operate at a loss regardless of the quantity sold.   **4.** $1300   **5.** (14, 42)

SECTION 2.1

1. 2   **2.** .96   **3.** .4   **4.** $-.25$

SECTION 2.3

1. (a) does not exist   **(b)** does not exist   **(c)** 1   **(d)** does not exist
    (e) 2.732 (approximately)   **2. (a)** 5   **(b)** does not exist   **(c)** does not exist

SECTION 2.4

1. (a) 1   **(b)** 1   **2. (a)** 3   **(b)** 3

SECTION 2.5

1. (a) $-\infty$ **(b)** ∞ **2. (a)** ∞ **(b)** $-\infty$ **(c)** $-\infty$ **(d)** ∞ **3. (a)** .2
    (b) .2   **(c)** $-\infty$ **(d)** ∞ **4.** $x = -5, x = 5$ **5.** $x = -6$ **6.** $x = 4$

SECTION 3.1

1. (a) positive   **(b)** zero   **(c)** positive   **(d)** negative   **2.** yes   **3.** no
4. yes

SECTION 3.3

1. (a) 2.3 seconds   **(b)** 88 feet

SECTION 3.4

1. 26   **2.** 21

SECTION 3.5

1. (a) negative   **(b)** negative   **(c)** negative

SECTION 3.6

1. positive   **2.** no
3. (a) Any x such that $x > 1.29$ **(b)** $x \approx 1.285$ **(c)** Any x such that $0 < x < 1.28$

SECTION 3.7

1. (a) positive   **(b)** undefined   **(c)** positive   **2. (a)** positive   **(b)** negative

SECTION 3.8

1. $y = .75x + 6.25$

SECTION 4.1

1. increasing on $(-\infty, 0)$; decreasing on $(0, \infty)$
2. increasing on $(0, 4)$ and $(4, \infty)$; decreasing on $(-\infty, 4)$ and $(-4, 0)$
3. increasing on $(0, 2)$; decreasing on $(-\infty, 0)$ and $(2, \infty)$
4. increasing on $(-\infty, 0)$ and $(2, \infty)$; decreasing on $(0, 2)$
5. increasing on $(-\infty, 0)$; decreasing on $(0, \infty)$

SECTION 4.2

1. $(-1, -3)$ relative minimum
2. $(0, 6)$ relative maximum; $(-2, 2)$ relative minimum
3. $(.6, 8.7)$ relative maximum; $(3.4, -2.7)$ relative minimum
4. $(-4.5, 20.5)$ relative maximum; $(1.1, 3.2)$ relative minimum
5. -1 is the x coordinate of the relative maximum point; 3 is the x coordinate of the relative minimum point.
6. f has no relative extreme points.

SECTION 4.3

1. concave up on $(-3, \infty)$; concave down on $(-\infty, -3)$
2. concave up on $(-\infty, -5)$ and $(5, \infty)$; concave down on $(-5, 5)$
3. 1.6 to 1.8   **4.** Your answer should be one of these: $(-3, -2)$ or $(-3, -1)$ or $(-2, 0)$ or $(-2, 1)$
5. **(a)** one   **(b)** none   **(c)** two   **(d)** two

SECTION 4.4

1. 9.5   **2. (a)** 5.9   **(b)** 3.9   **3. (a)** -1.8 **(b)** -7.0 **4. (a)** -2.7
(b) 1.0

SECTION 5.1

1. (a) 8   **(b)** 22   **(c)** 12   **(d)** $y = 20$ **2.** $x \approx 1.8$

SECTION 5.3

1. (a) $(4.0, 3.0)$ **(b)** nowhere   **2.** $(-\infty, 0)$ **3. (a)** negative   **(b)** positive
(c) $y = 0$

SECTION 5.4

1. (a) $(1.1, 5.0)$ **(b)** $(0, \infty)$ **2. (a)** increasing   **(b)** decreasing
(c) decreasing
3. (a) positive   **(b)** positive

SECTION 7.4

1. .1159119   **2.** .1159119   **3.** .2027326   **4.** .2027326   **5.** .3923840
6. .3923840

SECTION 7.5

1. The magnitude of the area between the curve and the x axis is the same on $(-\infty, 0)$ as it is on $(0, \infty)$. However, between $-\infty$ and 0, the curve is *below* the x axis. This means that the integral from $-\infty$ to 0 will be negative. Since the integral from 0 to ∞ is positive, the two integrals will add up to 0.

SECTION 8.1

1. $y = 1 - 2x$ **2.** $y = x + 4$ **3. (a)** y_1 **(b)** y_3

SECTION 8.5

1. (a) $(3, 7)$ **(b)** $(7, 8)$ **2. (a)** $(3, 6)$ **(b)** $(4, 6)$

SECTION 9.2

1. $\sin x = \cos\left(x - \dfrac{\pi}{2}\right)$ **2.** $\cos x = \sin\left(x + \dfrac{\pi}{2}\right)$

3. $\sin x = -\cos\left(x + \dfrac{\pi}{2}\right)$ **4.** $\cos x = -\sin\left(x - \dfrac{\pi}{2}\right)$

5. (a) When $\cos x$ approaches 0 through positive values, $\sec x$ approaches $+\infty$. When $\cos x$ approaches 0 through negative values, $\sec x$ approaches $-\infty$.
(b) $x = \pi/2, x = -\pi/2$

6. (a) When $\sin x$ approaches 0 through positive values, $\csc x$ approaches $+\infty$. When $\sin x$ approaches 0 through negative values, $\csc x$ approaches $-\infty$.
(b) $x = 0, x = \pi, x = -\pi$

7. (a) false   **(b)** true   **(c)** true   **(d)** false   **(e)** false   **(f)** false
(g) true   **(h)** false

SECTION 9.3

1. (a) down   **(b)** up   **(c)** up   **(d)** down   **(e)** up   **(f)** down
2. (a) down   **(b)** up   **(c)** down   **(d)** up   **(e)** up   **(f)** down
3. (a) negative   **(b)** positive   **(c)** positive   **(d)** negative   **(e)** negative
(f) negative
4. (1.57, 0)
5. The slope of the curve $y = 5 \sin x$ has the same sign as the slope of the curve $y = \sin x$; however, it is steeper (5 times the slope) when the slope is positive or negative. The concavity of the two graphs is the same. The relative extrema occur at the same x, but the y coordinates of the relative extreme points of $y = 5 \sin x$ are 5 times those of $y = \sin x$ (5 versus 1 and -5 versus -1).

ANSWERS TO ODD-NUMBERED EXERCISES

SECTION 1.1 (page 4)

1. [5, 9] **3.** [6, ∞) **5.** (−∞, 0) **7.** (−2, ∞) **9.** (−∞, π) **11.** $(-\sqrt{3}, \sqrt{3})$ **13.** $x \geq 0$

15. $1 \leq x \leq 75$ **17.** $x < -2$ **19.** $x > -5$ **21.** $\pi \leq x < 7$ **23.** $-4 < x \leq 19$

25. **27.** **29.**

31. **33.**

35. $x > 8$ **37.** $x \leq 6$ **39.** $x \geq 0$ **41.** $t < -7$ **43.** $x \geq 1/8$ **45.** $x < 2$ **47.** $y < 8/5$

SECTION 1.2 (page 13)

1. x^{17} **3.** b^{21} **5.** x^6 **7.** y^2 **9.** 2 **11.** $1/x^3$ **13.** $7x^2$ **15.** $1/t^9$ **17.** $1/y^4$ **19.** $1/9$ **21.** $1/125$

23. 7 **25.** 3 **27.** 1/4 **29.** 1 **31.** 8 **33.** 9 **35.** 1/27 **37.** 6 **39.** 64 **41.** 1/1000

43. 11.3137 **45.** 1142.4502 **47.** 1.0601 **49.** $4\sqrt{3}/3$ **51.** $2\sqrt{5}/5$ **53.** $2\sqrt{3}$ **55.** ±8 **57.** 0, 9

59. −7, −2 **61.** −2, 4 **63.** −2, 5 **65.** −2/3, 4 **67.** −2, 3/2 **69.** $\dfrac{-3 \pm \sqrt{5}}{2}$ **71.** $1 \pm \sqrt{5}$

73. $2 \pm \sqrt{6}$ **75.** $\dfrac{-5 \pm 3\sqrt{5}}{4}$ **77.** $\dfrac{-3 \pm \sqrt{15}}{2}$ **79.** $x - 3$ **81.** $\dfrac{x}{x + 4}$ **83.** 2/3 **85.** $\dfrac{1}{x + 5}$ **87.** $\dfrac{11}{3x}$

89. $\dfrac{8x - 1}{x(x - 1)}$ **91.** $\dfrac{6 - 7x}{2x(x + 2)}$ **93.** $\dfrac{x}{x - 1}$ **95.** $5x/2$ **97.** $\dfrac{5}{x + 4}$ **99.** $\dfrac{4x^2}{x - 1}$ **101.** $\dfrac{3x}{2 - x}$

103. $\dfrac{3h + h^2x}{hx^2 + x}$ **105.** 8 **107.** 3

109. (a) If $a = 0$, the equation is then linear ($bx + c = 0$) rather than quadratic.
(b) If $a = 0$, then division by 0 is indicated. But we cannot divide by 0.

SECTION 1.3 (page 19)

1. $f(0) = 7, f(1) = 12, f(2) = 17, f(-1) = 2$ **3.** $f(0) = 1, f(1) = 5, f(2) = 11, f(-1) = -1$

5. $f(0) = 5, f(1) = 4, f(2) = 1, f(-1) = 4$ **7.** $f(0) = 6, f(1) = 6, f(2) = 6, f(-1) = 6$

9. $f(0) = 1, f(1) = \sqrt{2}, f(2) = \sqrt{3}, f(-1) = 0$ **11.** $f(x + 2) = x^2 + x + 5, f(x - 3) = x^2 - 9x + 25$

13. $f(x + 2) = 4x^2 + 25x + 34, f(x - 3) = 4x^2 - 15x + 9$ **15.** $f(x + 2) = \dfrac{x + 7}{x - 5}, f(x - 3) = \dfrac{x + 2}{x - 10}$

17. $f(x + h) = 3x + 3h - 4, f(x + h) - f(x) = 3h$ **19.** $f(x + h) = -9x - 9h + 2, f(x + h) - f(x) = -9h$

21. \$256   **23.** (a) 505°F   (b) 65°F   **25.** \$69,500   **27.** (a) 1021   (b) 1300

29. (a) $D(c) = \dfrac{50(c + 1)}{3}$ (b) $D(8) = 150$ (c) The dosage for an 8-year-old child is 150 milligrams.

31. all the real numbers   **33.** $x \neq 0$ **35.** $x \geq 2$ **37.** $x \neq -3$ **39.** $x \neq 0, 1$ **41.** $x \geq 2/3$

43. $x \geq -1$ **45.** $x \neq 0$ **47.** $x \neq -1/2, 5$ **49.** -3 **51.** ± 3 **53.** 4, 5   **55.** $\dfrac{-5 \pm \sqrt{33}}{2}$

57. $(f \circ g)(x) = 21x + 1; (g \circ f)(x) = 21x + 7$ **59.** $(f \circ g)(x) = x^2 - 1; (g \circ f)(x) = x^2 + 2x - 1$

61. $(f \circ g)(x) = \dfrac{1}{3x}; (g \circ f)(x) = \dfrac{3}{x}$

63. f is the *name* of the function, whereas $f(x)$ is the value of f that corresponds to a particular x value.

SECTION 1.4   (page 31)

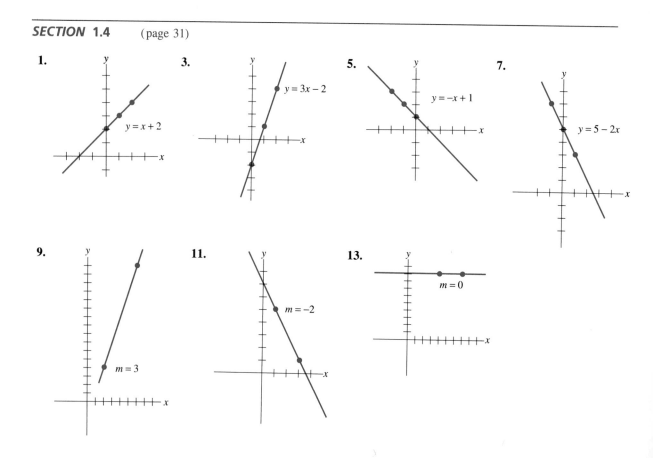

15. slope $= 5$; y intercept $= (0, 3)$ or 3   **17.** slope $= 1$; y intercept $= (0, -9)$ or -9

19. slope $= -7$; y intercept $= (0, 1)$ or 1   **21.** slope $= 0$; y intercept $= (0, 3)$ or 3

23. slope $= 8$; y intercept $= (0, 6)$ or 6   **25.** slope $= -5/2$; y intercept $= (0, 4)$ or 4   **27.** $y = -2x + 4$

29. $y = 5x - 3$ **31.** $y = -1$ **3.** $y = 3x + 5$ **35.** $y = -2x + 7$ **37.** $y = -x - 3$

39. $y = 5x - 18$ **41.** $y = -2x + 7$ **43.** $y = -x + 6$ **45.** $y = 3x - 10$ **47.** perpendicular

49. (a) When the temperature is 60°, a cricket will chirp at the rate of 80 chirps per minute.
    (b) $y = 4x - 160$ (for $x \geq 40°$)

51. (a) $y = -200t + 3400$ **(b)** \$1800

53. (a) m represents the annual appreciation, which is \$100 in this case. y represents the (appreciated) value after x years.
    (b) $y = 100x + 2000$ **(c)** \$2700   **(d)** 12 years

55. The domain of the function is all of the real numbers. For every real number x there is a corresponding real number y. (The y is computed as $2 \cdot x + 1$.)

SECTION 1.5   (page 42)

1. (a) 8   **(b)** 6   **(c)** -4 **(d)** 2   **(e)** 2, 6

3. (a) 30   **(b)** 5   **(c)** 40   **(d)** $m = 1/2$, $b = 0$ **(e)** $0 \leq t \leq 60$

5. (a) $t = 0$ **(b)** $t > 0$ (could also be $0 < t \leq 2$)
    (c) Memorized nonsense words are more readily forgotten than are the memorized words of a song.

7. (a) The northeastern city.
    (b) Yes. Since the point $(15, w(15))$ is above the point $(15, n(15))$, it follows that $w(15)$ is greater than $n(15)$.
    (c) No. Since the point $(5, n(5))$ is above the point $(5, w(5))$, it follows that $n(5) > w(5)$.

9. function   **11.** not a function   **13.** function

15.   **17.**   **19.**

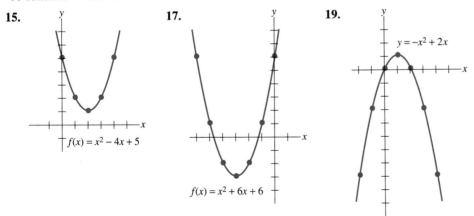

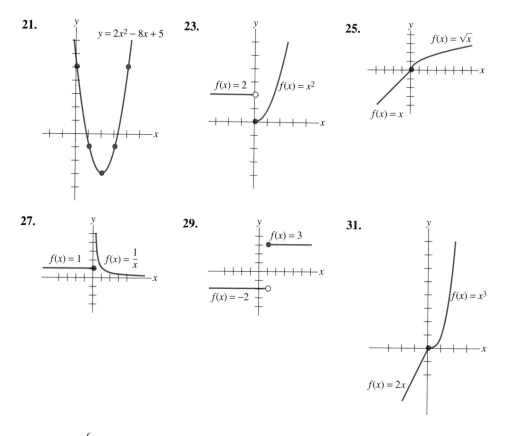

21. $y = 2x^2 - 8x + 5$

23. $f(x) = 2$ $f(x) = x^2$

25. $f(x) = \sqrt{x}$ $f(x) = x$

27. $f(x) = 1$ $f(x) = \dfrac{1}{x}$

29. $f(x) = 3$ $f(x) = -2$

31. $f(x) = x^3$ $f(x) = 2x$

33. $W(x) = \begin{cases} 5x & 0 \le x \le 40 \\ 7.5x - 100 & x > 40 \end{cases}$ **35.** When x is 2, y is 8.

37. $P(6) - P(0)$ is the change in the insect population over the next 6 months, that is, the amount by which the insect population will increase or decrease over the next 6 months.

SECTION 1.6   (page 49)

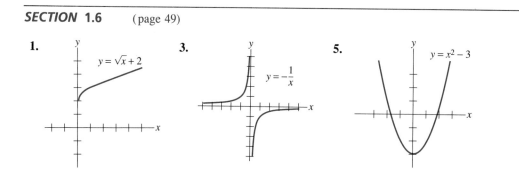

1. $y = \sqrt{x} + 2$

3. $y = -\dfrac{1}{x}$

5. $y = x^2 - 3$

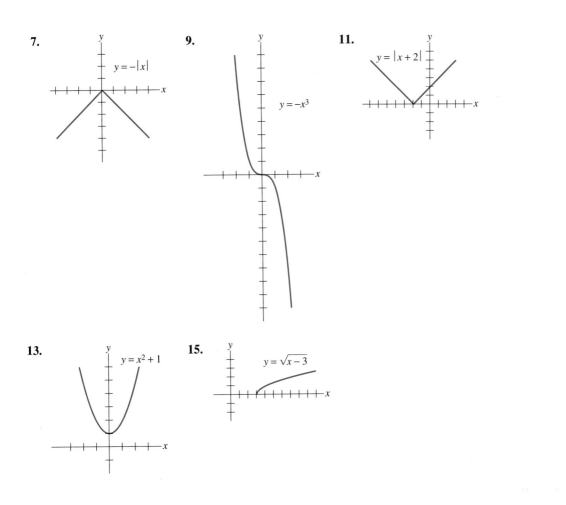

7. $y = -|x|$

9. $y = -x^3$

11. $y = |x + 2|$

13. $y = x^2 + 1$

15. $y = \sqrt{x - 3}$

SECTION 1.7   (page 58)

1. (a) $397.50   **(b)** $9.90   **(c)** $210   **3.** 23   **5. (a)** $2600   **(b)** $25.01

7. (a) $P(x) = -.1x^2 + 130x - 1000$ **(b)** $5250

9. (a) $C(x) = 29x + .02x^2$ **(b)** $R(x) = 54x$ **(c)** $P(x) = 25x - .02x^2$

11. $P(x) = 9x - .02x^2$ **13. (a)** $170   **(b)** $210   **(c)** $4100

15. First, determine the profit function $P(x)$ (as revenue minus cost). Then evaluate $P(x)$ at 75 and at 74. Finally, subtract $P(74)$ from $P(75)$.

17. 1900 units   **19.** $R(x) = 50x - .1x^2$; $1840   **21.** 160 units

23. (a) $C(x) = 24x + .4x^2$ **(b)** $640   **(c)** $31.60   **25. (a)** 30   **(b)** $11   **(c)** $(30, 11)$

27. (a) 480   **(b)** $52   **(c)** $(480, 52)$ **29. (a)** 200   **(b)** $7   **(c)** $(200, 7)$ **31.** 39; $121   **33.** 80; $29

35. $C(x) = .2x^2 + 11x$; $R(x) = 90x - .4x^2$; $P(x) = 79x - .6x^2$

37. $C(x) = 7x^2 + 10x; R(x) = 5x + 3; P(x) = -7x^2 - 5x + 3$ **39.** $388

41. Since the equilibrium quantity (x) is the quantity for which *supply and demand are equal*, it does not matter which equation (supply or demand) is used to compute the price that corresponds to the equilibrium quantity.

REVIEW EXERCISES FOR CHAPTER 1   (page 61)

1. $[1, 7)$ **3.** $(-\infty, 11/4]$ **5.** $f(0) = 0; f(2) = 4; f(-2) = -12$ **7.** $f(4) = 2; f(7) = 4; f(8) = 2\sqrt{5}$

9. $f(x + 1) = 3x^2 + 6x + 3; f(x + h) = 3x^2 + 6xh + 3h^2$ **11.** $x \neq 1/2$ **13.** $x \geq -9$ **15.** ± 3

17. $(f \circ g)(x) = x^2 + 2x - 2$

19.

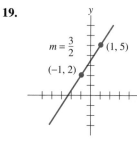

21. slope is 5; y intercept is $(0, -1)$ or -1

23. $y = 6x - 19$

25. $y = -3x + 6$

27.   **29.**   **31.**

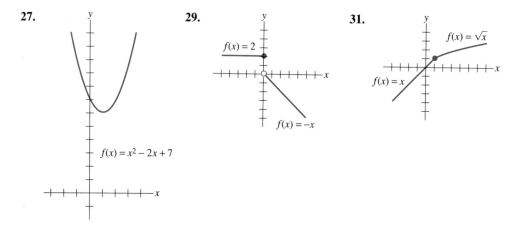

33. $y = .30x + 26$ (dollars)

35. (a) 300   **(b)** $n(5) = 385$. After 5 hours there will be 385 bacteria in the culture.

37. (a) $R(x) = 300x$ **(b)** $P(x) = -.1x^2 + 130x - 900$
(c) $P(5) = -252.50$. The manufacturer will lose $252.50 on the production and sale of 5 stereo units.
(d) $2610   **(e)** $171.90

39. $(200, 5)$

SECTION 2.1   (page 71)

1. 8   **3.** 20   **5.** 19   **7.** 9   **9.** 2/9   **11.** 3   **13.** 1/8   **15.** −2   **17.** 0

19. cannot determine by substitution; 4   **21.** 0   **23.** cannot determine by substitution; −3

25. cannot determine by substitution; 7   **27.** 0   **29.** cannot determine by substitution; −4   **31.** 7   **33.** −6

35. (a) 1000
    (b) As the time traveled approaches 2.5 hours, the distance traveled by the airplane approaches 1000 miles.

37. (a) $p(x) = .25$ (dollars)   *or* $p(x) = 25$ (cents)
    (b) .25 (dollars)   *or*   25 (cents)

39. 0   **41. (a)** *A*   **(b)** *B*

SECTION 2.2   (page 78)

1. 28   **3.** 37   **5.** 2/7   **7.** 0   **9.** −1/5   **11.** 5   **13.** 7   **15.** 0   **17.** −4   **19.** 1   **21.** 3

23. 0   **25.** 3/2

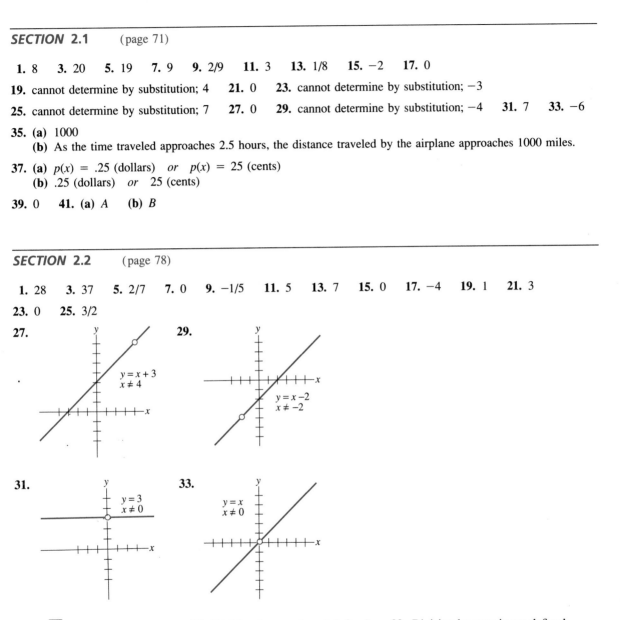

27. $y = x + 3$, $x \neq 4$

29. $y = x - 2$, $x \neq -2$

31. $y = 3$, $x \neq 0$

33. $y = x$, $x \neq 0$

35. $\sqrt{-1}$ is not a real number.   **37.** Division by zero is not defined.   **39.** Division by zero is not defined.

41. 3   **43.** 4

45. (a) At the beginning of each year, beginning in 1992.
    (b) At the beginning of each year (1992, 1993, and so on), new cases are added. This means that the workload is suddenly increased at that time.

47.

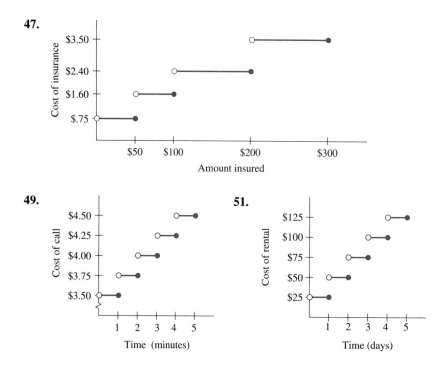

49.

51.

53. Not necessarily true. Only if f is continuous at 2 would it be true that $f(2)$ is equal to the limit as x approaches 2.

55. Determine the limit of $f(x)$ as x approaches 4. Also, compute $f(4)$. If the limit is equal to $f(4)$, then the function is continuous at 4. If the limit is not equal to $f(4)$, then the function is not continuous at 4. (Of course, if the limit fails to exist or if f is not defined at 4, then the function is not continuous at 4.)

57. continuous   **59.** discontinuous at 0   **61.** discontinuous at 0

63. discontinuous at all numbers in the interval.   **65.** 1   **67.** 1/12   **69.** 1/2   **71.** −1/9

SECTION 2.3   (*page* 85)

1. (a) 2   **(b)** 1   **3. (a)** limit does not exist   **(b)** 2   **5. (a)** −1   **(b)** −1

7. $\sqrt{2x}$ is not defined for numbers less than 0; that is, $f(x) = \sqrt{2x}$ is not defined for the values of x along the approach. As x approaches 0 from the left, $2x$ takes on negative values.

9. $\sqrt{1-x}$ is not defined for numbers less than 0; that is, $f(x) = \sqrt{1-x}$ is not defined for the values of x along the approach. As x approaches 1 from the right, $1-x$ takes on negative values.

11. In order for $\lim_{x \to 0} \sqrt{3x}$ to exist, both the left-hand and right-hand limits must exist and be equal. Although the right-hand limit exists and is 0, that is, $\lim_{x \to 0^+} \sqrt{3x} = 0$, the left-hand limit fails to exist (for the same reason as in Exercise 7).

13. In order for $\lim_{x \to -2} \sqrt{x + 2}$ to exist, both the left-hand and right-hand limits must exist and be equal. Although the right-hand limit exists and is 0, that is, $\lim_{x \to -2^+} \sqrt{x + 2} = 0$, the left-hand limit fails to exist (for the same reason as in Exercise 7).

15. (a) 2   (b) −1   17. (a) −1   (b) −1   19. (a) 0   (b) 0   21. (a) 0   (b) 3   23. does not exist

25. −1   27. 0   29. does not exist

31. (a) $2.40   (b) $1.60   (c) $1.60   (d) $.75   (e) does not exist   (f) no
    (g) Yes, since $\lim_{x \to 120} M(x) = M(120)$. Both the limit and the functional value are the same, 2.40.

33. (a) 10   (b) 10.5   (c) no   (d) 9.5   (e) 10   (f) no

35. f is continuous at 3 because $\lim_{x \to 3} f(x) = f(3)$; that is, 11 = 11.

37. f is continuous at 2 because $\lim_{x \to 2} f(x) = f(2)$; that is, 5 = 5.

39. f is not continuous at 2 because $\lim_{x \to 2} f(x) \neq f(2)$. Specifically, $\lim_{x \to 2} f(x)$ does not exist, since the left-hand limit is 11 and the right-hand limit is 16.

41. f is not continuous at 4 because $\lim_{x \to 4} f(x) \neq f(4)$. Specifically, $\lim_{x \to 4} f(x)$ does not exist, since the left-hand limit is 14 and the right-hand limit is 13.

43. No. Since $\lim_{x \to 3} f(x) = 5$, the left-hand and right-hand limits must be equal—and equal to 5. If $\lim_{x \to 3^+} f(x)$ were equal to 4, then $\lim_{x \to 3} f(x)$ would not exist.

SECTION 2.4   (page 94)

1.

x	$\dfrac{x + 1}{x}$
100	1.01
1000	1.001
10,000	1.0001
1,000,000	1.000001

The limit is 1.

3.

x	$\dfrac{1 + 3x}{2x}$
−100	1.495
−1000	1.4995
−10,000	1.49995
−1,000,000	1.4999995

The limit is 1.5.

5. 0   7. 0   9. 0   11. 0   13. 3/5   15. 2   17. −2   19. 0   21. 1/2   23. 1   25. −1/2

27. 0   29. $y = 2$ 31. $y = -1$ 33. $y = 1$ 35. $y = 0$ 37. $y = 1/3$ 39. $12   41. 5 feet

43. (a) 0
    (b) After the drug is injected, the amount remaining in the bloodstream will eventually tend toward zero. In other words, eventually there will be no presence of the drug.

45. b

47. A limit as $x \to \infty$ is a left-hand limit, since the approach is through values less than ∞. A limit as $x \to -\infty$ is a right-hand limit, since the approach is through values greater than $-\infty$.

SECTION 2.5   (page 101)

1. ∞ **3.** $-\infty$ **5.** ∞ **7.** $-\infty$ **9.** $-\infty$ **11.** ∞ **13.** $-\infty$ **15.** ∞ **17.** ∞ **19.** $-\infty$ **21.** ∞

23. ∞ **25.** ∞ **27.** ∞ **29.** $-\infty$ **31.** ∞ **33.** $-\infty$ **35.** ∞ **37.** $-\infty$ **39.** $x = 5$ **41.** $x = 0$

43. $x = -2$ **45.** $x = 1$ **47.** $x = -4$ **49.** ∞ **51.** ∞ **53.** 5/7

REVIEW EXERCISES FOR CHAPTER 2   (page 102)

1. 1/64   **3.** 5/13   **5.** 17   **7.** 5   **9.** 3   **11.** 0   **13.** 4   **15.** 0   **17.** 5/7   **19.** 0   **21.** $-\infty$

23. ∞ **25.** 3   **27.** 6   **29.** does not exist

31. (a) no   **(b)** yes   **(c)** yes   **(d)** no   **(e)** no   **(f)** no   **(g)** yes   **(h)** yes

33. 4 minutes

SECTION 3.1   (page 116)

1. $f'(x) = 5$ **3.** $f'(x) = 2x$ **5.** $f'(x) = 2x - 4$ **7.** $f'(x) = 6x + 7$ **9.** $f'(x) = 0$ **11.** $f'(x) = 3x^2$

13. $f'(x) = 3x^2 + 2x + 1$ **15.** $f'(x) = -\dfrac{2}{x^2}$ **17.** $f'(x) = -\dfrac{1}{3x^2}$ **19.** $f'(x) = \dfrac{-1}{(x+1)^2}$

21. $f'(x) = 2x + 6$; $f'(2) = 10$ **23.** $f'(x) = 3x^2$; $f'(4) = 48$ **25.** $f'(x) = 4$; $f'(3) = 4$

27. $y = 10x - 4$ **29.** $f'(x) = 4x^3$ **31.** $f'(x) = \dfrac{1}{2\sqrt{x}}$ **33.** $f'(x) = -\dfrac{1}{2x^{3/2}}$

35. This is not division by zero. The number Δx is not zero. Even after the reduction, when the limit as $\Delta x \to 0$ is considered, Δx is very close to zero but still not equal to zero.

SECTION 3.2   (page 123)

1. $f'(x) = 4x^3$ **3.** $f'(x) = -\dfrac{2}{x^3}$ **5.** $g'(x) = 0$ **7.** $y' = \dfrac{3}{2}x^{1/2}$ or $\dfrac{3\sqrt{x}}{2}$ **9.** $y' = -\dfrac{2}{3x^{5/3}}$

11. $f'(x) = \dfrac{-5}{x^6}$ **13.** $y' = \dfrac{1}{3x^{2/3}}$ **15.** $f'(x) = \dfrac{15}{\sqrt{x}}$ **17.** $y' = x^3$ **19.** $y' = \dfrac{-5}{t^{3/2}}$ **21.** $-3x^2$ **23.** $-\dfrac{1}{x^2}$

25. $1 + \dfrac{6}{x^3}$ **27.** $\dfrac{1}{2\sqrt{x}}$ **29.** $3x^2 + \dfrac{3}{x^4}$ **31.** $\dfrac{dy}{dx} = 2x - 5$ **33.** $\dfrac{dy}{dx} = \dfrac{3}{2}\sqrt{x} + 8x$ **35.** $\dfrac{dy}{dx} = \dfrac{4}{\sqrt{x}} + 16$

37. $\dfrac{dy}{dx} = -\dfrac{1}{x^{3/2}}$ **39.** $D_x y = 18x^5 - 4x^3$ **41.** $D_x y = \dfrac{5}{x^6}$ **43.** $D_x y = \dfrac{2}{x^{2/3}}$ **45.** $f'(x) = 3x^2 - 12x + 4$

47. $f'(x) = 14x^{3/4} + 10x^{2/3}$ **49.** $f'(x) = \dfrac{3}{2\sqrt{x}} + 10x$ **51.** $f'(x) = -\dfrac{4}{x^2} - 2$ **53.** 14   **55.** 34

57. 6   **59.** 1/6   **61.** 6   **63.** -12 **65.** 2   **67.** -9 **69.** $y = -32x - 48$

71. Here $f'(x) = \lim\limits_{\Delta x \to 0} \dfrac{c - c}{\Delta x} = \lim\limits_{\Delta x \to 0} \dfrac{0}{\Delta x} = \lim\limits_{\Delta x \to 0}(0) = 0$

73. The function is not defined at 0. Since the definition of $f'(x)$ includes $f(x)$ in its numerator, $f'(x)$ cannot exist if it includes an expression that is not defined.

75. Once the specific value (such as 3 or 1) is substituted, the expression becomes a constant. Differentiation at that point will always produce zero as a result.

77. (a) $f'(x_1)$ **(b)** $g'(x_2)$ **(c)** $f'(x_3)$

SECTION 3.3   (page 130)

1. 1.75 miles per hour   **3.** 19 birds per year   **5.** $20,100 per year   **7.** 10,500 people per year

9. 54 points per day   **11.** 64 feet per second downward   **13.** 2   **15.** 2/5   **17.** 96 feet per second

19. *After 5 hours*: 300 bacteria per hour (no inhibitor) *versus* 100 bacteria per hour (inhibitor). *After 10 hours*: 500 bacteria per hour (no inhibitor) *versus* 0 bacteria per hour (inhibitor).

21. (a) 5135 cones   **(b)** increasing at the rate of 28 cones per day
   (c) decreasing at the rate of 8 cones per day   **(d)** June 19

23. $2kr$ **25.** $2\pi r$

27. (a) $v(1) = 3$ feet per second; $v(5) = 11$ feet per second   **(b)** after 4 seconds   **(c)** 9 feet per second

29. (a) g_1 has the greater rate of increase. $g'_1(t_1) > g'_2(t_1)$; the slope of the tangent line is greater for g_1
   (b) No; g_1 has the greater rate of increase for the same reason as in part (a).

31. For any linear function, the instantaneous rate of change and the average rate of change are the same; in this case, 10.

SECTION 3.4   (page 138)

1. $C'(x) = -.2x$ **3.** $C'(x) = 150 - 2x$ **5.** $C'(x) = 90 + .04x$ **7.** $C'(x) = 3 - .003x^2$

9. (a) $20   **(b)** $19
   (c) $MC(10)$, the marginal cost when 10 units have been produced, is the approximate cost of producing the 11th unit.

11. 20 chairs   **13.** 14 hats   **15.** $R'(x) = 50 + .4x$ **17.** $R'(x) = 400 - .006x$ **19.** $R'(x) = .002x + .7$

21. (a) $R(1) = \$399.99$, $R(10) = \$3999$, $R(100) = \$39,900$ **(b)** $MR(x) = 400 - .02x$
   (c) $MR(1000) = \$380$. Once 1000 CD players have been sold, the revenue from the sale of the next CD player is approximately $380.

23. $P'(x) = .04x + 9$ **25.** $P'(x) = 40 - .02x$ **27.** $P'(x) = 200 + .6x - .003x^2$

29. (a) $60   **(b)** $108.75   **(c)** $P'(x) = .001x + 1$ **31.** $MP(x) = .02x + 85$ **33.** $MP(x) = 10$

35. 1500   **37.** 5000   **39.** 750   **41.** 500   **43.** 500   **45.** 30   **47. (a)** $MP(x) = 26 - .02x$ **(b)** 1300

49. \$59.60   **51. (a)** $MR(x) = 1000 - .08x$ **(b)** \$394   **(c)** \$393.98

53. The tax rates are marginal in the sense that they are the rates at which the "next income" will be taxed. For example, once your taxable income reaches \$8000, any income above that amount will be taxed at 16%. Once your taxable income reaches \$20,000, any income above that amount will be taxed at 26%.

SECTION 3.5   (page 145)

1. $\dfrac{dy}{dx} = 2x - 1$ **3.** $f'(x) = 20x - 1$ **5.** $\dfrac{dy}{dt} = 4t^3 + 14t$ **7.** $f'(x) = 16x^3 - 9x^2$ **9.** $f'(t) = 3t^2 + 2t$

11. $g'(x) = 4 - \dfrac{1}{x^2}$ **13.** $y' = 18 + \dfrac{4}{x^2}$ **15.** $y' = -4x^3 + 3x^2 - 2$ **17.** $\dfrac{dy}{dx} = 2x - 3x^2 - 5x^4$

19. $\dfrac{dy}{dx} = \dfrac{1}{(1 + x)^2}$ **21.** $\dfrac{dy}{dx} = \dfrac{7}{(x + 4)^2}$ **23.** $f'(x) = \dfrac{2}{(3x + 1)^2}$ **25.** $f'(t) = 0$ **27.** $y' = \dfrac{-2}{(1 + x)^2}$

29. $\dfrac{ds}{dt} = \dfrac{-4t}{(t^2 + 1)^2}$ **31.** $s'(t) = \dfrac{12t^2 + 8t + 1}{(1 + 3t)^2}$ **33.** $\dfrac{dy}{dx} = \dfrac{x^2 + 8x + 23}{(x + 4)^2}$

35. $\dfrac{dy}{dx} = \dfrac{2x^{3/2} - (4 + 2x)\frac{3}{2}x^{1/2}}{x^3}$, which simplifies to $-\dfrac{x + 6}{x^{5/2}}$ **37.** $y' = -1 - 2x$ **39.** $y' = 2x - 3x^2$

41. $f'(x) = 7 - \dfrac{2x}{3} - \dfrac{1}{x^2}$ **43.** $f'(t) = 5t^4 - 24t^2 + 2t$ **45.** $\dfrac{ds}{dt} = \dfrac{-t^2 + 2t + 8}{(1 - t)^2}$

47. 235   **49.** 7/12   **51.** 0   **53.** 7   **55.** 3.9 inches per second

57. (a) $-\$20$, a *loss* of \$20   **(b)** \$0   **(c)** \$45.45   **(d)** $MP(x) = \dfrac{10x^2 + 20x - 50}{(x + 1)^2}$

59. $\dfrac{d}{dx}\left(\dfrac{x^2}{9}\right) = \dfrac{(9)(2x) - (x^2)(0)}{(9)^2} = \dfrac{18x}{81} = \dfrac{2x}{9}$

61. Once the expression is written as x^{-2}, then the power rule can be used to obtain the derivative. Such an approach is much simpler than using the quotient rule. (Check it out if you aren't convinced.)

SECTION 3.6   (page 152)

1. $\dfrac{dy}{dx} = 10x(x^2 + 3)^4$ **3.** $\dfrac{dy}{dx} = 15(3x)^4$ **5.** $\dfrac{dy}{dx} = \dfrac{1}{(3x + 4)^{2/3}}$ **7.** $\dfrac{ds}{dt} = \dfrac{24t^3}{(1 - t^4)^7}$ **9.** $\dfrac{dy}{dx} = \dfrac{2}{\sqrt{4x + 1}}$

11. $\dfrac{dy}{dx} = -8(1 - 6x)^{1/3}$ **13.** $f'(x) = 8x(2x^2 + 1)(x^4 + x^2 + 1)^3$ **15.** $f'(t) = \dfrac{-12t}{(t^2 + 1)^7}$ **17.** $y' = \dfrac{-3}{(6x + 5)^{3/2}}$

19. $\dfrac{dy}{dx} = 20x^2(5x - 2)^3 + 2x(5x - 2)^4,\ or\ 2x(5x - 2)^3(15x - 2)$

21. $\dfrac{dy}{dx} = 6(x + 3)(2x + 1)^2 + (2x + 1)^3,\ or\ (2x + 1)^2(8x + 19)$

23. $\dfrac{dy}{dx} = (x^2 + 1)^4 + 8x(x - 2)(x^2 + 1)^3,\ or\ (x^2 + 1)^3(9x^2 - 16x + 1)$

25. $\dfrac{dy}{dx} = 6x(5x - 2)^2(x^2 + 7)^2 + 10(x^2 + 7)^3(5x - 2),\ or\ 2(5x - 2)(x^2 + 7)^2(20x^2 - 6x + 35)$

27. $\dfrac{dy}{dt} = 3(2t + 3)^4(t - 7)^2 + 8(t - 7)^3(2t + 3)^3,\ or\ (2t + 3)^3(t - 7)^2(14t - 47)$

29. $\dfrac{dy}{dx} = \dfrac{3(x + 1)(x + 4)^2 - (x + 4)^3}{(x + 1)^2}\ or\ \dfrac{(x + 4)^2(2x - 1)}{(x + 1)^2}$

31. $\dfrac{dy}{dt} = \dfrac{8(t - 2)(2t + 3)^3 - (2t + 3)^4}{(t - 2)^2}\ or\ \dfrac{(2t + 3)^3(6t - 19)}{(t - 2)^2}$

33. $\dfrac{dy}{dx} = \dfrac{10x(3x - 2)(1 + x^2)^4 - 3(1 + x^2)^5}{(3x - 2)^2}\ or\ \dfrac{(1 + x^2)^4(27x^2 - 20x - 3)}{(3x - 2)^2}$

35. $\dfrac{dy}{dx} = 3(2x - 5)(2x + 1)^{1/2} + 2(2x + 1)^{3/2},\ or\ (2x + 1)^{1/2}(10x - 13)$

37. $\dfrac{dy}{dx} = 2(x^3 + 1)^{4/3} + 4x^2(1 + 2x)(x^3 + 1)^{1/3},\ or\ 2(x^3 + 1)^{1/3}(5x^3 + 2x^2 + 1)$

39. $\dfrac{dy}{dx} = 2x(2x + 1)^{-1/2} + 2(2x + 1)^{1/2},\ or\ \dfrac{2(3x + 1)}{\sqrt{2x + 1}}$

41. $\dfrac{dy}{dx} = x^3(x^2 + 1)^{-1/2} + 2x(x^2 + 1)^{1/2},\ or\ \dfrac{x(3x^2 + 2)}{\sqrt{x^2 + 1}}$ **43.** $\dfrac{dy}{dx} = \dfrac{8(x - 1)^3}{(x + 1)^5}$

45. $\dfrac{dy}{dx} = \dfrac{1}{2}\left(\dfrac{2x + 1}{x - 1}\right)^{-1/2} \cdot \dfrac{-3}{(x - 1)^2},\ or\ \dfrac{-3}{2(2x + 1)^{1/2}(x - 1)^{3/2}}$ **47.** -4 **49.** $2/3$ **51.** 64 **53.** $1/2$

55. $MP(x) = \dfrac{100x}{\sqrt{x^2 - 1}}$ **57. (a)** $100\%, 95\%, 45\%$ **(b)** 0% per mile, -1.58% per mile, -13.4% per mile

59. The power rule, $D_x x^n = nx^{n-1}$, provides a rule for differentiating a power of a variable (say x) with respect to that variable. For example, $D_x x^3 = 3x^2$. But this power rule cannot be used to differentiate $(x^2 + 1)^3$ with respect to x, because the expression is not a power of x; it is a power of $x^2 + 1$. The chain rule is needed for such a differentiation.

61. $\dfrac{dy}{dx} = 16x(x^2 + 1)^2(2x - 3)^5(x^2 + 1)^7 + 10(x + 1)^2(x^2 + 1)^8(2x - 3)^4 + 2(x + 1)(2x - 3)^5(x^2 + 1)^8,\ or$
$2(x + 1)(2x - 3)^4(x^2 + 1)^7(23x^3 - 6x^2 - 17x + 2)$

63. $\dfrac{dy}{dx} = \dfrac{2(2x + 7)(x^2 + 3)^5(x + 1) + 10x(2x + 7)(x + 1)^2(x^2 + 3)^4 - 2(x^2 + 3)^5(x + 1)^2}{(2x + 7)^2}\ or$

$\dfrac{2(x + 1)(x^2 + 3)^4(11x^3 + 51x^2 + 38x + 18)}{(2x + 7)^2}$

SECTION 3.7 (page 156)

1. $f''(x) = 12x^2 - 20$ **3.** $y''' = 6$ **5.** $f^{(4)}(x) = 24$ **7.** $D_x^3 y = -\dfrac{10}{27}x^{-4/3}$ or $\dfrac{-10}{27x^{4/3}}$ **9.** $y'' = 0$

11. $y'' = 20x^{-6}$ or $\dfrac{20}{x^6}$ **13.** $y'' = \dfrac{-2}{(1 + x)^3}$ **15.** $\dfrac{d^2y}{dx^2} = 6x - 10$ **17.** $\dfrac{d^2y}{dx^2} = 180(3x - 1)^3$

19. $D_x^2 y = \dfrac{15}{4}x^{1/2}$ or $\dfrac{15\sqrt{x}}{4}$ **21.** $D_x^2 y = \dfrac{3}{4}x^{-5/2}$ or $\dfrac{3}{4x^{5/2}}$ **23.** $g''(x) = 6x - 16$ **25.** $\dfrac{d^2y}{dt^2} = \dfrac{200}{t^6}$

27. $\dfrac{d^2s}{dt^2} = -32$ **29.** 14   **31.** 1/32   **33.** $-5/16$ **35.** $6x - 14$ **37.** $\dfrac{20}{x^3}$ **39.** 4.5   **41.** $-.25$ **43.** 2

45. $-.12$ **47.** $-.2$ **49.** $-.02$ **51.** $-.005$ **53.** $a = -32$ **55.** $a = 6t - 2$

57. (a) $v = 2t - 2 - \dfrac{4}{t^2}$ **(b)** $a = 2 + \dfrac{8}{t^3}$ **(c)** 10 centimeters per second per second
    (d) 3 centimeters per second per second

59. $f''(x) = \dfrac{-2x^3 + 6x^2 + 6x - 2}{(x^2 + 1)^3}$ **61.** $y'' = 15x(1 + x^2)^{-1/2} - 5x^3(1 + x^2)^{-3/2}$ or $\dfrac{5x(2x^2 + 3)}{(1 + x^2)^{3/2}}$

63. (a) $v = -5.3t + 106$ **(b)** $a = -5.3$
    (c) Yes. On the moon there is less gravitational pull on the ball $(-5.3$ feet per second per second) than there is on the earth $(-32$ feet per second per second).
    (d) 40 seconds

65. $P''(x)$ is the rate of change of $P'(x)$, which means that $P''(x)$ is the rate of change of marginal profit.

SECTION 3.8 (page 163)

1. $\dfrac{dy}{dx} = -\dfrac{x}{y}$ **3.** $\dfrac{dy}{dx} = \dfrac{2x + 6}{3y^2}$ **5.** $\dfrac{dy}{dx} = -\dfrac{x^2}{y^2}$ **7.** $\dfrac{dy}{dx} = \dfrac{2x + 7}{2y}$ **9.** $\dfrac{dy}{dx} = \dfrac{1 - 2x}{2y - 3y^2}$ **11.** $\dfrac{dy}{dx} = -\dfrac{x^{1/2}}{y^{1/2}}$

13. $\dfrac{dy}{dx} = \dfrac{10x^{3/2} - 1}{2y}$ **15.** $\dfrac{dy}{dx} = \dfrac{1 - 2xy^2}{2x^2y - 3}$ **17.** $\dfrac{dy}{dx} = \dfrac{3 - y^3}{3xy^2 + 5}$ **19.** $\dfrac{dy}{dx} = -\dfrac{y}{x}$ **21.** $\dfrac{dy}{dx} = \dfrac{-1 - 8y}{8x}$

23. $\dfrac{dy}{dx} = \dfrac{-2x}{3y^2 - 1}$ or $\dfrac{2x}{1 - 3y^2}$; m at $(1, 2)$ is $-2/11$.   **25.** $\dfrac{dy}{dx} = \dfrac{3x + 3}{y}$; m at $(-1, 1)$ is 0.

27. $\dfrac{dy}{dx} = \dfrac{-10y - 1}{10x}$; m at $(5, -1)$ is 9/50.   **29.** $y = -\dfrac{3}{4}x + \dfrac{25}{4}$ **31.** $\dfrac{dy}{dx} = \dfrac{2(x + 5)}{3(y - 3)^2}$ **33.** $\dfrac{dy}{dx} = \dfrac{14 - 2y^{1/2}}{xy^{-1/2} + 2}$

35. (a) no   **(b)** yes   **(c)** yes   **(d)** no

37. This equation can be considered as $y = x^2 - 9x$, in which y is given *explicitly* in terms of x. Consequently, implicit differentiation is not needed.

SECTION 3.9 (page 169)

1. 4.5π (≈ 14) square miles per year   **3.** $\dfrac{1}{4\pi}$ $(\approx .08)$ meter per hour

5. 500π (≈ 1571) cubic millimeters per month   **7.** $\dfrac{1}{4\pi}$ ($\approx .08$) foot per minute

9. 144 square millimeters per second   **11.** \$400 per day   **13. (a)** $R = 12x - .00015x^2$ **(b)** \$22.50 per week

15. 1.5 feet per second   **17. (a)** $V = \dfrac{1}{3}\pi r^3$ **(b)** $\dfrac{2}{25\pi}$ ($\approx .025$) foot per minute   **19.** 1920 miles per hour

21. (a) $\dfrac{dA}{dt}$ and $\dfrac{dr}{dt}$

    (b) the rate of change of the area with respect to time; the rate of change of the radius with respect to time

    (c) $\dfrac{dA}{dt} = 2\pi r \dfrac{dr}{dt}$

SECTION 3.10   (page 175)

1. $dy = 3x^2\, dx$ **3.** $dy = 2x\, dx$ **5.** $dy = (2x + 5)\, dx$ **7.** $dy = 10x(x^2 - 3)^4\, dx$

9. $dy = -36(1 - 9x)^3\, dx$ **11.** .8   **13.** .008   **15.** .000125   **17.** .30   **19.** .0175   **21.** $-.045$

23. \$12,750   **25.** Computing $R(16) - R(15)$ leads to 12.25, which is \$12,250.   **27.** $-.02$ **29.** 3

REVIEW EXERCISES FOR CHAPTER 3   (page 177)

1. $f'(x) = 0$ **3.** $f'(x) = x^6$ **5.** $f'(t) = \dfrac{-1}{8t^{3/2}}$ **7.** $2x + 3$ **9.** $\dfrac{dy}{dx} = \dfrac{4}{x^{1/3}} - 14$ **11.** $\dfrac{dy}{dx} = \dfrac{-1}{x^2}$

13. $D_x y = 1 - 9x^8$ **15.** $D_x y = \dfrac{-1}{x^{4/3}}$ **17.** $f'(x) = \dfrac{5}{3}x^{2/3}$ **19.** $y' = \dfrac{4}{\sqrt{8x}}$ **21.** $f'(t) = \dfrac{2t}{3(1 + t^2)^{2/3}}$

23. $y' = \dfrac{-x^2}{(1 + x^3)^{4/3}}$ **25.** $\dfrac{dy}{dx} = \dfrac{8(x^2 + 4)(2x - 1)^3 - 2x(2x - 1)^4}{(x^2 + 4)^2}$ or $\dfrac{2(2x - 1)^3(2x^2 + x + 16)}{(x^2 + 4)^2}$

27. $\dfrac{dy}{dx} = 4(x^2 + 7)^3(x - 5)^3 + 6x(x - 5)^4(x^2 + 7)^2$ or $2(x - 5)^3(x^2 + 7)^2(5x^2 - 15x + 14)$

29. $y'' = \dfrac{-12}{(x + 3)^3}$ **31.** $\dfrac{dy}{dx} = \dfrac{-2x - y^4}{4xy^3 - 1}$ or $\dfrac{2x + y^4}{1 - 4xy^3}$ **33.** .0275   **35.** 972

37. (a) \$34.80   **(b)** \$34.79   **(c)** $-\$.02$ per item per item   **39.** $MP(x) = -4 + .002x$ **41.** 35 units

---

SECTION 4.1   (page 189)

1. increasing on $[-2, 1)$; decreasing on $(1, 5]$ **3.** increasing on $(-2, 3)$; decreasing on $[-5, -2)$ and $(3, 7]$

5. increasing on $(-3, 2)$; decreasing on $[-5, -3)$ **7.** increasing on $[-6, 0)$ and $(2, 6]$; decreasing on $(0, 2)$

9. never increasing; decreasing on $[-6, 0)$ and $(0, 6]$ **11.** increasing when $x > 3$, decreasing when $x < 3$

13. increasing when $x > -5/2$, decreasing when $x < -5/2$ **15.** increasing when $x < 0$, decreasing when $x > 0$

17. increasing when $x < 5$, decreasing when $x > 5$ **19.** increasing when $x < 150$, decreasing when $x > 150$

21. never increasing; decreasing when $x < 0$ and when $x > 0$ **23.** never decreasing; increasing when $x > 0$

25. never decreasing; increasing when $x > 0$ **27.** never increasing; decreasing when $x < 0$ and when $x > 0$

29. never decreasing; increasing when $x < -1$ and when $x > -1$

31. increasing when $x < -2$ and when $x > 2$, decreasing when $-2 < x < 2$ **33.** After 16 seconds (when $x > 16$)

35. **(a)** During the first 25 seconds of the flight (when $0 < t < 25$)
  (b) After 50 seconds (when $t = 50$)
  (c) During the last 25 seconds of the flight (when $25 < t < 50$)

37. When the number of TV antennas produced is less than 900   **39.** decreasing   **41.** increasing

43. decreasing   **45. (a)** $\overline{C}(x) = 36 - .02x$ **(b)** $\overline{C}(x)$ is decreasing for all $x > 0$ $(0 < x \le 1200)$.

47. increasing

49. **(a)** No. Since the graph of $y = B(t)$ is *below* the graph of $y = C(t)$ on $[0, 6]$, the $B(t)$ values are less than the $C(t)$ values on $[0, 6]$.
  (b) No. $B'(t)$ is the slope of the curve $y = B(t)$ and $C'(t)$ is the slope of the curve $y = C(t)$. From the graph we can see that the slope of $y = B(t)$ is not greater than the slope of $y = C(t)$ on $[0, 6]$.
  (c) Although crash dieting provides greater weight loss (and a faster rate of weight loss) early on, the crash dieter eventually gains back the lost weight. By contrast, the behavior modification dieter keeps the weight off.

51. 8   **53.** $-3/10$ **55.** 0   **57.** 2, 7   **59.** $-5/3$, 1   **61.** $\pm\sqrt{2}$ **63.** $-2 \pm \sqrt{3}$ **65.** 0   **67.** none

69. $-1 \pm \sqrt{6}$ **71.** $-3, -1$ **73.** none   **75.** 0, 2   **77.** $-3/2$, 1/36   **79.** $\pm 2, -1, 4/5$

81. never increasing; decreasing for $x < 2$ and for $x > 2$ **83.** increasing for $x > 0$; decreasing for $x < 0$

85. increasing for $x > -3$ (but not at $x = 0$); decreasing for $x < -3$ **87.** $0, -3 \pm \sqrt{7}$

89. **(a)** 2   **(b)** -1 **(c)** not defined

91. At a it is continuous and differentiable. At b it is continuous but not differentiable. At c it is continuous and differentiable. At d it is not continuous and not differentiable. At e it is not continuous and not differentiable.

93. Yes. If $f'(3) = 0$, then f must be defined at 3. (The derivative cannot exist at a number, such as 3, unless the function is defined there.)

95. Begin by finding the derivative of g. Then find all numbers for which g' is zero or undefined. Test those numbers to determine for which ones the function g is defined. The critical numbers of g are the numbers for which g' is zero or undefined *and* g is defined.

97. There are four possibilities for the derivative $f'(x)$. It can be positive, negative, zero, or undefined. When $f'(x)$ is positive, the function is increasing, so there can be no relative maximum or minimum point for any such x. When $f'(x)$ is negative, the function is decreasing, so there can be no relative maximum or minimum point for any such x. Thus, the only possible numbers that can lead to relative maximum or minimum points are numbers for which $f'(x)$ is zero or undefined (provided that f is defined); that is, critical numbers.

***SECTION* 4.2**   (page 200)

1. $(3, 24)$ relative maximum   **3.** $(10, -100)$ relative minimum

5. $(1, -4)$ relative minimum; $(-1, 0)$ relative maximum   **7.** $(0, 0)$ relative maximum; $(4, -32)$ relative minimum

9. $(-2, -16)$ relative minimum; $(2, 16)$ relative maximum

11. $(-3, 25)$ relative maximum; $(1, -7)$ relative minimum

13. $(-8, 711)$ relative maximum; $(1, -18)$ relative minimum

15. $(-3, 86)$ relative maximum; $(2, -39)$ relative minimum

17.

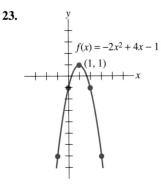

$f(x) = x^2 - 8x + 9$

$(4, -7)$

19.

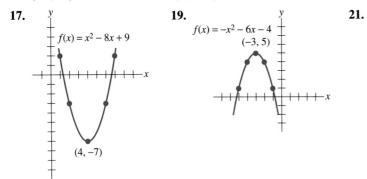

$f(x) = -x^2 - 6x - 4$

$(-3, 5)$

21.

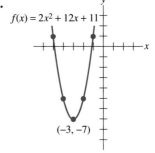

$f(x) = 2x^2 + 12x + 11$

$(-3, -7)$

23.

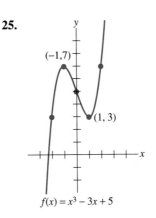

$f(x) = -2x^2 + 4x - 1$

$(1, 1)$

25.

$(-1, 7)$

$(1, 3)$

$f(x) = x^3 - 3x + 5$

27.

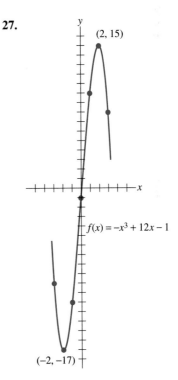

$(2, 15)$

$f(x) = -x^3 + 12x - 1$

$(-2, -17)$

29. **31.** **33.** $f(x) = 2x^3 - 3x^2 - 12x + 8$

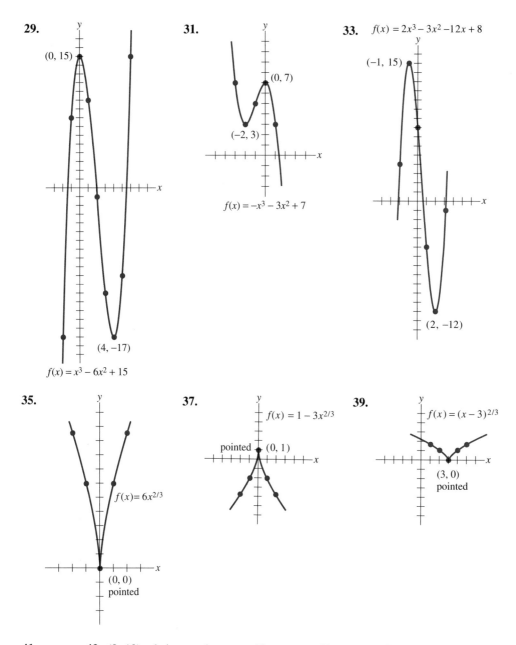

(0, 15)

(4, −17)

$f(x) = x^3 - 6x^2 + 15$

(0, 7)

(−2, 3)

$f(x) = -x^3 - 3x^2 + 7$

(−1, 15)

(2, −12)

35. **37.** **39.**

$f(x) = 6x^{2/3}$

(0, 0)
pointed

$f(x) = 1 - 3x^{2/3}$

pointed (0, 1)

$f(x) = (x - 3)^{2/3}$

(3, 0)
pointed

41. none   **43.** (0, 10) relative maximum   **45.** none   **47.** none   **49.** none

51. (−3, 0) relative maximum; (−1, −16) relative minimum   **53.** (0, 0) relative minimum

55. $f'(x_1)$ is zero; $f'(x_2)$ is positive; $f'(x_3)$ is undefined; $f'(x_4)$ is negative

57. $f'(x_1)$ is zero; $f'(x_2)$ is negative; $f'(x_3)$ is zero; $f'(x_4)$ is undefined

59.   **61.**

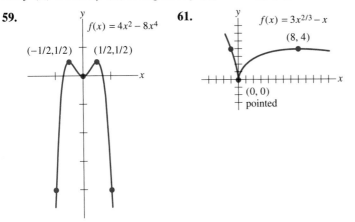

63. A relative maximum function value is the largest value the function takes on (in some interval)—the largest y value. A relative maximum point is a *point*, the highest point (in some interval) of the graph of the function.

65. (a) horizontal   **(b)** vertical

SECTION 4.3   (page 210)

1. concave up on $(6, 13)$; concave down on $(-5, 6)$ **3.** concave up on $(-\infty, 0)$ and $(0, \infty)$

5. concave up on $(0, 7)$; concave down on $(-9, 0)$ **7.** concave up on $(-\infty, \infty)$

9. concave down on $(-\infty, \infty)$ **11.** concave up on $(0, \infty)$; concave down on $(-\infty, 0)$

13. concave up on $(-1, \infty)$; concave down on $(-\infty, -1)$ **15.** concave up on $(-\infty, 1/2)$; concave down on $(1/2, \infty)$

17. concave up on $(0, \infty)$; concave down on $(-\infty, 0)$ **19.** concave up on $(-\infty, 0)$; concave down on $(0, \infty)$

21. $(-2, 9)$ relative maximum; $(0, 5)$ relative minimum

23. $(-4, 130)$ relative maximum; $(4, -126)$ relative minimum

25. $(-4, -73)$ relative minimum; $(2, 35)$ relative maximum

27. $(0, 0)$ relative maximum; $(5, -125)$ relative minimum   **29.** none   **31.** none

33. $(0, 5)$ relative minimum   **35.** 6   **37.** none   **39.** 0   **41.** $(0, -4)$ **43.** $(-2, 4)$

45. $(1, 2)$ **47.** $\left(\dfrac{1}{2}, 3\dfrac{1}{2}\right)$

49.

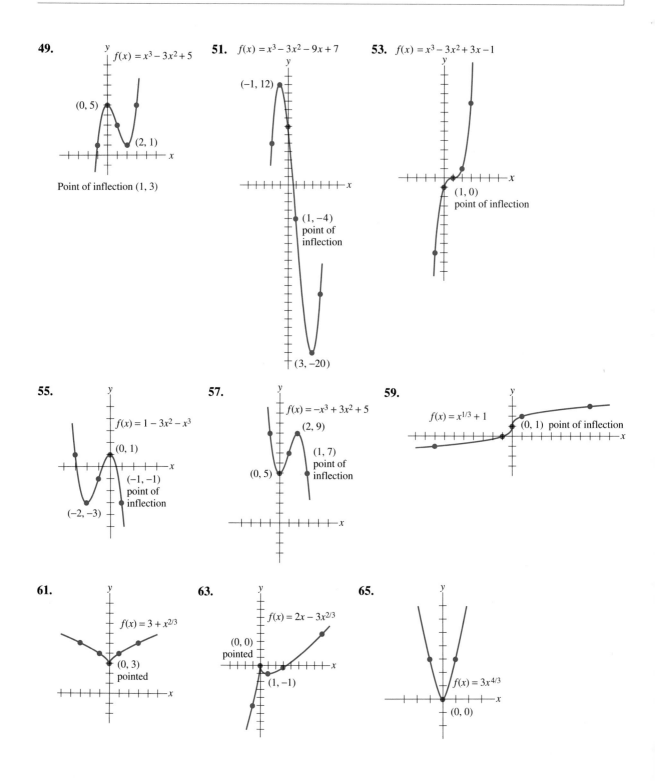

$f(x) = x^3 - 3x^2 + 5$

$(0, 5)$

$(2, 1)$

Point of inflection $(1, 3)$

51. $f(x) = x^3 - 3x^2 - 9x + 7$

$(-1, 12)$

$(1, -4)$
point of
inflection

$(3, -20)$

53. $f(x) = x^3 - 3x^2 + 3x - 1$

$(1, 0)$
point of inflection

55.

$f(x) = 1 - 3x^2 - x^3$

$(0, 1)$

$(-1, -1)$
point of
inflection

$(-2, -3)$

57.

$f(x) = -x^3 + 3x^2 + 5$

$(2, 9)$

$(1, 7)$
point of
inflection

$(0, 5)$

59.

$f(x) = x^{1/3} + 1$

$(0, 1)$ point of inflection

61.

$f(x) = 3 + x^{2/3}$

$(0, 3)$
pointed

63.

$f(x) = 2x - 3x^{2/3}$

$(0, 0)$
pointed

$(1, -1)$

65.

$f(x) = 3x^{4/3}$

$(0, 0)$

67.

69.

71. $f(x) = x^3 - 6x^2 + 12x - 8$

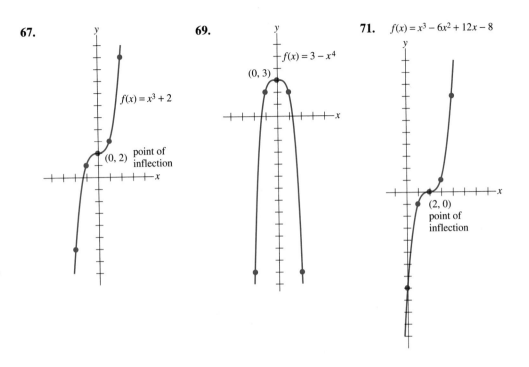

73. $f''(x) = -\dfrac{2}{9x^{4/3}}$. Since $x^{4/3} = (x^{1/3})^4$, $f''(x) < 0$ for $x > 0$ and for $x < 0$. Thus, it is concave down everywhere.

75. Each point $(a, f(a))$ and $(b, f(b))$ is both a relative extreme point *and* a point of inflection.

77. (a) zero   **(b)** zero   **(c)** negative   **(d)** negative   **(e)** negative   **(f)** positive   **(g)** positive
(h) negative

79. No. The second derivative can be zero at a relative maximum or minimum point.

81. The function is not defined at $x = 0$; that is, there is no point $(0, f(0))$ on the graph of f.   **83.** up

85. The slope of the curve (the derivative of the revenue function) is the marginal revenue. For R_1, the slope of the curve (the marginal revenue) is decreasing.

87.

89. $y = 2$ horizontal asymptote

91.

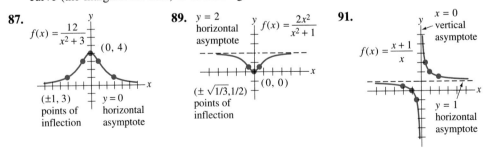

93.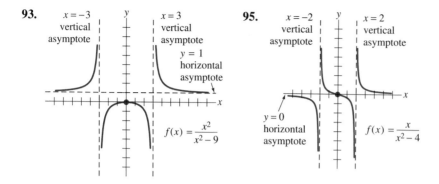

$x = -3$
vertical
asymptote

$x = 3$
vertical
asymptote

$y = 1$
horizontal
asymptote

$f(x) = \dfrac{x^2}{x^2 - 9}$

95.

$x = -2$
vertical
asymptote

$x = 2$
vertical
asymptote

$y = 0$
horizontal
asymptote

$f(x) = \dfrac{x}{x^2 - 4}$

SECTION 4.4   (page 215)

1. relative maximum 3; relative minimum 1; absolute maximum 5; absolute minimum 1

3. relative maximum 5; no relative minimum; absolute maximum 5; absolute minimum 1

5. relative maximum 6; relative minimum 5; absolute maximum 7; absolute minimum 2

7. absolute maximum 69; absolute minimum 5   **9.** absolute maximum -9; absolute minimum -41

11. absolute maximum 400; absolute minimum -32 **13.** absolute maximum 1344; absolute minimum -141

15. absolute maximum -34; absolute minimum -50 **17.** absolute maximum 205; absolute minimum 5

19. absolute maximum 109; absolute minimum -591

21. The only x values that can correspond to the absolute extrema of a function (on a closed interval) are the endpoints and any x values within the interval that correspond to relative extrema. The critical numbers are the only numbers that can possibly lead to relative extrema. Since our concern here is *absolute* extrema, it is faster to simply evaluate the function at all critical numbers in the interval (and at the endpoints) and see which numbers lead to absolute extrema. If some critical number does not correspond to a relative extremum, then it will not correspond to an absolute extremum, and we would have tested it for nothing. Still, this is far more efficient than testing each critical number first to see if it corresponds to a relative extremum.

SECTION 4.5   (page 224)

1. 370°F   **3.** 151 feet   **5.** (a) 650   (b) $422,500   **7.** 25   **9.** 1 hour after being swallowed

11. 50, 50   **13.** 400 feet by 400 feet   **15.** 120 feet   **17.** width 24 feet, length 72 feet

19. 60 inches long, 30 inches wide, 20 inches high   **21.** $6\frac{2}{3}$ cm   **23.** 7 cm

25. width 7 inches, length 14 inches   **27.** 22 trees   **29.** 48 feet (vertical sides) by 36 feet (horizontal sides)

31. 4 feet wide by 8 feet high   **33.** (a) 300   (b) 500   **35.** (a) 1050   (b) $24.50   **37.** 80   **39.** d

---

SECTION 4.6   (page 232)

1. 1.5   **3.** .125   **5.** $\dfrac{2p^2}{321 - p^2}$ **7.** 1   **9. (a)** 460,000   **(b)** 140,000   **(c)** .94   **11.** 25

13. .54% *or* .55%   **15.** $28

17. Raising the price will increase revenue, because demand is inelastic ($E = .315 < 1$).

19. The wholesaler should decrease the price in order to increase revenue, because demand is elastic ($E = 2.57 > 1$).

21. $1600   **23.** $5

REVIEW EXERCISES FOR CHAPTER 4   (page 234)

1. increasing when $x < 0$; decreasing when $x > 0$ **3.** never increasing; decreasing when $x < 1$ and when $x > 1$

5. $-2, 6$ **7.** 0   **9.** $-1, 1$

11.   **13.**

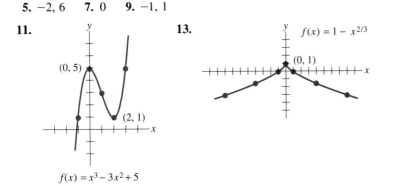

$f(x) = x^3 - 3x^2 + 5$

$f(x) = 1 - x^{2/3}$

15. $(6, -415)$ relative minimum   **17.** none   **19.** concave up on $(0, \infty)$; concave down on $(-\infty, 0)$ **21.** $(4, -64)$

23. (a) negative   **(b)** positive   **(c)** negative   **(d)** positive   **(e)** negative   **(f)** negative   **(g)** negative
(h) negative   **(i)** positive   **(j)** zero   **(k)** positive   **(l)** positive

25. absolute maximum 559; absolute minimum -305 **27.** 14 items   **29.** 60 feet by 60 feet   **31.** .25

33. Yes, this will happen when the graph is concave down (dy/dt decreasing) while the slope of the curve is positive (y increasing).

SECTION 5.1   (page 249)

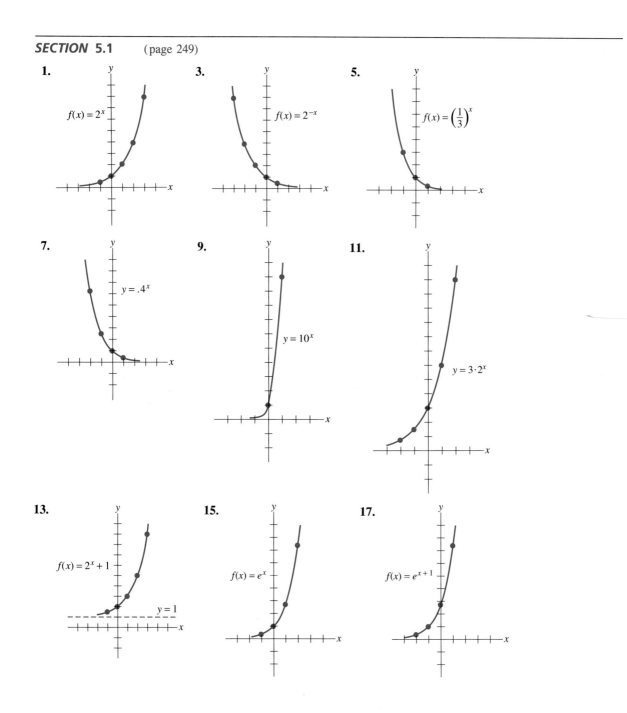

1. $f(x) = 2^x$

3. $f(x) = 2^{-x}$

5. $f(x) = \left(\frac{1}{3}\right)^x$

7. $y = .4^x$

9. $y = 10^x$

11. $y = 3 \cdot 2^x$

13. $f(x) = 2^x + 1$ $y = 1$

15. $f(x) = e^x$

17. $f(x) = e^{x+1}$

19. **21.**

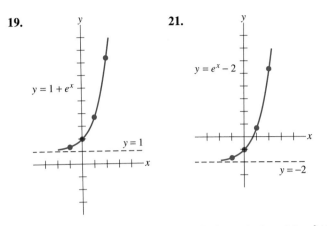

23. 4   **25.** −5   **27.** e^{x-2} **29.** e^{2x-1} **31.** 2   **33.** e^{3x+1} **35.** e^{7x} **37.** e^{6x} **39.** $2e^{3x+1}$ **41.** $1492

43. $2594   **45.** $8510   **47.** $29,447   **49. (a)** 50   **(b)** 305   **(c)** 1500   **51. (a)** 2000   **(b)** 10,000,000

53. (a) $y = 18(1 - e^{-.3(0)}) = 18(1 - e^0) = 18(1 - 1) = 18 \cdot 0 = 0$
(b) 5, 8, 11, 13, 14, 17   **(c)** 18
(d)

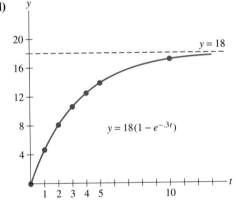

55. (a) $C(x) = 30xe^{-.02x}$ **(b)** $136

57. (a) $y = 0$ for both
(b) For $f(x) = 3^x$, $\lim_{x \to -\infty} 3^x = 0$. For $f(x) = .5^x$, $\lim_{x \to \infty} .5^x = 0$.

59.

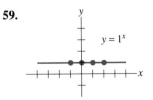

61. (a) The slopes of the tangent lines are increasing, which means that the function is always increasing.
(b) The graph is always concave up.

SECTION 5.2   (page 262)

1. $\log_2 32 = 5$ **3.** $\log_{10} 100 = 2$ **5.** $\ln 1 = 0$ **7.** $3^2 = 9$ **9.** $10^2 = 100$ **11.** $10^{-1} = .1$

13. $e^1 = e$ **15.** $\log 2$ **17.** $\ln 3$ **19.** $\dfrac{1}{3} \ln 2$ **21.** $-\ln 50$ **23.** $\dfrac{1}{5} \ln 14$ **25.** 81 **27.** e^{-2} **29.** $\dfrac{1}{4} e^{30}$

31. $1/3$ **33.** $\dfrac{1}{3} e^8$ **35.** 4 **37.** 10 **39.** 0 **41.** $3x^2$ **43.** 7 **45.** $1/2$ **47.** -1 **49.** $\ln x + \ln y$

51. $1 + \ln x$ **53.** $\ln 7 + 2 \ln x$ **55.** $\dfrac{1}{2} \ln x$ **57.** $3x \ln 2$ **59.** $x \ln x$ **61.** $\ln 2x$ **63.** $\ln \dfrac{4}{x}$

65. $\ln x^3$ **67.** $\ln 5x^2$ **69.** $\ln 3xy$ **71.** 9.9 years   **73.** 9.16%   **75.** .049   **77.** 8.6 hours   **79.** 43.8 years

81. (a) 19.87%   **(b)** 1995   **83.** .6 gram   **85.** Approximately 8748 or 8749 years old in 1970

87. (a) $-.025$ **(b)** 2056   **89. (a)** $-.081$ **(b)** 6%

91. (a) 15.06 units per volume   **(b)** .41 units per volume   **(c)** 57.6 minutes

93. (a) $R(x) = \dfrac{50x}{\ln(x + 3)}$ **(b)** \$429   **95. (a)** $D = 10^{a - b\log T}$ **(b)** $T = 10^{(a - \log D)/b}$

97. (a) $r = \dfrac{\ln 2}{t}$ **(b)** $r = $ 13.9%, 11.6%, 9.9%, 8.7%, 7.7%, 6.9%, 6.3%, 5.8%

99. (a) logistic growth   **(b)** learning curve   **(c)** exponential decay   **(d)** logistic growth
    (e) exponential growth   **(f)** logistic growth   **(g)** logistic growth   **(h)** logistic growth

101. (a) 315°   **(b)** $-.0047$ **(c)** between 81 and 82 minutes

103. When $t = 0$, $A = Ce^{k(0)} = Ce^0 = C \cdot 1 = C$

105. Yes, c is an exponent when $\log_b a = c$ is written in the alternative form $a = b^c$.

107. (a) The tangent lines all have positive slopes, which shows that $f(x) = \ln x$ is increasing.
    (b) The graph is concave down everywhere.

SECTION 5.3   (page 273)

1. $y' = x^2 e^x + 2xe^x = xe^x(x + 2)$ **3.** $y' = 5e^x$ **5.** $f'(x) = 4e^x(e^x + 2)^3$ **7.** $y' = \dfrac{xe^x - e^x - 1}{x^2}$

9. $f'(x) = -\dfrac{3(e^x + 1)}{(e^x + x)^4}$ **11.** $f'(x) = \dfrac{e^x}{(e^x + 1)^2}$ **13.** $f'(x) = 6e^{6x-1}$ **15.** $f'(x) = -2xe^{-x^2}$

17. $f'(x) = \dfrac{e^{\sqrt{x}}}{2\sqrt{x}}$ **19.** $\dfrac{dy}{dx} = -4xe^{-5x^2}$ **21.** $\dfrac{dy}{dx} = \dfrac{e^x}{4}$ **23.** $f'(x) = \dfrac{7e^{7x}}{2}$ **25.** $y' = e^{-x} - xe^{-x} = (1 - x)e^{-x}$

27. $y' = -10xe^{1-x^2}$ **29.** $f'(x) = \dfrac{3xe^{3x} + 2e^{3x}}{(x + 1)^2}$ or $\dfrac{(3x + 2)e^{3x}}{(x + 1)^2}$ **31.** $y' = \dfrac{2(x - 1)e^{1+2x}}{x^3}$ **33.** $y' = \dfrac{e^x - e^{-x}}{2}$

35. $f'(x) = \dfrac{5 - 5x}{e^x}$ **37.** $y' = 3x^2 e^{1+3x} + 2xe^{1+3x} = (3x + 2)xe^{1+3x}$ **39.** $y' = 50e^{5x}(1 + e^{5x})^9$

41. $f''(x) = 9e^{3x}$ **43.** $f''(x) = xe^x + 2e^x = (x + 2)e^x$ **45.** increasing for all x; never decreasing

47. never increasing; decreasing for all x **49.** increasing when $x > -1$; decreasing when $x < -1$

51. increasing when $x > 0$; decreasing when $x < 0$

53. (a) $-2, 0$ **(b)** relative maximum $\dfrac{4}{e^2}$ (when $x = -2$); relative minimum 0 (when $x = 0$)

55. (a) -2 **(b)** relative minimum $-\dfrac{2}{e}$ **57. (a)** $-1/2$ **(b)** relative minimum $-\dfrac{1}{2e}$

59. (a) $0, 2$ **(b)** relative minimum 0 (when $x = 0$); relative maximum $\dfrac{4}{e^2}$ (when $x = 2$)

61. (a) $-4/3$ **(b)** relative minimum $-\dfrac{1}{3e^4}$

63. The exponential function has no critical numbers (the only numbers that can possibly correspond to relative extrema), since $f'(x) = e^x$, and e^x is never zero or undefined.

65.   **67.**

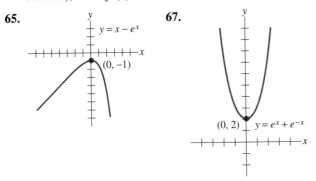

69. (a) $C(x) = 50xe^{-.04x}$ **(b)** $MC(x) = (1 - .04x)50e^{-.04x}$
(c) $C(10) = \$335.16$; $MC(10) = \$20.11$ **(d)** when $x = 25$

71. $MP(x) = 1.2xe^{.03x} + 40e^{.03x} + .8xe^{-.02x} - 40e^{-.02x}$ **73.** 10 units

75. (a) .001 grams per cubic centimeter (source), .000135 grams per cubic centimeter (1 mile downstream)
(b) $-.002e^{-2x}$ grams per cubic centimeter per mile

77. 4   **79.** $y = ex$ **81.** $\dfrac{dy}{dx} = \dfrac{-e^y - 1}{1 + xe^y}$ **83.** $\dfrac{dy}{dx} = \dfrac{ye^{xy} - 2}{1 - xe^{xy}}$ or $\dfrac{2 - ye^{xy}}{xe^{xy} - 1}$

85. $\dfrac{dy}{dx} = \dfrac{e^y - ye^x}{1 + e^x - xe^y}$ or $\dfrac{ye^x - e^y}{xe^y - e^x - 1}$ **87.** $\dfrac{dy}{dx} = \dfrac{-2x - e^{x+y}}{2y + e^{x+y}}$ **89.** $\dfrac{dy}{dx} = \dfrac{1 - 2e^{2x}}{1 - e^{-y}}$

91. $\dfrac{dy}{dx} = \dfrac{e^{y^2} + 2xe^y}{1 - x^2e^y - 2xye^{y^2}}$ **93.** $y' = 2^x \ln 2$ **95.** $y' = 7^{4x} \cdot 4 \ln 7$ **97.** $f'(x) = -3^{-x^2} \cdot 2x \ln 3$

99. $f'(x) = 9^{1+x^2} \cdot 4x \ln 9$ **101.** $y' = 2^{3x}(1 + 3x \ln 2)$ **103.** $y' = \dfrac{3^x(x \ln 3 - 1)}{2x^2}$

105. $f'(x) = \dfrac{10^{2x} - 1 - 10^{2x} \cdot 2x \ln 10}{3x^2}$ **107.** 0   **109.** e^2 **111.** $\dfrac{1}{1 + e}$ **113.** $(0, 1)$

SECTION 5.4   (page 285)

1. $y' = \dfrac{1}{x}$ **3.** $y' = 1 + \ln x$ **5.** $y' = \dfrac{1 - \ln x}{x^2}$ **7.** $y' = \dfrac{\ln x - 1 - \dfrac{1}{x}}{(\ln x)^2}$ or $\dfrac{x \ln x - x - 1}{x(\ln x)^2}$

9. $f'(x) = \dfrac{1}{2x\sqrt{\ln x}}$ **11.** $f'(x) = \dfrac{\ln x}{(1 + \ln x)^2}$ **13.** $y' = \dfrac{2x}{x^2 + 7}$ **15.** $y' = \dfrac{6}{2x + 1}$ **17.** $f'(x) = -\dfrac{1}{x}$

19. $f'(x) = \dfrac{30e^x}{1 + e^x}$ **21.** $f'(x) = \dfrac{1}{x(x + 1)}$ **23.** $f'(x) = \dfrac{1}{x \ln x}$ **25.** $y' = \dfrac{x + 1}{x}$ **27.** $y' = e^{-x}\left(\dfrac{2}{x} - \ln x^2\right)$

29. $y = 3x - 3$ **31.** $f''(x) = \dfrac{1}{x}$ **33.** $f''(x) = \dfrac{2 - 2\ln x}{x^2}$ **35.** $f''(x) = \dfrac{-3 + 2\ln x}{x^3}$

37. (a) $MR(x) = 70 + \dfrac{100}{x}$ **(b)** \$75   **39. (a)** $C(x) = 3x \ln(x + 1)$ **(b)** $MC(x) = \dfrac{3x}{x + 1} + 3 \ln(x + 1)$

41. increasing   **43.** increasing   **45.** increasing   **47.** decreasing   **49.** critical number: 1; relative minimum: 1

51. critical number: 1; relative minimum: 2   **53.** critical number: e; relative maximum: $\dfrac{2}{e}$

55. critical number: $\dfrac{1}{\sqrt{e}}$; relative minimum: $-\dfrac{1}{2e}$ **57.** critical number: 1; relative maximum: 0

59.

$y = x - \ln x$

$(1, 1)$

61.

$y = \ln(x^2 + 1)$

$(0, 0)$

points of inflection $(\pm 1, \ln 2)$

63. $\dfrac{dy}{dx} = \dfrac{-\dfrac{3}{x} - y}{x + 1} = \dfrac{-3 - xy}{x^2 + x}$ **65.** $\dfrac{dy}{dx} = \dfrac{-y - 2x^2 y}{2xy + x}$ **67.** $\dfrac{dy}{dx} = \dfrac{\dfrac{y}{x} - \ln y}{\dfrac{x}{y} - \ln x} = \dfrac{y^2 - xy \ln y}{x^2 - xy \ln x}$

69. $\dfrac{dy}{dx} = \dfrac{3y - 1 - e^x \ln y}{\dfrac{e^x}{y} - 3x} = \dfrac{3y^2 - y - ye^x \ln y}{e^x - 3xy}$ **71.** $\dfrac{dy}{dx} = \dfrac{2}{\dfrac{5}{y} + 3e^{3y}} = \dfrac{2y}{5 + 3ye^{3y}}$ **73.** $y' = \dfrac{1}{x \ln 10}$

75. $y' = \dfrac{1}{x \ln 2}$ **77.** $y' = \log_{10}(1 - 3x) - \dfrac{3x}{(1 - 3x) \ln 10}$ **79.** $y' = \dfrac{\dfrac{1}{\ln 10} - \log_{10} x}{x^2}$

81. $\dfrac{dy}{dx} = x^{2x}(2 + 2 \ln x)$ **83.** $\dfrac{dy}{dx} = (x + 1)^x\left[\dfrac{x}{x + 1} + \ln(x + 1)\right]$ **85.** $\dfrac{dy}{dx} = x^{x^2}(x + 2x \ln x)$

87. $\dfrac{dy}{dx} = (3x)^{x+1}\left(\dfrac{x + 1}{x} + \ln 3x\right)$ **89.** $\dfrac{dy}{dx} = x^{1/x}\left(\dfrac{1}{x^2} - \dfrac{1}{x^2} \ln x\right)$

91. As x gets larger and larger, the slope of the tangent line gets closer and closer to zero. However, the slope can never be zero. After all, the slope is $1/x$ and $1/x$ is never equal to zero.

93. 0   **95.** $\dfrac{dy}{dx} = \dfrac{-ye^{xy} - \dfrac{1}{2x}}{xe^{xy} + \dfrac{1}{2y}} = \dfrac{-2xy^2e^{xy} - y}{2x^2ye^{xy} + x} = -\dfrac{y}{x}$ **97.** $\dfrac{dy}{dx} = -\dfrac{1}{2} - \dfrac{1}{2}\ln x$ **99.** increasing

101. The only possible candidates for x values associated with relative extrema are the critical numbers—x such that $f'(x) = 0$ or $f'(x)$ is undefined when $f(x)$ is defined. Since $f'(x) = 1/x$, there are no x values for which $f'(x) = 0$. (The numerator is always 1.) Although $f'(x)$ is undefined when $x = 0$, so is the original function $f(x) = \ln x$ undefined when $x = 0$. So f has no critical numbers and therefore $f(x) = \ln x$ has no relative extrema.

103. $f^{(40)}(x) = -1 \cdot 2 \cdot 3 \cdot 4 \cdot 5 \cdot \cdots \cdot 39x^{-40}$

SECTION 5.5   (page 291)

1. 6.50%   **3.** 6.29%   **5. (a)** \$116.40   **(b)** \$130.27   **7.** 5.94%

9. (a) increases   **(b)** positive   **(c)** positive   **(d)** increasing, increasing   **11.** 10 years   **13.** 8 years

15. 11.7%   **17. (a)** 14 years   **(b)** 7 years

19. (a) 10.66 years   **(b)** 10.77 years
 (c) In this instance, the rule of 70 yields a slightly larger value.

21. \$38,553   **23.** \$1779

REVIEW EXERCISES FOR CHAPTER 5   (page 293)

1. $\dfrac{1}{7}\ln 6$ **3.** $\dfrac{1}{4}e^6$ **5.** $\dfrac{\ln .5}{.03}$ **7.** e^5 **9.** -7 **11.** -2 **13.** $\dfrac{dy}{dx} = 4e^4x^3$ **15.** $f'(x) = -7e^x(3 - e^x)^6$

17. $\dfrac{dy}{dx} = e^{1/x^2} - \dfrac{2}{x^2}e^{1/x^2} = e^{1/x^2}\left(1 - \dfrac{2}{x^2}\right)$ **19.** $f'(x) = 3x^2e^{-3x} - 3x^3e^{-3x} = 3x^2e^{-3x}(1 - x)$ **21.** $\dfrac{dy}{dx} = 5^x \ln 5$

23. $f'(x) = e^{3x}\left(\dfrac{1}{x} + 3\ln x\right)$ **25.** $y' = 2x\ln x + 2x(\ln x)^2$ **27.** $f'(x) = -\dfrac{2}{x}$ **29.** $y' = \ln x$

31. $y' = \dfrac{2}{x} + 1$ or $\dfrac{2 + x}{x}$ **33.** $\ln \dfrac{4}{3}$ **35.** 10   **37.** $\ln\dfrac{1}{x} = \ln 1 - \ln x = 0 - \ln x = -\ln x$ **39.** \$7408

41. approximately 24   **43.** $14{,}112 - 14{,}146$ **45. (a)** 100   **(b)** $5317 - 5321$ **(c)** 200,000

47. (a) $P = 10^{a - b\log A}$ **(b)** $A = 10^{(a - \log P)/b}$

49. The calculator indicates an error, because $\ln 0$ and $\ln(-1)$ are not defined. Specifically, $\ln x$ is only defined for $x > 0$.

SECTION 6.1 (page 303)

1. $4x^2 + C$ **3.** $2x^3 + C$ **5.** $\dfrac{t^4}{4} + C$ **7.** $\dfrac{5x^6}{3} + C$ **9.** $-\dfrac{1}{x} + C$ **11.** $-\dfrac{1}{x^3} + C$ **13.** $-\dfrac{1}{4x^4} + C$

15. $-\dfrac{4}{z^5} + C$ **17.** $\dfrac{2x^{3/2}}{3} + C$ **19.** $\dfrac{4x^{7/4}}{7} + C$ **21.** $3x^{1/3} + C$ **23.** $\dfrac{21t^{4/3}}{4} + C$ **25.** $3x + C$

27. $\dfrac{x^3}{3} + 3x^2 + C$ **29.** $\dfrac{2x^{3/2}}{3} - x^3 + C$ **31.** $2\sqrt{x} + 9x + C$ **33.** $\dfrac{t^3}{3} - 4t^2 + t + C$ **35.** $e^x + C$

37. $x^2 - e^x + C$ **39.** $e^x + x + C$ **41.** $\ln |z| + C$ **43.** $4 \ln |x| + 3x^2 + C$ **45.** $\dfrac{1}{5} \ln |x| + C$

47. $3x + \ln |x| + C$ **49.** $\dfrac{1}{7} e^{7x} + C$ **51.** $5e^x + C$ **53.** $4x - 10e^{.1x} + C$ **55.** $20e^{.05t} + t + C$

57. $-\dfrac{1}{6} e^{-6x} + C$ **59.** $800e^{.01x} + C$ **61.** $-100e^{-.01x} + C$ **63.** $\ln |x| + x^2 + C$ **65.** $\dfrac{x^5}{5} + 2 \ln |x| + C$

67. $\dfrac{2}{3} t^{3/2} + 2t^{1/2} + C$ **69.** $\dfrac{2}{5} x^{5/2} + \dfrac{2}{3} x^{3/2} + C$ **71.** $x - e^{-x} + C$ **73.** $\dfrac{x^3}{3} + \dfrac{x^4}{4} + C$

75. $2x^{1/2} + 3x^{1/3} + C$ **77.** $x + 2e^{-x} - \dfrac{1}{2} e^{-2x} + C$

79. (a) Yes.
(b) No. The integration leaves you with a "+ C" term. In particular, the function $f(x) + 5$ (after differentiation and integration) becomes $f(x) + C$, where C is *any* constant.

SECTION 6.2 (page 310)

1. $f(x) = x^3 - x^2 + 5x + 3$ **3.** $f(x) = \dfrac{1}{x^3} + 2$ **5.** $f(x) = x + 2x^{3/2} - 4$ **7.** $f(x) = -\sqrt{x} + 5$

9. $f(x) = e^x - x^2 + 5$ **11.** $f(x) = \ln |x| + 5$ **13.** $f(x) = \dfrac{1}{2} e^{2x} + 4x^2 + \dfrac{3}{2}$ **15.** $y = x^2 + 5$

17. $y = x^3 + 5x - 4$ **19.** $y = \dfrac{2}{3} x^{3/2} + 1$ **21.** $y = e^x$ **23.** $C(x) = 40x - .03x^2 + 200$

25. $C(x) = 20\sqrt{x} + 50$ **27.** \$1015   **29.** $R(x) = 50x - .2x^2$ **31.** $R(x) = 2\sqrt{x} - \dfrac{1}{10} x$

33. $R(x) = 10x - 20e^{.05x} + 20$ **35.** $P(x) = 40x - .4x^2 - 30$ **37.** $P(x) = 100x + .2x^2 - .02x^3$

39. $P(x) = 50x - .2x^{3/2} - 130$ **41.** $P(x) = 70x - 100e^{.01x} + 70$

43. (a) $v = -32t + 105$ **(b)** $-16t^2 + 105t$ feet

45. (a) $v(t) = -32t$ **(b)** $s(t) = -16t^2 + 576$ **(c)** 128 feet per second   **(d)** 6 seconds

47. (a) $v(t) = -5.3t + 120$ **(b)** $s(t) = -2.65t^2 + 120t$

49. (a) $\dfrac{dn}{dt} = 5t^{2/3} + 22$ **(b)** $n = 3t^{5/3} + 22t + 50$ **(c)** 322   **51. (a)** $n = t^2 + 30t + 2000$ **(b)** 2099

53. 102 milligrams   **55.** $P = 14.7e^{-.21x}$ **57.** 288π cubic centimeters

SECTION 6.3   (page 320)

1. 55   **3.** 48   **5.** 100   **7.** 29/12   **9.** $\sum_{i=1}^{9} i$ **11.** $\sum_{i=4}^{n} i$ **13.** $\sum_{i=1}^{49} \frac{i}{i+1}$ **15.** $\sum_{i=1}^{10} a_i x_i$ **17.** $\sum_{i=0}^{n} f(x_i)$

19. 36   **21.** 70   **23.** 9   **25.** 11.25   **27.** 8.25   **29.** 8.12

SECTION 6.4   (page 328)

1. the total revenue from the sale of x units

3. the total amount of gasoline consumed in the U.S. from 1964 to 1976

5. the total interest paid by a bank on its money market account from 1984 to 1992

7. the total amount of growth (in milligrams) of the yeast this week

9. 36   **11.** 21   **13.** 6   **15.** 95/6   **17.** 3   **19.** $e - 2$ **21.** 15/32   **23.** 3/10   **25.** 76   **27.** 7/8

29. $\ln 3$ **31.** $6 - 4e^2$ **33.** $20(e^5 - 1)$ **35.** $\frac{9}{2} + 4 \ln 2$ **37.** $\frac{1}{e} - \frac{5}{4} + \ln 4$ **39.** 64/3   **41.** 23/3

43. 9   **45.** 16/3   **47.** 10/3   **49.** 45/4   **51.** $e - 1$ **53.** $\ln 6$ **55.** $2e - 2$

57. Sample: $\int_{2}^{2} 6x\, dx = \left[3x^2\right]_{2}^{2} = 12 - 12 = 0$ **59.** There is no area (zero area) under a point.   **61.** $2 - e$

63. 9/2   **65.** 8

SECTION 6.5   (page 338)

1. $197.75   **3.** $145,634   **5.** No. There will be a loss of $3000 from the additional 300 computers.

7. 23.85 feet   **9.** 109   **11.** 378,000 cubic meters   **13.** 55,350 or 55,351   **15.** 736   **17.** 13/3   **19.** 10

21. 2/7   **23.** $e - 1$ **25.** 43 centimeters per second   **27.** 80 feet per second   **29.** 17.6 feet   **31.** $4596

33. The average value is 5, which can be determined by noting that f is 5 at every number in the interval.

35. 6π **37.** $124\pi/3$ **39.** $28\pi/15$ **41.** $127\pi/7$ **43.** 48π **45.** $\frac{\pi}{4}(e^{12} - 1)$ **47.** 3π **49.** $\pi(\ln 10 - 1)$

51. $\pi\left(\frac{3}{2} + 2 \ln 2\right)$ **53.** $V = \int_{0}^{h} \pi\left(\frac{r}{h}x\right)^2 dx = \left[\frac{\pi r^2}{h^2} \cdot \frac{x^3}{3}\right]_{0}^{h} = \frac{\pi r^2 h}{3}$

SECTION 6.6   (page 344)

1. (a) (11, 29)   **(b)** $60.50   **(c)** $121   **3. (a)** (13, 23.5)   **(b)** $42.25   **(c)** $126.75

5. (a) (28, 8)   **(b)** $98   **(c)** $98   **7. (a)** (5, 3.5)   **(b)** $25   **(c)** $8.33

9. (a) (16, 9)   **(b)** $21.33   **(c)** $42.67   **11. (a)** (3, 10)   **(b)** $4.50   **(c)** $18

SECTION 6.7   (page 352)

1. 66   **3.** 68/3   **5.** 47/6   **7.** 52/3   **9.** 9/2   **11.** 4/3   **13.** 1/12   **15.** 1/3   **17.** 4/3   **19.** $e - 2$

21. 34/3   **23.** 20   **25.** 59/3   **27.** 4/3   **29.** 2   **31.** 8   **33.** 1/2   **35.** 1/2   **37.** 9/2   **39.** 4/15

41. $\int_0^{x_e} [D(x) - p_e]\, dx$ **43.** 71/6   **45.** $\dfrac{2}{e} - \ln 2$ **47.** 7/3

REVIEW EXERCISES FOR CHAPTER 6   (page 354)

1. $4x^2 + 2x + C$ **3.** $2x^5 - \dfrac{1}{3}x^3 + C$ **5.** $x + \dfrac{2}{3}x^{3/2} + C$ **7.** $-\dfrac{3}{t} + C$ **9.** $2 \ln |x| + C$

11. $-50e^{-.02x} + C$ **13.** 10   **15.** 5/2   **17.** $\ln 6$ **19.** $25(e^4 - e^{.04})$ **21.** 15   **23.** $\dfrac{9}{2} - e$ **25.** 64/3

27. \$35   **29.** 110.6   **31.** 633/95   **33.** $2\pi\left(1 - \dfrac{1}{e^2}\right)$ **35.** (20, 11); \$40; \$80   **37.** $\displaystyle\sum_{i=1}^{n} x_i f(x_i)$

SECTION 7.1   (page 363)

1. $\dfrac{(x^2 + 3)^6}{6} + C$ **3.** $\dfrac{2(x^2 - 6)^{3/2}}{3} + C$ **5.** $\dfrac{(3x - 2)^7}{21} + C$ **7.** $\dfrac{(x^2 - 3)^{3/2}}{3} + C$ **9.** $\dfrac{(5x^3 + 1)^5}{75} + C$

11. $\dfrac{-1}{14(7x + 2)^2} + C$ **13.** $\dfrac{2}{3}\sqrt{3x + 2} + C$ **15.** $\dfrac{3}{8}(1 + 4x)^{2/3} + C$ **17.** $\dfrac{3}{8}(x + 2)^8 + C$

19. $\dfrac{(x^2 - 4x + 1)^6}{12} + C$ **21.** $\dfrac{2}{5}(x^{1/2} + 1)^5 + C$ **23.** 21/2   **25.** 39   **27.** 1/3   **29.** 104/3   **31.** 2

33. 38/3   **35.** $e^{x^3} + C$ **37.** $e^{x+1} + C$ **39.** $-e^{-x} + C$ **41.** $\dfrac{1}{6}e^{3t^2} + C$ **43.** $-e^{-x^2} + C$

45. $e^x - e^{-x} + C$ **47.** $\ln |x + 1| + C$ **49.** $\dfrac{1}{5}\ln |5x + 2| + C$ **51.** $\ln |1 + x^3| + C$

53. $\dfrac{2}{3}\ln |1 + x^{3/2}| + C$ **55.** $\dfrac{1}{3}\ln |x^3 + 6x^2 - 15| + C$ **57.** $\dfrac{1}{2}\ln |1 + x^2| + C$ **59.** $t + \dfrac{1}{3}e^{3t} + C$

61. $\ln |x| + x + C$ **63.** $-\dfrac{1}{2}\ln |1 - e^{2x}| + C$ **65.** $\dfrac{1}{5}(\ln x)^5 + C$ **67.** $-\dfrac{1}{\ln x} + C$ **69.** $x + \dfrac{1}{e^x} + C$

71. $(\ln x)^2 + C$ **73.** $\dfrac{1}{3}(\ln x)^3 + C$ **75.** $-\dfrac{1}{2}\ln |1 - 2x| + C$ **77.** $\ln 2$ **79.** $\ln(1 + e) - \ln 2$

81. $\dfrac{1}{2} - \dfrac{1}{2e}$ **83.** 1   **85.** 1/2   **87.** $\dfrac{1}{2}\ln 3$ **89.** $\dfrac{1}{3} - \dfrac{1}{3e}$ **91.** $e^2x + C$ **93.** $\dfrac{1}{4}(\ln x)^2 + C$

95. $\dfrac{x^2}{2} + x + 3\ln |x + 1| + C$ **97.** $3x + 3\ln |x - 1| + C$ **99.** $\dfrac{-1}{x \ln x} + C$ **101.** $y = 2e^{x^2} + 3$

103. 49/12   **105.** 45/28   **107.** $2\pi/77$ **109.** 10 feet

111. (a) The second integral can be evaluated by the methods studied thus far. Insert a factor of 2 into the integral (and 1/2 in front of the integral). The integrand then has the form $u^n du$, where $u = 2x + 1$ and $du = 2x\, dx$.

(b) The first integral cannot be evaluated by the methods studied thus far. The integrand is *almost* $u^n du$, but it is missing a factor of x (since $u = x^2 + 1$ and $du = 2x\, dx$). Unfortunately, we cannot insert *variable* factors (such as x) into the integrand.

SECTION 7.2   (page 370)

1. $\dfrac{1}{2}xe^{2x} - \dfrac{1}{4}e^{2x} + C$ **3.** $\dfrac{x^2}{2}\ln x - \dfrac{x^2}{4} + C$ **5.** $\dfrac{x^3}{3}\ln x - \dfrac{x^3}{9} + C$ **7.** $xe^x + C$ **9.** $-\dfrac{1}{x}\ln x - \dfrac{1}{x} + C$

11. $\dfrac{x(x+3)^5}{5} - \dfrac{(x+3)^6}{30} + C$ **13.** $\dfrac{-x}{2(x+7)^2} - \dfrac{1}{2(x+7)} + C$ **15.** $2x^{1/2}\ln x - 4x^{1/2} + C$

17. $x^3 e^x - 3x^2 e^x + 6xe^x - 6e^x + C$ **19.** $\dfrac{-x}{2(x+2)^2} - \dfrac{1}{2(x+2)} + C$ **21.** e^2 **23.** 1   **25.** 3/2   **27.** -1

29. 12,149   **31.** $\dfrac{1}{2}e^{x^2} + C$ **33.** $\dfrac{(\ln x)^2}{2} + C$ **35.** $x\ln 2x - x + C$ **37.** $\dfrac{2x}{3}(1+x)^{3/2} - \dfrac{4}{15}(1+x)^{5/2} + C$

39. $\dfrac{1}{3}(1+x^2)^{3/2} + C$ **41.** $-25xe^{-.04x} - 625e^{-.04x} + C$ **43.** $-150xe^{-.08x} - 1875e^{-.08x} + C$

45. $-20xe^{-.05x} - 400e^{-.05x} + C$ **47.** \$93,535 $-$ \$93,592   **49.** \$29,410   **51.** 1   **53.** $y = x\ln x - x$

55. $2e^{-1} - 4e^{-3}$ **57.** $\dfrac{\pi}{4}(e^2 - 1)$

59. The choice is not a good one, since it produces the identical integral that we are trying to evaluate. Nothing is accomplished by using $u = 1$ and $dv = \ln x\, dx$.

SECTION 7.3   (page 375)

1. $\dfrac{x}{2}\sqrt{x^2+16} + 8\ln\left|x + \sqrt{x^2+16}\right| + C$ **3.** $\dfrac{x}{2}\sqrt{9x^2+5} + \dfrac{5}{6}\ln\left|3x + \sqrt{9x^2+5}\right| + C$ **5.** $\dfrac{1}{14}\ln\left|\dfrac{x-7}{x+7}\right| + C$

7. $\dfrac{1}{90}\ln\left|\dfrac{5x-9}{5x+9}\right| + C$ **9.** $x - \dfrac{1}{7}\ln(1 + e^{7x}) + C$ **11.** $x + \dfrac{1}{4}\ln(1 + e^{-4x}) + C$

13. $\ln\left|x + \sqrt{x^2+25}\right| + C$ **15.** $\dfrac{1}{3}\ln\left|3x + \sqrt{9x^2+16}\right| + C$ **17.** $-\dfrac{\sqrt{x^2+1}}{x} + C$ **19.** $-\dfrac{\sqrt{4x^2+3}}{12x} + C$

21. $-\dfrac{\sqrt{4x^2+3}}{3x} + C$ **23.** $\dfrac{1}{6}\ln\left|\dfrac{x-3}{x+3}\right| + C$ **25.** $\dfrac{1}{6}\ln\left|\dfrac{3x-1}{3x+1}\right| + C$ **27.** $\dfrac{x^3}{3}\ln x - \dfrac{x^3}{9} + C$

29. $\dfrac{x^6}{6}\ln x - \dfrac{x^6}{36} + C$ **31.** $-\ln\left|\dfrac{1 + \sqrt{x^2+1}}{x}\right| + C$ **33.** $-\dfrac{1}{10}\ln\left|\dfrac{5 + \sqrt{4x^2+25}}{2x}\right| + C$

35. $-\dfrac{1}{2} \ln \left| \dfrac{2 + \sqrt{9x^2 + 4}}{3x} \right| + C$ **37.** $x(\ln x)^3 - 3x(\ln x)^2 + 6x \ln x - 6x + C$ **39.** $x \ln x - x + C$

41. $xe^x - e^x + C$ **43.** $x^2 e^x - 2xe^x + 2e^x + C$ **45.** $\ln \dfrac{e^u}{1 + e^u} = \ln e^u - \ln(1 + e^u) = u - \ln(1 + e^u)$

47. $\dfrac{81}{8} \ln 3 - \dfrac{5}{2}$ **49.** $3 \ln\left(7 + \sqrt{40}\right) - 3 \ln 9$

SECTION 7.4   (page 383)

1. $\Delta x = 1$; $x_0 = 0$, $x_1 = 1$, $x_2 = 2$, $x_3 = 3$, $x_4 = 4$

3. $\Delta x = 1$; $x_0 = 1$, $x_1 = 2$, $x_2 = 3$, $x_3 = 4$, $x_4 = 5$, $x_5 = 6$, $x_6 = 7$

5. $\Delta x = .5$; $x_0 = 0$, $x_1 = .5$, $x_2 = 1$, $x_3 = 1.5$, $x_4 = 2$

7. $\Delta x = .5$; $x_0 = 2$, $x_1 = 2.5$, $x_2 = 3$, $x_3 = 3.5$, $x_4 = 4$, $x_5 = 4.5$, $x_6 = 5$

9. $\Delta x = .5$; $x_0 = 2$, $x_1 = 2.5$, $x_2 = 3$, $x_3 = 3.5$, $x_4 = 4$, $x_5 = 4.5$, $x_6 = 5$, $x_7 = 5.5$, $x_8 = 6$

11. $\Delta x = .25$; $x_0 = 3$, $x_1 = 3.25$, $x_2 = 3.50$, $x_3 = 3.75$, $x_4 = 4$; $x_5 = 4.25$, $x_6 = 4.50$, $x_7 = 4.75$, $x_8 = 5$

13. 2.75   **15.** .5019   **17.** 5.6724   **19.** 2.0214 *or* 2.0215   **21.** 81.5657 to 81.5668   **23.** 3.9827 *or* 3.9828

25. 53.3333   **27.** 9.3004   **29.** 4.0299 to 4.0307   **31.** 1.0983 to 1.0987   **33.** 6.3912 to 6.3925

35. .6445 *or* .6446   **37. (a)** 22   **(b)** 64/3 *or* 21.3333   **(c)** 64/3   **39. (a)** 64.6875   **(b)** 63.75   **(c)** 63.75

41. (a) 4.6615   **(b)** 4.6665 *or* 4.6666   **(c)** 14/3 *or* 4.6667

43. (a) .8959   **(b)** .8610   **(c)** $\dfrac{e - 1}{2}$ *or* .8592 *or* .8591

45. (a) 2.5543   **(b)** 2.6249 *or* 2.6250   **(c)** 8/3 *or* 2.6667

47. (a) 2.9957 *or* 2.9958   **(b)** 3.0836   **(c)** 3.1416

49. The graph is a parabola, and Simpson's rule uses parabolas to approximate the area.

51. (a) 2   **(b)** .0167   **53. (a)** .0026   **(b)** .000033   **55. (a)** .0052   **(b)** .000022   **57.** 64   **59.** 566

SECTION 7.5   (page 391)

1. 0   **3.** 0   **5.** ∞ **7.** 0   **9.** ∞ **11.** ∞ **13.** ∞ **15.** 3   **17.** ∞ **19.** ∞ **21.** 2   **23.** 1/2

25. 1/3   **27.** divergent   **29.** divergent   **31.** divergent   **33.** 1/e   **35.** divergent   **37.** divergent

39. divergent   **41.** divergent   **43.** 2   **45.** 1   **47.** 1/8   **49.** 1   **51.** 50 pounds   **53.** $50,000

55. $71,000   **57.** $-1/2$ **59.** 1/3   **61.** divergent   **63.** divergent   **65.** $-1/8$ **67.** 1/6   **69.** 1/e

71. 1/3   **73.** 1/32   **75.** divergent   **77.** divergent   **79.** divergent   **81.** 0   **83.** 0   **85.** divergent

87. 1 square unit

SECTION 7.6   (page 398)

1. $x(10x) = 2(5x^2)$ leads to $10x^2 = 10x^2$.   **3.** $4e^{4x} - 4(e^{4x}) = 0$

5. $x(6x^{1/2}) - 1.5(4x^{3/2} - 2) = 3$ leads to $3 = 3$.

7. $2 - 3(2x + 6) + (x^2 + 6x + 16) = x^2$ leads to $x^2 = x^2$.

9. $(-2)^2 - 4(1 - 2x) - 8x = 0$ leads to $0 = 0$.   **11.** $y = 3x^2 + 19x + C$ **13.** $y = e^x + x + C$

15. $f(x) = 2x + \frac{2}{3}x^{3/2} + C$ **17.** $y = -\frac{1}{2}e^{-2t} + C$ **19.** $f(t) = t + 3 \ln |t| + C$ **21.** $y = 2x^3 - x^2 + 5$

23. $f(x) = \frac{x^8}{8} + 3x + 14$ **25.** $y = 5x^2 - e^x + 1$ **27.** $y = 2x^{3/2} - 5$ **29.** $y = 8 \ln |t| + 3$

31. $y = x^2 + C$ **33.** $\ln |y| = \frac{x^2}{2} + C$ **35.** $e^y = \frac{t^2}{2} + C$ **37.** $-\frac{1}{y} = 3x^2 + 5x + C$

39. $-\ln |y| = \ln |x| + C$ **41.** $e^y = e^x + C$ **43.** $C(x) = 50x - .03x^2 + 150$ **45.** $R(x) = 5x + .0001x^2$

47. $P(x) = 80x - .2x^{3/2} - 50$ **49.** $\frac{dy}{dx} = x - 3$ **51.** $\frac{dB}{dt} = .05B$ **53.** $\frac{dM}{dt} = -.02M$

55. (a) $\frac{dA}{dt} = .09A$ **(b)** $A = Ce^{.09t}$ **(c)** $C = 3000$

57. (a) $\frac{dA}{dt} = .03A; A(0) = 100,000,000$ **(b)** $A = 100,000,000e^{.03t}$

59. (a) $\frac{dA}{dt} = -.08A; A(0) = 10$ **(b)** $A = 10e^{-.08t}$

61. Yes. $y = 3$ is a solution of such differential equations as $y' = 0$ and $y' + 5y = 15$.

SECTION 7.7   (page 406)

1. (a) .3   **(b)** .4   **(c)** .7   **(d)** 1   **3. (a)** .4   **(b)** .5   **(c)** 1   **(d)** .6

5. $\frac{1}{30}x \geq 0$ for x in $[2, 8]$ and $\int_2^8 \frac{1}{30}x \, dx = 1$ **7.** $\frac{1}{12}(3x^2 + 2x) \geq 0$ for x in $[0, 2]$ and $\int_0^2 \frac{1}{12}(3x^2 + 2x) \, dx = 1$

9. $\frac{3}{38}\sqrt{x} \geq 0$ for x in $[4, 9]$ and $\int_4^9 \frac{3}{38}\sqrt{x} \, dx = 1$ **11.** $\frac{10}{9x^2} \geq 0$ for x in $[1, 10]$ and $\int_1^{10} \frac{10}{9x^2} \, dx = 1$

13. (a) 63/64   **(b)** 1/8   **15. (a)** 1/2   **(b)** 3/10   **17. (a)** $\ln 2$ **(b)** $1 - \ln 2$ **19.** $g(x) = \frac{2x}{63}$

21. $g(x) = 4x^3$ **23.** $g(x) = \frac{1}{12}(x - 5)$ **25.** .36   **27.** 1/5   **29. (a)** $1 - \frac{1}{e}$ **(b)** .2858

31. (a) 9/13   **(b)** 1/13

33. No, f cannot be a probability density function. For f to be a probability density function, it must be true that $\int_0^{10} f(x)\,dx = 1$, yet we already have 2.5 on the subinterval $[1, 4]$. The integrals over the remaining portions of the interval $[0, 10]$ cannot be negative, since $f(x) \geq 0$ on $[0, 10]$. Consequently, $\int_0^{10} f(x)\,dx$ will be greater than 2.5 and thus not equal to one as needed for a probability density function.

REVIEW EXERCISES FOR CHAPTER 7   (page 408)

1. $\frac{1}{3}(1 + x^2)^{3/2} + C$ **3.** $x(1 + 2x)^{1/2} - \frac{1}{3}(1 + 2x)^{3/2} + C$ or $\frac{x - 1}{3}\sqrt{1 + 2x} + C$ **5.** $\frac{(x - 2)^4}{4} + C$

7. $\frac{x(x - 2)^4}{4} - \frac{(x - 2)^5}{20} + C$ **9.** $\frac{1}{2}e^{x^2} + C$ **11.** $\frac{1}{3}xe^{3x} - \frac{1}{9}e^{3x} + C$ **13.** $\sqrt{x^2 + 1} + C$

15. $\ln\left|x + \sqrt{x^2 + 1}\right| + C$ **17.** $x \ln 2x - x + C$ **19.** $\frac{(\ln 2x)^2}{2} + C$ **21.** $\frac{1}{2}\ln(1 + e^{2x}) + C$

23. $x - \frac{1}{2}\ln(1 + e^{2x}) + C$ **25.** $-2000xe^{-.05x} - 40{,}000e^{-.05x} + C$ **27.** $2x + 5\ln|x - 1| + C$

29. (a) 5.5013   **(b)** 5.5196   **31. (a)** .323   **(b)** .3217   **33.** 1/3   **35.** divergent   **37.** divergent

39. divergent

SECTION 8.1   (page 420)

1. 20, 37, 4, 1, 1   **3.** 2/3, $-1/5$, 0, 1, 0   **5.** -5, -4, -1, 0, 0   **7.** 34, -39, -14, 3, -7

9. 5, $\sqrt{42}$, 3, $\sqrt{6}$, $\sqrt{6}$ **11.** 0, 0   **13.** 4, e **15.** 4   **17.** 1/2   **19. (a)** \$1173.51   **(b)** \$6041.26

21. 65,215   **23. (a)** 130   **(b)** 80   **(c)** The person's mental age and actual age are the same.

25. $P(x, y, z) = x + 2y + 3z$ **27. (a)** $4k, 64k$ **(b)** The resistance becomes 16 times as great.

29. (a) $R(x, y) = 6.5x + 5y$ **(b)** $P(x, y) = 2.5x + 2y$ **31. (a)** $V(x, y, z) = xyz$ **(b)** $V(x, y) = 2xy^2$

33. all (x, y) **35.** all (x, y) **37.** all (x, y) in which $y \geq x$ **39.** all (x, y) in which $y \neq 1$

41. all (x, y) in which $x \neq 0$ and $y \neq 0$ **43.** all (x, y) in which $x \neq y$

45. all (x, y) in which $x \geq 0$ and $y \neq -3/2$ **47.** all (x, y) **49.** all (x, y) in which $x > 1$

51.   **53.**

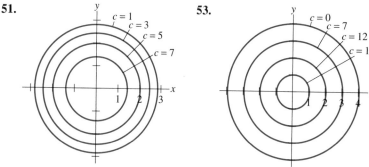

The radii are $\sqrt{8}$, $\sqrt{6}$, 2, $\sqrt{2}$.   The radii are 4, 3, 2, 1.

55.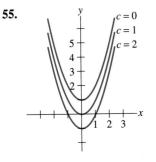

The parabolas $y = x^2 + 1$, $y = x^2$, $y = x^2 - 1$

57.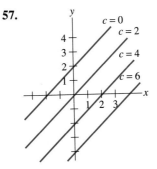

The lines $y = x + 2$, $y = x$, $y = x - 2$, $y = x - 4$

59.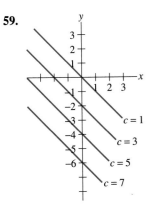

The lines $y = -x$, $y = -x - 2$, $y = -x - 4$, $y = -x - 6$

61.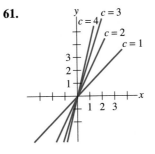

The lines $y = x$, $y = 2x$, $y = 3x$, $y = 4x$

63.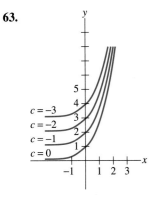

The graphs of $y = e^x + 3$, $y = e^x + 2$, $y = e^x + 1$, $y = e^x$

65.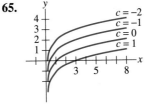

The graphs of $y = 2 + \ln x$, $y = 1 + \ln x$, $y = \ln x$, $y = -1 + \ln x$

SECTION 8.2   (page 429)

1. $\partial f/\partial x = 2$; $\partial f/\partial y = 5$ **3.** $\partial f/\partial x = 3x^2$; $\partial f/\partial y = -8y$ **5.** $\partial f/\partial x = 1/y$; $\partial f/\partial y = -x/y^2$

7. $\partial f/\partial x = ye^x$; $\partial f/\partial y = e^x$ **9.** $\partial f/\partial x = y/x$; $\partial f/\partial y = \ln x$ **11.** $\partial f/\partial x = 2x \ln y^3$; $\partial f/\partial y = 3x^2/y$

13. $\partial f/\partial x = 3ye^{3xy}$; $\partial f/\partial y = 3xe^{3xy}$ **15.** $\partial f/\partial x = \sqrt{y}$; $\partial f/\partial y = \dfrac{x}{2\sqrt{y}}$

17. $\partial f/\partial x = x/\sqrt{x^2 + y^2}$; $\partial f/\partial y = y/\sqrt{x^2 + y^2}$ **19.** $\partial f/\partial x = y/(x + y)^2$; $\partial f/\partial y = -x/(x + y)^2$

21. $f_x = 3x^2y^5$; $f_y = 5x^3y^4$ **23.** $f_x = -y^4$; $f_y = -4xy^3$ **25.** $f_x = \dfrac{2x}{x^2 + y^2}$; $f_y = \dfrac{2y}{x^2 + y^2}$

27. $f_x = e^{-y}$; $f_y = -xe^{-y}$ **29.** $f_x = \dfrac{y^3}{2\sqrt{xy^3}}$ or $\dfrac{1}{2}\sqrt{\dfrac{y^3}{x}}$; $f_y = \dfrac{3xy^2}{2\sqrt{xy^3}}$ or $\dfrac{3}{2}\sqrt{xy}$ **31.** y **33.** $2x$ **35.** 0

37. $2xy^2$ **39.** x **41.** 0   **43.** 3   **45.** $2x^2y$ **47. (a)** 157   **(b)** 18   **(c)** 47   **(d)** 2

49. (a) 12   **(b)** 1   **(c)** $1 + e$ **(d)** $2 + 2e$ **51.** $f_x = 2x$; $f_y = 2y$; $f_z = 2z$

53. $f_x = yz - 1$; $f_y = xz + 1$; $f_z = xy$ **55.** $f_x = ye^z$; $f_y = xe^z$; $f_z = xye^z$

57. $f_x = \dfrac{1}{\sqrt{2x + 2y + 2z}}$; $f_y = \dfrac{1}{\sqrt{2x + 2y + 2z}}$; $f_z = \dfrac{1}{\sqrt{2x + 2y + 2z}}$

59. $f_x = \dfrac{1}{y + z}$; $f_y = -x/(y + z)^2$; $f_z = -x/(y + z)^2$ **61.** $f_{xx} = 2$; $f_{xy} = 1$; $f_{yx} = 1$; $f_{yy} = 2$

63. $f_{xx} = -y/x^2$; $f_{xy} = 1/x$; $f_{yx} = 1/x$; $f_{yy} = 0$ **65.** $g_{xx} = 0$; $g_{xy} = 3e^y$; $g_{yx} = 3e^y$; $g_{yy} = 3xe^y$

67. $h_{xx} = 2 \ln y$; $h_{xy} = 2x/y$; $h_{yx} = 2x/y$; $h_{yy} = -x^2/y^2$

69. $f_{xx} = -1/(x + y^2)^2$; $f_{xy} = -2y/(x + y^2)^2$; $f_{yx} = -2y/(x + y^2)^2$; $f_{yy} = (2x - 2y^2)/(x + y^2)^2$

71. (a) \$148.40   **(b)** \$96.40   **73. (a)** $\partial f/\partial x = .8x^{-.6}y^{.6}$ **(b)** $\partial f/\partial y = 1.2x^{.4}y^{-.4}$ **75.** $135/64 \approx 2.1$

77. (a) $350°$

(b) The square of any $x \neq 0$ and the square of any $y \neq 0$ will be subtracted from $350°$. Thus, $T(x, y) < 350°$ for any point other than $(0,0)$.

(c) -20 **(d)** -12 **(e)** The temperature *decreases* as you move away from the origin.

79. πr^2

81. (a) k/V **(b)** $-kT/V^2$

(c) $\partial P/\partial V$ is the rate of change of pressure with respect to volume, assuming the temperature is kept constant.

83. (a) 1.98   **(b)** $S_w = .00306w^{-.575}h^{.725}$; $S_h = .00522w^{.425}h^{-.275}$ **(c)** .01, .008

(d) S_w is the rate of change of body surface area with respect to body weight, assuming the body height is kept constant. S_h is the rate of change of body surface area with respect to body height, assuming the body weight is kept constant.

SECTION 8.3   (page 439)

1. $(3, -1)$ **3.** $(-2, 5), (2, 5)$ **5.** $(-1, 4), (0, 4)$ **7.** $(-1, -3), (-1, 3), (1, -3), (1, 3)$ **9.** $(0, 0)$

11. $(-3, 2)$ **13.** $(-1, 0), (-1, 2), (3, 0), (3, 2)$ **15.** $(0, 0), (1, 1)$

17. relative minimum value is -12, at $(-4, 1)$ **19.** relative maximum value is 14, at $(2, -3)$

21. relative minimum value is -15, at $(1, 3)$ **23.** relative maximum value is 7, at $(-2, -2)$

25. relative minimum value is -7, at $(4, 2)$

27. relative maximum value is 9, at $(-1, 0)$; relative minimum value is -27, at $(3, 2)$ **29.** none

31. $1000 on radio, $4000 on newspaper   **33.** 13 at $1200 and 14 at $1500   **35.** 9, 9, 9

37. 14 inches wide, 28 inches long, 14 inches high

39. The shape indicated by the lowest cost approach would not be visually appealing, nor would it give the fish a long portion in which to swim.

SECTION 8.4   (page 446)

1. $f(x, y) = 25$, from $x = 5$ and $y = 5$ **3.** $f(x, y) = 8$, from $x = 2$ and $y = 1$

5. $f(x, y) = -16$, from $x = 4$ and $y = 2$ **7.** $f(x, y) = 4$, from $x = 2$ and $y = 0$

9. $f(x, y) = 50$, from $x = 5$ and $y = 7$ **11.** $f(x, y) = 300$, from $x = 10$ and $y = 10$

13. 500, 500   **15.** 2500 square meters   **17.** 70 units labor, 70 units capital; $21,000 minimum cost

19. 240 units labor, 60 units capital; $9600 minimum cost   **21.** 50 units labor, 50 units capital

23. radius $= \sqrt[3]{100/\pi} \approx 3.17$ inches, height $= 2\sqrt[3]{100/\pi} \approx 6.34$ inches

---

SECTION 8.5   (page 451)

1. $y = \dfrac{19}{14}x + 4$ **3.** $y = \dfrac{10}{7}x - \dfrac{73}{7}$ **5.** $y = \dfrac{6}{5}x + 3$ **7.** $y = 2x + \dfrac{5}{3}$ **9.** $y = \dfrac{53}{35}x + \dfrac{18}{5}$

11. $y = -\dfrac{3}{2}x + \dfrac{15}{2}$ **13. (a)** $y = -\dfrac{43}{35}x + \dfrac{69}{7}$ **(b)** 26/7   **(c)** $205/43 \approx 4.7674$

15. (a) $y = \dfrac{15}{113}x - \dfrac{1}{565}$ **(b)** 3.05   **17. (a)** $y = \dfrac{3}{5}x + 54$ **(b)** 75 grams   **(c)** 43.3°C

19. $b = \dfrac{\Sigma\, y_i - m\,\Sigma\, x_i}{n} = \dfrac{\Sigma\, y_i}{n} - \dfrac{m\,\Sigma\, x_i}{n} = \bar{y} - m \cdot \bar{x}$

SECTION 8.6   (page 455)

1. $dz = 2x\,dx + 2y\,dy$ **3.** $dz = (y^2 + 6x)\,dx + (2xy - 4y)\,dy$ **5.** $dz = (e^y + ye^x)\,dx + (xe^y + e^x)\,dy$

7. $dz = \left(-\dfrac{y}{x^2} - 8\right)dx + \left(\dfrac{1}{x} + 3\right)dy$ **9.** $dz = (\ln y - 2x)\,dx + \dfrac{x}{y}\,dy$ **11.** 777.6   **13.** .22   **15.** .1275

17. 163.22   **19.** 2   **21.** 15.56   **23.** $30,000

25. $R(21.5, 13) - R(20, 12) = 806.9375 - 784 = \$22,937.50$ **27.** $-4.4\pi \approx 13.8$ cubic inches
29. \$118.60   **31.** 44.4

SECTION 8.7   (page 465)

1. $104y$ **3.** $36x$ **5.** $3 + 9x$ **7.** $21 + 6y$ **9.** $2x^2 + 6$ **11.** 56   **13.** 120   **15.** 63   **17.** 462
19. 10/3   **21.** 76   **23.** $e^3 - e^2 - e + 1$ **25.** 32   **27.** $\int_0^2 \int_1^3 (2x + 6y)\, dy\, dx = \int_0^2 (4x + 24)\, dx = 56$
29. $\int_1^3 \int_2^4 (x + 2y)\, dx\, dy = 28$ **31.** 12   **33.** 4   **35.** 22/3   **37.** 40/3   **39.** 9   **41.** 20   **43.** 97.5
45. 52   **47.** 17/3 or 5.67   **49.** $e - 2$ **51.** 55   **53.** 25/3   **55.** $e^4 - 1$ **57.** 55.8   **59.** 1/12
61. $-4/21$

63.   **65.**   **67.**

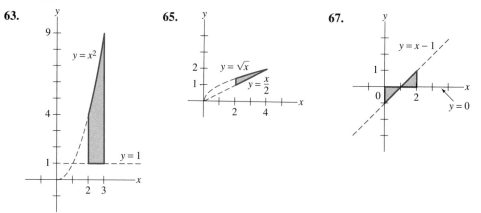

REVIEW EXERCISES FOR CHAPTER 8   (page 466)

1. $10, -3$ **3.** all (x, y) in which $x \neq 1$

5.   **7.**

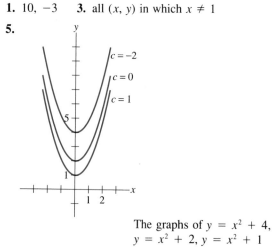

The graphs of $y = x^2 + 4$,
$y = x^2 + 2$, $y = x^2 + 1$

The graphs of $y = 2x - 1$,
$y = 2x - 3$, $y = 2x - 5$

9. $f_x = 4x^3 + 2xy^2$; $f_y = 2x^2y - 4y^3$　　**11.** $f_x = 3e^{3x} + \ln y$; $f_y = x/y$　　**13.** $f_x = e^y$; $f_y = xe^y$; $f_z = 15z^4$

15. $f_{xx} = 10$; $f_{yy} = 6y$; $f_{xy} = -1$; $f_{yx} = -1$　　**17.** $(2, -6)$　　**19.** none　　**21. (a)** 2.25　　**(b)** 0

23. 300 regular, 700 deluxe　　**25.** $f(4,3) = 31$　　**27.** $y = \dfrac{17}{12}x + \dfrac{595}{12}$　　$(952/672 = 17/12)$

29. 110,400 cubic centimeters　　**31.** -84　　**33.** $10(e - 1)$　　**35.** 72　　**37.** 90　　**39.** $-4/15$

SECTION 9.1　　(page 473)

1. .4067　　**3.** .7986　　**5.** 1.8807　　**7.** .2924　　**9.** .8387　　**11.** .8693　　**13.** 319 feet　　**15.** 10 feet

17. 217 feet

SECTION 9.2　　(page 485)

1. 30°　　**3.** 90°　　**5.** 150°　　**7.** 225°　　**9.** 240°　　**11.** $\pi/2$　　**13.** $\pi/4$　　**15.** $2\pi/3$　　**17.** $7\pi/6$　　**19.** $3\pi/4$

21. $\sqrt{3}$　　**23.** -1　　**25.** 0　　**27.** 0　　**29.** -1　　**31.** 2　　**33.** .8912　　**35.** .6967　　**37.** -6.2281

39. -2.4030　　**41.** 2.0858　　**43.** $-.3416$

45. (a) $\sin\dfrac{\pi}{4} = \dfrac{1}{\sqrt{2}} = \dfrac{1}{\sqrt{2}} \cdot \dfrac{\sqrt{2}}{\sqrt{2}} = \dfrac{\sqrt{2}}{2}$

　　　　$\cos\dfrac{\pi}{4} = \dfrac{1}{\sqrt{2}} = \dfrac{1}{\sqrt{2}} \cdot \dfrac{\sqrt{2}}{\sqrt{2}} = \dfrac{\sqrt{2}}{2}$

(b) $\sin\dfrac{\pi}{6} = \dfrac{1}{2}$;　$\cos\dfrac{\pi}{6} = \dfrac{\sqrt{3}}{2}$;　$\sin\dfrac{\pi}{3} = \dfrac{\sqrt{3}}{2}$;　$\cos\dfrac{\pi}{3} = \dfrac{1}{2}$

47. (a) 130　　**(b)** 80　　**(c)** 130　　**(d)** 129　　**49. (a)** 4000　　**(b)** 1000　　**(c)** 5000　　**(d)** 1000

51. The limits do not exist. These functions do not tend toward any number as x tends toward infinity.

53. -1 is the smallest x coordinate of any point on the unit circle.

SECTION 9.3　　(page 495)

1. $dy/dx = 2\cos 2x$　　**3.** $dy/dx = 7\cos x$　　**5.** $dy/dx = 2x\cos x^2$　　**7.** $dy/dx = 2\sin x\cos x$

9. $f'(x) = 15\cos 3x$　　**11.** $f'(x) = 4x\cos 4x + \sin 4x$　　**13.** $g'(t) = 4t\cos t + 4\sin t$

15. $dy/dx = (x\cos x - \sin x)/x^2$　　**17.** $dy/dx = -2\sin 2x$　　**19.** $dy/dx = -\pi\sin\pi x$

21. $dy/dt = \dfrac{\pi}{3}\cos\dfrac{\pi t}{3}$　　**23.** $dy/dt = -\dfrac{\pi}{4}\sin\dfrac{\pi t}{4}$　　**25.** $dy/dx = \dfrac{\cos x}{3}$　　**27.** $dy/dx = -4\sin 2x\cos 2x$

29. $g'(x) = \cos x - x\sin x$　　**31.** $y' = \dfrac{-2\sin 4x}{\sqrt{\cos 4x}}$　　**33.** $f'(x) = e^x(\cos x + \sin x)$

35. $y' = \dfrac{-x\sin x - \cos x}{x^2}$　　**37.** $y' = 4\sec^2 4x$　　**39.** $y' = -16\csc 2t\cot 2t$　　**41.** $y' = \pi\sec\pi t\tan\pi t$

43. $y' = -6x^2 \csc x^3 \cot x^3$ **45.** $f'(x) = \dfrac{-\csc^2 x}{2\sqrt{\cot x}}$ or $\dfrac{-\sqrt{\tan x}}{2\sin^2 x}$ **47.** $y' = 10\pi \csc^2 2\pi x$

49. $y' = 6\tan 3x \sec^2 3x$ **51.** $y' = e^x \sec x (\tan x + 1)$ **53.** $f'(x) = 2x \sec^2 x + 2\tan x$

55. (a) 1000   **(b)** 19,000   **(c)** $dS/dt = -1500\pi \sin \dfrac{\pi t}{6}$

     (d) $1500\pi \approx 4712$ snow shovels per month (increasing)

     (e) $-1500\pi \approx -4712$ snow shovels per month (decreasing)

57. (a) $H(3) = 6 + 4\sin \dfrac{\pi}{2} = 6 + 4 = 10$

        $H(15) = 6 + 4\sin \dfrac{5\pi}{2} = 6 + 4\sin \dfrac{\pi}{2} = 10$

     (b) $H(9) = 6 + 4\sin \dfrac{3\pi}{2} = 6 + 4(-1) = 2$

        $H(21) = 6 + 4\sin \dfrac{7\pi}{2} = 6 + 4\sin \dfrac{3\pi}{2} = 2$

     (c) $dH/dt = \dfrac{2\pi}{3}\cos \dfrac{\pi t}{6}$

     (d) $\left.\dfrac{dH}{dt}\right|_{t=3} = \dfrac{2\pi}{3}\cos \dfrac{\pi(3)}{6} = \dfrac{2\pi}{3}\cos \dfrac{\pi}{2} = \dfrac{2\pi}{3}\cdot 0 = 0$

        Similarly for $t = 15$, $t = 9$, and $t = 21$.

     (e) High tides are relative maximum points and low tides are relative minimum points.

59. (a) 1000   **(b)** 2000   **(c)** decreasing at approximately 131 per month

     (d) $Y' = 0$ at this relative minimum point.

61.

t	$\sin t$	$\dfrac{\sin t}{t}$
.1	.099833416	.998334166
.01	.009999833	.999983333
.001	.000999999	.999999833
.0001	.000099999	.999999998
−.1	−.099833416	.998334166
−.01	−.009999833	.999983333
−.001	−.000999999	.999999833
−.0001	−.000099999	.999999998

63.

x	$\dfrac{\pi}{2} - x$	$\sin\left(\dfrac{\pi}{2} - x\right)$	$\cos x$
.3	1.2708	.9553	.9553
.7	.8708	.7648	.7648
1.2	.3708	.3624	.3624
2.3	−.7292	−.6663	−.6663
3.5	−1.9292	−.9365	−.9365
4.1	−2.5292	−.5748	−.5748

65. $\dfrac{d}{dx}\left(\dfrac{\cos x}{\sin x}\right) = \dfrac{-(\sin^2 x + \cos^2 x)}{\sin^2 x} = \dfrac{-1}{\sin^2 x} = -\csc^2 x$

67. $\dfrac{d}{dx}\left(\dfrac{1}{\sin x}\right) = \dfrac{-\cos x}{\sin^2 x} = \dfrac{-1}{\sin x} \cdot \dfrac{\cos x}{\sin x} = -\csc x \cot x$ **69.** concave down (since $y'' = -2\sqrt{3} < 0$)

71. -16 **73.** $dy/dx = 2/\cos y$ or $2\sec y$ **75.** $\dfrac{dy}{dx} = \dfrac{1 + e^y \sin x}{e^y \cos x}$ **77.** $\dfrac{dy}{dx} = \dfrac{1 - y\sec^2 xy}{x\sec^2 xy}$

SECTION 9.4 (page 503)

1. $-\dfrac{1}{2}\cos 2x + C$ **3.** $\dfrac{1}{5}\sin 5x + C$ **5.** $\dfrac{1}{2}\sin x^2 + C$ **7.** $-\cos(x+1) + C$ **9.** $\dfrac{\sin^3 x}{3} + C$

11. $-\dfrac{\cos^2 x}{2} + C$ or $\dfrac{\sin^2 x}{2} + C$ **13.** $\dfrac{1}{\cos x} + C$ **15.** $-\dfrac{2}{3}(1 + \cos x)^{3/2} + C$ **17.** $-\ln|1 + \cos x| + C$

19. $1/2$ **21.** $-1/2$ **23.** $7/24$ **25.** $-\dfrac{1}{4}\ln|\cos 4x| + C$ **27.** $-\dfrac{1}{2}\ln|\cos x^2| + C$ **29.** $\dfrac{1}{2}\tan 2x + C$

31. $\dfrac{1}{3}\sec 3x + C$ **33.** $\ln|1 + \tan x| + C$ **35.** $\dfrac{1}{2}\ln|\sin 2x| + C$ **37.** $\dfrac{1}{6}\tan^6 x + C$ **39.** $1 + \dfrac{\sqrt{3}}{2}$

41. $1/2$ **43.** $1/3$ **45.** 0 **47.** $1400 + \dfrac{400}{\pi} \approx 1527$ **49.** predators: 1000; prey: 4000

51. $R(x) = 200x + \dfrac{15}{\pi}\sin \pi x$ **53.** (a) $540{,}000 + \dfrac{168{,}000}{\pi} \approx 593{,}476$ (b) $540{,}000 - \dfrac{168{,}000}{\pi} \approx 486{,}524$

55. $y = x^2 - \tan x + C$ **57.** $y = -\ln|\cos x| + C$ **59.** $-\cos y = \sin x + C$

61. Yes, it can be simplified. Since $\sin 2x$ is never less than -1, it follows that $2 + \sin 2x$ is never less than 1. Thus, $2 + \sin 2x > 0$ for all x and the absolute value is not needed.

SECTION 9.5 (page 506)

1. $-x\cos x + \sin x + C$ **3.** $-\dfrac{1}{2}\cos x^2 + C$ **5.** $\dfrac{x^2}{2} - \cos x + C$ **7.** $\dfrac{x}{4}\sin 4x + \dfrac{1}{16}\cos 4x + C$

9. $\dfrac{\sin^2 x}{2} + C$ **11.** $-x^2\cos x + 2x\sin x + 2\cos x + C$ **13.** $x + \tan x + C$

15. $x\tan x + \ln|\cos x| + C$ **17.** $\ln|x + \tan x| + C$

REVIEW EXERCISES FOR CHAPTER 9 (page 507)

1. $210°$ **3.** $5\pi/4$ **5.** $1/\sqrt{3}$ or $\sqrt{3}/3$ **7.** $\sqrt{2}$ **9.** (a) $2\sqrt{2}$ (b) 4 **11.** $y' = 2\cos(2x + 1)$

13. $y' = -6\cos 3x \sin 3x$ **15.** $f'(x) = x\sec x \tan x + \sec x$ or $(x\tan x + 1)\sec x$

17. $y' = \dfrac{x\sec^2 x - \tan x}{x^2}$ **19.** $y' = 3\sec^2 x(1 + \tan x)^2$ **21.** $f'(x) = e^x \cos e^x$ **23.** $y' = \cot x$

25. $\dfrac{1}{4}\sin 4x + C$ **27.** $-\cos e^x + C$ **29.** $-\dfrac{5}{3}\cos 3x + C$ **31.** $\sin(x + \pi) + C$ **33.** $1/4$ **35.** $\sqrt{3}$

37. $-x\cos x + \sin x - \cos x + C$ **39.** $\sqrt{3}/2$ **41.** $1/4$

INDEX

INDEX OF APPLICATIONS